First International Meeting on Microbial Phosphate Solubilization

Developments in Plant and Soil Sciences

VOLUME 102

First International Meeting on Microbial Phosphate Solubilization

Edited by

E. Velázquez and
C. Rodríguez-Barrueco

Partly reprinted from *Plant and Soil*, Vol 287, pages 1–84

Springer

A C.I.P. Catalogue record for this book is available from the Library of Congress.

ISBN-978-1-4020-4019-1 (HB)
ISBN-978-1-4020-5765-6 (e-book)

Published by Springer,
P.O. Box 17, 3300 AA Dordrecht, The Netherlands.

www.springer.com

Printed on acid-free paper

PLANT AND SOIL

Contents

First International Meeting on Microbial Phosphate Solubilization, Salamanca, Spain, July 16–19, 2002

Official logo of the First International Meeting on Microbial Phosphate Solubilization, Salamanca, Spain, July 16–19, 2002

Cover Photo: The compilation (clockwise) shows a strawberry flower, a *Prosopis* nodule, a bacterial plate culture showing phosphate solubilization zones, and a photomicrography showing sporulated *Bacillus*. Images courtesy of Encarna.

Preface

University of Salamanca and Consejo Superior de Investigaciones Científicas (CSIC), two famous and traditional scientific organizations have sponsored the First International Meeting on Microbial Phosphate Solubilization (MPS) held in Salamanca, Spain, on 16–19 July 2002.

The so called green revolution has provided us with grains to feed millions of humans and progress in medicine has increased longevity. Other moves of science have seen major advances of knowledge into cell biology and genetics and a threshold to success on what biosciences can make regarding a sustainable agricultural production can be envisaged at both short and long term. Besides Carbon and Nitrogen biogeochemical cycles, that of Phosphorus adds extra interest at increasing soil biological fertility. Second to none, phosphorus is involved in many essential metabolic processes of the living cell and free access of living beings to Phosphorus is a must not only due to P important role in itself but because of the enhancement effect on the role of other nutrients and processes, e.g. Biological Nitrogen Fixation, in the nutrition of the cultivated plants. Updating knowledge on the role of soil microorganisms in the solubilization of Phosphorus was the aim of the meeting. To the purpose sixty specialists from thirteen countries met in Salamanca to discuss the problems on the high P-unavailability as a soil nutrient for crops and the hazards of an increasing phosphate input to aquatic habitats from industrial and mining activities, sewage disposal, detergents, and other sources. Recommendations to enhance P-uptake by plants and crops, bioremediation potential in the rehabilitation of ecosystems, taxonomic characterization, interactions with mycorrhizae, the physiological and molecular basis of phosphate solubilizing microorganisms, possibilities of genetic modifications *ad hoc* of rhizospheric microorganisms, and trials on prospective inoculants were among the highlighted topics covered.

Emphasis was made on the fact that studies on phosphate solubilization shall always be on the line of contributing with extra available Phosphorus to plants, with no competition whatsoever with the important role of mycorrhizal associations with plants already widely recognized as self sufficient, and complementary under certain conditions to the use of P industrial fertilizers.

Let be this First Meeting on MPS also a first effort in the coordination of scientific internationally reputed groups, and let be the beginning of a continuing relation along the years to come, a wish that is extended to all those groups that were not given the opportunity to participate, such was the short notice under which the meeting was announced for which the Organizers apologize. Thank you to all sponsors and to attendants who made the meeting possible.

The Editors

E. Velázquez and C. Rodríguez-Barrueco (eds.),
First International Meeting on Microbial Phosphate Solubilization, 1

The taxonomy of rhizobia: an overview

Anne Willems[1]
Laboratory of Microbiology (WE10), Faculty of Sciences, Ghent University, Ledeganckstraat 35, B-9000, Gent, Belgium. [1]*Corresponding author**

Received: 31 May 2006

Key words: *Agrobacterium*, review, Rhizobia, taxonomy

Abstract

The taxonomy of rhizobia, bacteria capable of nodulating leguminous plants, has changed considerably over the last 20 years, with the original genus *Rhizobium*, a member of the alpha-Proteobacteria, now divided into several genera. The study of new geographically dispersed host plants, has been a source of many new species and is expected to yield many more. Here we provide an overview of the history of the rhizobia, but focus on the *Rhizobium–Allorhizobium–Agrobacterium* relationship. Finally, we review recent reports of nodulation and nitrogen fixation with legume hosts by bacteria that are outside the traditional rhizobial phylogenetic lineages. They include species of *Methylobacterium* and *Devosia* in the alpha-Proteobacteria and of *Burkholderia* and *Ralstonia* in the beta-Proteobacteria.

Introduction

The term "rhizobia", in the strictest sense, refers to members of the genus *Rhizobium*. Over the years, however, the term has come to be used for all the bacteria that are capable of nodulation and nitrogen fixation in association with legumes and that belong to a genus that was at one time part of the genus *Rhizobium* or closely related to it.

The family *Rhizobiaceae* in the 1984 edition of Bergey's Manual of Systematic Bacteriology, is composed of the rhizobia (at that time just including *Rhizobium* and *Bradyrhizobium*), *Agrobacterium* and *Phyllobacterium* (Jordan, 1984).

History

By the end of the 19th century, it was realized that atmospheric nitrogen was being assimilated through the root-nodules of legume plants. In 1888, Beijerinck reported isolation of the root-nodule bacteria and established that they were responsible for this process of nitrogen fixation. He named these bacteria *Bacillus radicicola* (Beijerinck, 1888). Later, Frank changed the name to *Rhizobium* with originally just one species, *R. leguminosarum* (Frank, 1889).

Extensive testing of nodulation of diverse legume hosts by different bacteria in the beginning of the 20th century, led to the establishment of cross-inoculation groups, with rhizobia from one plant in a cross-inoculation group supposed to nodulate all other plants in the group (Fred et al., 1932). This concept was also used in rhizobial taxonomy, but later it was abandoned as an unreliable taxonomic marker (Graham, 1964; Wilson, 1944), in part because of aberrant cross-infection among plant groups. Beginning in the early 1960s, bacteriologists started using a large diversity of morphological, nutritional and metabolic characters (Graham, 1964; Moffet and Colwell, 1968; 'tMannetje, 1967), as well as serology (Graham, 1963; Vincent and Humphrey, 1970) and simple DNA characteristics (De Ley and Rassel, 1965) in numerical taxonomy studies. This

* FAX No: +32-9-2645092.
E-mail: Anne.Willems@UGent.be

E. Velázquez and C. Rodríguez-Barrueco (eds.),
First International Meeting on Microbial Phosphate Solubilization, 3–14

demonstrated the relatedness of *Rhizobium* and *Agrobacterium* and led to a clear distinction between the fast and slow growing rhizobia (Graham, 1964), with the latter group subsequently placed in a separate genus, *Bradyrhizobium* (Jordan, 1982).

From the 80s on, with the introduction of more genetic characteristics (DNA–DNA and DNA–rRNA hybridizations, rRNA catalogues, rDNA sequencing) more diversity was discovered among the rhizobia and their relationships with other groups of bacteria became apparent. This led to a gradual increase in the number of genera (Table 1). In parallel, there has also been a significant increase in the number of validly published species (Table 1), with 48 species of rhizobia now recognized.

Two main reasons for this increase in the number of genera and species are:

(1) Many different legume species have now been studied. This in contrast to original efforts, which emphasized those legumes that were important food and pasture species crops, mostly from the Western world. As an example, consider Table 2, where the *Mesorhizobium* species are listed together with the host plants from which they were reported. Even now, only about 20% of the total of about 18,000 species and 57% of about 650 genera of legume plants have been studied for nodulation (Sprent, 1995). This leaves a large number of legume species to be studied and potentially many more species and genera of rhizobia to be described.

Table 1. Rising number of species in the genera of the rhizobia

Genus	Original publication	Number of species					
		Before 1980	81–85	86–90	91–95	96–00	01–06
Agrobacterium	Cohn (1942)	4	4	5	5	5	5
Rhizobium	Frank (1889)	4	5	5	10	10	16
Bradyrhizobium	Jordan (1982)		1	1	3	3	7
Sinorhizobium	Chen et al. (1988)			2	5	8	11
Azorhizobium	Dreyfus et al., (1988)			1	1	1	2
Mesorhizobium	Jarvis et al. (1997)					7	11
Allorhizobium	de Lajudie et al. (1998a)					1	1
Total		8	9	13	23	34	53

Table 2. Overview of the species of *Mesorhizobium* (Jarvis et al., 1997) and the plants they were isolated from

Name	Year[a]	Host plants	Reference
M. loti	1982	*Lotus, Lupinus, Anthyllis, Leucaena*	Jarvis et al. (1982)
M. huakuii	1991	*Astragalus* (China)	Chen et al. (1991)
M. ciceri	1994	*Cicer arietinum* (Spain, USA, India, Russia, Turkey, Morocco, Syria)	Nour et al. (1994)
M. tianshanense	1995	*Glycyrrhiza, Sophora, Caragana, Halimodendron, Swainsonia, Glycine* (China)	Chen et al. (1995)
M. mediterraneum	1995	*Cicer arietinum* (Spain, Syria, India, Lebanon, Syria, Tunesia)	Nour et al. (1995)
M. plurifarium	1998	*Acacia, Prosopis, Chamaecrista, Leucaena* (Senegal, Sudan, Brazil)	de Lajudie et al. (1998b)
M. amorphae	1999	*Amorpha fruticosa* (China)	Wang et al. (1999)
M. chacoense	2001	*Prosopis* (Argentina)	Velázquez et al. (2001)
M. septentrionale	2004	*Astragalus adsurgens* (China)	Gao et al. (2004)
M. temperatum	2004	*Astragalus adsurgens* (China)	Gao et al. (2004)
M. thiogangeticum	2006	Rhizosphere of *Clitoria ternatea* (India)	Gosh and Roy (2006)

[a]Year of first description.

(2) The other reason for increasing numbers of rhizobial species is the ongoing evolution of taxonomic research. Improvements and new developments in the methods to study cell DNA and RNA have led to a more detailed characterization resulting in phylogenetic and polyphasic classifications. Currently an increasing number of total bacterial genomes are becoming available. This will undoubtedly have a further major impact on bacterial taxonomy. Most recent taxonomic studies have made use of a polyphasic approach (Graham et al., 1991;Vandamme et al., 1996), with genetic, phenotypic, chemotaxonomic, phylogenetic data combined to establish a comprehensive picture of the relationships of the bacteria, and to propose a suitable classification.

Sinorhizobium

Chen et al. (1988) proposed a separate genus for the fast-growing soybean rhizobia, renaming *R. fredii* as *Sinorhizobium fredii*, and proposing a second species, *S. xinjiangense* (the original spelling *S. xinjiangensis* was later corrected [Euzéby, 1998]). This new genus was controversial at first since genetic evidence to justify its creation and to separate it from *R. fredii* was not presented at the time (Jarvis et al., 1992). Later, phylogenetic data were presented to support a third genus of rhizobia, not restricted to the fast-growing soybean rhizobia (de Lajudie et al., 1994) and the genus definition was emended. *R. meliloti* was transferred to *Sinorhizobium* as *S. meliloti* and two additional species, *S. saheli* and *S. terangae* (the original spelling *S. teranga* was later corrected [Trüper and De'Clari, 1997]), were proposed for isolates from *Acacia* and *Sesbania* from Senegal. The genus *Sinorhizobium* is now widely accepted and currently has 11 valid species (Table 3). New genetic evidence in support of the separation of *S. xinjiangense* and *S. fredii* has been presented (Peng et al., 2002). However, this strongly relies on DNA–DNA hybridizations performed with DNA from the *S. fredii* type strain USDA 205^T the quality of which was not validated by homologous hybridization with *S. fredii* strains.

It has recently become evident from 16S rDNA comparisons that *Ensifer adhaerens* is also phylogenetically a member of the *Sinorhizobium* lineage (Balkwill, 2005). This organism is a soil bacterium that can adhere to and lyse other soil bacteria, and that was initially described mostly on the basis of phenotypic data (Casida Jr, 1982). Our own polyphasic studies have shown that a small group of four diverse rhizobial isolates and two soil isolates cannot be distinguished clearly from *Ensifer adhaerens* on the basis of DNA–DNA hybridizations and phenotypic features and we should therefore include these rhizobia in *Ensifer*. Phylogenetically, *Ensifer* and *Sinorhizobium*

Table 3. Species of *Sinorhizobium*

Name	Year[a]	Host plants	Reference
S. meliloti	1926	*Melilotus, Medicago, Trigonella*	Dangeard (1926)
S. fredii	1984	*Glycine, Vigna, Cajanus*	Scholla and Elkan (1984)
S. xinjiangense	1988	*Glycine*	Chen et al. (1988)
S. saheli	1994	*Sesbania, Acacia* (Senegal)	de Lajudie et al. (1994)
S. terangae	1994	*Sesbania* (Senegal)	de Lajudie et al. (1994)
S. medicae	1996	*Medicago* (Syria, France)	Rome et al. (1996)
S. arboris	1999	*Acacia, Prosopis* (Sudan, Kenya)	Nick et al. (1999
S. kostiense	1999	*Acacia, Prosopis* (Sudan)	Nick et al. (1999)
S. kummerowiae	2002	*Kummerowia stipulacea*	Wei et al. (2002
S. morelense	2002	*Leucaena leucocephala* (Mexico)	Wang et al. (2002)
"*S. adhaerens*"[b]	2003	*Medicago sativa* (Spain), *Leucaena leucocephala* (Brazil), *Pithecellobium dulce* (Brazil)	Willems et al. (2003)
S. americanum	2003	*Acacia* spp. (Mexico)	Toledo et al. (2003)

[a]Year of first description.
[b]*Ensifer adhaerens* was proposed to belong *Sinorhizobium*, however the name "*Sinorhizobium adhaerens*" remains not valid pending a judicial opinion (Willems et al., 2003).

form a single group in the 16S rDNA dendrogram of the alpha-Proteobacteria and may therefore be regarded as a single genus. This has important nomenclatural consequences because the older name *Ensifer* would have precedence. There are several reasons why a change from *Sinorhizobium* to *Ensifer* may not be the best solution, and allowing an exception to Rule 38 may be more appropriate. We have therefore proposed the creation of the species *Sinorhizobium adhaerens* comb. nov. and submitted a request for Opinion on the conservation of *Sinorhizobium adhaerens* over *Ensifer adhaerens* (Willems et al., 2003). This proposal was regarded unjustified by Young (2003) who proposed that all *Sinorhizobium* species should be transferred to *Ensifer* instead. While the request for opinion is pending the combination "*Sinorhizobium adhaerens*" is not valid and *Ensifer adhaerens* remains the correct name.

Mesorhizobium

The genus *Mesorhizobium* was proposed for five rhizobial species (*R. loti*, *R. huakuii*, *R. ciceri*, *R. mediterraneum* and *R. tianshanense*) that are phylogenetically related and distinct from the large phylogenetic grouping that includes *Rhizobium*, *Agrobacterium* and *Sinorhizobium* (Jarvis et al., 1997). They are characterized by a growth rate intermediate between the fast- and slow-growing rhizobia. On the basis of 16S rDNA sequence data, *Mesorhizobium* is phylogenetically separated from the fast-growing rhizobia by the genera *Bartonella*, *Defluvibacter*, *Aquamicrobium* *Phyllobacterium*, *Aminobacter* and *Pseudaminobacter* (Figure 1).

Bradyrhizobium

Bradyrhizobium was created for the slow-growing species *Rhizobium japonicum* (Jordan, 1982). Originally the soybean-nodulating *B. japonicum* was the only species described, although it was recognized that slow-growing strains occur on various legume genera (Elkan and Bunn, 1992). To date, five additional species have been validly named in this genus, two of them nodulating *Glycine* (*B. elkanii* [Kuykendall et al., 1992] and *B. liaoningense* [Xu et al., 1995]), *B. yuanmingense* nodulating *Lespedeza* (Yao et al., 2002),

Figure 1. 16S rDNA phylogeny of rhizobia and relatives in the alpha-Proteobacteria. The tree was calculated with the neighbor joining method, using Kimura-2 corrections. A bootstrap analysis was performed on 500 replicates and the groupings that were recovered in 95 or more percent of trees are marked in the dendrogram by a black dot at the branching point. Numbers P1–P9 refer to the *Agrobacterium* DNA groups of Popoff et al. (1984).

B. betae from the roots of *Beta vulgaris* afflicted with tumor-like deformations (Rivas et al., 2004), *B. canariense* from genistoid legumes from the Canary Islands (Vinuesa et al., 2005). In addition to the species subdivision, a number of serogroups have been described among slow-growing soybean symbionts (Date and Decker, 1965). Many other slow-growing rhizobia have been isolated from other legume hosts and are commonly referred to as *Bradyrhizobium* sp., followed by the name of the legume host. A special feature of the *Bradyrhizobium*–legume symbiosis is that some bradyrhizobia can form stem nodules on some plant species, produce bacteriochlorophyll and perform photosynthesis (Alazard, 1985; Evans et al., 1990; Molouba et al., 1999). Some photosynthetic bradyrhizobia have also been reported as endophytes of African wild rice (Chaintreuil et al., 2000).

A major factor complicating the evaluation of the taxonomic status and interrelationships of bradyrhizobia is the high similarity of 16S rDNA gene sequences. Many strains have 16S rDNA sequence divergences of 0.1–2.0%. Only sequences for *B. elkanii* and related strains differ by up to 4% from those of other bradyrhizobia (Willems et al., 2001a). A further complicating factor is the very slow growth of these organisms, often precluding the use of the standard phenotypic test procedures (e.g. Biolog, API systems). As a consequence many bradyrhizobia have been characterized more thoroughly by genotypic methods. Our own work using AFLP, DNA–DNA hybridizations and 16S–23S internal transcribed spacer (ITS) analyses has resulted in the delineation of at least 11 *Bradyrhizobium* genospecies, including the named species (Willems et al., 2001c).

From 16S rDNA phylogeny, the genera *Afipia*, *Rhodopseudomonas* and *Nitrobacter* also appear closely related to the bradyrhizobia, with *B. ekanii* occupying a more peripheral phylogenetic position (Willems et al., 2001a). This is in contrast to ITS

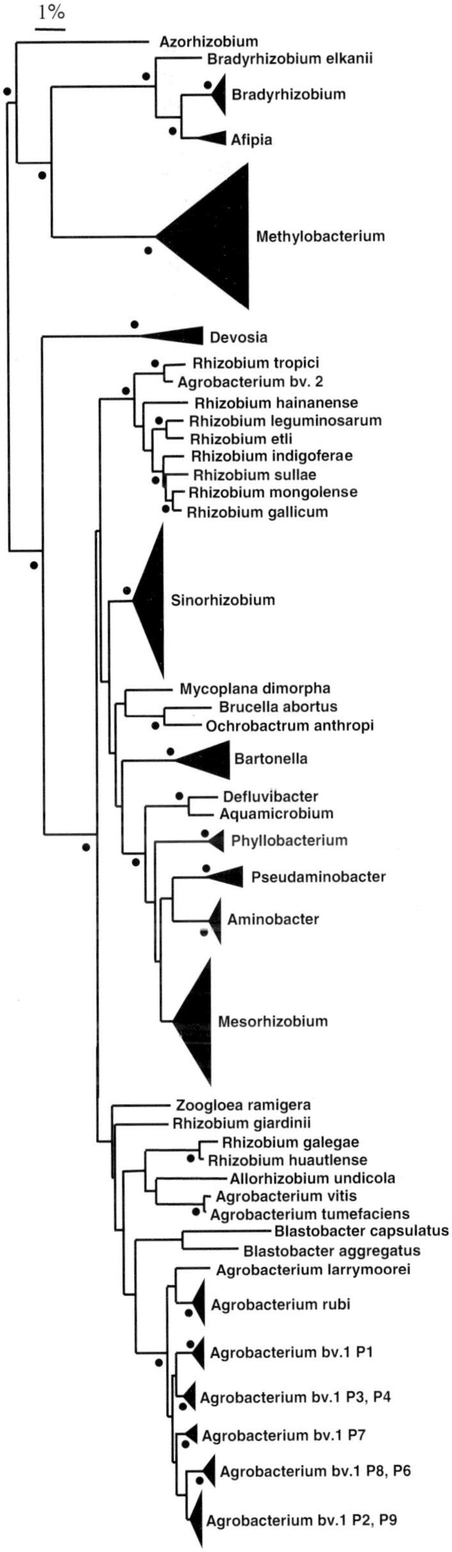

sequence data that show that *B. elkanii* is more closely related to the bradyrhizobia than are the three non-rhizobial genera. Based on ITS sequence data, the photosynthetic bradyrhizobia isolated from stem-nodules of *Aeschynomene*, form a distinct group closely related to *Blastobacter denitrificans* (Willems et al., 2001b; van Berkum and Eardly, 2002). As a result of a comprehensive study of both groups, van Berkum et al. (2006) recently proposed to transfer *Blastobacter denitrificans* to *Bradyrhizobium* and unite the species with the isolates from *Aeschynomene indica* as the species *Bradyrhizobium denitrificans*.

The Rhizobium–Allorhizobium–Agrobacterium issue

Figure 1 provides an overview of the phylogeny of the rhizobia and relatives in the alpha-Proteobacteria based on 16S rDNA sequence data. *Bradyrhizobium* and *Azorhizobium* are quite separate. *Sinorhizobium* and *Mesorhizobium* also form separate clusters, but it is clear that *Agrobacterium*, *Allorhizobium* and *Rhizobium* are rather more closely related. The recent proposal (Young et al., 2001) to abandon the genera *Agrobacterium* and *Allorhizobium* and incorporate them in *Rhizobium* has not met with universal approval (Farrand et al., 2003).

The genus *Allorhizobium* contains a single species, *Al. undicola*, for isolates from nodules of *Neptunia natans* from Senegal (de Lajudie et al., 1998a). Phylogenetically (Figure 1), it takes a separate position in the large *Agrobacterium–Rhizobium* 16S rDNA cluster, with *A. vitis* (96.3% 16S rDNA sequence similarity), *R. galegae* (95.1%) and *R. huautlense* (95.3%) as its nearest neighbors. In view of its remoteness from the *Rhizobium* type species, *R. leguminosarum*, and the confused taxonomic situation in *Agrobacterium* (see below) and because the *Neptunia* isolates can be distinguished phenotypically and genotypically from related taxa, it was considered most appropriate that they be placed in a separate genus (de Lajudie et al., 1998a). This genus may need emendation or revision in a future scheme to correct the classification and nomenclature of *Agrobacterium* and *Rhizobium* species, in particular *A. vitis*, *R. galegae* and *R. huautlense*.

The genus *Rhizobium* currently has 15 species, from various hosts (Table 4 – not including *Sinorhizobium* and *Agrobacterium*). *Agrobacterium*, a genus proposed in 1942 (Conn, 1942) that comprises bacteria responsible for various kinds of hypertrophies in plants, has six valid species. The oldest species were – for practical purposes – described mainly on the basis of phytopathological properties. For example, *A. tumefaciens* groups those strains that cause tumors on plants; *A. radiobacter* unites strains that are not pathogenic and *A. rhizogenes* comprises strains that cause hairy root growth (Conn, 1942). Of the other species, *A. rubi* is pathogenic on *Rubus* (Starr and Weiss, 1943), *A. vitis* on grapevine (Ophel and Kerr, 1990) and *A. larrymoorei* on *Ficus* (Bouzar and Jones, 2001).

Using a polyphasic approach, various authors have recognized three large groups or biovars among strains assigned to *A. tumefaciens*, *A. rhizogenes* and *A. radiobacter* in the official (phytopathology-based) classification (summarized in Kersters and De Ley, 1984). The taxonomic situation is complicated by the fact that these biovars do not correspond to the existing species, with biovar 1 containing strains of *A. tumefaciens*, *A. rhizogenes* and *A. radiobacter*. Among these strains are the type strains of *A. tumefaciens* and *A. radiobacter*. Biovar 2 also contains strains of all three species, including the type strain of *A. rhizogenes*, whereas biovar 3 contains *A. tumefaciens* and *A. vitis* strains. Furthermore, *Agrobacterium* biovar 1 has been shown to contain several genospecies by DNA–DNA hybridizations (De Ley, 1974; Popoff et al., 1984), some of which contain clinical isolates that do not have virulence genes but were originally named *Agrobacterium* because of biochemical features (Lautrop, 1967; Riley and Weaver, 1977). A further complicating factor is that the genus *Agrobacterium* was declared conserved by the Judicial Commission, with *A. tumefaciens* as the type species (Judicial Commission, 1970). However, as Young et al. (2006) recently pointed out, the species name *A. tumefaciens* is not a conserved name as some authors may previously have believed.

The phytopathology-based taxonomy and the polyphasic classification (Kersters and De Ley, 1984) of the genus *Agrobacterium* is shown in Table 5. No species names were proposed for the biovars described by Kersters and De Ley (1984) because the rules of the Bacteriological Code

Table 4. Species of *Rhizobium*

Name	Year[a]	Host plant(s)	Reference
R. leguminosarum	1879	*Pisum, Lathyrus, Vicia, Lens, Phaseolus, Trifolium*	Frank (1879)
R. lupinii[b]	1886	*Lupinus, Ornithopus*	Schroeter (1886)
R. galegae	1989	*Galega*	Lindström (1989)
R. tropici	1991	*Phaseolus vulgaris, Leucaena*	Martínez-Romero et al. (1991)
R. etli	1993	*Phaseolus vulgaris*	Segovia et al. (1993)
R. gallicum	1997	*Phaseolus vulgaris*	Amarger et al. (1997)
R. giardinii	1997	*Phaseolus vulgaris*	Amarger et al. (1997)
R. hainanense	1997	*Desmodium, Stylosanthes, Centrosema, Tephrosia, Acacia, Zornia, Macroptilium*	Chen et al. (1997)
R. mongolense	1998	*Medicago ruthenica*	van Berkum et al. (1998)
R. huautlense	1998	*Sesbania herbacea*	Wang et al., (1998)
R. yanglingense	2001	*Coronilla, Gueldenstaedtia, Amphicarpaea*	Tan et al. (2001)
R. sullae	2002	*Hedysarum coronarium*	Squartini et al. (2002)
R. indigoferae	2002	*Indigofera*	Wei et al. (2002)
R. loessense	2003	*Astragalus*	Wei et al. (2003)
R. daejeonense	2005	*Medicago*	Quan et al. (2005)

[a]Year of first description.

[b]The relationships of *R. lupinii* (Kuykendall et al., 2005) are unclear because of doubts on the purity of the type strain, but the species was included on the approved lists.

Table 5. Official and polyphasic classification of *Agrobacterium*

Official classification[a] (phytopathology)	Polyphasic classification[b] (Kersters and De Ley, 1984)
A. tumefaciens (tumors)	Biovar 1: tumorigenic (*A. tumefaciens*), rhizogenic (*A. rhizogenes*) and avirulent (*A. radiobacter*) strains, includes type strains of *A. tumefaciens* and *A. radiobacter*
A. radiobacter (no symptoms)	Biovar 2: tumorigenic (*A. tumefaciens*), rhizogenic (*A. rhizogenes*) and avirulent (*A. radiobacter*) strains, includes type strain of *A. rhizogenes*
A. rhizogenes (hairy roots)	Biovar 3: tumorigenic on *Vitis* (*A. tumefaciens* and *A. vitis*)
A. rubi (*Rubiaceae*)	*A. rubi*
A. vitis (*Vitis*)	
A. larrymoorei (*Ficus*)	

[a]The official classification includes the species that are in the Approved List of Bacterial Names (Skerman et al., 1980) or were subsequently published in International Journal of Systematic Bacteriology (now International Journal of Systematic and Evolutionary Microbiology). All *Agrobacterium* species are transferred to *Rhizobium* in the proposal of Young et al. (2001).
[b]The polyphasic classification is a consensus classification based on different methods as presented in Bergey's Manual of Systematic Bacteriology (Kersters and De Ley, 1984).

would have required biovar 1 to be named *A. tumefaciens* and biovar 2 *A. rhizogenes*. These names would then apply to strains with and without the phytopathological properties their name implies. This was regarded as unacceptable at the time and thus the official species classification and the polyphasic biovar system have been used in parallel for many years. This situation is clearly unsatisfactory and in 1993 Sawada et al. (1993) proposed that biovar 1 be named *A. radiobacter* and biovar 2 *A. rhizogenes*. This was thought to go against Opinion 33 of the Judicial Commission (Bouzar, 1994) and was not widely adopted.

From the 16S rDNA phylogeny (Figure 1), it is clear that *Rhizobium* and *Agrobacterium* are highly related and their species are interwoven. In particular, biovar 2 groups with the majority of *Rhizobium* species. Biovar 1 consists of several smaller groups representing different genospecies. *A. rubi* and *A. larrymoorei* are closely related to these biovar 1 genospecies. *A. vitis* is close to *Allorhizobium*. *Rhizobium giardinii* is the most peripheral of the whole group. Young et al. (2001) proposed the transfer all these taxa to *Rhizobium*. They proposed to unite *A. radiobacter* and *A. tumefaciens* in *R. radiobacter*, which thus represents biovar 1, while *A. rhizogenes* becomes *R. rhizogenes* and represents biovar 2. *A. rubi* and *A. vitis* are transferred to *Rhizobium* as distinct species and also *A. larrymoorei* is transferred as *R. larrymoorei* (Young, 2004). This proposal solves the species matching the biovars and the placement of *A. rhizogenes* in *Rhizobium* is clearly justified, but it is not widely accepted (Farrand et al., 2003) and several problems remain to be addressed:

(1) Many strains of *Agrobacterium* species have in the past been named on the basis of phytopathological effects they cause on plants and are listed in culture collection catalogues as such, often without their biovar status being known. It is not clear which *Rhizobium* species these should be classified as. For example, an *A. tumefaciens* strain may belong to biovar 1, 2 or 3 and depending on this should be classified as *R. radiobacter*, *R. rhizogenes* or *R. vitis*, respectively.
(2) The biovar 1 genospecies are ignored in the new scheme.
(3) The incomplete phenotypic differentiation of these genospecies.
(4) The proposed enlarged genus *Rhizobium* would be a large, widely defined and phylogenetically deep genus (Figure 1).
(5) The species *Blastobacter capsulatus*, *Blastobacter aggregatus* and *Zooglea ramigera*, that group with the *Agrobacterium–Rhizobium* phylogenetic cluster (Figure 1) and therefore would group in the proposed large genus *Rhizobium*, should be taken into account.

When considering the 16S rDNA phylogeny of part of the alpha-Proteobacteria (Figure 1), it is obvious that the new genus *Rhizobium* is rather large and represents a phylogenetically more divers (deeper) group then the other genera in its phylogenetic vicinity. However, to present an alternative proposal, additional data are essential. The current proposal was based mostly on 16S rDNA data. It is clear that data from other genes can provide useful insights to unravel relationships in these groups. With the new complete genome sequence data that are now becoming available, it may soon become possible to make a more comprehensive comparison and arrive at a suitable classification. Meanwhile, microbiologists should be aware that all validly published names can be used: it is the scientific community that decides on the value of any new proposal by either using it or, alternatively, by using a previously validly published classification.

Other nitrogen-fixing legume symbionts

Recently, a number of isolates have been reported from legume nodules, capable of nitrogen fixation but phylogenetically located outside the traditional groups of rhizobia in the alpha-Proteobacteria. New lines that contain nitrogen-fixing legume symbionts include *Methylobacterium*, *Devosia*, *Ochrobactrum* and *Phyllobacterium* in the alpha-Proteobacteria and *Burkholderia*, *Ralstonia* and *Cupriavidus* in the beta-Proteobacteria.

In *Burkholderia*, a genus that contains over 20 species of plant pathogens, soil and plant-associated bacteria and clinical isolates, the following symbiotic, nitrogen fixing strains have been identified: (1) two strains from *Mimosa* were found to belong to *B. caribensis*, a species that was first described for soil isolates from Martinique; (2) one strain from *Alysicarpus* was found to belong to *B. cepacia* genomovar VI (now *B. dolosa* [Vermis et al., 2004], a group previously only found in cystic fibrosis patients; (3) one strain from *Machaerium* was identified as a new species for which the name *B. phymatum* has been proposed and (4) one strain from *Aspalathus* was identified as a second new species for which the name *B. tuberum* was proposed (Moulin et al., 2001; Vandamme et al., 2003).

In *Ralstonia*, like *Burkholderia* a genus of plant pathogenic or plant associated, soil and clinical organisms, *Ralstonia taiwanensis* was proposed for strains from *Mimosa* species in Taiwan (Chen et al., 2001).

In the alpha-Proteobacteria, *Devosia neptuniae* was proposed for strains from *Neptunia natans* from India (Rivas et al., 2003) and *Methylobacterium nodulans* for strains from *Crotalaria* (Jourand et al., 2004; Sy et al., 2001). *Ochrobactrum lupinus* was described for nodule isolates from *Lupinus* sp. (Trujillo et al., 2005) and *Phyllobacterium lupinii* for isolates nodulating *Trifolium* and *Lupinus* (Valverde et al., 2005).

All these new nodulating bacteria are phylogenetically (16S rDNA) distinct from the rhizobia, but do carry nod genes similar to those of rhizobia. These genes encode for Nod factors, signal molecules in the bacterium–legume communication that accompanies nodulation. It is in fact from studies of nod gene diversity that some discoveries of new nodulating strains outside the rhizobia have originated. The nod genes were most probably obtained by these new nitrogen-fixing legume symbionts through lateral-gene transfer (Moulin et al., 2001; Sy et al., 2001). Most of the new nodulating bacteria belong to genera that have at least some plant-associated species and that are therefore likely to have the molecular strategies to overcome plant defenses. Recent reports confirm that it is quite likely that more such bacteria, capable of effective nodulation will be discovered outside the traditional rhizobia (Barret and Parker, 2006; Rasolomampianina et al., 2005; Zakhia et al., 2006).

Acknowledgements

The author is grateful to the Fund for Scientific Research – Flanders for a Postdoctoral Fellowship.

References

Alazard D 1985 Stem and root nodulation in *Aeschynomene* sp. Appl. Evironm. Microbiol. 50, 732–734.

Amarger N, Macheret V and Laguerre G 1997 *Rhizobium gallicum* sp. nov. and *Rhizobium giardinii* sp. nov., from *Phaseolus vulgaris*. Int. J. Syst. Bacteriol. 47, 996–1006.

Balkwill D L 2005 Genus VI. *Ensifer* Cassida 1982, 343^{VP}. *In* Bergey's Manual of Systematic Bacteriology 2nd ed. Vol. 2, part C. Ed. G M Garrity. pp. 354–358. Springer, New York.

Barrett C F and Parker M A 2006 Coexistence of *Burkholderia*, *Cupriavidus*, and *Rhizobium* sp nodule bacteria on two *Mimosa* spp. in Costa Rica. Appl. Environ. Microbiol. 72, 1198–1206.

Beijerinck M W 1888 Cultur des *Bacillus radicola* aus den Knöllchen. Bot. Ztg. 46, 740–750.

Bouzar H 1994 Request for a Judicial Opinion concerning the type species of *Agrobacterium*. Int. J. Syst. Bacteriol. 44, 373–374.

Bouzar H and Jones J B 2001 *Agrobacterium larrymoorei* sp. nov., a pathogen isolated from aerial tumours of *Ficus benjamini*. Int. J. Syst. Evol. Microbiol. 51, 1023–1026.

Casida L E Jr 1982 *Ensifer adhaerens*, gen. nov., sp. nov.: A bacterial predator of bacteria in soil. Int. J. Syst. Bacteriol. 32, 339–345.

Chantreuil C, Giraud E, Prin Y, Lorquin J, Bâ A, Gillis M, de Lajudie P and Dreyfus B 2000 Photosynthetic bradyrhizobia are natural endophytes of the African wild rice *Oryza breviligulata*. Appl. Environ. Microbiol. 66, 5437–5447.

Chen W-M, Laevens S, Lee T-M, Coenye T, De Vos P, Mergeay M and Vandamme P 2001 *Ralstonia taiwanensis* sp. nov., isolated from root nodules of *Mimosa* species and sputum of cystic fibrosis patients. Int. J. Syst. Evol. Microbiol. 51, 1729–1735.

Chen W X, Li G S, Qi Y L, Wang E T, Yuan H L and Li J L 1991 *Rhizobium huakuii* sp. nov. isolated from the root nodules of *Astragalus sinicus*. Int. J. Syst. Bacteriol. 41, 275–280.

Chen W-X, Tan Z-Y, Gao J-L, Li Y and Wang E T 1997 *Rhizobium hainanense* sp. nov., isolated from tropical legumes. Int. J. Syst. Bacteriol. 47, 870–873.

Chen W, Wang E, Wang S, Li Y, Chen X and Li Y 1995 Characterization of *Rhizobium tianshanense* sp. nov., a moderately and slow growing root nodule bacterium isolated from an arid saline environment in Xinjiang, People's Republic of China. Int. J. Syst. Bacteriol. 45, 153–159.

Chen W X, Yan G H and Li J L 1988 Numerical taxonomic study of fast-growing soybean rhizobia and proposal that *Rhizobium fredii* be assigned to *Sinorhizobium* gen. nov. Int. J. Syst. Bacteriol. 38, 392–397.

Conn H J 1942 Validity of the genus *Alcaligenes*. J. Bacteriol. 44, 353–360.

Dangeard P A 1926 Recherches sur les turbercles radicaux des Légumineuses. Botaniste (Paris) 16, 1–275.

Date R A and Decker A M 1965 Minimal antigenic constitution of 28 strains of *Rhizobium japonicum*. Can. J. Microbiol. 11, 1–8.

de Lajudie P, Laurent-Fulele E, Willems A, Torck U, Coopman R, Collins M D, Kersters K, Dreyfus B and Gillis M 1998a *Allorhizobium undicola* gen. nov., sp. nov., nitrogen-fixing bacteria that efficiently nodulate *Neptunia natans* in Senegal. Int. J. Syst. Bacteriol. 48, 1277–1290.

de Lajudie P, Willems A, Nick G, Moreira F, Molouba F, Hoste B, Torck U, Neyra M, Collins M D, Lindström K, Dreyfus B and Gillis M 1998b Characterization of tropical tree rhizobia and description of *Mesorhizobium plurifarium* sp. nov. Int. J. Syst. Bacteriol. 48, 369–382.

de Lajudie P, Willems A, Pot B, Dewettinck D, Maestrojuan G, Neyra M, Collins M D, Dreyfus B, Kersters K and Gillis M 1994 Polyphasic taxonomy of rhizobia: Emendation of the genus *Sinorhizobium* and description of *Sinorhizobium meliloti* comb. nov., *Sinorhizobium saheli* sp. nov., and *Sinorhizobium teranga* sp. nov. Int. J. Syst. Bacteriol. 44, 715–733.

De Ley J 1974 Phylogeny of Prokaryotes. Taxon 23, 291–300.

De Ley J and Rassel A 1965 DNA base composition, flagellation and taxonomy of the genus *Rhizobium*. J. Gen. Microbiol. 41, 85–91.

Dreyfus B, Garcia J L and Gillis M 1988 Characterization of *Azorhizobium caulinodans* gen. nov., sp. nov., a stem-nodulating nitrogen-fixing bacterium isolated from *Sesbania rostrata*. Int. J. Syst. Bacteriol. 38, 89–98.

Elkan G H and Bunn C R 1992 Chapter 107. The rhizobia. *In* The Prokaryotes. A handbook on the Biology of Bacteria: Ecophysiology, Isolation, Identification, Applications. Eds. A Balows, H G Trüper, M Dworkin, W Harder and K-H Schleifer. pp. 2197–2213. 2nd edition, Vol. III Springer-Verlag, New York.

Euzéby J P 1998 Taxonomic note: Necessary correction of specific and subspecific epithets according to Rules 12c and 13b of the International Code of Nomenclature of Bacteria (1990 Revision). Int. J. Syst. Bacteriol. 48, 1073–1075.

Evans W R, Fleischman D E, Calvert H E, Pyati R V, Alter G M and Subba Rao N S 1990 Bacteriochlorophyll and photosynthetic reaction centers in *Rhizobium* strain BTAi1. Appl. Environ. Microbiol. 56, 3445–3449.

Farrand S K, van Berkum P B and Oger P 2003 *Agrobacterium* is a definable member of the family *Rhizobiaceae*. Int. J. Syst. Evol. Microbiol. 53, 1681–1687.

Frank B, 1879 Ueber die Parasiten in den Wurzelanschwillungender Papilionaceen Ber. Dtsch. Bot. Ges. 37, 376–387 and 394–399.

Frank B 1889 Ueber die Pilzsymbiose der Leguminosen. Ber. Deut. Bot. Ges. 7, 332–346.

Fred E B, Baldwin I L and McCoy E 1932 Root Nodule Bacteria and Leguminous Plants. University of Wisconsin Studies in Science, number 5. University of Wisconsin Press, Madison.

Gao J-L, Turner S L, Kan F L, Wang E T, Tan Z Y, Qiu Y H, Gul J, Terefework Z, Young J P W, Lindström K and Chen W X 2004 *Mesorhizobium septentrionale* sp. nov. and *Mesorhizobium temperatum* sp. nov., isolated from *Astragalus adsurgens* growing in the northern regions of China. Int. J. Syst. Evol. Microbiol. 54, 2003–2012.

Ghosh W and Roy P 2006 *Mesorhizobium thiogangeticum* sp. nov., a novel sulfur-oxidizing chemolithoautotroph from rhizosphere soil of an Indian tropical leguminous plant. Int. J. Syst. Evol. Microbiol. 56, 91–97.

Graham P H 1963 Antigenic affinities of the root-nodule bacteria of legumes. Antonie van Leeuwenhoek J. Microbiol. Serol. 29, 281–291.

Graham P H 1964 The application of computer techniques to the taxonomy of the root-nodule bacteria of legumes. J. Gen. Microbiol. 35, 511–517.

Graham P H, Sadowsky M J, Keyser H H, Barnet Y M, Bradley R S, Cooper J E, De Ley J, Jarvis B D W, Roslycky E B, Stijdom B W and Young J P W 1991 Proposed minimal standards for the description of new genera and species of root- and stem-nodulating bacteria. Int. J. Syst. Bacteriol. 41, 582–587.

Jarvis B D W, Downer H L and Young J P W 1992 Phylogeny of fast-growing soybean-nodulating rhizobia supports synonymy of *Sinorhizobium* and *Rhizobium* and assignment to *Rhizobium fredii*. Int. J. Syst. Bacteriol. 42, 93–96.

Jarvis B D W, Pankhurst C E and Patel J J 1982 *Rhizobium loti*, a new species of legume root nodule bacteria. Int. J. Syst. Bacteriol. 32, 378–380.

Jarvis B D W, van Berkum P, Chen W X, Nour S M, Fernandez M P, Cleyet-Marel J C and Gillis M 1997 Transfer of *Rhizobium loti*, *Rhizobium huakuii*, *Rhizobium ciceri*, *Rhizobium mediterraneum*, and *Rhizobium tianshanense* to *Mesorhizobium* gen. nov. Int. J. Syst. Bacteriol. 47, 895–898.

Jordan D C 1982 Transfer of *Rhizobium japonicum* Buchanan 1980 to *Bradyrhizobium* gen. nov., a genus of slow-growing, root nodule bacteria from leguminous plants. Int. J. Syst. Bacteriol. 32, 136–139.

Jordan D C 1984 Family III. Rhizobiaceae Conn 1938. *In* Bergey's Manual of Systematic Bacteriology. Eds. N Krieg and R G Holt. pp. 234–235. 1st edition, Vol. 1 The Williams & Wilkins Co, Baltimore.

Jourand P, Giraud E, Béna G, Sy A, Willems A, Gillis M, Dreyfus B and de Lajudie P 2004 *Methylobacterium nodulans* sp. nov., for a group of aerobic, facultatively methylotrophic, legume root-nodule-forming and nitrogen-fixing bacteria. Int. J. Syst. Evol. Microbiol. 54, 2269–2273.

Judicial Commission 1970 Opinion 33. Conservation of the generic name *Agrobacterium* Conn 1942. Int. J. Syst. Bacteriol 20, 10.

Kersters K and De Ley J 1984 Genus III. *Agrobacterium* Cohn 1942. *In* Bergey's Manual of Systematic Bacteriology. Eds. N Krieg and R G Holt. pp. 244–254. 1st edition, Vol. 1 The Williams & Wilkins Co, Baltimore.

Kuykendall L D, Saxena B, Devine T E and Udell S E 1992 Genetic diversity in *Bradyrhizobium japonicum* Jordan 1982 and a proposal for *Bradyrhizobium elkanii* sp. nov. Can. J. Microbiol. 38, 501–505.

Kuykendall L D, Young J M, Martínez-Romero E, Kerr A and Sawada H 2005 Genus I. Rhizobium Frank 1889, 338[AL]. *In* Bergey's Manual of Systematic Bacteriology, 2nd ed., Vol. 2, part C. Ed. G M Garrity. pp. 325–340. Springer, New York.

Lautrop H 1967 *Agrobacterium* spp. isolated from clinical specimens. Acta Pathol. Microbiol. Scand. 187, 63–64.

Lindström K 1989 *Rhizobium galegae*, a new species of legume root nodule bacteria. Int. J. Syst. Bacteriol. 39, 365–367.

Martínez-Romero E, Segovia L, Martins Mercante F, Franco A A, Graham P and Pardo M A 1991 *Rhizobium tropici*, a novel species nodulating *Phaseolus vulgaris* L. Beans and *Leucaena* sp. trees. Int. J. Syst. Bacteriol. 41, 417–426.

Moffet M L and Colwell R R 1968 Adansonian analysis of the *Rhizobiaceae*. J. Gen. Microbiol. 51, 245–266.

Molouba F, Lorquin J, Willems A, Hoste B, Giraud E, Dreyfus B, Gillis M, de Lajudie P and Masson-Boivin C 1999 Photosynthetic bradyrhizobia from *Aeschynomene* spp. are specific to stem-nodulated species and form a separate 16S ribosomal DNA restriction fragment length polymorphism group. Appl. Environ. Microbiol. 65, 3084–3094.

Moulin L, Munive A, Dreyfus B and Boivin-Masson C 2001 Nodulation of legumes by members of the β-subclass of Proteobacteria. Nature 411, 948–950.

Nick G, de Lajudie P, Eardly B D, Suomalainen S, Paulin L, Zhang X, Gillis M and Lindström K 1999 *Sinorhizobium arboris* sp. nov. and *Sinorhizobium kostiense* sp. nov., isolated from leguminous trees in Sudan and Kenya. Int. J. Syst. Bacteriol. 49, 1359–1368.

Nour S M, Cleyet-Marel J-C, Normand P and Fernandez M P 1995 Genomic heterogeneity of strains nodulating chickpeas (*Cicer arietinum* L.) and description of *Rhizobium mediterreaneum* sp. nov. Int. J. Syst. Bacteriol. 45, 640–648.

Nour S M, Fernandez M P, Normand P and Cleyet-Marel J-C 1994 *Rhizobium ciceri* sp. nov., consisting of strains that nodulate chickpeas (*Cicer arietinum* L.). Int. J. Syst. Bacteriol. 44, 511–522.

Ophel K and Kerr A 1990 *Agrobacterium vitis* sp. nov. for strains of *Agrobacterium* biovar 3 from grapevines. Int. J. Syst. Bacteriol. 40, 236–241.

Peng G X, Tan Z Y, Wang E T, Reinhold-Hurek B, Chen W F and Chen W X 2002 Identification of isolates from soybean nodules in Xinjiang region as *Sinorhizobium xinjiangense* and genetic differentiation of *S. xinjiangense* from *Sinorhizobium fredii*. Int. J. Syst. Evol. Microbiol. 52, 457–462.

Popoff M Y, Kersters K, Kiredjian M, Miras I and Coynault C 1984 Position taxonomique de souches de *Agrobacterium* d'origine hospitalière. Ann. Microbiol. (Inst. Pasteur) 135A, 427–442.

Quan Z-X, Bae H-S, Baek J-H, Chen W-F, Im W-T and Lee S-T 2005 *Rhizobium daejeonense* sp. nov. isolated from a cyanide treatment bioreactor. Int. J. Syst. Evol. Microbiol. 55, 2543–2549.

Rasolomampianina R, Bailly X, Fetiarison R, Rabevohitra R, Bena G, Ramaroson L, Raherimandimby M, Moulin L, De Lajudie P, Dreyfus B and Avarre J C 2005 Nitrogen-fixing nodules from rose wood legume trees (*Dalbergia* spp.) endemic to Madagascar host seven different genera belonging to alpha- and beta-Proteobacteria. Mol. Ecol. 14, 4135–4146.

Riley P S and Weaver R E 1977 Comparison of thirty-seven strains of Vd-3 bacteria with *Agrobacterium radiobacter*: Morphological and physiological observations. J. Clin. Microbiol. 5, 172–177.

Rivas R, Willems A, Palomo J L, García-Benavides P, Mateos P F, Martínez-Molina E, Gillis M and Velázquez E 2004 *Bradyrhizobium betae* sp. nov., isolated from roots of *Beta vulgaris* affected by tumour-like deformations. Int. J. Syst. Evol. Microbiol. 54, 1271–1275.

Rivas R, Willems A, Subba-Rao N, Mateos P F, Dazzo F B, Martínez-Molina E, Gillis M and Velàzquez E 2003 Description of *Devosia neptunia* sp. nov. that nodulates and fixes nitrogen in symbiosis with *Neptunia natans*, an aquatic legume from India. Syst. Appl. Microbiol. 26, 47–53.

Rome S, Fernandez M P, Brunel B, Normand P and Cleyet-Marel J-C 1996 *Sinorhizobium medicae* sp. nov., isolated from annual *Medicago* spp. Int. J. Syst. Bacteriol. 46, 972–980.

Sawada H, Ieki H, Oyaizu H and Matsumoto S 1993 Proposal for rejection of *Agrobacterium tumefaciens* and revised descriptions for the genus *Agrobacterium* and for *Agrobacterium radiobacter* and *Agrobacterium rhizogenes*. Int. J. Syst. Bacteriol. 43, 694–702.

Scholla M H and Elkan G H 1984 *Rhizobium fredii* sp. nov., a fast-growing species that effectively nodulates soybean. Int. J. Syst. Bacteriol. 34, 484–486.

Schroeter J 1886 Schizomycetes. *In* Kryptogamenflora von Sclesien, Bd. 3, Heft 3, Pilze. Ed. Cohn. pp. 1–814. J U Kern's Verlag, Breslau.

Segovia L, Young J P W and Martinez-Romero E 1993 Reclassification of American *Rhizobium leguminosarum* biovar *phaseoli* type I strains as *Rhizobium etli*. Int. J. Syst. Bacteriol. 43, 374–377.

Skerman V B D, McGowan V and Sneath P H A 1980 Approved lists of bacterial names. Int. J. Syst. Bacteriol. 30, 225–420.

Sprent J I 1995 Legume trees and shrubs in the tropics: N_2 fixation in prespective. Soil Biol. Biochem. 27, 401–407.

Squartini A, Struffi P, Döring H, Selenska-Pobell S, Tola E, Giacomini A, Vendramin E, Velàzquez E, Mateos P F, Martínez-Molina E, Dazzo F B, Casella S and Nuti M P 2002 *Rhizobium sullae* sp. nov. (formerly '*Rhizobium hedysari*'), the root-nodule microsymbiont of *Hedysarum coronarium* L. Int. J. Syst. Evol. Microbiol. 52, 1267–1276.

Starr M P and Weiss J E 1943 Growth of phytopathogenic bacteria in a synthetic asparagine medium. Phytopathology 33, 314–318.

Sy A, Giraud E, Jourand P, Garcia N, Willems A, de Lajudie P, Prin Y, Neyra M, Gillis M, Boivin-Masson C and Dreyfus B

2001 Methylotrophic *Methylobacterium* bacteria nodulate and fix nitrogen in symbiosis with legumes. J. Bacteriol. 183, 214–220.

Tan Z Y, Kan F L, Peng G X, Wang E T, Reinholdt-Hurek B and Chen W X 2001 *Rhizobium yanglingense* sp. nov., isolated from arid and semi-arid regions in China. Int. J. Syst. Evol. Microbiol. 51, 909–914.

'tMannetje L 1967 A re-examination of the taxonomy of the genus *Rhizobium* and related genera using numerical analysis. Antonie van Leeuwenhoek J. Microbiol. Serol. 33, 477–491.

Toledo I, Lloret L and Martínez-Romero E 2003 *Sinorhizobium americanus* sp. nov., a new *Sinorhizobium* species nodulating native *Acacia* spp. in Mexico. Syst. Appl. Microbiol. 26, 54–64.

Trujillo M E, Willems A, Abril A, Planchuelo A-M, Rivas R, Ludeña D, Mateos P F, Martínez-Molina E and Velázquez E 2005 Nodulation of *Lupinus* by strains of *Ochrobactrum lupini* sp. nov. Appl. Environ. Microbiol. 71, 1318–1327.

Trüper H G and De'Clari L 1997 Taxonomic note: necessary correction of specific epithets formed as substantives (nouns) "in apposition". Int. J. Syst. Bacteriol. 47, 908–909.

Valverde A, Velázquez E, Fernández-Santos F, Vizcaíno N, Rivas R, Gillis M, Mateos P F, Martínez-Molina E, Igual J M and Willems 2005 *Phyllobacterium trifolii* sp. nov. nodulating *Trifolium* and *Lupinus* in Spanish soils. Int. J. Syst. Evol. Microbiol. 55, 1985–1989.

van Berkum P, Beyene D, Bao G, Campbell T A and Eardly B D 1998 *Rhizobium mongolense* sp. nov. is one of three rhizobial genotypes identified which nodulate and form nitrogen-fixing symbioses with *Medicago ruthenica* [(L.) Ledebour]. Int. J. Syst. Bacteriol. 48, 13–22.

van Berkum P and Eardly B D 2002 The aquatic budding bacterium *Blastobacter denitrificans* is a nitrogen-fixing symbiont of *Aeschynomene indica*. Appl. Environ. Microbiol. 68, 1132–1136.

van Berkum P, Leibold J M and Eardly B D 2006 Proposal for combining *Bradyrhizobium* spp. (*Aeschynomene indica*) with *Blastobacter denitrificans* and to transfer *Blastobacter denitrificans* (Hirsch and Muller, 1985) to the genus *Bradyrhizobium* as *Bradyrhizobium denitrificans* (comb. nov.). Syst. Appl Microbiol 29, 207–215.

Vandamme P, Goris J, Chen W-M, De Vos P and Willems A 2003 *Burkholderia tuberum* sp. nov. and *Burkholderia phymatum* sp. nov., nodulate the roots of tropical legumes. Syst. Appl. Microbiol. 25, 507–512.

Vandamme P, Pot B, Gillis M, De Vos P, Kersters K and Swings J 1996 Polyphasic taxonomy, a consensus approach to bacterial systematics. Microbiol. Rev. 60, 407–438.

Velázquez E, Igual J M, Willems A, Fernández M P, Muñoz E, Mateos P F, Abril A, Toro N, Normand P, Cervantes E, Gillis M and Martínez-Molina E 2001 Description of *Mesorhizobium chacoense* sp. nov., a novel species that nodulates *Prosopis alba* in the Chaco Arido region (Argentina). Int. J. Syst. Evol. Microbiol. 51, 1011–1021.

Vermis K, Coenye T, LiPuma J J, Mahenthiralingam E, Nelis H J and Vandamme P 2004 Proposal to accommodate *Burkholderia cepacia* genomovar VI as *Burkholderia dolosa* sp. nov. Int. J. Syst. Evol. Microbiol. 54, 689–691.

Vincent J M and Humphrey B A 1970 Taxonomically significant group antigens in *Rhizobium*. J. Gen. Microbiol. 63, 379–382.

Vinuesa P, León-Barrios M, Silva C, Willems A, Jarabo-Lorenzo A, Pérez-Galdona R, Werner D and Martínez-Romero E 2005 *Bradyrhizobium canariense* sp. nov., an acid-tolerant endosymbiont that nodulates endemic genistoid legumes (*Papilionoideae*: *Genisteae*) from the Canary Islands, along with *Bradyrhizobium japonicum* bv. *genistearum*, *Bradyrhizobium* genospecies alpha and *Bradyrhizobium* genospecies beta. Int. J. Syst. Evol. Microbiol. 55, 569–575.

Wang E T, van Berkum P, Beyene D, Sui X H, Dorado O, Chen W X and Martínez-Romero E 1998 *Rhizobium huautlense* sp. nov., a symbiont of *Sesbania herbacea* that has a close phylogenetic relationship with *Rhizobium galegae*. Int. J. Syst. Bacteriol. 48, 687–699.

Wang E, Tan Z Y, Willems A, Fernández-López M, Reinhold-Hurek B and Martínez-Romero E 2002 *Sinorhizobium morelense*, sp. nov. a *Leucaena leucocephala*-associated bacterium that is highly resistant to multiple antibiotics. Int. J. Syst. Evol. Microbiol. 52, 1687–1693.

Wang E T, van Berkum P, Sui X H, Beyene D, Chen W X and Martínez-Romero E 1999 Diversity of rhizobia associated with *Amorpha fruticosa* isolated from Chinese soils and descritpion of *Mesorhizobium amorphae* sp. nov. Int. J. Syst. Bacteriol. 49, 51–65.

Wei G H, Tan Z Y, Zhu M E, Wang E T, Han S Z and Chen W X 2003 Characterization of rhizobia isolated from legume species within the genera *Astragalus* and *Lespedeza* grown in the Loess Plateau of China and description of *Rhizobium loessense* sp. nov. Int. J. Syst. Evol. Microbiol. 53, 1575–1583.

Wei G H, Wang E T, Tan Z Y, Zhu M E and Chen W X 2002 *Rhizobium* indigoferae sp. nov. and *Sinorhizobium kummerowiae* sp. nov., respectively isolated from *Indigofera* spp. and *Kummerowia stipulaceae*. Int. J. Syst. Evol. Microbiol. 52, 2231–2239.

Willems A, Coopman R and Gillis M 2001a Phylogenetic and DNA: DNA hybridization analyses of *Bradyrhizobium* species. Int. J. Syst. Evol. Microbiol. 51, 111–117.

Willems A, Coopman R and Gillis M 2001b Comparison of sequence analysis of 16S–23S spacer regions, AFLP analysis and DNA–DNA hybridizations in *Bradyrhizobium*. Int. J. Syst. Evol. Microbiol. 51, 623–632.

Willems A, Doignon-Bourcier F, Goris J, Coopman R, de Lajudie P, De Vos P and Gillis M 2001c DNA–DNA hybridization study of *Bradyrhizobium* strains. Int. J. Syst. Evol. Microbiol. 51, 1315–1322.

Willems A, Fernández-López M, Muñoz-Adelantado E, Goris J, De Vos P, Martínez-Romero E, Toro N and Gillis M 2003 Description of New *Ensifer* strains from nodules and proposal to transfer *Ensifer adhaerens* Cassida 1982 to *Sinorhizobium* as *Sinorhizobium adhaerens* comb. nov. Request for an Opinion. Int. J. Syst. Evol. Microbiol. 53, 1207–1217.

Wilson J K 1944 Over five hundred reasons for abandoning the cross inoculation groups of legumes. Soil Sci. 58, 61–69.

Xu M L, Ge C, Cui Z, Li J and Fan H 1995 *Bradyrhizobium liaoningense* sp. nov., isolated from the root nodules of soybeans. Int. J. Syst. Bacteriol. 45, 706–711.

Yao Z Y, Kan F L, Wang E T, Wei G H and Chen W X 2002 Characterization of rhizobia that nodulate legume species within the genus Lespedeza and description of *Bradyrhizobium yuanmingense* sp. nov. Int. J. Syst. Evol. Microbiol. 52, 2219–2230.

Young J M 2003 The genus name *Ensifer* Casida 1982 takes priority over *Sinorhizobium* Chen et al. 1988, and *Sinorhizobium morelense* Wang et al. 2002 is a later synonym of *Ensifer adhaerens* Casida 1982. Is the combination '*Sinorhizobium adhaerens*' (Casida 1982) Willems et al. 2003 legitimate? Request for an Opinion. Int. J. Syst. Evol. Microbiol. 53, 2107–2110.

Young J M 2004 Renaming of *Agrobacterium larrymoorei* Bouzar and Jones 2001 as *Rhizobium larrymoorei* (Bouzar and Jones 2001) comb. nov. Int. J. Syst. Evol. Microbiol. 54, 149.

Young J M, Kuykendall L D, Martínez-Romero E, Kerr A and Sawada H 2001 A revision of *Rhizobium* Frank 1889, with an emended description of the genus, and the inclusion of all species of *Agrobacterium* Conn 1942 and *Allorhizobium undicola* de Lajudie et al. 1998 as new combinations: *Rhizobium radiobacter*, *R. rhizogenes*, *R. rubi*, *R. undicola* and *R. vitis*. Int. J. Syst. Evol. Microbiol. 51, 89–103.

Young J M, Pennycook S R and Watson D R W 2006 Proposal that *Agrobacterium radiobacter* has priority over *Agrobacterium tumefaciens*. Request for an Opinion. Int. J. Syst. Evol. Microbiol. 56, 491–493.

Zakhia F, Jeder H, Willems A, Dreyfus B, de Lajudie P 2006 Diverse bacteria associated with root nodules of spontaneous legumes in Tunisia and first report for nifH-like gene within the genera *Microbacterium* and *Starkeya*. Microb. Ecol. 51, 375–393.

Genetics of phosphate solubilization and its potential applications for improving plant growth-promoting bacteria

H. Rodríguez[1,2,*], R. Fraga[1], T. Gonzalez[1] & Y. Bashan[2]
[1]*Dept. of Microbiology, Cuban Research Institute on Sugar Cane By-Products (ICIDCA), 4026, CP 11 000, Havana, Cuba.* [2]*Environmental Microbiology Group, Center for Biological Research of the Northwest (CIB), La Paz B.C.S., Mexico.* [1]*Corresponding author**

Received: 31 May 2006

Key words: genetically modified microorganisms, organic acids, phosphatases, phosphorus solubilization, phytases, plant growth promoting bacteria

Abstract

Plant growth-promoting bacteria (PGPB) are soil and rhizosphere bacteria that can benefit plant growth by different mechanisms. The ability of some microorganisms to convert insoluble phosphorus (P) to an accessible form, like orthophosphate, is an important trait in a PGPB for increasing plant yields. In this mini-review, the isolation and characterization of genes involved in mineralization of organic P sources (by the action of enzymes acid phosphatases and phytases), as well as mineral phosphate solubilization, is reviewed. Preliminary results achieved in the engineering of bacterial strains for improving capacity for phosphate solubilization are presented, and application of this knowledge to improving agricultural inoculants is discussed.

Introduction

Plant growth-promoting bacteria (PGPB) are soil and rhizosphere bacteria that can benefit plant growth by different mechanisms (Glick, 1995). Given the negative environmental impact of chemical fertilizers and their increasing costs, the use of PGPB as natural fertilizers is advantageous for the development of sustainable agriculture.

There are two components of P in soil, organic and inorganic phosphates. A large proportion is present in insoluble forms, and therefore, not available for plant nutrition. Inorganic P occurs in soil, mostly in insoluble mineral complexes, some of them appearing after the application of chemical fertilizers. These precipitated forms cannot be absorbed by plants. Organic matter, on the other hand, is an important reservoir of immobilized P that accounts for 20–80% of soil P (Richardson, 1994). To convert insoluble phosphates (both organic and inorganic) to a form accessible to the plants, like orthophosphate, is an important trait for a PGPB for increasing plant yields.

Molecular biology techniques are an advantageous approach for obtaining and characterizing improved PGPB strains (Igual et al., 2001; Rodríguez and Fraga, 1999). Release of genetically modified organisms is controversial. While some countries encourage it, others prohibit the use of the technology and require labeling of products containing genetically modified food ingredients. However, studies carried out so far have shown that following appropriate regulations, genetically modified microorganisms can be applied safely in agriculture (Armarger, 2002; Morrissey et al., 2002). Chromosomal insertion of the genes is one of the tools to avoid horizontal transfer of the introduced genes within the rhizosphere.

* Fax No: + 53-7-338236.
E-mail: hildadiaxi03@yahoo.com

E. Velázquez and C. Rodríguez-Barrueco (eds.),
First International Meeting on Microbial Phosphate Solubilization, 15–21

Some barriers should be overcome first to achieve successful gene insertions using this approach, such as the dissimilarity of metabolic machinery and different regulating mechanism between the donor and recipient strains. Despite the difficulties, significant progress has been made in obtaining genetically engineering microorganisms for agricultural use (Armarger, 2002).

There are several advantages of developing genetically-modified PGPB over transgenic plants for improving plant performance: (1) With current technologies, it is far easier to modify a bacterium than complex higher organisms, (2) Several plant growth-promoting traits can be combined in a single organism, and (3) Instead of engineering crop by crop, a single, engineered inoculant can be used for several crops, especially when using a nonspecific genus like *Azospirillum*.

Introduction or over-expression of genes involved in soil phosphate solubilization (both organic and inorganic) in natural rhizosphere bacteria is a very attractive approach for improving the capacity of microorganisms to work as inoculants. Insertion of phosphate-solubilizing genes into microorganisms that do not have this capability may avoid the current need of mixing two populations of bacteria, when used as inoculants (nitrogen fixers and phosphate-solubilizers) (Bashan et al., 2000). We report on recent advances in the manipulation of genes related to microbial phosphate-solubilization and its relationship to rhizobacteria, as improved inoculants.

Organic phosphate solubilization

Phosphorus can be released from organic compounds in soil by three groups of enzymes: (1) Nonspecific phosphatases, which perform dephosphorylation of phospho-ester or phosphoanhydride bonds in organic matter, (2) Phytases, which specifically cause P release from phytic acid, and (3) Phosphonatases and C–P Lyases, enzymes that perform C–P cleavage in organophosphonates. The main activity apparently corresponds to the work of acid phosphatases and phytases because of the predominant presence of their substrates in soil.

Availability of organic phosphate compounds for plant nutrition could be a limitation in some soils resulting from precipitation with soil particle ions. Therefore, the capability of enzymes to perform the desired function in the rhizosphere is a crucial aspect for their effectiveness in plant nutrition. Nevertheless, the efficiency of plant and microbial phosphatases on organic P depletion in the rhizosphere and P uptake by plants has been well documented (Tarafdar and Jungk, 1987; Tarafdar and Claassen, 1988).

Nonspecific acid phosphatases

Bacterial nonspecific acid phosphatases (phosphohydrolases) (NSAPs) are formed by three molecular families, which have been designated as molecular class A, B, and C (Thaller et al., 1995a). From their cellular location, these enzymes seem to function as organic phospho-ester scavengers, providing the cell with essential nutrients (releasing inorganic phosphates from nucleotides and sugar phosphates, for example, while the organic by-products are incorporated into the cell) (Beacham, 1980; Wanner, 1996).

Interest in these enzymes has increased during the last decade because of their potential biotechnological applications. Macaskie et al. (1997) reported on the successful use of Class A NSAPs as tools for environmental bioremediation of uranium-bearing wastewater, and Baskanova and Macaskie (1997) and Bonthrone et al. (1996) on heavy metal biomineralization (particularly Ni^{2+}). A new biotechnological application for NSAPs would be to transfer and express these genes in PGPB to obtain improved phosphate-solubilizing strains using recombinant DNA technology.

Several acid phosphatase genes from Gram-negative bacteria have been isolated and characterized (Rossolini et al., 1998). These cloned genes represent an important source of material for the genetic transfer of this trait to PGPB strains. Some of them code for acid phosphatase enzymes that are capable of performing well in soil. For example, the *acpA* gene isolated from *Francisella tularensis* expresses an acid phosphatase with optimum action at pH 6, with a wide range of substrate specificity (Reilly et al., 1996). Also, genes encoding nonspecific acid phosphatases class A (PhoC) and class B (NapA) isolated from *Morganella morganii* are very promising, since the biophysical and functional properties of the encoded enzymes were extensively studied

(Thaller et al., 1994; Thaller et al., 1995b). Besides, they are P-irrepressible enzymes showing broad substrate action and high activity around pH 6 and 30°C.

Among rhizobacteria, a gene from *Burkholderia cepacia* that facilitates phosphatase activity was isolated (Rodríguez et al., 2000a). This gene codes for an outer membrane protein that enhances synthesis in the absence of soluble phosphates in the medium, and could be involved in P transport to the cell. Besides, cloning of two nonspecific periplasmic acid phosphatase genes (*nap*D and *nap*E) from *Rhizobium (Sinorhizobium) meliloti* was accomplished (Deng, et al., 1998, 2001).

Heterologous expression of these genes in agriculturally important bacterial strains would be the next step in programs of improving organic phosphate mineralization in PGPB. The *napA* phosphatase gene from the soil bacterium *Morganella morganii* was transferred to *Burkholderia cepacia* IS-16, a strain used as a biofertilizer, using the broad-host range vector pRK293 (Fraga et al, 2001). An increase in extracellular phosphatase activity of the recombinant strain was achieved.

Insertion of the transferred genes into the bacterial chromosome is advantageous for stability and ecological safety. In our lab, a plasmid for the stable chromosomal insertion of the *phoC* phosphatase gene from *Morganella morganii* was constructed, based on the delivery system developed by de Lorenzo et al. (1990). This plasmid was transferred to *Azospirillum* spp. Preliminary results indicate that strains with increased phosphatase activity were obtained.

Phytases

Most phytases (myo-inositol hexakisphosphate phosphohydrolases) belong to high molecular weight acid phosphatases. In its basic form, phytate is the primary source of inositol and the major stored form of phosphate in plant seeds and pollen. Nevertheless, monogastric animals are incapable of using the P bound in phytate because their gastrointestinal tracts have low levels of phytase activity. Thus, nearly all the dietary phytate phosphorus ingested by these species is excreted, resulting in phosphorus pollution in areas of intensive animal production, and why phytases have emerged as very attractive enzymes for industrial and environmental applications. Genetic studies of phytases began in 1984, and the first commercial phytase, produced by genetically modified microorganisms, appeared on the market in the mid 1990s (Yanming et al., 1999).

Most genetic engineering studies have focused on the search for phytases that are optimal for improving animal nutrition. Another attractive application of these enzymes that is not currently exploited is solubilization of soil organic phosphorus through phytate degradation. Phytate is the major component of organics forms of P in soil (Richardson, 1994). The ability of plants to obtain phosphorus directly from phytate is very limited. However, the growth and phosphorus nutrition of *Arabidopsis* plants supplied with phytate was improved significantly when they were genetically transformed with the phytase gene (*phyA*) from *Aspergillus niger* (Richardson et al., 2001a). This resulted in improved P nutrition, such that the growth and P content of the plant was equivalent to control plants supplied with inorganic phosphate.

The enhanced utilization of inositol phosphate by plants by the presence of soil microorganisms has also been reported (Richardson et al., 2001b). Therefore, developing agricultural inoculants with high phytase production would be of great interest for improving plant nutrition and reducing P pollution in soil. Although phytase genes have been cloned from fungi, plants, and bacteria (Lei and Stahl, 2001), we will discuss only bacteria because they are the most feasible for the genetic improvement of rhizobacteria.

Thermally stable phytase genes (*phy*) from *Bacillus* sp. DS11 (Kim et al., 1998a) and from *B. subtilis* VTT E-68013 (Kerovuo et al., 1998) has been cloned. Acid phosphatase/phytase genes from *E. coli* (*appA* and *appA2 genes)* have also been isolated and characterized (Golovan et al., 2000; Rodríguez et al., 1999). The bi-functionality of these enzymes makes them attractive for solubilization of organic P in soil. Also, neutral phytases have great potential for genetic improvement of PGPB. Neutral phytase genes have been recently cloned from *B. subtilis* and *B. licheniformis* (Tye et al., 2002), A *phyA* gene has been cloned from the FZB45 strain of *B. amyloliquefaciens*. This strain was isolated from a group of several *Bacillus* having plant-growth-promoting activity (Idriss et al., 2002). It showed the highest extracellular phytase activity, and diluted culture filtrates of these strains stimulated growth of maize seedlings under limited phosphate in the presence of phytate. Culture filtrates obtained from a phytase-negative mutant strain, whose *phyA* gene was disrupted, did not stimulate plant growth. In addition, growth of maize seedlings was enhanced in the presence of purified phytase and the absence of culture filtrate.

These experiments provide strong evidence that phytase activity can be important for stimulating growth under limited P in soil, and supports the potential of using phytase genes to improve or transfer the P-solubilizing trait to PGPB strains used as agricultural inoculants.

Inorganic phosphate solubilization

Isolation of mineral phosphate-solubilizing (mps) genes

In most bacteria, mineral phosphate-dissolving capacity has been shown to be related to the production of organic acid (Rodríguez and Fraga, 1999). Goldstein (1996) proposed direct glucose oxidation to gluconic acid (GA) as a major mechanism for mineral phosphate solubilization (MPS) in Gram-negative bacteria. GA biosynthesis is carried out by the glucose dehydrogenase (GDH) enzyme and the co-factor, pyrroloquinoline quinone (PQQ). Some genes involved in MPS in different species have been isolated (Table 1).

Goldstein and Liu (1987) were the first to clone a gene involved in MPS from the Gram-negative bacteria *Erwinia herbicola*. Expression of this gene allowed production of GA in *E. coli* HB101 and conferred the ability to solubilize hydroxyapatite. *E. coli* can synthesize GDH, but not PQQ, thus it does not produce GA. The cloned 1.8 kb locus encodes a protein similar to the gene III product of a *pqq* synthesis gene complex from *Acinetobacter calcoaceticus*, and to *pqqE* of *Klebsiella neumonie* (Liu et al., 1992). These authors suggested that the *E. herbicola* DNA fragment functions as a PQQ synthase gene, and that probably, some *E. coli* strains contain some cryptic PQQ synthase genes that could be complemented by this single open reading frame (ORF) isolated by them.

Coincidentally, nucleotide sequence analysis of a 7.0 kb fragment from *Rhanella aquatilis* genomic DNA that induced hydroxyapatite solubilization in *E. coli*, showed two complete ORFs and a partial ORF. One of the cloned proteins showed similarity to *pqq E* of *E. herbicola, K. neumoniae*, and *A. calcoaceticus* (Kim et al.,

Table 1. Cloning of genes involved in mineral phosphate solubilization (MPS)

Microorganism	Gene or plasmid	Features	Reference
Erwinia herbicola	*mps*	Produces gluconic acid and solubilizes mineral P in *E. coli* HB101 Probably involved in PQQ[1] synthesis	Goldstein and Liu (1987)
Pseudomonas cepacia	*gab*Y	Produces gluconic acid and solubilizes mineral P in *E. coli* JM109 No homology with PQQ genes	Babu-Khan et al. (1995)
Enterobacter agglomerans	pKKY	Solubilizes P in *E. coli* JM109 Does not lower pH	Kim et al. (1997)
Rahnella aquatilis	pK1M10	Solubilizes P and produces gluconic acid in *E. coli* DH5α Probably related to PQQ synthesis	Kim et al. (1998b)
Serratia marcescens	pKG3791	Produces gluconic acid and solubilizes mineral P	Krishnaraj and Goldstein (2001)
Synechococcus PCC 7942	*pcc* gene	Synthesizes phosphoenol pyruvate carboxylase	N. Kumar (pers. comm.)

PQQ: pyrroloquinoline quinone.

1998b), while the partial ORF is similar to the *pqq C* of *K. neumoniae*. These authors also report that these genes complement cryptic *pqq E. coli* genes, thus allowing GA production.

Another type of gene (*gabY*) involved in GA production and MPS was cloned from *Pseudomonas cepacia* (Babu-Khan et al., 1995). The deduced amino acid sequence showed no homology with previously cloned direct oxidation pathway (GA synthesis) genes, but was similar to histidine permease membrane-bound components. In the presence of *gabY*, GA is produced only if the *E. coli* strain expresses a functional glucose dehydrogenase (*gcd*) gene. The authors (Babu-Khan et al., 1995), speculated that this ORF could be related to the synthesis of PQQ by an alternative pathway, or the synthesis of a *gcd* co-factor different from PQQ. The reported synergistic effect of exogenous PQQ and this gene supports this alternative, in our opinion. Also, a DNA fragment from *Serratia marcenses* induces GA synthesis in *E. coli*, but showed no homology to *pqq* or *gcd* genes (Krishnaraj and Goldstein, 2001). They suggested that this gene acted by regulating GA production under cell-signal effects.

Other isolated genes involved in the MPS phenotype seem not to be related with *pqq* DNA or *gcd* biosynthetic genes. A genomic DNA fragment from *Enterobacter agglomerans* showed MPS activity in *E. coli* JM109, although the pH of the medium was not altered (Kim et al., 1997). These results indicate that acid production is an important way, but not the only mechanism, of phosphate solubilization by bacteria (Illmer and Shinnera, 1995). More recently, a phosphoenol pyruvate carboxylase (*pcc*) gene from *Synechococcus PCC* 7942 appears to be involved in MPS (Kumar Naresh, pers. comm.). All these findings demonstrate the complexity of MPS in different bacterial strains, but at the same time, offer a basis for better understanding of this process.

Manipulation of mps genes for PGPB improvement

Expression in *E. coli* of the *mps* genes from *Ranella aquatilis* supported a much higher GA production and hydroxyapatite dissolution in comparison with the donor strain (Kim et al., 1998b). The authors suggested that different genetic regulation of the *mps* genes might occur in both species. MPS mutants of *Pseudomonas* spp. showed pleiotropic effects, with apparent involvement of regulatory *mps* loci in some of them (Krishnaraj et al., 1999). This suggests a complex regulation and various metabolic events related to this trait. Expression of a *mps* gene in a different host could be influenced by the genetic background of the recipient strain, the copy number of plasmids present, and metabolic interactions. Thus, genetic transfer of any isolated gene involved in MPS to induce or improve phosphate-dissolving capacity in PGPB strains, is an interesting approach.

An attempt to improve MPS in PGPB strains, using this approach, was carried out (Rodríguez et al., 2000b) with a PQQ synthetase gene from *Erwinia herbicola*. This gene, isolated by Goldstein and Liu (1987), was subcloned in a broad-host range vector (pKT230). The recombinant plasmid was expressed in *E. coli*, and transferred to PGPB strains of *Burkholderia cepacia* and *Pseudomonas aeruginosa*, using tri-parental conjugation. Several of the exconjugants that were recovered in the selection medium showed a larger clearing halo in medium with tricalcium phosphate as the sole P source. This indicates the heterologous expression of this gene in the recombinant strains, which gave rise to improved MPS ability of these PGPBs. More recently, a genomic integration of the *pcc gene of Synechococcus PCC* in *P. fluorescent* 7942 allowed phosphate solubilization in the recipient strain (Kumar Naresh, pers. comm.).

In other work, a bacterial citrate synthase gene was reported to increase exudation of organic acids and P availability to the plant when expressed in tobacco roots (López-Bucio et al., 2000). Citrate overproducing plants yielded more leaf and fruit biomass when grown under P-limiting conditions, and required less P-fertilizer to achieve optimal growth. This shows the putative role of organic acid synthesis genes in P uptake in plants.

Concluding remarks

Although knowledge of the genetics of phosphate solubilization is still scanty, some genes involved in mineral and organic phosphate solubilization have been isolated and characterized.

Initial achievements in the manipulation of these genes open a promising perspective for obtaining PGPB strains with enhanced phosphate solubilizing capacity, and thus, a more effective use of these microbes as agricultural inoculants.

Acknowledgements

H.R. received support from Consejo Nacional de Ciencia y Tecnología of Mexico (CONACyT Catedra Patrimonial de Excelencia grant EX-000580). We thank Ira Fogel at CIB for editing the English text.

References

Armarger N 2002 Genetically modified bacteria in agriculture. Biochimie 84, 1061–1072.

Babu-Khan S, Yeo C, Martin W L, Duron M R, Rogers R and Goldstein A 1995 Cloning of a mineral phosphate-solubilizing gene from *Pseudomonas cepacia*. Appl. Environ. Microbiol. 61, 972–978.

Bashan Y, Moreno M and Troyo E 2000 Growth promotion of the seawater-irrigated oil seed halophyte *Salicornia bigelovii* inoculated with mangrove rhizosphere bacteria and halotolerant *Azospirillum* spp. Biol. Fertil. Soils 32, 265–272.

Baskanova G and Macaskie L E 1997 Microbially-enhanced chemisorption of nickel into biologically-synthesized hydrogen uranyl phosphate: a novel system for the removal and recovery of metals from aqueous solutions. Biotechnol. Bioeng. 54, 319–329.

Beacham I R 1980 Periplasmic enzymes in Gram-negative bacteria. Int. J. Biochem. 10, 877–883.

Bonthrone K M, Baskanova G, Lin F and Macaskie L E 1996 Bioaccumulation of nickel by intercalation into polycrystalline hydrogen uranyl phosphate deposited via an enzymatic mechanism. Nat. Biotechnol. 14, 635–638.

de Lorenzo V, Herrero M, Jakubzik U and Timmis K N 1990 Mini-Tn5 transposon derivatives for insertion mutagenesis, promoter probing and chromosomal insertion of cloned DNA in Gram-negative Eubacteria. J. Bacteriol. 172, 6568–6572.

Deng S, Summers M L, Kahn M L and McDermontt T R 1998 Cloning and characterization of a *Rhizobium. meliloti* nonspecific acid phosphatase. Arch. Microbiol. 170, 18–26.

Deng S, Elkins J G, Da L H, Botero L M and McDermott T R 2001 Cloning and characterization of a second acid phosphatase from *Sinorhizobium meliloti* strain 104A14. Arch. Microbiol. 176, 255–263.

Fraga R, Rodríguez H and Gonzalez T 2001 Transfer of the gene encoding the Nap A acid phosphatase from *Morganella morganii* to a *Burkholderia cepacia* strain. Acta. Biotechnol. 21, 359–369.

Glick B R 1995 The enhancement of plant growth by free living bacteria. Can. J. Microbiol. 41, 109–117.

Goldstein A H and Liu S T 1987 Molecular cloning and regulation of a mineral phosphate solubilizing gene from *Erwinia herbicola*. Biotechnology 5, 72–74.

Goldstein A H 1996 Involvement of the quinoprotein glucose dehydrogenase in the solubilization of exogenous phosphates by Gram-negative bacteria. *In* Phosphate in Microorganisms: Cellular and Molecular Biology. Eds. A Torriani-Gorini, E Yagil and S Silver. pp. 197–203. ASM Press, Washington, DC.

Golovan S, Wang G, Zhang J and Forsberg C W 2000 Characterization and overproduction of the *Escherichia coli* appA encoded bifunctional enzyme that exhibits both phytase and acid phosphatase activities. Can. J. Microbiol. 46, 59–71.

Iddris E E, Makarewicz O, Farouk A, Rosner K, Greiner R, Bochow H, Richter T and Borris R 2002 Extracellular phytase activity of *Bacillus amyloliquefaciens* FZB45 contributes to its plant-growth-promoting effect. Microbiology 148, 2097–2109.

Igual J M, Valverde A, Cervantes E and Velázquez E 2001 Phosphate-solubilizing bacteria as inoculants for agriculture: use of updated molecular techniques in their study. Agronomie 21, 561–568.

Illmer P and Shinnera F 1995 Solubilization of inorganic calcium phosphates. Solubilization mechanisms. Soil Biol. Biochem. 27, 257–263.

Kerovuo J, Lauraeus M, Nurminen P, Kalkinen N and Apajalahti J 1998 Isolation, characterization, molecular gene cloning, and sequencing of a novel phytase from *Bacillus subtilis*. Appl. Environ. Microbiol. 64, 2079–2085.

Kim K Y, McDonald G A and Jordan D 1997 Solubilization of hydroxypatite by *Enterobacter agglomerans* and cloned *Escherichia coli* in culture medium. Biol. Fert. Soils 24, 347–352.

Kim Y O, Lee J K, Kim H K, Yu J H and Oh T K 1998a Cloning of the thermostable phytase gene (*phy*) from *Bacillus* sp. DS11 and its overexpression in Escherichia coli. FEMS Microbiol. Lett. 162, 185–191.

Kim K Y, Jordan D and Krishnan H B 1998b Expression of genes from *Rahnella aquatilis* that are necessary for mineral phosphate solubilization in *Escherichia coli*. FEMS Microb. Lett. 159, 121–127.

Krishnaraj P U and Goldstein A H 2001 Cloning of a *Serratia marcescens* DNA fragment that induces quinoprotein glucose dehydrogenase-mediated gluconic acid production in *Escherichia coli* in the presence of stationary phase *Serratia marcescens*. FEMS Microbiol. Lett. 205, 215–220.

Krishnaraj P U, Sadasivam K V and Khanuja S PS 1999 Mineral phosphate soil defective mutants of *Pseudomonas sp.* express pleiotropic phenotypes. Curr. Sci. (Bangalore, India) 76, 1032–1034.

Lei X G and Stahl C H 2001 Biotechnological development of effective phytases for mineral nutrition and environmental protection. Appl. Microbiol. Biotechnol. 57, 474–481.

Liu S T, Lee L Y, Taj C Y, Hung C H, Chang Y S, Wolfrang J H, Rogers R and Goldstein A H 1992 Cloning of an *Erwinia herbicola* gene necessary for gluconic acid production and enhanced mineral phosphate solubilization in *Escherichia coli* HB101: nucleotide sequence and probable involvement in biosynthesis of the coenzyme Pyrroloquinoline Quinone. J. Bacteriol. 174, 5814–5819.

López-Bucio J, de la Vega O M, Guevara-García A and Herrera-Estrella L 2000 Enhanced phophorus uptake in transgenic tobacco plants that overproduce citrate. Nat. Biotechnol. 18, 450–453.

Macaskie L E, Yong P, Doyle T C, Roig M G, Díaz M and Manzano T 1997 Bioremediation of uranium-bearing wastewater: biochemical and chemical factors affecting bioprocess application. Biotechnol. Bioeng. 53, 100–109.

Morrissey J P, Walsh O'Donnell U F; A, Moenne-Loccoz Y and O'Gara F 2002 Exploitation of genetically modified inoculants for industrial ecology applications. Antonie van Leeuwenhoek 81, 599–606.

Reilly T J, Baron G S, Nano F and Kuhlenschmidt M S 1996 Characterization and sequencing of a respiratory burst-inhibiting acid phosphatase from *Francisella tularensis*. J. Biol. Chem. 271, 10973–10983.

Richardson A E 1994 Soil microorganisms and phosphorous availability. *In* Soil Biota: Management in Sustainable Farming Systems. Eds. CE Pankhurst, BM Doube and VVSR Gupta. pp. 50–62. CSIRO, Victoria, Australia.

Richardson A E, Hadobas P A and Hayes J E 2001a Extracellular secretion of *Aspergillus* phytase from *Arabidopsis* roots enables plants to obtain phosphorous from phytate. Plant J 25, 641–649.

Richardson A E, Hadobas P A, Hayes J E, O'Hara C P and Simpson R J 2001b Utilization of phosphorus by pasture plants supplied with *myo*-inositol hexaphosphate is enhanced by the presence of soil micro-organisms. Plant Soil 229, 47–56.

Rodríguez E, Han Y and Lei X G 1999 Cloning, sequencing and expression of an *Escherichia. coli* acid phopshatase/phytase gene (*appA2*) isolated from pig colon. Biochem. Biophys. Res. Comm. 257, 117–123.

Rodríguez H and Fraga R 1999 Phosphate solubilizing bacteria and their role in plant growth promotion. Biotechnol. Adv. 17, 319–339.

Rodríguez H, Gonzalez T and Selman G 2000b Expression of a mineral phosphate solubilizing gene from *Erwinia herbicola* in two rhizobacterial strains. J. Biotechnol. 84, 155–161.

Rodríguez H, Rossolini G M, Gonzalez T, Jiping L and Glick B R 2000a Isolation of a gene from *Burkholderia cepacia* IS-16 encoding a protein that facilitates phosphatase activity. Curr. Microbiol. 40, 362–366.

Rossolini G M, Shipa S, Riccio M L, Berlutti F, Macaskie L E and Thaller M C 1998 Bacterial non-specific acid phosphatases: physiology, evolution, and use as tools in microbial biotechnology. Cell Mol. Life Sci. 54, 833–850.

Tarafdar J C and Jung A 1987 Phosphatase activity in the rhizosphere and its relation to the depletion of soil organic phosphorus. Biol. Fertil. Soils 3, 199–204.

Tarafdar J C and Claassen N 1988 Organic phosphorus compounds as a phosphorus source for higher plants through the activity of phosphatases produced by plant roots and microorganisms. Biol. Fertil. Soils 5, 308–312.

Thaller M C, Berlutti F, Schippa S, Lombardi G and Rossolini G M 1994 Characterization and sequence of PhoC, the principal phosphate-irrepressible acid phosphatase of *Morganella morganii*. Microbiology 140, 1341–1350.

Thaller M C, Berlutti F, Schippa S, Iori P, Passariello C and Rossolini G M 1995a Heterogeneous patterns of acid phosphatases containing low-molecular-mass polypeptides in members of the family *Enterobacteriaceae*. Int. J. Syst. Bacteriol. 4, 255–261.

Thaller M C, Lombardi G, Berlutti F, Schippa S and Rossolini G M 1995b Cloning and characterization of the NapA acid phosphatase/phosphotransferase of *Morganella morganii*: identification of a new family of bacterial acid phosphatase encoding genes. Microbiology 140, 147–151.

Tye A J, Siu F K, Leung T Y and Lim B L 2002 Molecular cloning and the biochemical characterization of two novel phytases from Bacillus subtilis 168 and Bacillus licheniformis. Appl. Microbiol. Biotechnol. 59, 190–197.

Wanner B L 1996 Phosphorus assimilation and control of the phosphate regulon. *In* Escherichia Coli and Salmonella, Cellular and Molecular Biology,. Eds. FC Niedhardt, R Curtiss III, JL Ingraham, EC Lin, KB Low, B Magasanik, WS Reznikoff, M Riley, M Schaechter and HE Umbarger. pp. 1357–1381. 2nd edition, 1ASM Press, Washington, DC.

Yanming H, Wilson D B and Lei X G 1999 Expression of an *Aspergillus niger* phytase gene (*phyA*) in *Saccharomyces cerevisiae*. Appl. Environ. Microbiol. 65, 15–18.

Biodiversity of populations of phosphate solubilizing rhizobia that nodulates chickpea in different Spanish soils

R. Rivas[1], A. Peix[2], P. F. Mateos[3,*], M. E. Trujillo[1], E. Martínez-Molina[3] & E. Velázquez[1]
[1]*Departamento de Microbiología y Genética, Lab. 209, Edificio Departamental de Biología, Universidad de Salamanca, Pl. Doctores de la Reina s/n, 37007, Salamanca, Spain.* [2]*Instituto de Recursos Naturales y Agrobiología, CSIC, Salamanca, Spain.* [3]*Departamento de Microbiología y Genética, Instituto Hispano Luso de Investigaciones Agrarias, Universidad de Salamanca, Salamanca, Spain.* [1]*Corresponding author**

Received: 31 May 2006

Key words: Phosphate solubilizing bacteria, rhizobia, *Mesorhizobium*, chickpea

Abstract

Within rhizobia, two species nodulating chickpea, *Mesorhizobium ciceri* and *Mesorhizobium mediterraneum*, are known as good phosphate solubilizers. For this reason, we have analysed the ability to solubilize phosphate of a wide number of strains isolated from *Cicer arietinum* growing in several soils in Spain. The aim of this work was to analyse microbial populations nodulating chickpea, that are able to solubilize phosphates, using molecular techniques. In the present work we analyzed 19 strains isolated from effective nodules of *C. arietinum* growing in three soils from the North of Spain. Nineteen strains showed ability to solubilize phosphate in YED-P medium. These strains were separated into 4 groups according to the results obtained by 879F-RAPD fingerprinting. The 16S rDNA sequencing of a representative strain from each group allowed the identification of strains as belonging to the genus *Mesorhizobium*. Strains from groups I and II showed a 99.4% and 99.2% similarity with *M. mediterraneum* UPM-CA142^T, respectively. The strains from group III were related to *M. tianshanense* USDA 3592^T at a 99.4% similarity level. Finally, the strain from group IV was related to *M. ciceri* USDA 3383^T with a 99.3% similarity. The LMW RNA profiles confirmed these results. Strains from groups I and II showed an identical LMW RNA profile to that of *M. mediterraneum* UPM-CA142^T, the profile of strains from group III was identical to that of *M. tianshanense* USDA 3592^T and the profile of strains from group IV was identical to that of *M. ciceri* USDA 3383^T. Different 879F-RAPD patterns were obtained for strains of the group I, group II and the *M. mediterraneum* type strain (UPM-CA142^T). The 879-RAPD patterns obtained for group III also differed from the pattern shown by *M. tianshanense* USDA 3592^T. Finally, the patterns between group IV and *M. ciceri* USDA 3383^T were also different. These results suggest that groups I and II may be subspecies of *M. mediterraneum*, group III a subspecies of *M. tianshanense* and group IV a subspecies of *M. ciceri*. Nevertheless, more studies are needed to establish the taxonomic status of strains isolated in this study.

Introduction

The solubilization of phosphates has been found in several species of rhizobia that nodulate different legumes. Within them, species nodulating chickpea are the most powerful P solubilizers (Halder et al. 1990, Peix et al. 2001). Currently these species, that were previously classified in the genus *Rhizobium*, are included in the genus *Mesorhizobium* (Jarvis et al. 1998). The species from this genus have a lower growth rate than those of genus *Rhizobium* and form a group phylogenetically separated from this genus. Currently the

* FAX No: +34923224876.
E-mail: pfmg@usal.es

E. Velázquez and C. Rodríguez-Barrueco (eds.),
First International Meeting on Microbial Phosphate Solubilization, 23–33

genus *Mesorhizobium* includes several species, many of them recently described. *Mesorhizobium mediterraneum* and *Mesorhizobium ciceri* have been described as chickpea endosymbionts (Nour et al. 1994, 1995), separating them from the species *Mesorhizobium loti* that initially included the strains nodulating chickpea (Jarvis et al. 1982). The species *M. loti* is a very complex group of strains, and many of them are currently included in other species. *Mesorhizobium huakuii* nodulates *Astragalus* (Chen et al. 1991), *M. tianshanense* nodulates *Glycyrrhiza pallidiflora* (Chen et al. 1995), *M. amorpha* nodulates *Amorpha fruticosa* (Wang et al. 1999), *M. plurifarium* nodulates tropical trees (de Lajudie et al. 1998) and *M. chacoense*, isolated in the Chaco Arido (Argentina), nodulates *Prosopis* (Velázquez et al. 2001a).

Nevertheless, it is not possible to relate an exclusive bacterial species with a legume species, because most of the rhizobial endosymbionts and legumes are promiscuous and even non-rhizobial species have been recently described as endosymbionts of several legumes. Some of them belong to alpha subclass of Proteobacteria as *Methylobacterium* (Sy et al. 2001) or *Devosia* (Rivas et al. 2002b) and other to beta subclass of Proteobacteria as *Burkholderia* (Moulin et al. 2001) and *Ralstonia* (Chen et al. 2001). Nevertheless, all species of legumes nodulated by non-rhizobial strains are tropical legumes and only few studies have been made on biodiversity of species nodulating temperate legumes. Within them, the chickpea endosymbionts have not been exhaustively studied, except when *M. ciceri* and *M. mediterraneum* were described.

To perform taxonomic studies of endosymbiont populations of legumes we have already applied the LMW RNA profiles obtained by using staircase electrophoresis (Cruz-Sánchez et al. 1997) which comprise 5S rRNA and class 1 and 2 tRNA in bacteria. These profiles can be applied to a large number of isolates and allowed the differentiation among microbial genera, based on the 5S rRNA zone, and species, based on tRNA profiles. The LMW RNA profiles have been used to differentiate species of rhizobia (Velázquez et al. 1998b), to detect new species of *Mesorhizobium* (Velázquez et al. 2001a) and to identify strains isolated from several legumes, including chickpea, in diverse geographical locations (Peix et al. 2001; Velázquez et al. 2001b; Jarabo-Lorenzo et al. 2000). These profiles have also been applied to Gram positive bacteria (Palomo et al. 2000) including endosymbionts of non-legumes as *Frankia* (Velázquez et al. 1998a) and eukaryotic microorganisms (Velázquez et al. 2000). From all these works it is possible to conclude that LMW RNA are molecular signatures of both prokaryotic and eukaryotic microorganisms (Velázquez et al. 2001c).

Therefore, we have used LMW RNA profiling to identify the phosphate solubilizing strains isolated from chickpea in different geographical locations in Spain and the biodiversity within each species was analysed using RAPD patterns. Finally, we have compared the ability of these strains to solubilize phosphate and to movilize it to chickpea plants.

Material and methods

Bacterial strains

The reference strains, the new isolates and their host plants are listed in Table 1. A total of 19 new rhizobial isolates were obtained from effective nodules of chickpea growing in different soils from Spain. Isolations were made according to Vincent (1970) using yeast manitol agar, YMA (Bergersen 1961). The same medium was used to grow all strains tested.

Evaluation of bicalcium phosphate solubilization of rhizobial strains

The ability to solubilize bicalcium phosphate of the type strains of species from genus *Mesorhizobium* and those of isolates from this study was tested in Petri dishes containing YED (yeast extract 0.5%; glucose 1% and agar 2%) supplemented with a 0.2% of bicalcium phosphate (YED-P). The inoculated plates were incubated for 7 days until the solubilization zone surrounding the colonies was observed.

Analysis of 879F-RAPD and RAPD patterns

Total genomic DNA from the isolates was extracted according to the method employed by Rivas et al. (2001). The primer 879F (5′-GCC TGGGGAGTACGGCCGCA-3′) was used to

Table 1. Characteristics of strains used in this study

Strain	Host	Soil	Phylogenetic group (16S rRNA sequence)	LMW RNA[1]	879F-RAPD pattern[2]	RAPD pattern[3]
PECA03	*Cicer arietinum* (var. "pedrosillano")	Spain (Salamanca)	*Mesorhizobium mediterraneum*	A	I	Ia
PECA11	*Cicer arietinum* (var. "pedrosillano")	Spain (Salamanca)	*Mesorhizobium mediterraneum*	A	I	Ib
PECA12	*Cicer arietinum* (var. "pedrosillano")	Spain (Salamanca)	*Mesorhizobium mediterraneum*	A	I	Ib
PECA15	*Cicer arietinum* (var. "pedrosillano")	Spain (Salamanca)	*Mesorhizobium mediterraneum*	A	I	Ib
PECA19	*Cicer arietinum* (var. "pedrosillano")	Spain (Salamanca)	*Mesorhizobium mediterraneum*	A	I	Ic
PECA21	*Cicer arietinum* (var. "pedrosillano")	Spain (Salamanca)	*Mesorhizobium mediterraneum*	A	I	Ib
PECA23	*Cicer arietinum* (var. "pedrosillano")	Spain (Salamanca)	*Mesorhizobium mediterraneum*	A	I	Ib
PECA09	*Cicer arietinum* (var. "pedrosillano")	Spain (Salamanca)	*Mesorhizobium mediterraneum*	A	II	IIa
PECA10	*Cicer arietinum* (var. "pedrosillano")	Spain (Salamanca)	*Mesorhizobium mediterraneum*	A	II	IIb
PECA13	*Cicer arietinum* (var. "pedrosillano")	Spain (Salamanca)	*Mesorhizobium mediterraneum*	A	II	IIa
PECA14	*Cicer arietinum* (var. "pedrosillano")	Spain (Salamanca)	*Mesorhizobium mediterraneum*	A	II	IIa
PECA16	*Cicer arietinum* (var. "pedrosillano")	Spain (Salamanca)	*Mesorhizobium mediterraneum*	A	II	IIa
PECA18	*Cicer arietinum* (var. "pedrosillano")	Spain (Salamanca)	*Mesorhizobium mediterraneum*	A	II	IIa
PECA20	*Cicer arietinum* (var. "pedrosillano")	Spain (Salamanca)	*Mesorhizobium mediterraneum*	A	II	IIb
PECA22	*Cicer arietinum* (var. "pedrosillano")	Spain (Salamanca)	*Mesorhizobium mediterraneum*	A	II	IIb
PECA30	*Cicer arietinum* (var. "pedrosillano")	Spain (Salamanca)	*Mesorhizobium mediterraneum*	A	II	IIb
RCAN03	*Cicer arietinum* (var. "pedrosillano")	Spain (León)	*Mesorhizobium tianshanense*	B	III	IIIa
RCAN08	*Cicer arietinum* (var. "pedrosillano")	Spain (León)	*Mesorhizobium tianshanense*	B	III	IIIa
FCA08	*Cicer arietinum* (var. "fuentesauco")	Spain (Salamanca)	*Mesorhizobium ciceri*	C	IV	IVa
Mesorhizobium mediterraneum UPM-CA142^T	*Cicer arietinum*	Spain	*Mesorhizobium mediterraneum*	A	V	Va
Mesorhizobium tianshanense USDA 3592^T	*Glyzyrrhiza pallidiflora*	China	*Mesorhizobium tianshanense*	B	VI	VIa
Mesorhizobium ciceri USDA 3383^T	*Cicer arietinum*	Spain	*Mesorhizobium ciceri*	C	VII	VIIa

[1]Each letter corresponds to a different group of strains which present a different 16S rRNA sequence.
[2]Each roman number corresponds to a different pattern obtained using 879F-RAPD. Strains from the same 16SrRNA sequence group may belong to different 879F-RAPD groups.
[3]A combination of roman numbers and letters indicates that within a 879F-RAPD pattern group there are different RAPD patterns.

obtain groups within the species isolated in this study. The M13 primer (5′-GAGGGTGGCGGTTCT-3′) was used to differentiate among the strains from the same subspecies. Crude DNA (2 μl) was used as template for PCR amplifications. PCR was performed using AmpliTaq Gold reagent kit (Perkin-Elmer Biosystems, California, USA) following the manufacturer's instructions: 2.5 μl of GeneAmp 10× buffer, 1 μl of BSA 0.1%, 1.5 mM $MgCl_2$, 200 μM of each dNTP, 2 U of AmpliTaq Gold DNA polymerase, and 2 μM of primer for 25 μl of final reaction volume. PCR conditions were as follow: pre-heating at 95°C for 9 min; 35 cycles of denaturing at 95°C for 1 min; annealing at 50°C for 1 min and extension at 75°C for 2 min, and a final extension at 72°C for 7 min. Eight microliters of amplified PCR product was separated by electrophoresis on 1.5% agarose gels, in TBE buffer (100 mM Tris, 83 mM boric acid, 1 mM EDTA; pH 8.5) for 2 h at 6 V cm^{-1}, stained in a solution containing 0.5 μl of ethidium bromide ml^{-1}, and photographed on a UV transilluminator. Standard VI (Boehringer-Roche, Indianapolis, IN, USA) was used as a size marker.

Amplification and determination of nucleotide sequences of the 16S rRNA gene and analysis of the sequence data

DNA extraction was carried out as previously described (Rivas et al. 2001). PCR was performed using an AmpliTaq reagent kit (Perkin-Elmer Biosystems, California, USA) following the manufacturer's instructions (1.5 mM $MgCl_2$, 200 μM of each dNTP and 2 U of Taq polymerase for 25 μg final volume of reaction). The PCR amplification of 16S rDNA was carried out using the following primers: 5′-AGAGTTTGAT CTGGCTCAG-3′ (*Escherichia coli* positions 8–27) and 5′-AAGGAGGTGATCCANCCRCA-3′ (*Escherichia coli* positions 1509–1522) at a final concentration of 0.2 μM. PCR conditions were as follows: pre-heating at 95°C for 9 min; 35 cycles of denaturing at 95°C for 1 min; annealing at 59°C for 1 min and extension at 72°C for 2 min, and a final extension at 72°C for 7 min. The PCR product (25 μg) was electrophoresed on 1% agarose gel in TBE buffer (100 mM Tris, 83 mM boric acid, 1 mM EDTA, pH: 8.5) at 6 V cm^{-1}, stained in a solution containing 0.5 μl ethidium bromide ml^{-1}. Standard VI (Boehringer-Roche, USA) was used as a size marker. Three microliters of 6× loading solution (40% sucrose and 0.25% bromophenol blue) were added to each sample. The band corresponding to the 16S rDNA was purified directly from the gel by centrifugation in Eppendorff tubes with a special filter (Millipore Co., Illinois, USA) for 10 min at 5000 × *g* at room temperature according to the manufacturer's instructions.

The sequence reaction was performed on an ABI377 sequencer (Applied Biosystems Inc.) using a BigDye terminator v3.0 cycle sequencing kit as supplied by the manufacturer. The following primers were used: 5′-AACGCTGGCGGCR KGCYTAA-3′, 5′-ACTCCTACGGGAGGCAG CAG-3′, 5′-CTGCTGCCTCCCGTAGGAGT-3′, 5′-CGTGCCAGCAGCCGCGGTAA-3′, 5′-CAG GATTAGATACCCTGGTAG-3′ and 5′-GAGG AAGGTGGGGATGACGTC-3′, which correspond to *E. coli* small-subunit rDNA sequence positions 32–52, 336–356, 356–336, 512–532, 782–803 and 1173–1194, respectively. The sequence obtained was compared with those from the GenBank using the FASTA program (Pearson and Lipman 1988).

Sequences were aligned using the Clustal W software (Thompson et al. 1997). The distances were calculated according to Kimura's two-parameter method (Kimura 1980). Phylogenetic trees were inferred using the neighbour-joining method (Saitou and Nei 1987). Bootstrap analysis was based on 1000 resamplings. The MEGA2 package (Kumar et al. 2001) was used for all analyses. The trees were rooted using *Bradyrhizobium japonicum* as outgroup.

LMW RNA extraction and SCE LMW RNA profiling

LMW RNA extraction was accomplished following the phenol/chloroform method described by Höfle (1988), using cells grew in tryptone-yeast agar, TY (Beringer 1974). The following commercial molecules from Boehringer Manheim (Manheim, Germany) and Sigma (St. Louis, MO, USA) were used as reference: 5S rRNA from *Escherichia coli* MRE 600 (120 and 115 nucleotides) (Bidle and Fletcher 1995), tRNA specific for tyrosine from *E. coli* (85 nucleotides) and tRNA specific for valine from *E. coli* (77 nucleotides)

(Sprinzl et al. 1985). Samples containing 3 μg were added to 5 μg of loading solution (300 mg/ml of sucrose, 460 mg/ml of urea, 10 μl/ml 20% SDS, 1 mg/ml xylene cyanol) and, after 10 min of heating at 70°C, applied to each well. LMW RNA profiles were obtained using staircase electrophoresis (Cruz-Sánchez et al. 1997) which was performed in 400 × 360 × 0.4 mm gels in a vertical slab unit (Poker Face SE 1500 Sequencer, Hoeffer Scientific Instruments, San Francisco, CA, USA). The separating gel contained 14% acrylamide/Bis (acrylamide: *N*, *N*-methylene bis-acrylamide 29:1 (w/w), 7 M urea in TBE buffer: 100 mM Tris, 83 mM boric acid, 1 mM EDTA, pH: 8.5) in TBE buffer, pH: 8.5. Before running the pre-electrophoresis (30 min at 100 V), the system was stabilized at 50°C. The running buffer (TBE, x1.2) was recycled at a flow rate of 300 ml/min with a peristaltic pump (MasterFlex, Cole Parmer Instruments, Chicago, Illinois, USA). After electrophoresis, gels were silver-stained according to Haas et al. (1994).

Mobilization of phosphorous in plants

Experiments for studying the phosphorous mobilization in plants were made on chickpea and were conducted in pots containing vermiculite as sterile support amended with 0.2% (w/w) bicalcium phosphate. The pots were placed in a plant growth chamber with mixed incandescent and fluorescent lighting (400 microeinsteins m^{-2} s^{-1}; 400–700 nm), programmed for a 16 h photoperiod, day–night cycle, with a constant temperature varying from 15–27° C (night–day), and 50–60% relative humidity. Fifteen pots were used for each treatment. The seeds were placed in each pot at a depth of 2 cm.

For inoculation, each strain was grown in Petri dishes with YMA (Bergersen 1961) for 7 days. After that, sterile water was added to the plates to obtain a suspension with ca. 10^8 cells ml^{-1}. For inoculation we added 1 ml of the suspension of each strain to each seed placed in Petri dishes. The seeds were dried overnight at room temperature.

At harvest (30 days) the dry weight of the aerial part of the inoculated plants was determined. Plant nitrogen, phosphorous, potassium, calcium and magnesium content was measured according to the A.O.A.C. methods (Johnson 1990). The data obtained were analyzed by one-way analysis of variance, with the mean values compared using the Fisher's Protected LSD (Low Significative Differences) (P = 0.05).

Results and discussion

Analysis of 879-F patterns

According to the results obtained in other microbial groups, primers targeting 16S rDNA sequences, when used at relatively high annealing temperatures (typically 50°C or 55°C), yield DNA patterns that allow to discriminate at species or subspecies levels (Igual et al. in press). For that reason we have used this primer to obtain groups within the isolates of this study. Figure 1 shows that these strains are separated in four groups. The groups I (lanes 4–10) and II (lanes 11–19) include strains isolated in a soil from Salamanca (Spain). The group III (lanes 20 and 21) includes strains isolated from a soil in León (Spain) and the group IV (lane 22) a strain isolated in the soil of Salamanca (Spain). In lanes 1, 2 and 3 the 879-F patterns of *M. mediterraneum* UPM-CA142^T , *M. ciceri* USDA 3383^T and *M. tianshanense* USDA 3592^T are shown.

16S rDNA sequencing and analysis

The complete sequences of 16S rDNA genes from strains PECA03 (group I), PECA20 (group II), RCAN03 (group III) and FCA08 (group IV) were obtained and compared with those from databanks using the FASTA program (Pearson and Lipman 1988). Strain PECA03 sequence showed a 99.4% similarity with that of *M. mediterraneum* UPM-CA142^T. Strain PECA20 sequence showed a 99.2% similarity with that of *M. mediterraneum* UPM-CA142^T. The sequence of strain RCAN03 showed a 99.4% similarity with that of *M. tianshanense* USDA 3592^T. Finally, the strain FCA08 sequence showed a 99.3% similarity with that of *M. ciceri* USDA 3383^T. Therefore, the phylogenetic analysis (Figure 2) of 16S rRNA sequences places the strains from this study in the genus *Mesorhizobium*.

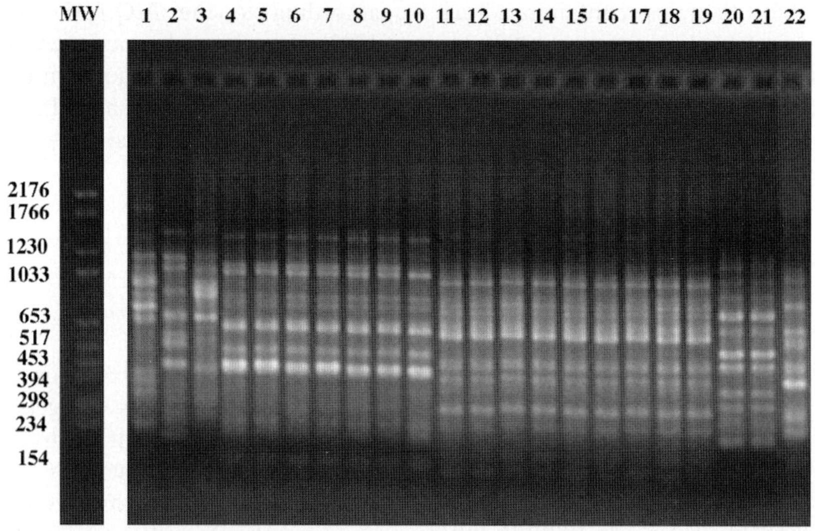

Figure 1. Patterns obtained using the primer 879-F: *M. ciceri* USDA 3383^T (lane 1), *M. mediterraneum* UPM-CA142^T (lane 2), *M. tianshanense* USDA 3592^T (lane 3), PECA03 (lane 4), PECA11 (lane 5), PECA12 (lane 6), PECA15 (lane 7), PECA19 (lane 8), PECA21 (lane 9), PECA23 (lane 10), PECA09 (lane 11), PECA10 (lane 12), PECA13 (lane 13), PECA14 (lane 14), PECA16 (lane 15), PECA18 (lane 16), PECA20 (lane 17), PECA22 (lane 18), PECA30 (lane 19), RCAN03 (lane 20), RCAN08 (lane 21) and FCA08 (lane 22).

LMW RNA profiles

In order to identify bacteria at the species level, 16S rDNA sequences were complemented with the LMW RNA profile analysis of our strains comparing them with those of the type strains *M. mediterraneum* UPM-CA142^T, *M. tianshanense* USDA 3592^T and *M. ciceri* USDA 3383^T. The LMW RNA profile of *M. mediterraneum* UPM-CA142^T (Figure 3, lane 1) is identical to that of strains from groups I and II (represented in Figure 3, lane 2), *M. ciceri* USDA 3383^T (Figure 3, lane 3) shows the same LMW RNA profile than strains from group IV (represented in Figure 3, lane 4) and *M. tianshanense* USDA 3592^T shows the same LMW RNA profile (Figure 3, lane 5) than strains from group III (represented in lane 6, Figure 3). In previous studies, we have demonstrated that LMW RNA profiles are molecular signatures of eukaryotic and prokaryotic microorganisms at genus and at species level (Velázquez et al. 2001c). Therefore, strains from groups I and II belong to *M. mediterraneum* species. Group III strains belong to *M. tianshanense* species and group IV belong to species *M. ciceri.* These results confirm the identification obtained by means of 16S rDNA sequence.

According to the results the 879F-RAPD pattern of strains from group I and II do not coincide between them and neither with that of type strain of *M. mediterraneum* UPM-CA142^T. The 879F-RAPD pattern of strains from group III do not coincide with the type strain of *M. tianshanense* USDA 3592^T and the pattern of strains from group IV does not coincide with the type strain of *M. ciceri* USDA 3383^T. Therefore, these results point out the existence of more than one genomic group within the three species of this study. These results confirm those obtained in *C. michiganensis* subspecies using primers targeting 16S rDNA sequence (Rivas et al. 2002a). The taxonomic status of the strains isolated in this study must be established in further studies, but it is possible that the species *M. mediterraneum*, *M. ciceri* and *M. tianshanense* contain several subspecies. These results coincide with those obtained by other authors that have recently described new subspecies in the species *Bacillus subtilis* (Nakamura et al. 1999) and *Photorhabdus* (Fischer-Le Saux et al. 1999). Moreover, our results are in agreement with those of other authors that have proposed a new subspecies within *M. huakuii* (Nuswantara et al. 1999).

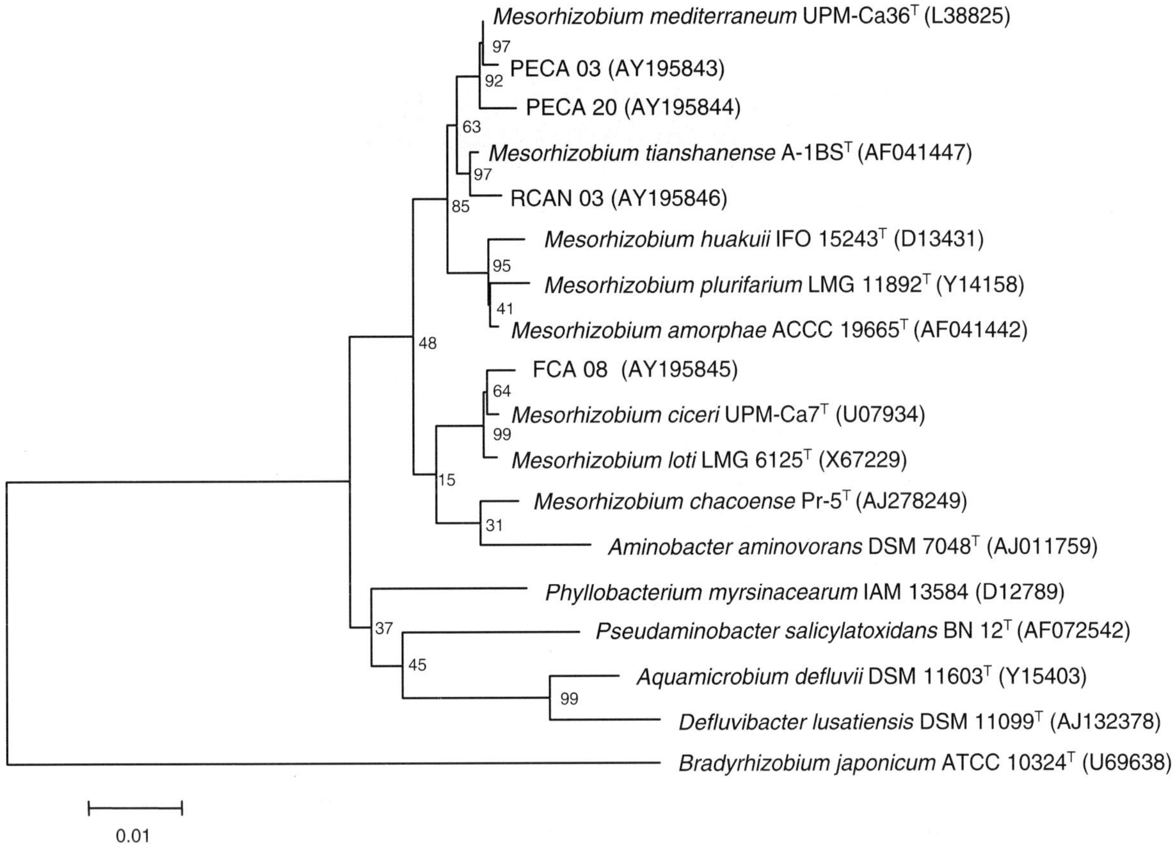

Figrue 2. Comparative sequence analysis of 16S rDNA from the strains PECA03, PECA20, FCA08 and RCN03 and representative strains from the GenBank. The significance of each branch is indicated by a bootstrap value calculated for 1,000 subsets. Bar, 1 nt substitutions per 100 nt.

Analysis of the intraspecific biodiversity using RAPD patterns

To analyse the intraspecific biodiversity from species of this study we also used the primer M13 to detect strain specific patterns (Table 1). The results are shown in Figure 4. Within strains from group I of 879-F patterns, three RAPD patterns have been found using M13 primer: Ia (lane 4), Ib (lanes 5, 6, 7, 9 and 10) and Ic (lane 8). Most of the strains from this group showed the RAPD pattern type Ib. The strains from group II of 879-F patterns showed two types of RAPD pattern: IIa (lanes 11, 13, 14, 15, 16, 18 and 19) and IIb (lanes 12 and 17). The strains from group III showed identical RAPD pattern (lanes 18 and 19). The RAPD pattern of strain from group IV is shown in lane 20. Finally, the type strains of *M. mediterraneum* UPM-CA142^T (lane 1), *M. tianshanense* USDA 3592^T (lane 2) and *M. ciceri* USDA 3383^T (lane 3) showed different RAPD patterns among them and with respect to the other strains from this study.

The conventional RAPD patterns are strain dependent and usually vary among the strains from the same subspecies (de la Puente-Redondo et al. 2000; Wieser and Busse 2000). The results of this work are in agreement with those reported in the literature and confirm that the 879F-RAPD patterns are strain non-dependent and that probably the strains showing different pattern belong to different subspecies. Nevertheless, to demonstrate that a taxonomic polyphasic study must be performed on the strains isolated in this study.

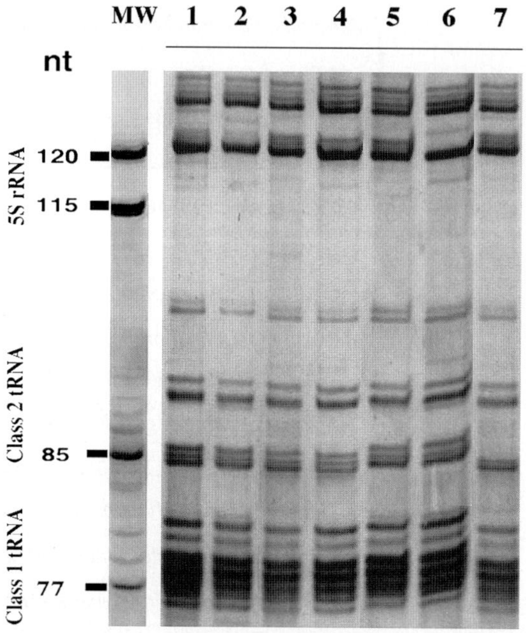

Figure 3. LMW RNA profiles displayed by the strains of this study. Lane 1, *M. mediterraneum* UPM-CA142^T. Lane 2 shows the profile of strains PECA03, PECA11, PECA12, PECA15, PECA19, PECA21, PECA23, PECA09, PECA10, PECA13, PECA14, PECA16, PECA18, PECA20, PECA22, PECA30. Lane 3, *M. ciceri* USDA 3383^T. Lane 4, strain FCA08. Lane 5, *M. tianshanense* USDA 3592^T. Lane 6 shows the profile of strains RCAN03, RCAN08. Lane 7, *M. loti* DSM 2626^T.

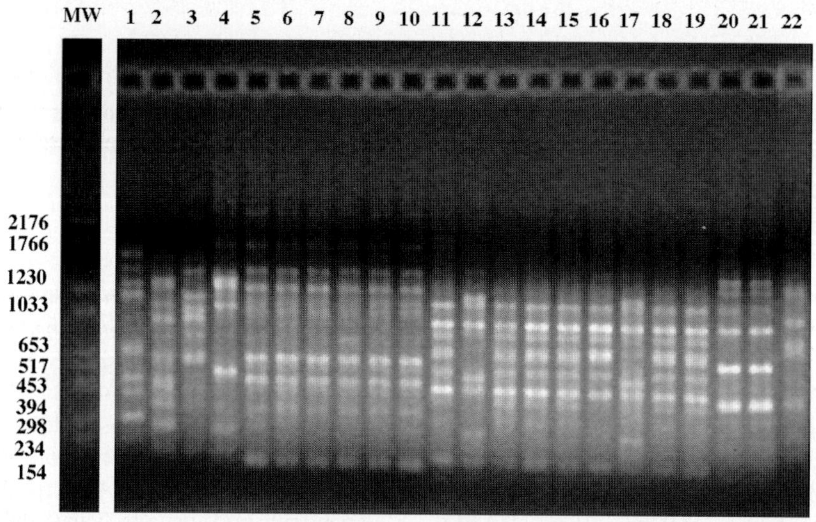

Figure 4. RAPD patterns obtained using the primer M13: *M. ciceri* USDA 3383^T (lane 1), *M. mediterraneum* UPM-CA142^T (lane 2), *M. tianshanense* USDA 3592^T (lane 3), PECA03 (lane 4), PECA11 (lane 5), PECA12 (lane 6), PECA15 (lane 7), PECA19 (lane 8), PECA21 (lane 9), PECA23 (lane 10), PECA09 (lane 11), PECA10 (lane 12), PECA13 (lane 13), PECA14 (lane 14), PECA16 (lane 15), PECA18 (lane 16), PECA20 (lane 17), PECA22 (lane 18), PECA30 (lane 19), RCAN03 (lane 20), RCAN08 (lane 21) and FCA08 (lane 22).

Evaluation of bicalcium phosphate solubilization of rhizobial strains

The strains isolated in this study showed differences in their ability to solubilize phosphate in plates (Table 2). All strains nodulating chickpea in the soils studied were able to solubilize phosphate. The strains belonging to species *M. ciceri* presented the smallest clearing halo on YED-P plates followed by those of *M. mediterraneum*

Table 2. Symbiotic characteristics of strains nodulating chickpea used in this study

Strain	Solubilization "halo" (mm)*	Number of nodules	Dry weight per plant (mg)	Total N (mg)	Total P (mg)	Total Ca (μg)	Total Mg (μg)	Total K (mg)
PECA03	10	8^{a}	120^{a}	2.4^{ab}	0.20^{a}	43.5^{c}	86.0^{b}	0.48^{b}
PECA20	5	5^{a}	90^{a}	1.8^{a}	0.15^{a}	32.6^{b}	89.0^{b}	0.34^{a}
RCAN03	15	7^{a}	120^{a}	3.0^{c}	0.6^{b}	50.0^{d}	106.2^{c}	1.30^{c}
FCA08	2	9^{a}	120^{a}	2.5^{ab}	0.1^{a}	22.0^{a}	76.3^{a}	0.25^{a}

Values followed by the same letter are no significantly different from each other at $P = 0.05$ according to Fisher's Protected LSD (Least Significant Differences). * After seven days of incubation at 28°C in YED-P plates with bicalcium phosphate as P source.

(Table 2). As we show in the present study some strains nodulating chickpea were identified as *M. tianshanense* and to our knowledge, this is the first report of the nodulation of *C. arietinum* by this species. The type strain of *M. tianshanense* USDA 3592^{T} showed low ability (solubilization halo lower than 2 cm) to solubilize phosphate but the strain RCAN03 belonging to this species showed the highest clearing halo on YED-P plates (15 cm). These results are in agreement with those obtained in previous studies (Halder et al. 1990; Peix et al. 2001) in which it is shown that strains nodulating chickpea are the best phosphate solubilizers. In this work we also report for the first time phosphate solubilization produced by strains of *M. tianshanense*.

Mobilization of phosphorous in plants

We analysed the content of P, N, Ca, Mg and K in plants inoculated with representative strains of each group of 879F-RAPD (Table 2). According to our results the ability to mobilize phosphorous to plants is directly related to that to solubilize phosphates in vitro. In this way, the highest P content was measured in the plants inoculated with the strain RCAN03 belonging to *M. tianshanense* and the lowest content in the plants inoculated with the strain FCA08 belonging to *M. ciceri*. The effect of the inoculation with different strains on the dry matter and nitrogen content was not related to the ability to solubilize phosphate. The content in K, Ca and Mg was the highest in the plants inoculated with the strain RCAN03 which is the best phosphate solubilizer and the lowest in plants inoculated with the strain FCA08 that showed a low ability to solubilize phosphate. The phosphate solubilizing strains form low number of nodules per plant and this fact is related to the N content and dry weight. This fact was already observed for the strain PECA21 belonging to *M. mediterraneum* (Peix et al. 2001). Although this strain was able to promote the growth of chickpea, the dry matter and fixed nitrogen were less increased than the phosphorous content. Therefore, to obtain an optimal growth promotion of chickpea it could be necessary to inoculate with strains that show a great ability to solubilize phosphate but also a high effectiveness in nodulation and nitrogen fixation. Taking into account that the type strain of *M. tianshanense* was not able to nodulate chickpea and it shows a low ability to solubilize phosphate it is possible that both nodulation and phosphate solubilization are related to determinate subspecies within the same species. Nevertheless, more strains of different subspecies, species and genera must be analyzed to confirm this hypothesis.

In conclusion, this study shows that chickpea can be nodulated by several species, not only by *M. ciceri* and *M. mediterraneum* but also species that were originally described as endosymbionts of other legumes such as *M. tianshanense*. These results support the findings of several authors concerning the nodulation of the same host by several species of rhizobia (Herrera Cervera et al. 1999; Velázquez et al. 2001b). Nevertheless, until the moment rhizobia associated with plants from tribe *Cicerae* were thought to belong to concrete rhizobial species (Perret et al. 2000). In this way, until now only two species have been described as effective endosymbionts of *Cicer arietinum*, *M. ciceri* (Nour et al. 1994) and *M. mediterraneum* (Nour et al. 1995). However, in this work it has been demonstrated that *M. tianshanense* can nodulate either species from tribe *Cicerae* as chickpea and species from other tribes as *Galegae* (*Glycyrrhiza pallidiflora*). Moreover, in this study it is shown that the chickpea isolates from

M. tianshanense have a higher ability to solubilize phosphate and mobilize phosphorous to the plants than those of *M. ciceri* and *M. mediterraneum* that are reported at present as the most powerful phosphate solubilizing rhizobia (Peix et al. 2001).

Acknowledgements

This work was supported by the Junta de Castilla y León and the DGICYT (Dirección General de Investigación Científica y Técnica).

References

Bergersen FJ 1961 The growth of *Rhizobium* in synthetic media. Aust J Biol Sci 14, 349–360.

Beringer JE 1974 R factors transfer in *Rhizobium leguminosarum*. J Gen Microbiol 84, 188–198.

Bidle KD and Fletcher M 1995 Comparison of free-living and particle-associated bacterial communities in the Chesapake Bay by stable low-molecular-weight RNA analysis. Appl Environ Microbiol 61, 944–952.

Chen WX, Li GS, Qi YL, Wang ET, Yuan HL and Li JL 1991 *Rhizobium huakuii* sp. nov. isolated from the root nodules of *Astragalus sinicus*. Int J Syst Bacteriol 41, 275–280.

Chen W, Wang E, Wang S, Li Y, Chen X and Li Y 1995 Characteristics of *Rhizobium tianshanense* sp. nov., a moderately and slowly growing root nodule bacterium isolated from an arid saline environment in Xinjiang, People's Republic of China. Int J Syst Bacteriol 45, 153–159.

Chen WM, Laevens S, Lee TM, Coenye T, de Vos P, Mergeay M and Vandamme P 2001 *Ralstonia taiwanensis* sp. nov., isolated from root nodules of *Mimosa* species and sputum of a cystic fibrosis patient. Int J Syst Evol Microbiol 51, 1729–1735.

Cruz-Sánchez JM, Velázquez E, Mateos P and Martínez-Molina E 1997 Enhancement of resolution of low molecular weight RNA profiles by staircase electrophoresis. Electrophoresis 18, 1909–1911.

de Lajudie P, Willems A, Nick G, Moreira F, Molouba F, Hoste B, Torck U, Neyra M, Collins MD, Lindström K, Dreyfus B and Gillis M 1998 Characterization of tropical tree rhizobia and description of *Mesorhizobium plurifarium* sp. nov. Int J Syst Bacteriol 48, 369–382.

de la Puente-Redondo VA, García del Blanco N, Gutiérrez Martín CB, García-Peña FJ and Rodríguez Ferri EF 2000 Comparison of different PCR approaches for typing of *Francisella tularensis* strains. J Clin Microbiol 38, 1016–1022.

Fischer-Le Saux M, Viallard V, Brunel B, Normand P and Boemare NE 1999 Polyphasic classification of the genus *Photorhabdus* and proposal of new taxa: *P. luminescens* subsp. *luminescens* subsp. nov., *P. luminescens* subsp. *akhurstii* subsp. nov., *P. luminescens* subsp. *laumondii* subsp. nov., *P. temperata* sp. nov., *P. temperata* subsp. *temperata* subsp. nov. and *P. asymbiotica* sp. nov. Int J Syst Bacteriol 49, 1645–1656.

Haas H, Budowle B and Weiler G 1994 Horizontal polyacrylamide gel electrophoresis for the separation of DNA fragments. Electrophoresis 15, 153–158.

Halder AK, Mishra AK and Chakrabartty PK 1990 Solubilization of phosphatic compounds by *Rhizobium*. Indian J Microbiol 30, 311–314.

Herrera-Cervera JA, Caballero-Mellado J, Laguerre G, Tichy HV, Requena N, Amarger N, Martínez-Romero E, Olivares J and Sanjuán J 1999 At least five rhizobial species nodulate *Phaseolus vulgaris* in a Spanish soil. FEMS Microbiol Ecol 30, 87–97.

Höfle MG 1988 Identification of bacteria by low molecular weight RNA profiles: a new chemotaxonomic approach. J Microbiol Methods 8, 235–248.

Igual JM, Valverde A, Rivas R, Mateos PF, Rodríguez-Barrueco C, Martínez-Molina E, Cervantes E, Velázquez E 2003 Genomic fingerprinting of *Frankia* strains by PCR-based techniques. Assessment of a primer based on the sequence of 16S rRNA gene of *Escherichia coli*. Plant Soil 254, 115–123.

Jarabo-Lorenzo A, Velázquez E, Pérez-Galdona R, Vega-Hernández MC, Martínez-Molina E, Mateos PF, Vinuesa P, Martínez-Romero E and León-Barrios M 2000 Restriction fragment length polymorphism analysis of PCR-amplified 16S rDNA and low molecular weight RNA profiling in the characterisation of rhizobial isolates from shrubby legumes endemic to the Canary Islands. Syst Appl Microbiol 23, 4–18.

Jarvis BDW, Pankhurst CE and Patel JJ 1982 *Rhizobium loti*, a new species of legume root nodule bacteria. Int J Syst Bacteriol 32, 378–380.

Jarvis BDW, van Berkum P, Chen WX, Nour SM, Fernandez MP, Cleyet-Marel JC and Gillis M 1998 Transfer of *Rhizobium loti*, *Rhizobium huakuii*, *Rhizobium ciceri*, *Rhizobium mediterraneum* and *Rhizobium tianshanense* to *Mesorhizobium* gen. nov. Int J Syst Bacteriol 47, 895–898.

Jonhson FJ 1990 Detection method of nitrogen (total) in fertilizers. *In* Methods of analysis of the Association of Official Analytical Chemists. Ed. K Elrich. pp. 17–19. Association of Official Analytical Chemists, USA.

Kimura M 1980 A simple method for estimating evolutionary rates of base substitutions through comparative studies of nucleotide sequences. J Mol Evol 16, 111–120.

Kumar S, Tamura K, Jakobsen IB and Nei M 2001 Molecular evolutionary genetics analysis software. Arizona State University, Tempe, Arizona. USA.

Moulin L, Munive A, Dreyfus B and Boivin-Masson C 2001 Nodulation of legumes by members of β–subclass of Proteobacteria. Nature 411, 948–950.

Nakamura LK, Roberts MS and Cohan FM 1999 Relationship of *Bacillus subtilis* clades associated with strains 168 and W23: a proposal for *Bacillus subtilis* subsp. *subtilis* subsp. nov. and *Bacillus subtilis* subsp. *spizizenii* subsp. nov. Int J Syst Bacteriol 49, 1211–1215.

Nour SH, Fernández MP, Normand P and Cleyet-Marel JC 1994 *Rhizobium ciceri*, sp. nov., consisting of strains that nodulate chickpeas (*Cicer arietinum* L.). Int J Syst Bacteriol 44, 511–522.

Nour SH, Cleyet-Marel JC, Normand P and Fernandez MP 1995 Genomic heterogeneity of strains nodulating chickpeas (*Cicer arietinum* L.) and description of *Rhizobium mediterraneum* sp. nov. Int J Syst Bacteriol 45, 640–648.

Nuswantara S, Fujie M, Yamada T, Malek W, Inaba M, Kaneko Y and Murooka Y 1999 Phylogenetic position of

Mesorhizobium huakuii subsp. *rengei*, a symbiont of *Astragalus sinicus* cv. Jpn J Biosci Bioeng 87, 49–55.

Palomo JL, Velázquez E, Mateos PF, García-Benavides P and Martínez-Molina E 2000 Rapid identification of *Clavibacter michiganensis* subspecies *sepedonicus* based on the stable low molecular weight RNA (LMW RNA) profiles. Eur J Plant Pathol 106, 789–793.

Pearson WR and Lipman DJ 1988 Improved tools for biological sequence comparison. Proc Natl Acad Sci USA 85, 2444–2448.

Peix A, Rivas-Boyero AA, Mateos PF, Rodríguez-Barrueco C, Martínez-Molina E and Velázquez E 2001 Growth promotion of chickpea and barley by a phosphate solubilizing strain of *Mesorhizobium mediterraneum* under growth chamber conditions. Soil Biol Biochem 33, 103–110.

Perret X, Staehelin C and Broughton WJ 2000 Molecular basis of symbiotic promiscuity. Microbiol Mol Biol Rev 64, 180–201.

Rivas R, Velázquez E, Valverde A, Mateos PF and Martínez-Molina E 2001 A two primers random amplified polymorphic DNA procedure to obtain polymerase chain reaction fingerprints of bacterial species. Electrophoresis 22, 1086–1089.

Rivas R, Velázquez E, Palomo JL, Mateos P, García-Benavides P and Martínez-Molina E 2002a Rapid identification of *Clavibacter michiganensis* subspecies *sepedonicus* using two primers random amplified polymorphic DNA (TP-RAPD) fingerprints. Eur J Plant Pathol 108, 179–184.

Rivas R, Velázquez E, Willems A, Vizcaíno N, Subba-Rao NS, Mateos PF, Gillis M, Dazzo FB and Martínez-Molina E 2002b A new species of *Devosia* that forms a nitrogen-fixing root-nodule symbiosis with the aquatic legume *Neptunia natans* (L. f.) Druce. Appl Environ Microbiol 68, 5217–5222.

Saitou N and Nei M 1987 A neighbour-joining method: a new method for reconstructing phylogenetics trees. Mol Biol Evol 44, 406–425.

Sprinzl M, Moll J, Meissner F and Hatmann T 1985 Compilation of tRNA sequences. Nucleic Acid Res 13, 1–49.

Sy A, Giraud E, Jourand P, García N, Willems A, de Lajudie P, Prin Y, Neyra M, Gillis M, Boivin-Masson C and Dreyfus B 2001 Methylotrophic *Methylobacterium* bacteria nodulate and fix nitrogen in symbiosis with legumes. J Bacteriol 183, 214–220.

Thompson JD, Gibson TJ, Plewniak F, Jeanmougin F and Higgins DG 1997 The clustalX windows interface: flexible strategies for multiple sequence alignment aided by quality analysis tools. Nucleic Acid Res 24, 4876–4882.

Velázquez E, Cervantes E, Igual JM, Peix A, Mateos PF, Benamar S, Moiroud A, Wheeler CT, Dawson J, Labeda D, Rodríguez-Barrueco C and Martínez-Molina E 1998a Analysis of LMW RNA profiles of *Frankia* strains by staircase electrophoresis. Syst Appl Microbiol 21, 539–545.

Velázquez E, Cruz-Sánchez JM, Mateos PF and Martínez-Molina E 1998b Analysis of stable low molecular weight RNA profiles of members of the family Rhizobiaceae. Appl Environ Microbiol 64, 1555–1559.

Velázquez E, Calvo O, Cervantes E, Mateos PF, Tamame M and Martínez-Molina E 2000 Staircase electrophoresis profiles of stable low molecular weight RNA as yeast fingerprinting. Int J Syst Evol Microbiol 50, 917–923.

Velázquez E, Igual JM, Willems A, Fernández MP, Muñoz E, Mateos PF, Abril A, Toro N, Normand P, Cervantes E, Gillis M and Martínez-Molina E 2001a Description of *Mesorhizobium chacoense* sp. nov. that nodulates *Prosopis alba* in the Chaco Arido region (Argentina). Int J Syst Evol Microbiol 51, 1011–1021.

Velázquez E, Martínez-Romero E, Rodríguez-Navarro DN, Trujillo ME, Daza A, Mateos PF, Martinez-Molina E and Van Berkum P 2001b Characterization of rhizobial isolates of *Phaseolus vulgaris* by staircase electrophoresis of low molecular weight RNA. Appl Environ Microbiol 67, 1008–1010.

Velázquez E, Trujillo ME, Peix A, Palomo JL, García-Benavides P, Mateos P, Ventosa A and Martínez-Molina E 2001c Stable low molecular weight RNA analyzed by staircase electrophoresis, a molecular signature for both prokaryotic and eukaryotic microorganisms. Syst Appl Microbiol 24, 490–499.

Vincent JM 1970 The cultivation, isolation and maintenance of rhizobia. *In* A manual for the practical study of root-nodule. Ed. JM Vincent. pp. 1–13. Blackwell Scientific Publications, Oxford.

Wang ET, van Berkum P, Sui XH, Beyene D, Chen WX and Martínez-Romero E 1999 Diversity of rhizobia associated with *Amorpha fruticosa* isolated from chinese soils and description of *Mesorhizobium amorphae* sp. nov. Int J Syst Bacteriol 49, 51–65.

Wieser M and Busse HJ 2000 Rapid identification of *Staphylococcus epidermidis*. Int J Syst Evol Microbiol 50, 1087–1093.

Original Paper

Phosphate solubilization activity of rhizobia native to Iranian soils

H.A. Alikhani[1], N. Saleh-Rastin[1] & H. Antoun[2]

[1]*Department of Soil Science, College of Agriculture, Tehran University, Tehran, Iran.* [2]*Département des Sols et de Génie Agroalimentaire, Université Laval, 2110 Pavillon de l'Envirotron, G1K 7P4, Québec, Qc, Canada. Corresponding author**

Received: 31 May 2006

Key words: calcium phosphate, legumes, inositol hexaphosphate, PGPR, pH, *Rhizobium*

Abstract

Agricultural soils in Iran are predominantly calcareous with very low plant available phosphorus (P) content. In addition to their beneficial N_2-fixing activity with legumes, rhizobia can improve plant P nutrition by mobilizing inorganic and organic P. Isolates from different cross-inoculation groups of rhizobia, obtained from Iranian soils were tested for their ability to dissolve inorganic and organic phosphate. From a total of 446 rhizobial isolates tested for P solubilization by the formation of visible dissolution halos on agar plates, 198 (44%) and 341(76%) of the isolates, solubilized $Ca_3(PO_4)_2$ (TCP) and inositol hexaphosphate (IHP), respectively. In the liquid Sperber TCP medium, phosphate-solubilizing bacteria (*Bacillus* sp. and *Pseudomonas fluorescens*) used as positive controls released an average of 268.6 mg L^{-1} of P after 360 h incubation. This amount was significantly ($P < 0.05$) higher than those observed with all rhizobia tested. The group of *Rhizobium leguminosarum* bv. *viciae* mobilized in liquid TCP Sperber medium significantly ($P < 0.05$) more P (197.1 mg L^{-1} in 360 h) than other rhizobia tested,. This group also showed the highest dissolution halo on the TCP solid Sperber medium. The release of soluble P was significantly correlated with a drop in the pH of the culture filtrates indicating the importance of acid production in the mobilization process. None of the 70 bradyrhizobial isolates tested was able to solubilize TCP. These results indicate that many rhizobia isolated from soils in Iran are able to mobilize P from organic and inorganic sources and this beneficial effect should be tested with crops grown in Iran.

Introduction

Microorganisms play an important role in effecting the availability of soil P to plant roots, and increasing P mobilization in soil, though the development of effective microbial inoculants remains a major scientific challenge (Richardson, 2001). Agricultural soils in Iran are predominately calcareous and are characterized by a high pH and low amounts of plant available phosphorus (P). The P deficiency can severely limit plant growth and productivity, in particular in legumes, where both the plants and their symbiotic bacteria are affected, and this may have a deleterious effect on nodule formation, development and function (Robson et al., 1981). Up to 75% of the soluble P fertilizers added to crops may be converted to sparingly soluble forms by reacting with the free Ca^{2+} ions in high pH soils or with Fe^{3+} or Al^{3+} in low pH soils (Goldstein, 1986). Organic P represents from 50% to 80% of the total soil P, and most plants are unable to utilize these sources of P (Richardson, 2001). Several bacterial and fungal species are phosphate-solubilizing microorganisms (PSM) and evaluation of their potential to mobilize soil P has been the subject of intensive investigations

*FAX No: +1-418-656-7871. E-mail: Hani.Antoun@sga.ulaval.ca

E. Velázquez and C. Rodríguez-Barrueco (eds.),
First International Meeting on Microbial Phosphate Solubilization, 35–41

(Rodriguez and Fraga, 1999; Whitelaw, 2000). Rhizobia, the beneficial N_2-fixing symbiotic partners of legumes, like other plant growth promoting rhizobacteria (PGPR), are also able to colonize the roots of non-legumes (Chabot et al., 1996; Schloter et al., 1997) and stimulate plant growth (Antoun et al., 1998; Yanni et al., 2001). Rhizobia are able to solubilize both organic (Abd-Alla, 1994) and inorganic phosphates (Antoun et al., 1998). The main advantage of using rhizobia, as PSM will be their dual beneficial nutritional effect resulting both from P mobilization and N_2-fixation (Peix et al., 2001) and their well-documented synergistic interactions with arbuscular mycorhizal fungi (Barea et al., 2002).

The current study was designed to determine the ability of 446 strains of rhizobia to mobilize inorganic and organic P, in order to identify strains with high activity to be tested as PGPR with crops cultivated in Iran.

Material and methods

The rhizobia

The 446 isolates of rhizobia tested in this study belong to the following groups:

Bradyrhizobium sp. (13); *Bradyrhizobium japonicum* (57); *Mesorhizobium ciceri* and *Mesorhizobium mediterraneum* (83); *Sinorhizobium meliloti* (168); *Rhizobium leguminosarum* bv. *phaseoli* (57); *Rhizobium leguminosarum* bv. *trifolii* (9) and *Rhizobium leguminosarum* bv. *viciae* (59). All of the isolates originated from fields under legume cultivation in different parts of Iran, and several isolates belong to the Soil and Water Research Institute of Iran. The selective medium, yeast extract mannitol agar (YMA) with congo red (Vincent, 1970), was used for isolation of rhizobia and a pure culture of each isolate was prepared after sub-culturing on the same medium. Pure cultures were authenticated as rhizobia through laboratory procedures and plant infection tests described by Somasegaran and Hoben (1994).

Inocula preparation

In order to prepare fresh inocula containing the same number of bacterial population for all rhizobia under study, a colony of each isolate was transferred to a 100 mL Erlenmeyer flask containing 15 mL of yeast extract mannitol broth (YMB). Inoculated flasks were incubated at 27 °C on a rotary shaker (100 rpm) for 96 h. All bacterial suspensions were adjusted to approximately 5×10^8 cfu mL^{-1}, with a sterile 0.5% NaCl solution and by using standard curves relating numbers of bacteria (cfu) to optical densities measured with a spectrophotometer at 570 nm.

Phosphate solubilization in solid media

The basal Sperber (1958) medium used contained (in g L^{-1} of distilled water): glucose 10.0, yeast-extract 0.5, $CaCl_2$ 0.1, $MgSO_4 \cdot 7H_2O$ 0.25 and agar 15.0. The medium was supplemented with 2.5 g L^{-1} of $Ca_3(PO_4)_2$ (TCP) or inositol hexaphosphate (IHP) as P source to appraise the ability of the strains to mobilize respectively inorganic or organic P sources. The pH of the medium was adjusted to 7.2 before autoclaving. The media were distributed in 9 cm diameter Petri plates and marked in four equal parts after solidification. Using the drop plate method, each part was inoculated with 7 μL of inocula. All tests were performed with four replications. Inoculated plates were incubated in dark at 27 °C and the diameter of clear zone (halo) surrounding the bacterial growth as well as the diameter of colony were measured after 10, 20 and 30 days. All assays were replicated four times and the results are shown as the ratio of halo/colony.

Phosphate solubilization in liquid medium

On solid media three isolates of *Rhizobium leguminosarum* biovar phaseoli, 12 of *R. leguminosarum* biovar viciae, 69 of *Sinorhizobium meliloti*, and 64 of *Mesorhizobium ciceri* and *M. mediterraneum* produced large halo zones (ratio of halo diameter/colony diameter >1.2), and were used to measure P solubilization in liquid medium. The 70 bradyrhizobia used were unable to solubilize TCP on the solid medium and were also tested in liquid medium. The following P-solubilizing bacteria isolated from Iranian soils were used as positive controls: an isolate of *Bacillus* sp., and three *Pseudomonas fluorescens* isolates. Erlenmeyer flasks (200 mL) containing 90 mL. of the liquid Sperber medium were inoculated with 200 μL of bacterial suspension (5×10^8 cfu mL^{-1}).

The flasks were incubated on rotary shaker (120 rpm) at 27 °C. After 72, 120, 240 and 360 h of incubation, aliquots of cultures were aseptically taken from each flask. The supernatant was separated from the bacterial cells by successive filtration through Whatman paper # 42 followed by 0.2 μm Millipore membrane and was used for the determination of the pH and the soluble P released into the solution. P was measured with the water-soluble phosphorus method using ammonium paramolybdate and ascorbic acid as described by Olsen and Sommers (1982). Control flasks were not inoculated, and had a pH of 7.20 and a water-soluble P content of 1.8 mg L^{-1}after autoclaving. After 360 h incubation the control flasks had a pH of 6.06 and contained 7.5 mg L^{-1} soluble P. Values obtained with the uninoculated controls were always substracted from their repective treatments. All experiments were performed in triplicates.

Statistical analysis

The experimental design used to analyse the P solubilization results obtained in solid and liquid media was a split plot in time based on completely randomized design (bacterial groups as main plot and time of measurements as subplot). Variance homogeneity determination (ANOVA) was conducted with the General Linear Models of SAS by using the type II sum of squares, and means were compared according to the Duncan test (SAS, 1990).

Results and discussion

Preliminary assays with culture media

In preliminary studies, we modified the well-known rhizobia YMA medium (Vincent, 1970) by replacing the soluble source of P (K_2HPO_4) with 2.5 g L^{-1} TCP or IHP and by adding 0.1 g L^{-1} of KCl as a source of K. On modified media, no clear P solubilization halos were observed in solid media, and P release in liquid media from TCP was negligible. These results indicated that mannitol was not a good C source for P mobilization studies in rhizobia, and therefore all tests were performed with Sperber medium (1958) containing glucose as C source. Phosphate solubilization in solid media was greatly affected by the C source used, and generally the larger calcium phosphate solubilization halos were obtained with glucose (Silva Filho and Vidor, 2000).

P mobilization in the solid Sperber medium

From the 446 strains of the Iranian rhizobia used in this study, 198 (44%) and 341 (76%) were able to mobilize TCP and IHP respectively. Antoun et al. (1998), tested 266 strains obtained from different laboratories in Australia, Columbia, Egypt and North America on the solid Goldstein (1986) medium supplemented with vitamins (Vincent, 1970), and found that 144 (54%) were dicalcium phosphate (DCP) solubilizers. The differences observed can be explained by the different calcium phosphate and nitrogen sources used. In the present work, yeast extract was used as nitrogen and vitamin sources while Antoun et al. (1998) used NH_4Cl as a nitrogen source. In developing efficient growth medium for screening PSM, yeast extract was avoided because of its inhibitory effect at concentration higher than 0.5 g L^{-1} (Nautiyal, 1999). However, Halder and Chakrabartty (1993) also observed that the inorganic P solubilization activity of some *Rhizobium* strains was better in a medium without NH_4^+, containing 0.4 g L^{-1} of yeast extract as the nitrogen source. Rhizobia have different vitamin requirements (Vincent, 1970) that are better satisfied by yeast extracts. In some studies, the plate screening method has produced contradictory results between plate halo detection and P solubilization in liquid cultures. However this method can be regarded as generally reliable for isolation and preliminary characterization of PSM (Rodriguez and Fraga, 1999). In our study the plate method was very practical for screening a very large number of rhizobial isolates, however the procedure developed by Gupta et al. (1994) using bromophenol blue to improve detection of acid production and its adaptation to liquid media (Mehata and Nautiyal, 2001) should be further evaluated in future screening work.

None of the 57 strains of *Bradyrhizobium japonicum* and of the 13 strains of *Bradyrhizobium* sp. tested were able to mobilize P from TCP in Sperber solid or liquid medium. This observation suggests that *B. japonicum* strains are not good inorganic P-solubilizers. In fact, Antoun et al. (1998) reported that only 1 out of the 18 strains of *B. japonicum* tested was able to mobilize P from DCP on a solid medium. The analysis of variance indicated that on the solid Sperber medium, the different groups of rhizobia mobilized P from TCP or IHP in a different manner (Table 1). Within each group a significant ($P < 0.001$) strain effect was also observed, indicating that the activity of the strains may vary significantly. Overall, strains of *Rhizobium leguminosarum* bv. *viciae* mobilized significantly ($P < 0.05$) more P from TCP than strains of *Mesorhizobium*, *Sinorhizobium* and *R. leguminosarum* bv. *phaseoli* (Table 2). The solubilization activity of the groups exhibited different trends at different time, as indicated by the significant ($P < 0.001$) group x time interactions observed (Table 1). However, in general for all strains tested, the solubilization activity of TCP and IHP by the strains significantly ($P < 0.05$) increased with time (results not shown).

Soils may contain a substantial quantity of organic P (Richardson 2001), and phosphatases from microorganisms may carry out mineralization of most organic phosphorus compounds. From 30 up to 63% of culturable soil bacteria can mineralize organic P in soils (Rodriguez and Fraga, 1999). More rhizobia were able to mobilize P from IHP than from TCP. In fact 341 (76%) of the 446 Iranian rhizobia were able to mineralize IHP. With the exception of the *Bradyrhizobium* group (5–7%), 70% or more of the other rhizobial isolates were able to mineralize IHP. The only strain of *Bradyrhizobium* spp. able to mobilize IHP, had the highest observed mineralization halo/colony ratio. This solubilization activity was comparable to that observed with the strains of the *M. ciceri* and *M. mediterraneum* group, and was significantly ($P < 0.05$) higher than that of the other groups tested. As observed with TCP, the strain effect within each group on IHP solubilization is very significant ($P < 0.001$).

Table 1. ANOVA of the solubilization of inorganic ($Ca_3(PO_4)_2$) and organic (inositol hexaphosphate) phosphorus on the solid Sperber medium by rhizobia isolated from Iranian soils

Source of variation	Inorganic P		Organic P	
	Degree of freedom	Mean square	Degree of freedom	Mean square
Rhizobial group (G)	3	56.16***	6	404.33***
Strains	194	4.81***	333	9.61***
Time (T)	2	37.89***	2	270.52***
Interaction G × T	6	12.43***	12	51.19***
Error	2170	0.06	3726	0.12

***Significant at $P < 0.001$.

Table 2. Solubilization of inorganic ($Ca_3(PO_4)_2$) and organic (inositol hexaphosphate) phosphorus on the solid Sperber medium by rhizobia isolated from Iranian soils.

Inorganic P		Organic P	
Rhizobial group	DH/CD	Rhizobial group	DH/CD
Rhizobium leguminosarum bv. *viciae*	2.48a	*Bradyrhizobium* spp.	4.68a
Mesorhizobium ciceri & *M. mediterraneum*	1.42b	*Mesorhizobium ciceri* & *M. mediterraneum*	3.65a
Sinorhizobium meliloti	1.40b	*Sinorhizobium meliloti*	2.29b
Rhizobium leguminosarum bv. *phaseoli*	0.96b	*Rhizobium leguminosarum* bv. *viciae*	2.06bc
		Rhizobium leguminosarum bv. *trifolii*	1.39bc
		Rhizobium leguminosarum bv. *phaseoli*	1.19bc
		Bradyrhizobium japonicum	0.83c

Values are the ratio of dissolution halo (DH)/colony diameter (CD).
Results are mean of three replicates, and three measurements made after 10, 20 and 30 days of incubation. Means followed by the same letter are not significantly different at $P < 0.05$.

P mobilization in the liquid Sperber medium

Strains producing halo/colony ratios higher than 1.2 on TCP plates (3, *Rhizobium leguminosarum* bv. *phaseoli*; 12, *R. leguminosarum* bv. *viciae*, 69, *Sinorhizobium meliloti*; and 64 *Mesorhizobium ciceri* & *M. mediterraneum*) were further investigated in liquid medium. All strains tested solubilized some P from DCP and produced acid in liquid culture. As observed on the solid medium, bacterial groups and strains within each group had significantly ($P < 0.001$) different solubilization and acid production activities (Table 3). The soluble P released by the strains significantly ($P < 0.05$) increased with time (Figure 1).

Table 3. ANOVA of phosphorus mobilized from $Ca_3(PO_4)_2$ and of the change in pH of the liquid Sperber medium inoculated with rhizobia isolated from Iranian soils and with phosphate-solubilizing bacteria (one isolate of *Bacillus* sp. and three isolates of *Pseudomonas fluorescens*) used as positive controls

Source of variation	Degree of freedom	Mean squares	
		Phosphorus	PH
Bacterial group (G)	4	557658.20***	21.64***
Strains	147	35085.25***	2.31***
Time (T)	3	2478319.23***	33.18***
Interaction G × T	12	25294.82***	1.51***
Error	1657	1492.10	0.1

***Significant at $P < 0.001$.

The four isolates of PSM (*Bacillus* sp. and *Pseudomonas fluorescens*) used in this study as positive controls released an average of 268.6 mg mL^{-1} of P from TCP. This quantity was significantly ($P < 0.05$) higher than the 197.1 mg mL^{-1} of P mineralized by strains of the group *R. leguminosarum* bv. *viciae*. The other three groups of rhizobia released the following comparable amounts of soluble P which are significantly lower ($P < 0.05$) than those obtained with PSM and *R. leguminosarum* bv. *viciae*: *S. meliloti*, 112.8 mg mL^{-1}; *M. ciceri* and *M. mediterraneum*, 102.3 mg mL^{-1}; and *R. leguminosarum* bv. *phaseoli*, 88.66 mg mL^{-1}. As revealed by statistical analyses, the results obtained in the TCP liquid medium corroborate those observed with the solid medium, indicating that strains of the group *R. leguminosarum* bv. *viciae* isolated from Iranian soils are the more effective TCP solubilizers. Halder and Chakrabartty (1993) previously reported that strains of *R. leguminosarum* bv. *viciae* can achieve high inorganic P solubilization.

Significant drops in pH accompanied the release of soluble P from TCP, in the culture supernatants (Table 3 and Figure 2). This confirms the implication of organic acid production in P solubilization by rhizobia (Halder and Chakrabartty, 1993). For all groups of rhizobia tested, strong significant ($P \leq 0.01$) inverse correlations ($r = -0.66$ to -0.89) were observed between the pH of the culture supernatants and

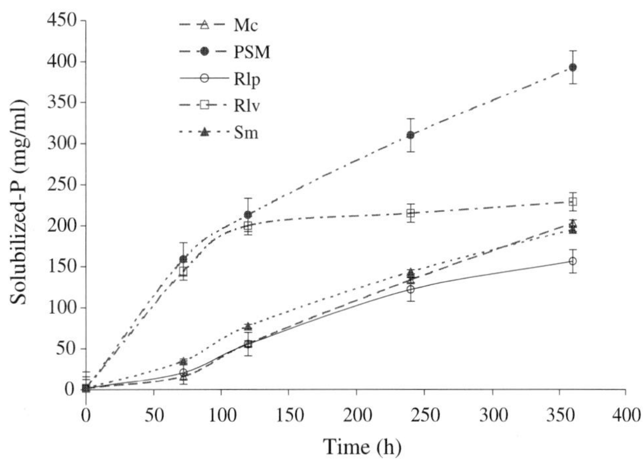

Figure 1. Solubilization of $Ca_3(PO_4)_2$ in the liquid Sperber medium by bacterial isolates belonging to the following groups: Mc, *Mesorhizobium ciceri* and *M. mediterraneum*; PSM, phosphate-solubilizing *Bacillus* sp. and *Pseudomonas fluorescens* used as controls; Rlp, *Rhizobium leguminosarum* bv. *phaseoli*; Rlv, *R. leguminosarum* bv. *viciae*; Sm, *Sinorhizobium meliloti*. Error bars are ± standard error ($n = 3$).

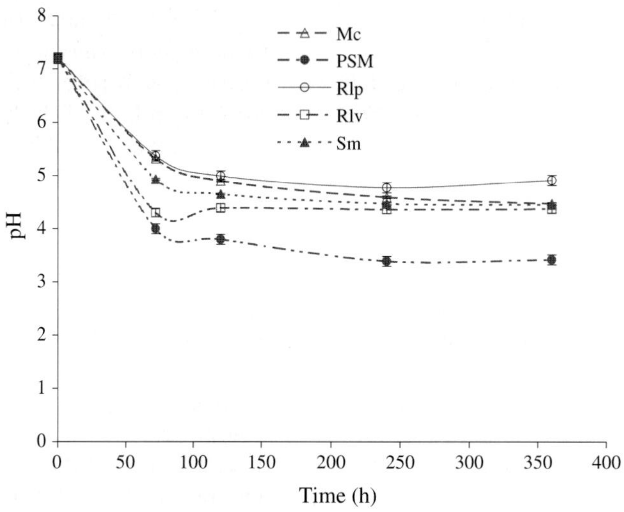

Figure 2. Changes of the pH of the culture filtrates of the liquid Sperber medium during the solubilization of $Ca_3(PO_4)_2$ by the different bacterial groups tested. For abbreviations see Figure 1. Error bars are ± standard error (n = 3), and are smaller than the symbols.

their soluble P content, corroborating similar observations made with rhizobia (Halder and Chakrabartty, 1993), and other bacteria mobilizing P from rock phosphate (Nahas, 1996).

These results indicate that many rhizobia are able to mobilize P from inorganic and organic sources. These rhizobia also have proved to be good plant growth PGPR with non-legumes (Antoun et al., 1998; Yanni et al., 2001). In developing inoculants that improve plant P nutrition and allow plants to use soil stocks of organic and inorganic P, rhizobia may present many advantages. In fact, in addition to their beneficial effects on legume and non-legume plants which will be an advantage in crop rotation systems, inoculation and inoculants production technologies are already available, and rhizobia are generally perceived as environmentally friendly, since they have been used with legumes for many years without causing harm to the environment or to farmers (Antoun et al., 1998).

Acknowledgements

The authors are grateful to professor Malakouti head of the Soil and Water Research Institute in Tehran and to the staff of the Department of Soil Biology of the same Institute for their help with the rhizobial collection and for their constructive discussions. Many thanks to professor Ghannadha for his help in the statistical analyses and the interpretation of the results, and to Miss Pashaky for her excellent technical assistance. Financial support for this investigation was provided by grants from the Iranian Government "Studies and Researches Between Universities" program for collaboration between the Universities of Tehran and Rafsanjan.

References

Abd-Alla M H 1994 Use of organic phosphorus by *Rhizobium leguminosarum* bv. *viciae* phosphatases. Biol. Fertil. Soils 8, 216–218.

Antoun H A, Beauchamp C J, Goussard N, Chabot R and Lalande R 1998 Potential of *Rhizobium* and *Bradyrhizobium* species as plant growth promoting rhizobacteria on non-legumes: effect on radishes (*Raphanus sativus* L.). Plant Soil 204, 57–67.

Barea J M, Azcón R and Azcón-Aguilar C 2002 Mycorhizosphere interactions to improve plant fitness and soil quality. Antonie van Leeuwenhoek 81, 343–351.

Chabot R, Antoun H, Kloepper J W and Beauchamp C J 1996 Root colonization of maize and lettuce by bioluminescent *Rhizobium leguminosarum* biovar phaseoli. Appl. Environ. Microbiol. 62, 2767–2772.

Goldstein A H 1986 Bacterial solubilization of mineral phosphates: historical perspectives and future prospects. Am. J. Altern. Agric. 1, 51–57.

Gupta R, Singal R, Shankar A, Kuhad R C and Saxena R K 1994 A modified plate assay for screening phosphate solubilizing microorganisms. J. Gen. Appl. Microbiol. 40, 255–260.

Halder A K and Chakrabartty P K 1993 Solubilization of inorganic phosphate by *Rhizobium*. Folia Microbiol. 38, 325–330.

Mehata S and Nautiyal C S 2001 An efficient method for qualitative screening of phosphate-solubilizing bacteria. Curr. Microbiol. 43, 51–56.

Nahas E 1996 Factors determining rock phosphate solubilization by microorganisms isolated from soil. World J. Microbiol. Biotechnol. 12, 567–572.

Nautiyal C S 1999 An efficient microbiological growth medium for screening phosphate solubilizing microorganisms. FEMS Microbiol. Lett. 170, 265–270.

Olsen S R and Sommers L E 1982 Phosphorus. *In* Methods of Soil Analysis, Part 2-chemical and Microbiological Properties, 2nd edn., Ed. Page AL. Am Soc. Agron. and Soil Sci. Soc. A. Madison, Wisconsin, USA.

Peix A, Rivas-Boyero A A, Mateos P F, Rodriguez-Barrueco C, Martinez-Molina E and Velazquez E 2001 Growth promotion of chickpea and barley by a phosphate solubilizing strain of *Mesorhizobium mediterraneum* under growth chamber conditions. Soil Biol. Biochem. 33, 103–110.

Richardson A E 2001 Prospects for using soil microorganisms to improve the acquisition of phosphorus by plants. Aust. J. Plant Physiol. 28, 897–906.

Robson A D, O'Hara G W and Abbott L K 1981 Involvement of phosphorus in nitrogen fixation by subterranean clover (*Trifolium subterraneum* L.). Aust. J. Plant Physiol. 8, 427–436.

Rodriguez H and Fraga R 1999 Phosphate solubilizing bacteria and their role in plant growth promotion. Biotech. Adv. 17, 319–339.

SAS, Institute Inc 1990 SAS procedure guide version 6 edn. SAS Institute Inc Cary, NC, 705 p.

Schloter M, Wiehe W, Assmus B, Steindl H, Beke H, Höflich G and Hartmann A 1997 Root colonization of different plants by plant-growth-promoting *Rhizobium leguminosarum* bv. *trifolii* R39 studied with monosporic polyclonal antisera. Appl. Environ. Microbiol. 63, 2038–2046.

Silva Filho G N and Vidor C 2000 Phosphate solubilization by microorganisms in the presence of different carbon sources. R. Bras. Ci. Solo. 24, 311–319.

Sperber J I 1958 The incidence of apatite solubilizing organisms in the rhizosphere and soil. Aust. J. Agric. Res. 9, 778–781.

Somasegaran P and Hoben H J 1994 Handbook for Rhizobia – Methods in Legume–*Rhizobium* Technology. Springer-Verlag, New York.

Vincent J M 1970 A Manual for the Practical Study of Root Nodule Bacteria. IBP handbook 15 Blackwell Scientific Publications, Oxford.

Whitelaw M A 2000 Growth promotion of plants inoculated with phosphate-solubilizing fungi. Adv. Agron. 69, 99–151.

Yanni Y G, Rizk R Y, Abd El-Fattah F K, Squartini A, Corich V, Giacomini A, de Bruin F, Rademaker J, Mayra-Flores J, Ostrom P, Vega-Hernandez M, Hollingsworth R I, Martinez-Molina E, Mateos P, Velazquez E, Wopereis J, Triplett E, Umali-Garcia M, Anarna J A, Rolfe B G, Ladha J K, Hill J, Mujoo R, Ng P K and Dazzo F B 2001 The beneficial plant growth-promoting association of *Rhizobium leguminosarum* bv. *trifolii* with rice roots. Aust. J. Plant Physiol. 28, 845–870.

Differential effects of coinoculations with *Pseudomonas jessenii* PS06 (a phosphate-solubilizing bacterium) and *Mesorhizobium ciceri* C-2/2 strains on the growth and seed yield of chickpea under greenhouse and field conditions

Angel Valverde[1], Araceli Burgos[2], Tiziana Fiscella[1,4], Raúl Rivas[3], Encarna Velázquez[3], Claudino Rodríguez-Barrueco[1], Emilio Cervantes[1], Manuel Chamber[2] & José-Mariano Igual[1,*]

[1]*Instituto de Recursos Naturales y Agrobiología-CSIC, Apartado 257, 37071, Salamanca, Spain.* [2]*CIFA Las Torres-Tomejil, 41200, Alcalá del Rio, Sevilla, Spain.* [3]*Departamento de Microbiología y Genética, Universidad de Salamanca, 37007, Salamanca, Spain.* [4]*Agriculture Faculty, University of Catania, Via Valdisaboia 5, 95123, Catania, Italy.* **Corresponding author*

Received: 31 May 2006

Key words: chickpea, *Mesorhizobium*, PGPR, phosphate-solubilizing bacteria, plant yield, *Pseudomonas*

Abstract

In the course of a project carried out in two regions of Spain, Castilla y León and Andalucía, aiming to find useful biofertilizers for staple grain-legumes, an efficient rhizobia nodulating chickpea (termed as C-2/2) and a powerful *in vitro* phosphate-solubilizing bacterial strain (termed as PS06) were isolated. Analyses of their 16S rDNA sequence indicated that they belong to the bacterial species *Mesorhizobium ciceri* and *Pseudomonas jessenii*, respectively. Greenhouse and field experiments were carried out in order to test the effect of single and dual inoculations on chickpea (ecotype ILC-482) growth. Under greenhouse conditions, plants inoculated with *Mesorhizobium ciceri* C-2/2 alone had the highest shoot dry weight. The inoculation treatment with *P. jessenii* PS06 yielded a shoot dry weight 14% greater than the uninoculated control treatment, but it was not correlated with shoot P contents. However, the co-inoculation of C-2/2 with PS06 resulted in a decrease in shoot dry weight with respect to the inoculation with C-2/2 alone. Under field conditions, plants inoculated with *M. ciceri* C-2/2, in single or dual inoculation, produced higher nodule fresh weight, nodule number and shoot N content than the other treatments. Inoculation with *P. jessenii* PS06 had no significant effect on plant growth. However, the co-inoculation treatment ranked the highest in seed yield (52% greater than the uninoculated control treatment) and nodule fresh weight. These data suggest that *P. jessenii* PS06 can act synergistically with *M. ciceri* C-2/2 in promoting chickpea growth. The contrasting results obtained between greenhouse and field experiments are discussed.

Introduction

A substantial number of bacterial species, mostly those associated with the plant rhizosphere, may exert a beneficial effect upon plant growth (Glick, 1995). This group of bacteria has been termed "plant growth promoting rhizobacteria" or PGPR (Kloepper and Schroth, 1978). Although plant growth promoting rhizobacteria occur in soil, usually their numbers are not high enough to compete with other bacteria commonly established in the rhizosphere. Therefore, for agronomic utility, inoculation of plants with target microorganisms

* FAX No: +34-923-219609.
E-mail: igual@usal.es

E. Velázquez and C. Rodríguez-Barrueco (eds.),
First International Meeting on Microbial Phosphate Solubilization, 43–50

at a much higher concentration than those normally found in soil is necessary to take advantages of their beneficial properties for plant yield enhancement. An increasing number of PGPR are successfully used as commercial biofertilizers for agricultural improvement (Subba Rao, 1993).

Among PGPR, phosphate-solubilizing bacteria (PSB) are considered as promising biofertilizers since they can supply plants with phosphorus (P) from sources otherwise poorly available. Beneficial effects of the inoculation with PSB to many crop plants have been described by numerous authors (Antoun et al., 1998; Chabot et al., 1993, 1996, 1998; Pal, 1998; Peix et al., 2001a, b; Sarawgi et al., 1999; Tomar et al., 1996). Moreover, synergistic interactions on plant growth have been observed by co-inoculation of PSB with other bacteria, such as *Azospirillum* (Alagawadi and Gaur, 1992; Belimov et al., 1995) and *Azotobacter* (Kundu and Gaur, 1984), or with vesicular arbuscular mycorrhizae (Kim et al., 1998; Piccini and Azcón, 1987; Ray et al., 1981; Toro et al., 1997, 1998).

Ectorhizospheric strains from pseudomonads and bacilli, and endosymbiotic rhizobia have been described as effective phosphate solubilizers (Igual et al., 2001; Rodríguez and Fraga, 1999). In addition to the phosphate-solubilizing capability of many *Pseudomonas* strains, they also can promote plant growth by mechanisms such as the production of plant growth regulators and vitamins, enhancement of plant nutrient uptake and suppression of pathogenic or deleterious organisms (Davison, 1998; Glick, 1995; O'Sullivan and O'Gara, 1992). Moreover, the tripartite association composed of legume plant, rhizobia and *Pseudomonas* spp. has been reported to increase root and shoot weight, plant vigour, nitrogen (N) fixation and grain yield in various legumes (Bolton et al., 1990; Dashti et al., 1998; Sindhu et al., 1999).

In the course of a project carried out in two regions of Spain, Castilla y León and Andalucía, aiming to find useful biofertilizers for staple grain-legumes, we isolated an efficient rhizobia nodulating chickpea and a powerful *in vitro* phosphate-solubilizing bacterial strain. We also identified them and evaluated their performance under greenhouse and field conditions in promoting the growth and yield of chickpea (*Cicer arietinum*, ecotype ILC-482).

Materials and methods

Isolation and selection of phosphate-solubilizing and rhizobial strains

Soil samples for isolation of PSB were taken from a soil in Pajares de la Laguna (Salamanca, Spain), which has been traditionally cultivated with autochthonous varieties of *Cicer arietinum* (chickpea, variety "Pedrosillano") and *Lens culinaris* (lentil, variety "Pardina") in rotation with barley (*Hordeum vulgare*). PSB were isolated by plating serial dilutions of this soil in the medium described by Gupta et al. (1994), containing poorly soluble tri-calcium phosphate and bromophenol blue. After an incubation period of 7 days at 28 °C, colonies showing large solubilization halos were selected and the persistence of their phosphate-solubilizing capacity checked by five successive subcultures in the same medium. The most efficient P solubilizing strain (PS06), with the largest solubilization halo, was selected for further studies.

Soil samples for isolation of rhizobia nodulating chickpea were taken from a soil in Carmona (Seville, Spain), a zone where this legume is traditionally cultivated. Isolations were made, also using chickpeas as trap plants, on YMA plates (Bergersen, 1961) according to Vincent (1970). The cultures were purified from single colonies after 10 days incubation at 28 °C. Several strains were isolated from different plants and greenhouse experiments to test their symbiotic capacities (nodulation and N fixation) were carried out with *Cicer arietinum* (ecotype ILC-482, ICARDA, Turkey) plants growing in N-free medium. The one (termed C-2/2) that yielded the highest shoot growth and N content was chosen for further studies.

Identification of the selected strains by 16S rDNA gene sequencing

Total genomic DNA from the bacterial isolates was extracted as described by Rivas et al. (2001). Cells were harvested by centrifugation at 9000*g* in a microspin centrifuge for 10 min at room temperature. DNA was extracted with 100 μl of 0.05 *M* NaOH (DNA-free) by heating at 100 °C for 4 min. Samples were then placed on ice and 900 μL of water was added to each microtube and mixed thoroughly. After an additional centrifugation at 9000*g*, 700 μL of the supernatant were removed and stored at −20 °C.

Polymerase chain reaction was performed using an AmpliTaq reagent kit (Perkin-Elmer Biosystems) following the manufacturer's instructions (1.5 m*M* $MgCl_2$, 200 μM of each dNTP and 2 U of *Taq* polymerase for 25 μL of final volume of reaction). The PCR amplification of 16S rDNA was carried out using the following primers: 5′-AGAGTTTGATCTGGCTCAG-3′ (*Escherichia coli* positions 8–27) and 5′-AAGGAGGTGATCCANCCRCA-3′ (*E. coli* positions 1502–1522) at a final concentration of 0.2 μM. PCR conditions were as follows: preheating at 95 °C for 9 min; 35 cycles of denaturing at 95 °C for 1 min; annealing at 59 °C for 1 min and extension at 72 °C for 2 min, and a final extension at 72 °C for 7 min. PCR products were electrophoresed in a 1% agarose gel at 6 V cm^{-1} and visualized by ethidium bromide staining. The band corresponding to the 16S rDNA was purified directly from the gel by centrifugation using Ultrafree®-DA tubes (Millipore) for 10 min at 5000*g* at room temperature according to the manufacturer's instructions.

Sequencing reactions were performed on an ABI377 sequencer (Applied Biosystems) using a BigDye terminator v3.0 cycle sequencing kit as supplied by the manufacturer. The following primers were used: 5′-AACGCTGGCGGCRKGCYTAA-3′, 5′-ACTCCTACGGGAGGCAGCA G-3′, 5′-CTGCTGCCTCCCGTAGGAGT-3′, 5′-CGTGCCAGCAGCCGCGGTAA-3′, 5′-CAGG ATTAGATACCCTGGTAG-3′ and 5′-GAGGA AGGTGGGGATGACGTC-3′, which correspon d to *E. coli* small-subunit rDNA sequence positions 32–52, 336–356, 356–336, 512–532, 782–803 and 1173–1194, respectively. The sequence obtained was compared with those from the GenBank using the FASTA program (Pearson and Lipman, 1988). Sequences were aligned using the Clustal W software (Thompson et al., 1997).

Greenhouse experiment

Seeds of *Cicer arietinum* ecotype ILC-482 obtained from ICARDA (Turkey), and adapted and selected at C.I.F.A. "Las Torres" (Seville), were surface-sterilized for 10 min in 5% sodium hypochlorite, and then repeatedly washed with sterile, distilled water. After sterilization, eight seeds were planted in 2-L pots filled with autoclaved perlite supplemented with $Ca_3(PO_4)_2$ to obtain a concentration of 400 mg P kg^{-1} (approximately 0.7 mg of available P) and the inoculation treatments applied. The experimental design consisted of four inoculation treatments: uninoculated seeds, seeds inoculated with the phosphate-solubilizing strain PS06, seed inoculated with the rhizobial strain C-2/2, and seed co-inoculated with PS06 and C-2/2 bacterial strains.

For inoculation, strain PS06 was grown in Petri dishes with YED for 2 days, and strain C-2/2 was grown in Petri dishes with YMA for 5 days. Sterile water was added to the plates and cells scraped to obtain a suspension, which was adjusted to 10^8 colony forming units (cfu) mL^{-1}, as corroborated by plating dilutions on the corresponding YED or YMA medium.

One mL of the appropriate suspension was added to each seed placed in the pots. Approximately five days after emergence seedlings were thinned to five per pot. The experiment was arranged in a randomized block design with three replicates per treatment.

Pots were watered weekly with the nutrient solution described by Rigaud and Puppo (1975) but devoid of combined N and soluble phosphate. Five weeks after sowing, the plants were harvested and the number of nodules and shoot dry weights per pot recorded. Shoot P concentrations were measured by molybdovanado phosphate method after calcination of materials in an electric muffle furnace and digestion in a mixture of water, nitric and hydrochloric acids (8:1:1). Shoot N concentrations were determined using an Orion Research Ioanalyzer 901 equipped with an ammonia electrode after digestion of plant material by the Kjeldahl method.

Field experiment

The field trial was conducted at the C.I.F.A. "Las Torres y Tomejil" (Consejería de Agricultura y Pesca, Junta de Andalucía) Research Centre at Alcalá del Rio (Seville, Spain) between December 2000 and July 2001. The principal soil properties are summarized in Table 1.

The experimental site was divided in plots, each 5 m by 2 m (10 m^2) containing four rows planted 0.5 m apart. Inoculation treatments

Table 1. Characteristics of the soil at the experimental site at CIFA Las Torres-Tomejil[a]

pH	Organic matter (%)	Total N (%)	C/N	Available P (mg kg^{-1})	Exchangeable Ca (mg kg^{-1})	Exchangeable K (mg kg^{-1})
7.71 (H_2O)						
7.05 (KCl)	1.16	0.076	8.81	20	4510	245

[a]Organic matter was determined by the Walkley–Black wet combustion method, total N by the Kjeldahl method, available P by the Olsen method, and exchangeable Ca and K by atomic absorption spectrophotometry after extraction with ammonium acetate. All soil analyses were performed as described by Tan (1996).

were the same as those described in the greenhouse trial. The experiment was arranged in a randomized block design with three replicates per treatment. No fertilization was applied to the soil.

For the inoculation of the seeds, peat base inoculants were prepared for both strains using peat adjusted to neutral by adding $CaCO_3$ and sterilized by autoclaving before hand (Subba Rao, 1993). Strains PS06 and C-2/2 were grown in liquid YED and YMA medium, respectively, for 2 days with continuous shaking at 25 °C. The cells were harvested by centrifugation at about 5000*g* for 15 min and washed 3 times and resuspended in sterile water. The resultant bacterial suspensions were added aseptically to trays containing 50 g of peat and mixed (so that the final moisture became 40% of the water-holding capacity). For uninoculated control, equal volume of sterile water was added to peat. For the co-inoculation treatment, 10 g of each PS06 and C-2/2 peat base inoculum were mixed. Seeds of the *Cicer arietinum* ecotype ILC-482 having their surface wetted with a 40% gum arabic-water solution (Subba Rao, 1993) were mixed well with the corresponding peat culture, at a rate of 1 g of peat culture per 400 g of seeds, before they were sown. At the time of application, the population of bacteria in each formulation was checked, by plating dilutions on the corresponding YED or YMA medium, and was approximately 10^6 cfu g^{-1} of peat.

At flowering (13 weeks after sowing), ten plants were taken from each plot (five from each lateral row) and the number and fresh weight of nodules were recorded. Seed yields, taken from the two central rows of each plot (5 m^2 per plot), was determined at maturity (28 weeks after sowing). Seed N and P concentrations were determined as described above.

Statistical analysis

Statistical analyses were conducted using one-way ANOVA according to Snedecor and Cochran (1989) using the Statistix v.4.0 software. Comparisons of means were performed by the Fisher's Protected LSD test at $P \leq 0.05$.

Results and discussion

Isolation and selection of phosphate-solubilizing and rhizobial strains

Indigenous PSB were isolated from a soil traditionally cultivated with cereals (mainly barley) in rotation with leguminous crops (chickpea and lentils). The number of PSB was 2×10^4 cells g^{-1} moist soil. We selected the strain PS06 because, among all the other PSB isolated, it produced the largest halos, of approximately 20 mm within 4 days of incubation. According to de Freitas et al. (1997), good phosphate-solubilizers produce halos around their colonies with diameters higher than 15 mm. Since it has been reported that some strains loose their phosphate-solubilizing capability after several cycles of inoculation (Halder et al., 1990; Illmer and Schinner, 1992), we corroborated the persistence of this trait in strain PS06 by successive subcultures.

Rhizobia were isolated using chickpea as trap plants. The symbiotic performance of each isolate with the chickpea ecotype ILC-482 growing in a N-free medium was evaluated in greenhouse experiments. At harvest, 8 weeks after inoculation, shoot dry weight of plants inoculated with strain C-2/2 ranked the highest and it was 111% greater than that of the uninoculated plants (data not shown). Therefore, strain C-2/2 was selected for further experiments.

Identification of the bacterial isolates by 16S rDNA sequence analyses

The complete 16S rDNA sequences of the PSB isolate PS06 and the rhizobial isolate C-2/2 were obtained.

A comparison with the 16S rDNA sequences available in the GenBank database indicated that the PS06 strain and the rhizobial strain C-2/2 are phylogenetically related to *Pseudomonas jessenii* CIP105274 and *Mesorhizobium ciceri* UPM-Ca^T, respectively. Strain PS06 showed a 99.6% of similarity with *P. jessenii* CIP105274 and, therefore, it can be considered as belonging to this *Pseudomonas* species. Strain C-2/2 constitutes, together with *M. ciceri* UPM-Ca^T, a separate group from the other species of the genus *Mesorhizobium*. The 16S rDNA sequence of strain C-2/2 showed a 100% similarity with respect to *M. ciceri* UPM-Ca^T, indicating that this strain belongs to *M. ciceri* species.

Greenhouse experiment

Five weeks after sowing, plants inoculated with C-2/2 alone had the highest shoot dry weight, which was 24% greater than that of uninoculated control plants (Table 2). Shoot dry weight of plants inoculated with PS06 alone or co-inoculated with C-2/2 and PS06 did not differ significantly with respect to that of control plants. Co-inoculated plants showed a significant decrease in shoot dry weight when compared to those exclusively inoculated with C-2/2. However, when compared to the shoot dry weight of the uninoculated, no deleterious effects of PS06 on plant growth was observed.

Inoculation treatments including strain C-2/2 yielded higher shoot N contents than the other treatments. The co-inoculation with PS06 did not affect significantly the nodulation. On the other hand, no differences in P contents were observed between treatments, indicating that PS06 did not improve P uptake under these experimental conditions.

Field experiment

At flowering, nodulation rates over a 10 plants samples per plot was almost three times higher in the two treatments inoculated with C-2/2 than in those not inoculated with this strain (Table 3), but no further differences were observed between treatments. Compared to plants inoculated with

Table 2. Effect of inoculation treatments with *P. jessenii* PS06 and *M. ciceri* C-2/2 on growth, number of nodules and shoot P and N contents of *Cicer arietinum* ecotype ILC-482 under greenhouse conditions

Inoculation treatment	Shoot dry weight (mg pot^{-1})	Nodulation (nodules pot^{-1})	Shoot P content (g kg^{-1})	Shoot N content (g kg^{-1})
Uninoculated	633 ± 46 a	0 a	0.60 ± 0.08 a	34.4 ± 3.0 a
P. jessenii PS06	720 ± 44 ab	0 a	0.60 ± 0.08 a	36.2 ± 1.4 a
M. ciceri C-2/2	787 ± 47 b	48 ± 21 b	0.58 ± 0.07 a	41.2 ± 2.0 b
P. jessenii PS06 + *M. ciceri* C-2/2	677 ± 49 a	32 ± 7 b	0.66 ± 0.09 a	42.5 ± 1.5 b

Data are average values of three replicates ± SD*.
*Means with different letters in the same column differ significantly at $P \leq 0.05$ according to Fisher's Protected LSD.

Table 3. Effect of inoculation treatments with *P. jessenii* PS06 and *M. ciceri* C-2/2 on shoot N and P contents and nodulation parameters of *Cicer arietinum* ecotype ILC-482 under field conditions at flowering

Inoculation treatment	Nodulation (nodules 10 plants^{-1})	Nodule fresh weight (mg 10 plants^{-1})	Shoot P content (g kg^{-1})	Shoot N content (g kg^{-1})
Uninoculated	101 ± 32 a	12.7 ± 2.1 a	1.2 ± 0.4 a	27.3 ± 2.6 a
P. jessenii PS06	102 ± 20 a	13.9 ± 1.9 a	1.2 ± 0.2 a	25.5 ± 1.8 a
M. ciceri C-2/2	274 ± 107 b	19.2 ± 1.8 b	1.1 ± 0.3 a	33.5 ± 1.8 b
P. jessenii PS06 + *M. ciceri* C-2/2	291 ± 44 b	27.3 ± 0.8 c	1.2 ± 0.3 a	40.0 ± 2.3 c

Data are average values of three replicates ± SD*.
*Means with different letters in the same column differ significantly at $P \leq 0.05$ according to Fisher's Protected LSD

C-2/2 alone, co-inoculated plants did not differ in number of nodules per 10 plants but had a higher nodule fresh weight. As the number of nodules per 10 plants was not different, the higher nodule fresh weight in the co-inoculated plants was due to a greater nodule size compared to the plants inoculated with C-2/2 alone. This suggests that PS06 acted synergistically with *M. ciceri* C-2/2 in promoting growth of the nodules. Similar results were obtained by Sindhu et al. (2002) in chickpea plants co-inoculated with some *Pseudomonas* strains and *Mesorhizobium*. At present we are unable to explain the precise basis of the role of PS06 in the observed nodule growth-promoting effect. It is well documented the involvement of plant hormones in legume nodule development (Hirsch et al., 1997), and that pseudomonas produce phytohormones (Glick, 1995; Persello-Cartieaux et al., 2003) and even are able to induce complex changes in the hormonal balance within the affected plant (Schmelz et al., 2003). Therefore, it is attractive to speculate that some or the coordination of these mechanisms may act to stimulate nodule growth. Certainly, exhaustive studies will be needed to confirm this hypothesis.

Treatments inoculated with C-2/2 also yielded higher N shoot content than those not inoculated with this strain and, similarly to the nodule fresh weight values, co-inoculated plants had the highest shoot N contents. There were no differences between treatments in shoot P content indicating that PS06 did not improve P uptake under the field conditions assayed. If any PS06-mediated P solubilization arose, the released P might have been further immobilized within the soil microbiota, since soil microorganisms are highly efficient in obtaining P from the surrounding medium (McLaughlin et al., 1988; Oberson et al., 2001; Oehl et al., 2001), or by physico-chemical reactions of P in soil (Umrit and Friesen, 1994), becoming unavailable for plants.

At maturity, all C-2/2-inoculated plants produced greater seed yield than the uninoculated control plants. There were no significant differences in seed yield between the single inoculation treatments, however co-inoculated plants produced more seeds than those inoculated with PS06 alone (Table 4). There were no significant differences neither in seed P nor N contents between treatments. Therefore, under the specific field conditions tested, PS06 acted synergistically with C-2/2 in promoting growth of chickpea. Other reports have also shown positive effects of the inoculation with *Pseudomonas* spp. on nodulation, plant growth and seed yield of legumes (Bolton et al., 1990; Dashti et al., 1998; Sindhu et al., 1999).

In conclusion, all *M. ciceri* C-2/2-inoculated plants produced higher seed yield and nodule number than the uninoculated control plants, indicating that this strain could be exploited as inoculant for improved chickpea productivity in the region where the field experiment was established. Moreover, our results from the field experiment indicate benefits on chickpea by combined inoculation of PS06 and C-2/2, and stress the suitability of using such mixed inoculants for the improvement of crop productivity.

The decrease in plant growth and nodulation due to the co-inoculations treatment under greenhouse conditions are in disagreement with the nodulation results obtained in the field experiment. Although there is no conclusive explanation for that, we speculate that it may be due to a differential survival or activity of PS06 to the chemical, physical and biological differences of the two substrates (perlite vs. soil) and/or to the environmental conditions (greenhouse vs. field

Table 4. Effect of inoculation treatments with *P. jessenii* PS06 and *M. ciceri* C-2/2 on seed yield and P and N content of *Cicer arietinum* ecotype ILC-482 under field conditions

Inoculation treatment	Seed yield (g plot^{-1})	Seed P content (g kg^{-1})	Seed N content (g kg^{-1})
Uninoculated	1742 ± 162 a	3.2 ± 0.6 a	34.2 ± 1.4 a
P. jessenii PS06	1887 ± 293 ab	3.4 ± 0.4 a	35.2 ± 1.4 a
M. ciceri C-2/2	2364 ± 202 bc	3.8 ± 0.6 a	34.6 ± 0.4 a
P. jessenii PS06 + *M. ciceri* C-2/2	2654 ± 396 c	3.7 ± 0.5 a	34.5 ± 2.5 a

Data are average values of three replicates ± SD*.
*Means with different letters in the same column differ significantly at $P \leq 0.05$ according to Fisher's Protected LSD

conditions). In these regards, chickpea plants grown in perlite (greenhouse experiment) were watered with a nutrient solution devoid of combined N. It has been shown a positive effect of N on root colonization by pseudomonads (Marschner et al., 1999) and other soil bacteria that may be explained by a increased carbon supply in the rhizosphere due to the greater exudation from the root of plants with sufficient N supply (Liljeroth et al., 1990a,b). Moreover, the beneficial effect of two pseudomonas strains (*P. fluorescens* ANP15 and *P. aeruginosa* 7NSK2) on plant growth showed to be more pronounced when plants were subjected to suboptimal conditions such as a soil with high microbial activity, unfavourable climatological conditions, or the presence of plant pathogens (Höfte et al., 1991). Seong et al. (1991) found that maize root colonization by these two pseudomonas strains and their pyoverdin (a fluorescent siderophore) production were strongly influenced by the ambient temperature. In the presence of soil phytopathogenic fungi the co-inoculation of chickpea with *Mesorhizobium* and two *Pseudomonas* strains resulted in synergistic effect on the symbiotic effectiveness (Sindhu et al., 2002) attributed to the production by pseudomonads of siderophores as well as antibiotics against phytopathogenic fungi (Sindhu et al., 1999).

In conclusion, the effects of inoculation with PGPR should be always corroborated under different field conditions in order to obtain accurate conclusions, since differences in the surrounding medium can deeply affect the outcome of such inoculations.

Acknowledgements

This work was supported by the Junta de Castilla y León and the Junta de Andalucía. The authors thank the soil analyses service staff from IRNA for their help in this work.

References

Alagawadi A R and Gaur A C 1992 Inoculation of *Azospirillum brasilense* and phosphate-solubilizing bacteria on yield of sorghum [*Sorghum bicolor* (L.) Moench] in dry land. Trop. Agric. 69, 347–350.

Antoun H, Beauchamp C J, Goussard N, Chabot R and Lalande R 1998 Potential of *Rhizobium* and *Bradyrhizobium* species as growth promoting bacteria on non-legumes: effect on radishes (*Raphanus sativus* L.). Plant Soil 204, 57–67.

Belimov A A, Kojemiakov A P and Chuvarliyeva C V 1995 Interaction between barley and mixed cultures of nitrogen fixing and phosphate-solubilizing bacteria. Plant Soil 173, 29–37.

Bergersen F J 1961 The growth of *Rhizobium* in synthetic media. Aust. J. Biol. Sci. 14, 349–360.

Bolton H, Elliot L F, Turco R F and Kennedy A C 1990 Rhizoplane colonization of pea seedlings by *Rhizobium leguminosarum* and a deleterious root colonizing *Pseudomonas* sp. and effects on plant growth. Plant Soil 123, 121–124.

Chabot R, Antoun H and Cescas M P 1993 Growth stimulation of corn and romiane lettuce by microorganisms solubilizing inorganic phosphorous. Can. J. Microbiol. 39, 941–947.

Chabot R, Antoun H and Cescas M P 1996 Growth promotion of maize and lettuce by phosphate-solubilizing *Rhizobium leguminosarum* biovar *phaseoli*. Plant Soil 184, 311–321.

Chabot R, Beauchamp C J, Kloepper J W and Antoun H 1998 Effect of phosphorous on root colonization and growth promotion of maize by bioluminescent mutants of phosphate-solubilizing *Rhizobium leguminosarum* biovar. *phaseoli*. Soil Biol. Biochem. 30, 1615–1618.

Dashti N, Zhang F, Hynes R and Smith D L 1998 Plant growth-promoting rhizobacteria accelerate nodulation and increase nitrogen fixation activity by field grown soybean [*Glycine max* (L.) Merr.] under short season conditions. Plant Soil 200, 205–213.

Davison J 1988 Plant beneficial bacteria. Biotechnology 6, 282–286.

de Freitas J R, Banerjee M R and Germida J J 1997 Phosphate-solubilizing rhizobacteria enhance the growth and yield but not phosphorous uptake in canola (*Brassica napus* L.). Biol. Fertil. Soils 24, 358–364.

Glick B R 1995 The enhancement of plant growth by free-living bacteria. Can. J. Microbiol. 41, 109–117.

Gupta R, Singal R, Sankar A, Chander R M and Kumar R S 1994 A modified plate assay for screening phosphate solubilizing microorganisms. J. Gen. Appl. Microbiol. 40, 255–260.

Halder A K, Mishra A K, Bhattacharyya P and Chakrabartty P K 1990 Solubilization of rock phosphate by *Rhizhobium* and *Bradyrhizobium*. J. Gen. Appl. Microbiol. 36, 81–92.

Hirsch A M, Fang Y, Asad S and Kapulnik Y 1997 The role of phytohormones in plant-microbe symbioses. Plant Soil 194, 171–184.

Höfte M, Boelens J and Verstraete W 1991 Seed protection and promotion of seedling emergence by the plant growth beneficial *Pseudomonas* strains 7NSK2 and ANP15. Soil Biol. Biochem. 23, 407–410.

Igual J M, Valverde A, Cervantes E and Velázquez E 2001 Phosphate-solubilizing bacteria as inoculants for agriculture: use of updated molecular techniques in their study. Agronomie 21, 561–568.

Illmer P and Schinner F 1992 Solubilization of inorganic phosphates by microorganisms isolated from forest soil. Soil Biol. Biochem. 24, 389–395.

Kim K Y, Jordan D and McDonald G A 1998 Effect of phosphate-solubilizing bacteria and vesicular-arbuscular mycorrhizae on tomato growth and soil microbial activity. Biol. Fertil. Soils 26, 79–87.

Kloepper J W and Schroth M N 1978 Plant growth-promoting rhizobacteria on radishes. *In* Proceedings of the IV International Conference on Plant Pathogenic Bacteria, vol. 2. Eds. Gibert-Clarey and Tours. pp. 879–882. Station de Phatologie Végétale et Phytobactériologie, INRA, Angers, France .

Kundu B S and Gaur A C 1984 Rice response to inoculation with N_2-fixing and P-solubilizing microorganisms. Plant Soil 79, 227–234.

Liljeroth E, Bååth E, Mathiasson I and Lundborg T 1990a Root exudation and rhizoplane bacterial abundance of barley (*Hordeum vulgare* L) in relation to nitrogen fertilization and root growth. Plant Soil 127, 81–89.

Liljeroth E, Van Veen J A and Miller H J 1990b Assimilate translocation to the rhizosphere of two wheat lines and subsequent utilization by rhizosphere microorganisms at two nitrogen concentrations. Soil Biol. Biochem. 22, 1015–1021.

Marschner P, Gerendás J and Sattelmacher B 1999 Effect of N concentration and N source on root colonization by *Pseudomonas fluorescens* 2-79RLI. Plant Soil 215, 135–141.

McLaughlin M J, Alston A M and Martin J K 1988 Phosphorus cycling in wheat-pasture rotations II The role of the microbial biomass in phosphorus cycling. Aust. J. Soil Res. 26, 333–342.

Oberson A, Friesen D K, Rao I M, Bühler S and Frossard E 2001 Phosphorus transformations in an Oxisol under contrasting land-use systems: The role of the soil microbial biomass. Plant Soil 237, 197–210.

Oehl F, Oberson A, Probst M, Fliessbach A, Roth H R and Frossard E 2001 Kinetics of microbial phosphorus uptake in cultivated soils. Biol. Fertil. Soils 34, 31–41.

O'Sullivan D J and O'Gara F 1992 Traits of fluorescent *Pseudomonas* spp. involved in suppression of plant root pathogens. Microbiol. Rev. 56, 662–676.

Pal S S 1998 Interaction of an acid tolerant strain of phosphate solubilizing bacteria with a few acid tolerant crops. Plant Soil 198, 169–177.

Pearson W R and Lipman D J 1988 Search for DNA homologies was performed with the FASTA program. Proc. Natl. Acad. Sci. USA 85, 2444–2448.

Peix A, Rivas-Boyero A A, Mateos P F, Rodríguez-Barrueco C, Martínez-Molina E and Velázquez E 2001a Growth promotion of chickpea and barley by a phosphate solubilizing strain of *Mesorhizobium mediterraneum* under growth chamber conditions. Soil Biol. Biochem. 33, 103–110.

Peix A, Mateos P F, Rodríguez-Barrueco C, Martínez-Molina E and Velázquez E 2001b Growth promotion of common bean (*Phaseolus vulgaris* L.) by a strain of *Burkholderia cepacia* under growth chamber conditions. Soil Biol. Biochem. 33, 1927–1935.

Persello-Cartieaux F, Nussaume L and Robaglia C 2003 Tales from the underground: molecular plant-rhizobacteria interactions. Plant Cell Environ. 26, 189–199.

Piccini D and Azcón R 1987 Effect of phosphate-solubilizing bacteria and vesicular arbuscular mycorrhizal (VAM) on the utilization of bayoran rock phosphate by alfalfa plants using a sand-vermiculite medium. Plant Soil 101, 45–50.

Ray J, Bagyaraj D J and Manjunath A 1981 Influence of soil inoculation with versicular arbuscular mycorrhizal (VAM) and a phosphate dissolving bacteria on plant growth and ^{32}P uptake. Soil Biol. Biochem. 13, 105–108.

Rigaud J and Puppo A 1975 Indole-3-acetic catabolism by soybean bacteroids. J. Gen. Microbiol. 88, 223–228.

Rivas R, Velázquez E, Valverde A, Mateos P F and Martínez-Molina E 2001 A two primers random amplified polymorphic DNA procedure to obtain polymerase chain reaction fingerprints of bacterial species. Electrophoresis 22, 1086–1089.

Rodríguez H and Fraga R 1999 Phosphate solubilizing bacteria and their role in plant growth promotion. Biotechnol. Adv. 17, 319–339.

Sarawgi S K, Tiwari P K and Tripathi R S 1999 Uptake and balance sheet of nitrogen and phosphorus in gram (*Cicer arietinum*) as influenced by phosphorus, biofertilizers and micronutrients under rainfed condition. Indian J. Agron. 44, 768–772.

Schmelz E A, Engelberth J, Alborn H T, O'Donnell P, Sammons M, Toshima H and Tumlinson J H III 2003 Simultaneous analysis of phytohormones, phytotoxins, and volatile organic compounds in plants. Proc. Natl. Acad. Sci. USA 100, 10552–10557.

Seong K Y, Hofte M, Boelens J and Verstraete W 1991 Growth, survival and root colonization of plant growth beneficial *Pseudomonas fluorescens* ANP15 and *Pseudomonas aeruginosa* 7NSK2 at different temperatures. Soil Biol. Biochem. 23, 423–428.

Sindhu S S, Gupta S K and Dadarwal K R 1999 Antagonistic effect of *Pseudomonas* spp. on pathogenic fungi and enhancement of growth of green gram (*Vigna radiata*). Biol. Fertil. Soils 29, 62–68.

Sindhu S S, Suneja S, Goel A K, Parma N and Dadarwal K R 2002 Plant growth promotion effects of *Pseudomonas* sp. on coinoculation with *Mesorhizobium* sp. *Cicer* strain under sterile and "wilt sick" soil conditions. Appl. Soil Ecol. 19, 57–64.

Snedecor G W and Cochran W G 1989 Statistical Methods. Iowa State University Press, Ames, Iowa 503 pp.

Subba Rao N S 1993 Biofertilizers in Agriculture and Forestry. Oxford and IBH Publishing Co. Pvt. Ltd, New Delhi 242 pp.

Tan K H 1996 Soil Sampling, Preparation, and Analysis. Marcel Dekker, Inc, New York 408 pp.

Thompson J D, Gibson T J, Plewniak F, Jeanmougin F and Higgins D G 1997 The clustalX windows interface: Flexible strategies for multiple sequence alignment aided by quality analysis tools. Nucleic Acid Res. 24, 4876–4882.

Tomar R K S, Namdeo K N and Ranghu J S 1996 Efficacy of phosphate solubilizing bacteria biofertilizers with phosphorus on growth and yield of gram (*Cicer arietinum*). Indian J. Agron. 41, 412–415.

Toro N, Azcón R and Barea J M 1997 Improvement of arbuscular mycorrhiza development by inoculation of soil with phosphate-solubilizing rhizobacteria to improve rock phosphate bioavailability (^{32}P) and nutrient cycling. Appl. Environ. Microbiol. 63, 4408–4412.

Toro N, Azcón R and Barea J M 1998 The use of isotopic dilution techniques to evaluate the interactive effects of *Rhizobium* genotype, mycorrhizal fungi, phosphate-solubilizing rhizobacteria and rock phosphate on nitrogen and phosphorus acquisition by *Medicago sativa*. New Phytol. 138, 265–273.

Umrit G and Friesen D K 1994 The effect of C:P ratio of plant residues added to soils of contrasting phosphate sorption capacities on P uptake by *Panicum maximum* (Jacq.). Plant Soil 158, 275–285.

Vincent J M 1970 The cultivation, isolation and maintenance of rhizobia. *In* A Manual for the Practical Study of Root-Nodule. Ed. J M Vincent. pp. 1–13. Blackwell Scientific Publications, Oxford.

Effect of Tilemsi phosphate rock-solubilizing microorganisms on phosphorus uptake and yield of field-grown wheat (*Triticum aestivum* L.) in Mali

A. H. Babana[1,2] & H. Antoun[2,3]

[1]*Faculté des Sciences et Techniques, Université du Mali, BPE 3204, Bamako, Mali.* [2]*Département des Sols et de Génie Agroalimentaire, Faculté des Sciences de l'Agriculture et de l'Alimentation, Université Laval, 2110 Pavillon de l' Envirotron, G1K 7P4, Québec, Qc, Canada.* [3]*Corresponding author**

Received: 31 May 2006

Key words: *Aspergillus awamori*, *Glomus intraradices*, Plant growth promotion, *Penicillium chrysogenum*, *Pseudomonas* sp., *Triticum aestivum*

Abstract

With the broad aim of biologically improving P uptake by wheat fertilized with Tilemsi phosphate rock (TPR), we investigated the effect of inoculation with TPR-solubilizing microorganisms isolated from Malian soils and with a commercial isolate of the arbuscular mycorrhizal (AM) fungus *Glomus intraradices* (Gi). AM root length colonization, and growth yield and P concentration of the cultivar Tetra of wheat were measured under field conditions in Mali. Experimental plots were established in Koygour (Diré) during the 2001–2002 cropping season. Inoculation treatments included two fungal isolates, *Aspergillus awamori* (C1) and *Penicillium chrysogenum* (C13), and an isolate of *Pseudomonas* sp. (BR2), used alone or in fungus-bacterium combinations in the presence or absence of the AM fungus Gi. In fertilized treatments, 0 or 30 kg P ha^{-1} was applied as TPR or diammonium phosphate (DAP). In 45-day-old wheat plants, the highest root length AM colonization (62%) was observed with TPR fertilized wheat inoculated with Gi and BR2. Our results suggest that BR2 is a mycorrhizal-helper bacteria and a good plant growth-promoting rhizobacteria. In fact, inoculation of wheat Tetra fertilized with TPR with a combination of Gi, BR2 and C1 produced the best grain yield with the highest P concentration. This work shows that by inoculating seeds with TPR-solubilizing microorganisms and AM fungi under field conditions in Mali it is possible to obtain wheat grain yields comparable to those produced by using the expensive DAP fertilizer.

Introduction

Phosphorus (P) deficiency is one of the major constraints to crop production in West Africa, and in Mali imported fertilizers are very expensive. However the Tilemsi phosphate rock (TPR) deposits are estimated to be between 20 and 25 million tonnes, and are a potential inexpensive source of P for farmers (Bationo et al., 1997). In fact, economic evaluation of TPR under farmers' operating conditions for three cropping rotations (groundnut/pearl millet; cotton/sorghum and cotton/maize) clearly indicated that the direct application of TPR could be profitable in comparison with recommended imported P fertilizers (Bationo et al., 1997).

Many bacteria (Rodriguez and Fraga, 1999) and fungi (Whitelaw, 2000) are able to improve plant growth by solubilizing sparingly soluble inorganic and organic phosphates in the soil. Production and release of organic acids is an important mechanism involved in inorganic P solubilization (Richardson, 2001). *Penicillium rugulosum* IR-94MF1 isolated from a soil in Venezuela mobilizes inorganic phosphates by producing

*Fax No: +1-418-656 7871.
E-mail: Hani.Antoun@sga.ulaval.ca

E. Velázquez and C. Rodríguez-Barrueco (eds.),
First International Meeting on Microbial Phosphate Solubilization, 51–58

gluconic and citric acids (Reyes et al., 1999). These organic acids are also involved in PR solubilization (Reyes et al., 2001), and inoculation of maize that had been fertilized with Venezuelan Navay PR with IR-94MF1 significantly increased shoot yield and P-uptake by the plants as compared to the uninoculated control (Reyes et al., 2002).

Most crop plants are colonized by arbuscular mycorrhizal (AM) fungi. Besides improving uptake of poorly mobile nutrients, AM symbioses benefit plant growth by other mechanisms of action such as improving drought tolerance, protecting the plant against pathogens or channeling carbon to the soil, thus improving soil aggregation (Sylvia and Chelleni, 2001). Many recent reports show synergistic interactions between phosphate-solubilizing microorganisms and AM, under different experimental conditions. Using transformed carrot (*Daucus carota* L.) roots, Villegas and Fortin (2001) observed that the combination of the P solubilizing *Pseudomonas aeruginosa* with *Glomus intraradices* enhanced the solubilization of sparingly soluble sources of P above the levels reached with each culture alone. In field trials performed in southern Egypt, the highest significant effect on wheat (*Triticum aestivum* L.) yield and phosphorus content, was observed when seeds were inoculated with a mixture of the AM fungus *Glomus constrictum* with two Egyptian fungal isolates *Aspergillus niger* and *Penicillium citrinum* that solubilize phosphate rock (Omar, 1998).

The aim of the present work was to evaluate TPR as a P source for wheat cultivated under field conditions in Mali, and to explore the possibility of improving its value by seed inoculation with TPR-solubilizing organisms. *Pseudomonas* sp. BR2, *Aspergillus awamori* C1, and *Penicillium chrysogenum* C13 were isolated from Malian soils and selected for their high TPR-solubilizing activity. These organisms were used alone or in combination, with or without a commercial AM isolate of *Glomus intraradices* to inoculate wheat seeds.

Material and methods

Microorganisms

Glomus intraradices, was supplied as a suspension of concentrated pure spores (PremierTech, Rivière-du-Loup, Quebec, Canada). TPR-solubilizing activity was measured in agar cultures as described by Chabot et al. (1993) by using Goldstein (1986) medium containing TPR as sole source of P. After screening of a large number of rhizosphere isolates obtained from wheat grown in Malian soil, the following organisms were selected for their high solubilization activity and were used in field inoculation trials: *Pseudomonas* sp. BR2, *Aspergillus awamori* Nakazawa C1, and *Penicillium chrysogenum* Thom C13.

Inoculant preparation and seed inoculation

BR2 was cultivated in 50 mL liquid NBRIP medium (Nautiyal, 1999) containing 5 g L^{-1} TPR as P source, for 48 h on a rotary shaker at 28 °C. Cells were collected and washed three times by centrifugation and suspended in sterile saline. Fungi C1 and C13 were grown on solid TPR-NBRIP medium for 3 and 5 days, respectively. Mycelia and spores from several plates were harvested in 100 mL sterile saline, homogenized in a domestic blender, and washed three times by centrifugation. Seeds of the wheat (*Triticum aestivum* L.) cv. Tetra were surface-sterilized (Chabot et al. 1996), and 200 seeds were soaked in 100 mL of microbial suspension for 4 h at room temperature. The wet seeds were then transferred to sterile plastic bags and mixed by the sequential addition of 20 mL a sterile 1% carboxymethylcellulose solution and 10 g of talc powder. Coated seeds were dried overnight in a laminar flow hood at room temperature. Uninoculated control seeds were treated similarly but without microorganisms.

At sowing each coated wheat seed contained an average of 1.8×10^5 CFU BR2, 1.5×10^2 CFU C1 and 2.1×10^3 CFU C13. The concentrated suspension of spores of *Glomus intraradices* was diluted to 200 spores mL^{-1} in sterile saline, and 7 mL of this suspension was directly applied in the seed bed in the AM treatments.

Phosphate rock

The TPR deposits contain between 23 and 32% of P_2O_5 and their solubility in neutral ammonium citrate is 4.2% (Bationo et al., 1997). The fine TPR powder used had the following composition (in mg g^{-1}): P, 150; Ca, 329; Al, 20; F, 29.

The extractability of P from TPR determined according to Bolland and Gilkes (1997) was 16.2 mg g^{-1} in 2% citric acid and 73.4 mg g^{-1} in 2% formic acid.

Field experiments

Experimental plots were established in Koygour (Diré) during the 2001–2002 cropping season. The 0–15 cm of the silty clay soil at the site had a pH of 6.37 (0.01 M $CaCl_2$, 1:1 v) and contained 0.17% organic matter. The Mehlich 3 (Mehlich, 1984) available elements (kg ha^{-1}) were as follows: P, 6.3; K, 240; Ca, 804; Mg, 217; Fe, 43 and Al, 255. A split–split plot design was used. The main plots were phosphate fertilization treatments with TPR or diammonium phosphate (DAP) applied at a rate of 30 kg P ha^{-1} and a non-fertilized control, arranged in randomized complete blocks. The additional N added with the DAP was calculated and compensated for in all other treatments. Sub-plots were inoculated with AM fungus *Glomus intraradices* or uninoculated control. Sub-subplots included the following treatments with TPR-solubilizing microorganims: *Pseudomonas* sp. BR2, *Aspergillus awamori* C1, and *Penicillium chrysogenum* C13, BR2 + C1, BR2 + C13, and an uninoculated control. The main plots (P fertilization) were 5 m wide and 15 m long. They were divided in two subplots (AM treatments) 2 m wide and 15 m long separated by a 1 m wide buffer zone. Sub-subplots were 2 × 2 m separated by a 60 cm buffer zone, and contained four rows 50 cm apart. Two wheat seeds were planted in each row every 20 cm. Only the two central rows received seeds inoculated with TPR-solubilizing microorganisms. All treatments were replicated four times. Planting was done on November 20, 2001. After emergence plants were thinned to one every 20 cm of row. Nitrogen was applied as 50 kg N ha^{-1} urea, 2 and 7 weeks after planting which corresponded to stage 2 and stage 5 of Feekes scale (Large, 1954), and a final application of 120 kg N ha^{-1} urea at stage 10.1 (11 weeks). All plots received 80 kg K ha^{-1} as KCl. The plots were irrigated 10 times during the growing season (each of approximately 500 m^3 ha^{-1}). Plant height was measured 8 weeks (Feekes scale 10) after planting on five randomly chosen plants in the two central rows. Wheat was harvested at maturity on February 18, 2002 (88 days after planting), from a 1-m^2 area in the center of each sub-subplot.

AM colonization of roots

In the central rows of each sub-subplot, three plants randomly chosen at 45 days after planting, were carefully excavated and their root washed free of soil and stained according to the ink and vinegar technique of Vierheilig et al. (1998), to measure the root length colonization by AM.

Soil and plant analysis

Soil was air-dried and sieved (2 mm) and treated with the Mehlich 3 extractant (Mehlich, 1984) for the determination of available elements. Soil organic matter was estimated by the modified Walkley and Black method (McKeague, 1978). Plant shoots and grain were air dried and weighed, grounded and digested in 15 mL $HClO_4$ and 5 mL HNO_3. The spectrophotometric vanado-molybdate method was used to measure P (Tandon et al., 1968). Other minerals were determined in plant tissues and soil extracts by atomic absorption spectrophotometry (Gaines and Mitchell, 1979).

Statistical analysis

A three-factor analysis of variance (P fertilization, *Glomus intraradices*, phosphate-solubilizing microorganims) for each parameter was performed using the general linear models procedure of SAS (1990).

Results and discussion

AM root colonization

P-fertilization and inoculation with phosphate-solubilizing microorganisms (PSM) and *G. intraradices* (Gi) significantly affected root colonization of the cv. Tetra of wheat by indigenous AM fungi (Table 1). All interactions between P-fertilization and inoculation with PSM and Gi were highly significant ($P < 0.001$). This indicates for example, that the colonization by indigenous AM will be affected differently by P-fertilization, according to the applied PSM or the Gi inoculation

Table 1. Summary from the analyses of variance for root arbuscular mycorrhizal colonization (% AM), plant height, grain and shoot yields and P concentrations of wheat cv. Tetra fertilized with Tilemsi phosphate rock or diammonium phosphate (P) and inoculated with different P-solubilizing microorganisms (PSM) in the presence or absence of the AM fungus *Glomus intraradices* (Gi)

Source of variations	Means squares						
	df	% AM	Plant height	Grain yield	Grain P	Shoot Yield	Shoot P
Main plots P	2	2132.4***	448.1***	2.8***	2.6***	3.9***	2.0***
Replications	3	5.7 NS[a]	95.3**	0.005 NS	0.03NS	0.006 NS	0.01NS
Main plots error	6	2.04	25.2	0.0003	0.03	0.0004	0.01
Subplots Gi	1	996.2***	3145.3***	8.5***	0.1*	10.3***	1.5***
P × Gi	2	674.4***	1.36 NS	0.05**	0.3***	0.03**	0.04NS
Subplots error	9	1.6	27.6	0.001	0.03	0.0003	0.01
Sub-subplots PSM	5	2291.1***	1143.6***	0.9***	0.5***	1.9***	0.2***
P × PSM	10	418.2***	100.3***	0.1***	0.06*	0.1***	0.03*
Gi × PSM	5	31.9***	150.5***	0.4***	0.03 NS	0.3***	0.006 NS
P × Gi × PSM	10	93.8***	96.8 ***	0.06***	0.07*	0.03***	0.007 NS
Sub-subplots error	90	3.54	18.9	0.007	0.03	0.005	0.01

*, **, *** Significant at $P < 0.05$, $P < 0.01$ and $P < 0.001$, respectively.
[a]NS: Statistically not significant.

treatments. In the uninoculated non-fertilized treatments, indigenous AM fungi colonized only 5.5% of wheat root length, 45 days after planting (Table 2). In semi-arid ecosystems, soil disturbance (grazing, erosion) results in loss of AM propagules and low numbers of viable spores, thus decreasing the mycorrhizal soil infectivity (Diop et al., 1994; McGee, 1989). Inoculation with Gi in absence of any other treatment did not improve the observed 5.5% AM colonization (Table 2). In pot experiments, Singh and Kapoor (1999) obtained a significant increase in wheat root colonization by inoculation with the AM *Glomus* sp. 88, applied as chopped mycorrhizal root fragments of 10-week-old pearl millet (*Pennisetum typhoides*) and soil. This inoculation

Table 2. Plant height 8 weeks after planting and root colonization by arbuscular mycorrhizal fungi (AM) 45 days after planting of wheat cv. Tetra as influenced by single or dual inoculation with P-solubilizing microorganisms *Pseudomonas* sp. (BR2), *Aspergillus awamori* (C1) and *Penicillium chrysogenum* (C13) in the presence or absence of *Glomus intraradices* (Gi) and by P fertilisation with Tilemsi phosphate rock (TPR) and diammonium phosphate (DAP)

Inoculation treatments	AM (% colonization)			Plant Height (cm)		
	Control	TPR	DAP	Control	TPR	DAP
Uninoculated	5.5 c	8.0 d	6.5 b	69.0 c	73.8 d	75.0 d
BR2	43.5 a	37.8 a	11.3 a	83.8 b	99.5 a	89.3 bc
C1	7.3 c	11.5 bc	6.0 b	86.3 ab	95.0 b	92.8 b
C13	9.0 c	10.3 cd	6.8 b	82.0 b	86.0 c	84.3 c
BR2 + C1	26.0 b	14.0 b	9.5 a	91.5 a	87.8 c	100.5 a
BR2 + C13	24.3 b	13.0 b	10.8 a	85.3 b	86.0 c	85.8 c
Gi	5.5 d	25.5 d	8.8 c	79.0 d	85.0 e	89.3 c
Gi + BR2	38.5 a	62.3 a	14.3 a	97.3 ab	106.0 a	115.0 a
Gi + C1	11.0 c	12.5 e	10.0 c	90.8 c	98.0 c	90.8 bc
Gi + C13	11.0 c	12.0 e	9.0 c	102.3 a	93.3 d	94.0 bc
Gi + BR2 + C1	24.0 b	34.5 b	12.5 ab	93.0 bc	100.3 bc	102.0 b
Gi + BR2 + C13	22.3 b	29.8 c	10.8 bc	89.0 c	102.8 ab	93.8 bc

For each AM treatment (uninoculated or Gi) within each column means followed by the same letter are not statistically different according to the Fisher protected Lsd test ($P < 0.05$).

procedure probably provided some nutrients not present in the pure spore suspension used in this study. In fact, addition of TPR increased the AM colonization from 5.5 to 8% in the uninoculated control treatment, but a more substantial increase (from 5.5 to 25.5%) was observed in the Gi treated plants (Table 2). In general, inoculation with Gi significantly increased root colonization with AM (Table 3). In general, in the presence of DAP wheat root length colonized with either indigenous or introduced AM fungi was lower than observed in TPR amended or in the unfertilized control plots (Tables 2 and 4). Our results corroborate the observations made by Graham and Abbott (2000) that application of a high rate of soluble P to soil reduces the percentage of root length colonization by AM in 42 day-old wheat plants. The results also agree with the findings of Barea et al. (1980) that phosphate rock does not reduce the level of mycorrhizal infection. Regardless of the phosphorus fertilization treatment, inoculation with the TPR-solubilizing bacterium *Pseudomonas* sp. BR2 significantly enhanced root colonization by indigenous or introduced AM fungi. The highest root length colonization (62%) was obtained with wheat fertilized with TPR and inoculated with Gi and BR2 (Table 2). Inoculation with the TPR-solubilizing *Aspergillus awamori* C1 or *Penicillium chrysogenum* C13 caused less pronounced colonization enhancement of the roots as compared to BR2 (Table 2). The results suggest that BR2 is a mycorrhizal-helper bacterium. Such synergistic interaction between bacteria and AM fungi is well documented in the literature (Barea et al. 2002).

Plant height

After 8 weeks of growth P-fertilization and inoculation with PSM and Gi significantly influenced the plant height. With the exception of the non-significant P-fertilization × Gi interaction, all other interactions between P-fertilization and inoculation with PSM and Gi were highly significant (Table 1). For all treatments combined, inoculation with Gi and P-fertilization with TPR or DAP significantly enhanced plant height (Tables 3 and 4). In non-fertilized treatments the highest plant height was recorded when wheat was inoculated with Gi and *Pseudomonas* sp. BR2 or *P. chrysogenum* C13 (Table 2). Plant height also was significantly correlated with grain ($r = 0.70^{**}$, $P < 0.01$) and straw ($r = 0.70^{**}$) yields of mature Tetra wheat.

Grain and shoot yields and P concentrations

Grain and shoot yields and P concentrations were significantly affected by P-fertilization, inoculation with Gi and PSM (Table 1). Interactions between the three treatments were significant for grain and shoot yields. For grain P concentration, the Gi × PSM interaction was not significant and all interactions involving Gi were not significant for shoot P concentration (Table 1). For all

Table 3. Effect of inoculation with *Glomus intraradices* (Gi) on wheat cv. Tetra height 8 weeks after planting, root arbuscular mycorrhizal colonization (AM), grain and shoot yields and P concentrations

	−Gi	+Gi
AM % colonization	14.5 b	19.7 a
Plant height (cm)	86.3 b	95.6 a
Grain yield (t/ha)	2.18 b	2.67 a
Grain P (mg/g dry matter)	2.30 b	2.36 a
Shoot yield (t/ha)	2.45 b	2.99 a
Shoot P (mg/g dry matter)	1.16 b	1.36 a

Values are means of P fertilisation and inoculation with P-solubilizing microorganisms treatments.
In each line, means followed by the same letter are not statistically different according to the Fisher protected Lsd test ($P < 0.05$).

Table 4. Effect of fertilisation with 30 kg P $^{-1}$ applied as Tilemsi phosphate rock (TPR) or diammonium phosphate (DAP) on wheat cv. Tetra height 8 weeks after planting, root arbuscular mycorrhizal colonization (AM), grain and shoot yields and P concentrations

	Control	TPR	DAP
AM % colonization	19.0 b	22.6 a	9.7 c
Plant height (cm)	87.4 b	92.8 a	92.7 a
Grain yield (t/ha)	2.14 b	2.55 a	2.57 a
Grain P (mg/g dry matter)	2.08 c	2.36 b	2.55 a
Shoot yield (t/ha)	2.44 b	2.71 b	3.00 a
Shoot P (mg/g dry matter)	1.08 b	1.21 b	1.48 a

Values are means of all inoculation treatments (P-solubilizing microorganisms and *Glomus intraradices*).
In each line means followed by the same letter are not statistically different according to the Fisher protected Lsd test ($P < 0.05$).

treatments combined, inoculation with Gi increased significantly plant and shoot yields and their P concentrations (Table 3). P-solubilizing microorganisms may also directly increase P uptake by changing root morphology. Root hairs can substantially increase root–soil contact, and play a determinant role in P acquisition. Gahoonia et al. (1997) found that the number, length and surface area of root hairs are very variable in wheat cultivars. Gulden and Vessey (2000) also reported that inoculation of field pea (*Pisum sativum* L.) with *Penicillium bilaii* resulted in a 22% increase in the proportion of root containing root hairs and a 33% increase in the mean root-hair length in seedlings. Future work should investigate the effects of inoculation with PSM and AM fungi on root hair development in different cultivars of wheat. Fertilization with DAP increased the four parameters studied as compared to the non-fertilized treatments (Table 4). Except for grain yield, as expected DAP was always superior to TPR. Straw yield and P concentration were not different in the non-fertilized control and the TPR amended plots (Table 4).

In the absence of any P fertilization treatment, inoculation of wheat with the AM fungus Gi produced lower grain yield and P concentration as compared to the uninoculated control (Table 5). In low P soils, inoculation with aggressive and non-aggressive AM fungi reduced the growth of wheat (Graham and Abbott, 2000). In our study this non-beneficial effect was eliminated by fertilization with TPR or DAP or by inoculation with the P-solubilizing microorganisms tested. On average, inoculation with the AM fungus Gi caused significant increases in grain (0.49 t ha^{-1}) and shoot (0.54 t ha^{-1}) yields for all P-fertilization and PSM inoculation treatments (Table 3). When inoculated with TPR-solubilizing microorganisms, grain yields obtained with TPR treatment were comparable to those produced with DAP (Table 5). When all Gi and PSM treatments are considered the addition of 30 kg P ha^{-1} as TPR or DAP produced 0.42 t ha^{-1} more grain than the unfertilized control (Table 4). Wheat grain yield was always improved by inoculation with PSM in the non-fertilized control and TPR treatments. No grain yield response to inoculation with PSM was observed when DAP was added in the absence of Gi (Table 5). In general grain and shoot yields of wheat inoculated with *A. awamori* C1 or *P. chrysogenum* C13 were always higher when plants were inoculated with Gi as compared to the uninoculated control

Table 5. Wheat cv. Tetra grain and shoot dry matter yields and P concentrations as influenced by single or dual inoculation with P-solubilizing microorganisms *Pseudomonas* sp. (BR2), *Aspergillus awamori* (C1) and *Penicillium chrysogenum* (C13) in the presence or absence of *Glomus intraradices* (Gi) and by P fertilisation with Tilemsi phosphate rock (TPR) and diammonium phosphate (DAP)

Inoculation treatments	Grain Yield (t/ha)			Grain P (mg/g dry matter)			Shoot yield (t/ha)			Shoot P (mg/g dry matter)		
	Control	TPR	DAP	Control	TPR	DAP	Control	TPR	DAP	Control	TPR	DAP
Uninoculated	1.94 d	2.14 c	2.23 a	1.96 d	1.99 e	2.09 c	2.10 e	2.08 f	2.25 b	0.89 e	1.09 b	1.19 d
BR2	2.04 b	2.35 ab	2.39 a	2.01 c	2.37 b	2.32 ab	2.28 b	2.60 b	2.94 a	1.02 bc	1.12 b	1.54 b
C1	2.00 bc	2.28 b	2.17 a	1.99 cd	2.22 d	2.28 ab	2.33 a	2.37 e	2.76 a	0.97 d	1.03 c	1.38 c
C13	1.55 e	2.32 ab	2.17 a	1.97 d	2.19 d	2.23 b	1.89 f	2.50 d	2.72 a	0.99 cd	1.02 c	1.18 d
BR2 + C1	2.12 a	2.38 a	2.40 a	2.17 a	2.48 a	2.35 a	2.15 d	2.55 c	2.79 a	1.08 a	1.16 a	1.61 a
BR2 + C13	1.98 cd	2.39 a	2.36 a	2.11 b	2.31 c	2.30 ab	2.23 c	2.72 a	2.87 a	1.05 ab	1.09 b	1.51 b
Gi	1.51 e	2.14 e	2.58 c	1.86 d	1.20 e	2.35 b	2.07 d	2.10 d	2.46 d	1.09 e	1.30 b	1.43 a
Gi + BR2	2.62 b	2.94 bc	2.90 a	2.18 b	2.57 b	2.53 a	2.93 a	3.22 a	3.57 a	1.12 d	1.37 b	1.65 a
Gi + C1	2.60 bc	2.86 d	2.98 a	2.20 b	2.51 c	2.52 a	2.95 a	2.99 c	3.59 a	1.15 cd	1.29 b	1.45 a
Gi + C13	2.12 d	2.90 cd	2.72 b	2.14 c	2.48 c	2.49 a	2.53 c	3.12 b	3.17 c	1.16 c	1.28 b	1.48 a
Gi + BR2 + C1	2.7 a	3.04 a	2.95 a	2.28 a	2.69 a	2.51 a	2.82 b	3.20 a	3.46 b	1.28 a	1.47 a	1.71 a
Gi + BR2 + C13	2.56 c	2.96 b	2.93 a	2.15 c	2.37 d	2.51 a	2.96 a	3.09 b	3.49 b	1.22 b	1.37 b	1.68 a

For each AM treatment (uninoculated or Gi) within each column means followed by the same letter are not statistically different according to the Fisher protected Lsd test ($P < 0.05$).

(Table 5). The positive interaction observed between Gi and PSM like C1 and C13 for shoot yield (Table 1) is comparable to that found with wheat cultivated under field conditions in Egypt, fertilized with phosphate rock and inoculated with *Glomus constrictum* and the two P-solubilizing fungi *A. niger* and *Penicillium citrinum* (Omar, 1998).

Without P-fertilization, the highest grain yields were observed with the BR2 + C1 treatment in the presence or absence of Gi (Table 5). This combination of PSM was also the best for grain yield in TPR fertilized treatments. In addition to its potential as mycorrhizal helper bacteria, *Pseudomonas* sp. BR2 like other PSM used in this study is probably a good PGPR. As the highest grain P concentrations were also observed with the Gi + BR2 + C1 combination in the non-fertilized and the TPR treatments, P-solubilization is probably an important mechanism involved in the observed growth promotion. In fact, a significant increase in P solubilization was observed when an isolate of P-solubilizing *Pseudomonas aeruginosa* was added with the AM fungus Gi to transformed carrot roots (Villegas and Fortin 2001). Similar *in vitro* studies between AM and the phosphate solubilizing fungi deserve to be investigated. *Pseudomonas* strains can also stimulate mycelial development from *Glomus mosseae* spores germinating in soil and tomato root colonization (Barea et al., 1998). Further work should be conducted to evaluate the performance of the Gi + BR2 + C1 combination in different soils and with different wheat cultivars. In fact some PGPR inoculants can adversely affect mutualistic associations between wheat and AM fungi under certain field conditions (Germida and Walley, 1996). The species and type of indigenous AM fungi involved (Graham and Abbott, 2000), wheat genotype (Zhu et al., 2001), and seed P contents (Zhu and Smith, 2001) are additional factors that can also significantly influence this symbiosis.

Concluding comments

This work shows that under field conditions in Mali it is possible to obtain wheat grain yields comparable to those produced by using the expensive DAP fertilizer, by using the less expensive locally available fertilizer TPR, combined with TPR-solubilizing microorganisms and AM fungi. To make TPR economically profitable to farmers, future work should be oriented towards the development and production of inexpensive formulations of Gi and PSM inoculants, by using locally available material. More field assays in different agricultural regions in Mali are necessary to test the efficacy of the inoculants in the presence of different indigenous soils microbial communities.

Acknowledgements

Babana Amadou Hamadoun was the recipient of a doctoral fellowship from Le Programme de bourses de la Francophonie (PCBF). The authors are grateful to the following persons in Mali who helped in the logistics and the establishment of the field plots: A. Allaye and the farmers of Koygour (Diré); D Maiga, B. Diallo and the personnel of PACCEM Diré, the director of the IER and Anne-Marie Marcotte. This work was supported by NSERC.

References

Barea J M, Azcón R and Azcón-Aguilar C 2002 Mycorhizosphere interactions to improve plant fitness and soil quality. Antonie van Leeuwenhoek 81, 343–351.

Barea J M, Andrade G, Bianciotto V, Dowling D, Lohrke S, Bonfante P, O'Gara F and Azcón-Aguilar C 1998 Impact on arbuscular mycorrhiza formation of *Pseudomonas* strains used as inoculants for biocontrol of soil borne fungal plant pathogens. Appl. Environ. Microbiol. 64, 2304–2307.

Barea J M, Escudero J L and Azcón-Aguilar C 1980 Effects of introduced and indigenous VA mycorrhizal fungi on nodulation, growth and nutrition of *Medicago sativa* in phosphate-fixing soils as affected by P-fertilizers. Plant and Soil 54, 283–296.

Bationo A, Ayuk E, Ballo D and Koné M 1997 Agronomic and economic evaluation of Tilemsi phosphate rock in different agroecological zones of Mali. Nutr. Cycling Agrosyst. 48, 179–189.

Bolland M D A and Gilkes R J 1997 The agronomic effectiveness of reactive phosphate rocks. 2. Effect of phosphate rock reactivity. Aust. J. Exp. Agric. 37, 937–946.

Chabot R, Antoun H and Cescas M P 1993 Stimulation de la croissance du maïs et de la laitue romaine par des microorganismes dissolvant le phosphore inorganique. Can. J. Microbiol. 39, 941–947.

Chabot R, Antoun H and Cescas M P 1996 Growth promotion of maize and lettuce by phosphate-solubilizing *Rhizobium leguminosarum* biovar *phaseoli*. Plant and Soil 184, 311–321.

Diop T, Gueye M, Dreyfus B, Plenchette C and Strullu D G 1994 Indigenous arbuscular mycorrhizal fungi associated with *Acacia albida* Del. in different areas of Senegal. Appl. Environ. Microbiol. 60, 3433–3436.

Gahoonia T S, Care D and Nielsen N E 1997 Root hairs and phosphorus acquisition of wheat and barley cultivars. Plant and Soil 191, 181–188.

Gaines P T and Mitchell A G 1979 Chemical methods for Soil and Plant Analysis. University of Georgia, Coastal Plain Station, Tifton, USA 105 pp.

Germida J J and Walley F L 1996 Plant growth-promoting rhizobacteria alter rooting patterns and arbuscular mycorrhizal fungi colonization of field-grown spring wheat. Biol. Fertil. Soils 23, 113–120.

Goldstein A H 1986 Bacterial solubilization of mineral phosphates: Historical perspective and future prospects. Am. J. Altern. Agric. 1, 51–57.

Graham J H and Abbott L K 2000 Wheat responses to aggressive and non-aggressive arbuscular mycorrhizal fungi. Plant and Soil 220, 207–218.

Gulden R H and Vessey J K 2000 *Penicillium bilaii* inoculation increases root-hair production in field pea. Can. J. Plant Sci. 80, 801–804.

Large E C 1954 Growth stages in cereals: Illustration of the Feekes scale. Plant Pathol. 3, 128–129.

McGee P 1989 Variations in propagules number of vesicular–arbuscular mycorrhizal fungi in a semi-arid soil. Mycol. Res. 92, 28–33.

McKeague J A Ed. 1978 Manual of Soil Sampling and Methods of Analysis. Canadian Soil Survey Committee, Canadian Society of Soil Science, Ottawa, 223 pp.

Mehlich A 1984 Mehlich 3 soil test extractant: A modification of Mehlich 2 extractant. Commun. Soil Sci. Plant Anal. 15, 1409–1416.

Nautiyal C S 1999 An efficient microbiological medium for screening phosphate solubilizing microorganisms. FEMS Microbiol. Lett. 170, 265–270.

Omar S A 1998 The role of rock-phosphate-solubilizing fungi and vesicular–arbuscular mycorrhiza (VAM) in growth of wheat plants fertilized with rock phosphate. World J. Microbiol. Biotech. 14, 211–218.

Reyes I, Bernier L, Simard R R and Antoun H 1999 Effect of nitrogen source on the solubilization of different inorganic phosphates by an isolate of *Penicillium rugulosum* and two UV-induced mutants. FEMS Microbiol. Ecol. 28, 281–290.

Reyes I, Baziramakenga R, Bernier L and Antoun H 2001 Solubilization of phosphate rocks and minerals by a wild-type strain and two UV-induced mutants of *Penicillium rugulosum*. Soil Biol. Biochem. 33, 1741–1747.

Reyes I, Bernier L and Antoun H 2002 Rock phosphate solubilization and colonization of maize rhizosphere by wild and genetically modified strains of *Penicillium rugulosum*. Microb. Ecol. 44, 39–48.

Richardson A E 2001 Prospects for using soil microorganisms to improve the acquisition of phosphorus by plants. Aust. J. Plant Physiol. 28, 897–906.

Rodriguez H and Fraga R 1999 Phosphate solubilizing bacteria and their role in plant growth promotion. Biotech. Adv. 17, 319–339.

SAS, Institute Inc 1990 SAS Procedure Guide Version 6. 3rd ed. SAS Institute Inc, Cary, NC 705 pp.

Singh S and Kapoor K K 1999 Inoculation with phosphate-solubilizing microorganisms and a vesicular–arbuscular mycorrhizal fungus improves dry matter yield and nutrient uptake by wheat grown in a sandy soil. Biol. Fertil. Soils 28, 139–144.

Sylvia D M and Chelleni D O 2001 Interactions among root-inhabiting fungi and their implications for biological control of root pathogens. Adv. Agron. 73, 1–33.

Tandon H L S, Cescas M P and Tyner E H 1968 An acid-free vanadate–molybdate reagent for the determination of total phosphorus in soils. Soil Sci. Soc. Am. Proc. 32, 48–51.

Vierheilig H, Goughlan A P, Wyss U and Piché Y 1998 Ink and vinegar, a simple technique for arbuscular-mycorrhizal fungi. Appl. Environ. Microbiol. 64, 5004–5007.

Villegas J and Fortin J A 2001 Phosphorus solubilization and pH changes as a result of the interactions between soil bacteria and arbuscular mycorrhizal fungi on a medium containing NH_4^+ as nitrogen source. Can. J. Bot. 79, 865–870.

Whitelaw M A 2000 Growth promotion of plants inoculated with phosphate-solubilizing fungi. Adv. Agron. 69, 99–151.

Zhu Y-G and Smith S E 2001 Seed phosphorus (P) content affects growth, and P uptake of wheat plants and their association with arbuscular mycorrhizal (AM) fungi. Plant and Soil 231, 105–112.

Zhu Y-G, Smith S E, Barritt A R and Smith F A 2001 Phosphorus (P) efficiencies and mycorrhizal responsiveness of old and modern wheat cultivars. Plant and Soil 237, 249–255.

Screening for PGPR to improve growth of *Cistus ladanifer* seedlings for reforestation of degraded mediterranean ecosystems

Ramos Solano B.[1], M. T. Pereyra de la Iglesia[1], A. Probanza[1], J. A. Lucas García[1], M. Megías[2] & F. J. Gutierrez Mañero[1]

[1]*Facultad Farmacia, Universidad San Pablo CEU, P.O. Box 67, Boadilla del Monte, 28668, Madrid, Spain.*
[2]*Fac. Farmacia, Universidad de Sevilla, C/Profesor García González s/n C.P., 41012, Sevilla, Spain.*
[1]*Corresponding author**

Received: 31 May 2006

Key words: genetic diversity, PGPR, phosphate solubilisation, rhizosphere, reforestation

Abstract

A screening for PGPRs was carried out in the rhizosphere of wild populations of *Cistus ladanifer*. Two hundred and seventy bacteria were isolated, purified and grouped by morphological criteria. Fifty percent of the isolates were selected and tested for aminocyclopropanecarboxylic acid (ACC) degradation, auxin and siderophore production and phosphate solubilisation. Fifty-eight percent of the isolates showed at least one of the evaluated activities, with phosphate solubilisation and siderophore production being the most abundant traits. After PCR-RAPDs (Randomly amplified polymorphic DNA) analysis, 11 groups appeared with 85% similiarity, revealing the low diversity in the system. One strain of each group was tested in a biological assay, and those that enhanced *Cistus* growth were identified by 16S rDNA sequencing.Although seven of the 11 assayed strains were phosphate solubilisers and able to produce siderophores, only one was really effective in increasing all biometric parameters in *Cistus ladanifer* seedlings, the lack of effect of the other six probably being due to the rich substrate used. This suggests that other mechanisms apart from nutrient mobilisation might be involved in growth promotion by this strain. However, the low diversity together with the high redundancy detected by PCR-RAPDs and the predominance of strains able to mobilise nutrients in the rhizosphere of *Cistus* reveals that the plant selects for bacteria that can help to supply scarce nutrients. This type of plant growth promoting rhizobacteria (PGPR) strains should be succesful in reforestation practices.

Introduction

Today it is a widely accepted fact that rhizobacteria play a key role in plant health and nutrition. Knowledge of the rhizosphere and its implications on plant physiology have dramatically changed traditional crop management practices regarding plant nutrition and defensive mechanisms (Ramamoorthy et al., 2001; Richardson, 2001). Those bacteria that are beneficial for plant growth are called Plant Growth Promoting Rhizobacteria (PGPR) (Kloepper et al., 1980). This term includes those bacteria that induce plant growth, improving plant nutrition or producing plant growth regulators (Gutierrez Mañero et al., 2001) as well as those that prevent the attack of pathogenic microorganisms (Bowen and Rovira, 1999; Van Loon et al., 1998).

The use of PGPRs on tree species is gaining interest since the usual pathogens in agricultural crops are also common to tree nurseries (Enebak et al., 1997). Inoculation of PGPRs in forest-tree nurseries has proved to be crucial in enhancing

* Phone: +1-34-913724733. FAX No: +1-34-913510496.
E-mail: bramsol@ceu.es

E. Velázquez and C. Rodríguez-Barrueco (eds.),
First International Meeting on Microbial Phosphate Solubilization, 59–68

survival of young seedlings when transplanted to the field. Inoculated seedlings with a more developed root system achieve better nutrition and survival after transplanting (Probanza et al., 2001). Recovering degraded mediterranean ecosystems directly into the mature stage with the dominant species *Quercus* is difficult due to the high percentage of individuals that do not survive. A different approach could be to introduce species that belong to immature stages of this ecosystem such as *Cistus*, which may prepare soil for a better establishment of *Quercus* species.

Plants select those bacteria that are more beneficial for their health by releasing organic compounds through exudates (Lynch, 1990), creating a very selective environment where diversity is low (Marilley and Aragno, 1999). Therefore, the rhizosphere of wild plant species seems to be the best source to isolate plant growth promoting rhizobacteria (Gutierrez Mañero et al., 2002; Lucas García et al., 2001). Consequently, our rationale was that the rhizosphere of wild populations of *Cistus ladanifer* would be a good source for putative PGPRs. A screening of 270 strain in the rhizosphere of *Cistus* was carried out to identify PGPRs associated to this genus. A subset of 144 isolates were characterised based on metabolic activities regarded as putative PGPR traits (auxin production, ACC degradation, siderophore production and phosphate solubilisation). The 83 isolates that tested positive for any of the evaluated traits were analysed by PCR-RAPDs to reduce genetic redundancy by selecting genetically different bacterial strains. Eleven groups appeared and one strain of each group was tested for growth promotion of *Cistus* seedlings. Those that showed a positive effect were identified by partial sequencing of 16S rDNA.

Materials and methods

Origin of bacteria

A bacterial screening was carried out in the rhizosphere of wild populations of *Cistus ladanifer* in Sierra de Aracena southwest Spain (Huelva, coordinates UTM 7° 6°33″ W 37°55″ N) in March 2000. The potential climax vegetation in the studied area would belong to the silicylic mesomediterranena series of *Quercus suber* (*Sanguisorbo agrimoniodi – Querceto suberis sigmetum*) and to the silicylic mesomediterranean series of *Quercus illex* (*Pyro bourgeanae – Querceto rotundifoliae sigmetum*). Three sampling areas (A, B and C) were determined to achieve the maximum edaphic and environmental variability. Physicochemical characterisation of soil was carried out in the three sampling areas. Zone A: texture 45%–15%–40% sand:clay:silt with acidity of pH 5.94, organic C 2.71%, nitrogen 1440.29 ppm, available phosphate 10.28 ppm, iron 37.3 ppm. Zone B: texture 45%–17%–37% sand:clay:silt with acidity of pH 6.02, organic C 1.32%, nitrogen 489.48 ppm, available phosphate 1.79 ppm, iron 40.99 ppm. Zone C: texture 45%–15%–40% sand:clay:silt with acidity of pH 6.78, organic C 1.52%, nitrogen 589.95 ppm, available phosphate 5.05 ppm, iron 62.5 ppm.

Nine plants were sampled in each area. The soil intimately adhered to roots and the thinner roots (diameter 1–2 mm) of three plants were pooled at random and constituted a working unit, all of which were brought to the lab in plastic bags at 4 °C.

Two grams of rhizosphere soil and thinner roots were suspended in 2 mL sterile distilled water and homogenised for 1 min in an omnimixer. One hundred microlitres of the soil suspension was used to prepare serial 10-fold dilutions in a final volume of 1 mL; 500 μL were plated on Standard Medium Agar (Pronadisa SPAIN) and incubated for 4 days at 28 °C. Individual colonies were selected after 36 h and after 4 days of incubation to select fast and slow growing strains. To avoid duplication, isolated colonies were marked on the plate after selection.

Fifteen colony forming units (cfu) were selected from each serial-dilution series, that is, from each working unit (3), in each area (A, B and C) at 36 h and at 4 days, constituting 270 cfu. All were purified and grouped according to Gram staining, morphological characteristics and sporulating capacity into five parataxonomic groups: Gram positive filamentous isolates, Gram positive endospore forming bacilli, Gram positive non-endospore forming bacteria, Gram positive cocci and Gram negative bacteria.

In vitro *tests*

Fifty percent of the isolates from each working unit, sampling area and sampling moments (36 h and 4 days), constituting 144 isolates, were selected for *in vitro* testing. All isolates are kept at −20 °C on glycerol:water (1:4). The following biochemical tests generally associated with PGPR traits were made on the 144 isolates: auxin production (Benizri et al., 1998), aminocyclopropanecarboxylic acid (ACC) degradation (Glick et al., 1995), siderophore production (Alexander and Zuberer, 1991) and phosphate solubilisation (de Freitas et al., 1997).

DNA extraction and RAPD-PCR analysis

A genetic analysis by PCR-RAPDs was carried out on those isolates that showed at least one of the phenotypic traits. Each strain was inoculated on nutrient broth (Pronadisa, Spain) at 28 °C overnight with shaking. DNA extraction was done with the Ultraclean™ Microbial DNA isolation Kit (MOBIO, USA) according to manufacturer instructions.

DNA amplification was carried out on a PE Cetus DNA Thermal cycler, in the following conditions: 5 min 95 °C followed by 45 cycles of 1 min at 94 °C, 2.5 min at 35 °C, 2 min at 72 °C, to end with 6 min at 72 °C. Six random primers were used, four of them (1, 2, 3 and 5) were common to all parataxonomic groups and other two were different for each one: for Gram positive endospore forming bacilli, primers 4 and 6; for Gram positive non-endospore forming bacteria, primers 6 and 8; for Gram negative bacteria, 6 and 7; and for filamentous isolates, 4 and 6. Sequences were: primer 1 (GTT TCG CTC C), primer 2 (GGA CTG GAG T), primer 3 (GGT GAC GCA G), primer 4 (TGG GGG ACT C), primer 5 (CTG CTG GGA C), primer 6 (CCT TGA CGC A), primer 7 (TTC CCC CGC T) and primer 8 (AGG GAA CGA G). The reproducibility of the unique RAPD-PCR patterns produced by single isolates was tested by performing three independent DNA amplifications followed by PCR-RAPD analysis of several randomly chosen strains (data not shown). Two microlitres of each amplification mixture was analysed by agarose (1.2% wt/vol) gel electrophoresis in Tris–acetate–EDTA (TAE) buffer containing 0.5 μg of ethidium bromide per mL.

Data from each parataxonomic group were treated individually. The amplification patterns were analysed with a scanner-densitometer Gel-Doc2000™ 170-8126 (Biorad, CA, USA), elaborating a dendroGram with Pearson's coefficient and UPGAMA method. Eighty-five percent similarity was defined to determine groups.

Short-term plant growth tests

The biological effect on the growth of *Cistus* seedlings was determined with one representative of each group defined by PCR-RAPDs.

After 24 h at 4 °C in 0.4% purified agar (Pronadisa, Spain), seeds were boiled for 10 min and germinated on 100 mL pots filled with peat:vermiculite (tremite no.3) (1:1, v/v). When plants showed the cotyledons and the first two true leaves (approximately 5 weeks), inoculation was carried out by soil drench with 1 mL of a bacterial suspension (10E8 cfu/mL). Six weeks after inoculation, seedlings were harvested and shoot fresh weight, shoot length and leaf number were determined.

Plants were kept under natural photoperiod (16 h light/8 h dark) in the greenhouse under controlled conditions of relative humidity (HR) and temperature (35/20 °C); watering was done twice a day (27.5 mL/m^2 min) for 5 min.

Statistics of growth parameters

One way analysis of variance with replicates was used to evaluate the effect of treatments on plant growth parameters (Harmann, 1967). When significant differences appeared a Fisher test was used.

PCR amplification of bacterial 16s-rDNA and sequencing

Those strains that demonstrated a positive effect on plant growth were identified by 16S rDNA sequencing and sequence phylogenetic analyses.

Each bacterial strain was amplified with 16s rDNA specific primers: P1F (AGA GTT TGA TCC TGG CTC AG) *E. coli* and P2R (AAG GAG GTG ATC CAG CCG CA. Amplification reactions were made with 5 μL DNA (20 ng/μL), 3 ud *Taq* polymerase (Roche Expand High-Fidelity™ PCR system), 5 μL 10× PCR buffer, primers 1 and 2 at 0.5 μM and ultrapure water at a 50 μL volume. The reaction mixtures were incubated in a thermocycler (PE Cetus DNA thermal cycler) at 95 °C for 5 min and then subjected to 30 cycles consisting of 95 °C for 60 s, the annealing temperature 64 °C for 60 s, and 72 °C for 2 min. Finally, the mixtures were incubated at 72 °C for 6 min. Two microlitres of each amplification mixture were analysed by agarose (1.2% wt/vol) gel electrophoresis in Tris–acetate–EDTA (TAE) buffer containing 0.5 μg of ethidium bromide per mL.

Phylogenetic analysis

The 16S rDNA sequences were aligned with Bioedit Secquence Alignment editor 5.0.3. (Hall TA 1999). The alignment was checked manually, corrected, and then analysed by BLAST (NCBI BLAST^R Home page. Basic Local Alignement Search Tool). The tree and its robustness was created and evaluated using PAUP 4.0 beta. The neighbour option was used to build the dendroGram and the distance was calculated with the HKY85 algorithm that evaluates a transition/transversion ratio and base frequencies.

Nucleotide sequence accession numbers

The 16S rDNA nucleotide sequences obtained were deposited in the GenBank database under the following accession numbers: strains A39: AY178859, B85: AY178857, B38: AY178858 and C50: AY178856.

Results

Table 1 shows frequency of bacterial isolates that belong to each parataxonomic group. Gram positive strains predominated (68.47%), with endospore-forming bacilli being the most numerous (46.57%). Over 50% of isolates tested positive for the putative PGPR traits evaluated. The highest frequency of isolates that tested positive for any of the evaluated traits was found in the most abundant group, the Gram positive endospore-forming bacilli, while the lowest was found on the least abundant group, Gram positive cocci. This group was not further studied due to the low number of representatives. Three Gram negative and two Gram positive non-spore forming bacteria did not survive cryoconservation.

Siderophore production was the best represented putative PGPR trait (Table 2), both in strains that only showed this trait (39.40%) and those associated with phosphate solubilisation (27.84%). The same trend applies to predominant groups (Gram positive endospore-forming and non-endospore forming bacteria): individuals that belonged to the less abundant group (Gram negative bacteria) showed siderophore production only associated to phosphate solubilisation (8.86%).

Isolates that were able to degrade ACC and those only able to solubilise phosphate showed similar frequencies and represent all parataxonomic groups to the same extent (Table 2). With regard to auxin production, all parataxonomic groups had representatives with this trait, but again, predominant groups (Gram positive endospore-forming and non-endospore forming bacteria) showed higher frequencies.

Table 1. Frequency of isolates that tested positive for at least one of the evaluated traits, or negative to all in each parataxonomic group

%	Gram positive endospore-forming bacilli	Gram positive non-endospore-forming bacteria	Gram positive cocci	Gram positive filamentous isolates	Gram negative bacteria	Total
Negative	26.70	10.27	0.68	0.68	4.97	41.78
Positive	19.86	11.71	1.37	14.38	10.95	58.22
Total	46.57	21.90	2.05	15.06	14.38	100

Table 2. Frequency of isolates and biochemical activities of PGPR traits in each morphological group isolated from the rhizosphere of *Cistus ladanifer*

	Gram positive endospore-forming bacilli	Gram positive non-endospore-forming bacteria	Gram positive cocci	Gram positive filamentous isolates	Gram negative bacteria	Total %
ACC	2.53	2.53	0	0	2.53	7.59
CAS	21.68	11.39	0	6.33	0	39.40
PDYA	2.53	2.53	2.53	0	0	7.59
AUX	3.79	2.53	0	1.26	1.26	8.86
AUX + CAS	1.26	0	0	3.79	0	5.06
AUX + PDYA	3.79	0	0	0	0	3.73
CAS + PDYA	1.26	3.79	0	13.92	8.86	27.83
Total %	36.70	22.78	2.54	25.32	12.65	100

AUX, auxin producers; CAS, siderophore producers; PDYA, phosphate solubilisers; ACC, aminocyclopropanecarboxylic degraders.

Those isolates that tested positive for at least one of the putative PGPR traits were analysed by PCR-RAPDs to reduce genetic redundance while retaining the maximum genetic diversity. Figure 1 shows the dendroGrams for each parataxonomic group. For our purposes, 85% similarity was determined to define groups within each parataxonomic group. The 12 Gram positive non-spore forming isolates segregated into 2 groups, of 6 isolates each 29; Gram positive endospore forming bacilli strains separated into 3 groups constituted by 2, 1 and 26 isolates; the 13 Gram negative bacteria split into 2 groups of 1 and 12 individuals and the 20 filamentous isolates segregated into 4 groups of 3, 3, 2 and 12 isolates.

One isolate of each of the above groups was selected at random to test its plant growth promoting capacity when inoculated on *Cistus ladanifer* seedling growth. Fresh weight, shoot length and leaf number were the evaluated parametres (Table 3). Isolates A39, B85, B38 and C50 were found to be the most effective for enhancing plant growth. Isolate C50 showed exceptional growth increase for all four parameters, while the other three isolates each affected only one parameter.

These four isolates were identified by partial sequencing of their 16s rDNA (Figure 2). C50 showed low genetic distances with *Burkholderia caryophylli U91570* and *B. Graminis U96941*, B38 was identified as *Bacillus senegalensis AF519468*, and A39 showed high similarity with *Streptomyces galileus AB045878*, as did B85 with *Arthrobacter oxydans AJ243423*.

Discussion

Traditionally, a search for PGPRs involves screening a large number of isolates and identifying a desired phenotypic trait. Once isolates are purified, the main goal is to keep the maximum genetic diversity in the minimum number of isolates, for further biological assays. This goal may be achieved through ITS-PCR, AFLP, AP-PCR/PCR-RAPDs, techniques that define differences at the strain level (Louws et al., 1999). PCR-RAPDs has proved to be a very efficient tool to define strains within the same bacterial species (Gutierrez Mañero et al., 2002; Lucas García et al., 2001). The International Comitee for Bacterial Systematics determined that 30% divergence in DNA estimated by DNA hybridisation defines the taxonomic range of bacterial species (Wayne et al., 1987), while divergences of 40–50% define bacterial genera. The arrangement obtained by PCR-RAPDs and UPGAMA and Pearson coefficient may be understood as genetic distances obtained by nucleotidic substitutions (Clark and Lanigan, 1993; Nei and Miller, 1990). We defined 15% genetic divergence to define bacterial strains within the same species, therefore, all isolates within the same group at 85% similarity are considered members of the same bacterial species. After PCR-RAPDS analysis of the 83 strains, only 11 genetically different strains appeared: four filamentous isolates, two Gram negative bacteria, two Gram positive non-spore forming bacteria and three Gram positive endospore forming bacilli. It should be noted that in each parataxonomic group, at least one of the

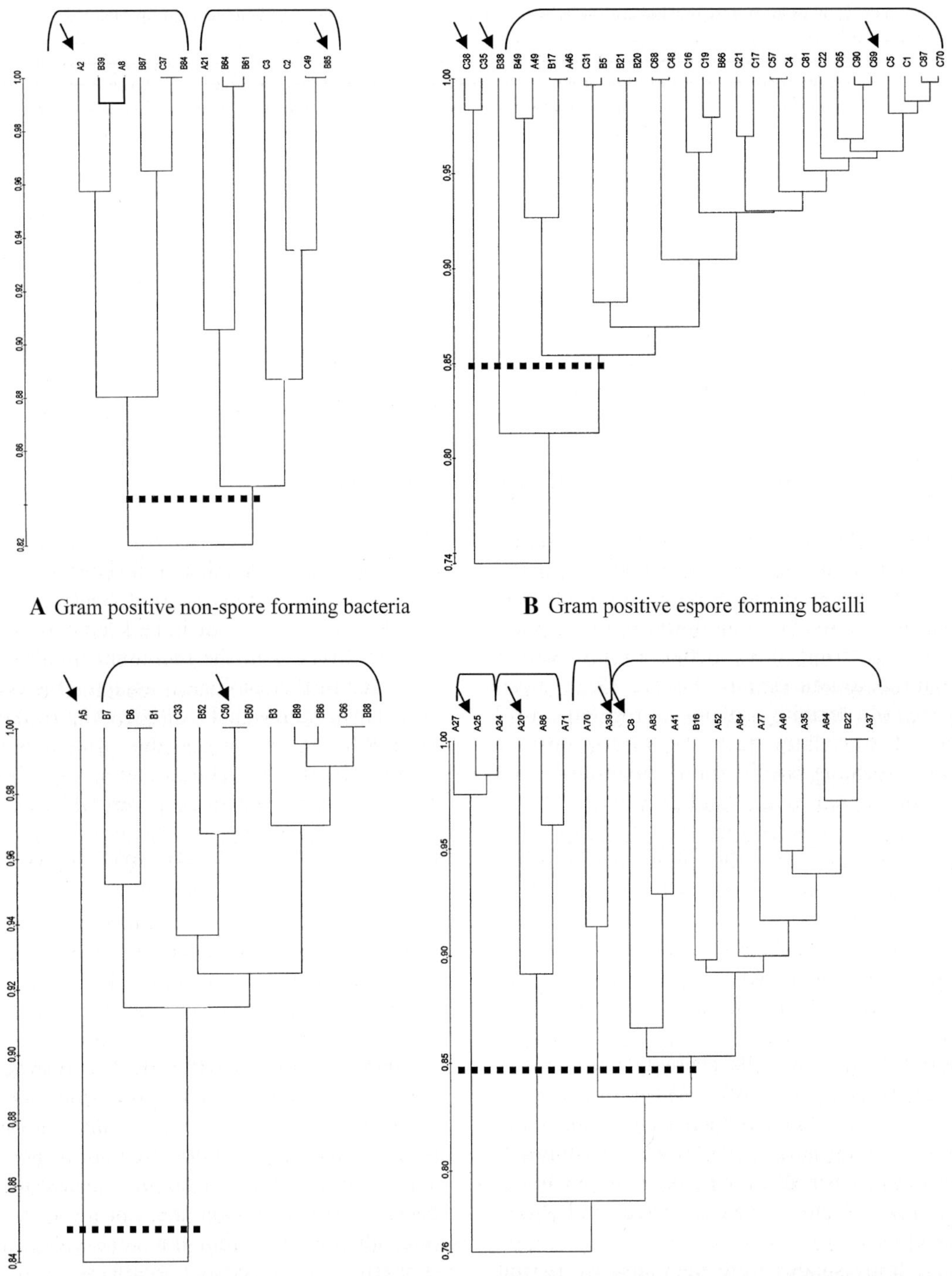

Figure 1. Genetic divergence among (A) Gram positive non-endospore-forming bacteria; (B) Gram positive endospore-forming bacilli; (C) Gram negative bacteria; (D) Gram positive filamentous isolates from the rhizosphere of *Cistus ladanifer* according to Pearson coefficient and UPGAMA, that integrates information of the six primers used.

Table 3. Growth parameters of *Cistus ladanifer* seedlings inoculated with the selected strains from PCR-RAPDs

Bacterial strain	PGPR activity	Parataxonomic group	Heigth (cm)	Fresh weight (g)	Leaf number
A20	CAS	Filamentous microorganism	2.38 ± 0.32	0.048 ± 0.029	7.6 ± 1.26
A25	CAS PDYA	Filamentous microorganism	1.88 ± 0.37	0.035 ± 0.021	7.2 ± 1.39
A39	CAS PDYA	Filamentous microorganism	2.36 ± 0.28	0.076 ± 0.122*	7.6 ± 1.26
C8	CAS PDYA	Filamentous microorganism	1.99 ± 0.31	0.027 ± 0.012	7.5 ± 1.99
A5	CAS PDYA	Gram negative bacteria	2.18 ± 0.27	0.045 ± 0.017	7.8 ± 1.13
C50	CAS PDYA	Gram negative bacteria	3.01 ± 0.65*	0.094 ± 0.068*	9.0 ± 1.94*
A8	CAS PDYA	Gram positive non-endospore-forming bacteria	1.78 ± 0.29	0.022 ± 0.007	6.6 ± 1.8
B85	ACC	Gram positive non-endospore-forming bacteria	2.69 ± 0.42*	0.049 ± 0.031	7.8 ± 1.47
C69	CAS PDYA	Gram positive endospore-forming bacilli	2.26 ± 0.42	0.041 ± 0.023	7.6 ± 1.26
C38	CAS	Gram positive endospore-forming bacilli	2.22 ± 0.21	0.042 ± 0.019	7.9 ± 1.37
B38	AUX	Gram positive endospore-forming bacilli	2.36 ± 0.62	0.067 ± 0.038	8.4 ± 1.83*
Control			2.21 ± 0.32	0.038 ± 0.023	6.8 ± 1.36

AUX, auxin producers; CAS, siderophore producers; PDYA, phosphate solubilisers; ACC, aminocyclopropanecarboxylic degraders. Data is the average of 6 plants ± SE. Asterisks indicate significant differences ($P < 0.05$) with controls.

groups defined by PCR-RAPDs had a high number of representatives, indicating a great deal of genetic redundance that was more marked in the most abundant groups. The low number of different strains reveals the low diversity existing in such a selective environment as the rhizosphere, a fact that has been demonstrated by several authors in different species (di Cello et al., 1997; Gutierrez Mañero et al., 2002; Lucas García et al., 2001; Marilley and Aragno, 1999).

The evaluated phenotypic traits have been previously proposed as good indicators of putative PGPRs (Cattelan et al., 1999). However, any phenotypic trait shown *in vitro* reveals that the information is contained within the bacterial genome, but it is not constitutively expressed. Because of this, we carried out the phenotypic screening first, to define genetic differences by PCR-RAPDs afterwards.

On the basis of laboratory screening assays, it has been shown that P-solubilising microorganisms may constitute up to 40% of the culturable population of soil microorganisms, and a significant proportion can be isolated from the rhizosphere (Kucey, 1983) as was found in this study, considering those isolates that showed this activity combined with others (Table 1). Additionally, 70% of the strains can produce siderophores, and the intersection of both groups constitutes one of the best represented phenotypic groups (CAS-PDYA). Considering these results together with the low diversity shown by PCR-RAPDs and with the high number of representatives in some of the groups, it may be concluded that the plant preferentially selects rhizobacteria with nutrient related mechanisms instead of those able to affect the plant's hormonal balance. Interpretation of these results is relevant for an effective use of *Cistus* to recuperate degraded ecosystems, since these soils usually show low nutritional status, and hence, a successful growth of *Cistus* would be ensured with nutrient-helper PGPRs. Nevertheless, all putative PGPR traits were found in the 11 strains, although with different proportions: 7 were CAS-PDYA, 2 CAS, 1 ACC and 1 AUX (Table 3).

Among the 11 strains inoculated on *Cistus* seedlings, only C50 significantly increased all the biometric parametres evaluated, and another 3 (A39, B85 and B38) significantly increased at least one of them. Despite the fact that seven out of 11 were CAS-PDYA, only two demonstrated a positive effect on *Cistus* growth (Table 3). All these PGPR strains belong to different bacterial genera according to 16s rDNA sequencing (Figure 2): *Burkholderia*, *Bacillus*, *Arthrobacter* and *Streptomyces*. Representatives of all of these genera have been reported in the literature as PGPRs showing different mechanisms of action (Burdman et al., 2000; Marten et al., 2001; Rodríguez et al., 2000).

Irrespective of the underlying mechanism controlling the desired activity, if the introduced strain does not survive and colonise in the rhizosphere, it will not be effective in promoting plant growth (Goddard et al., 2001; Wiehe and

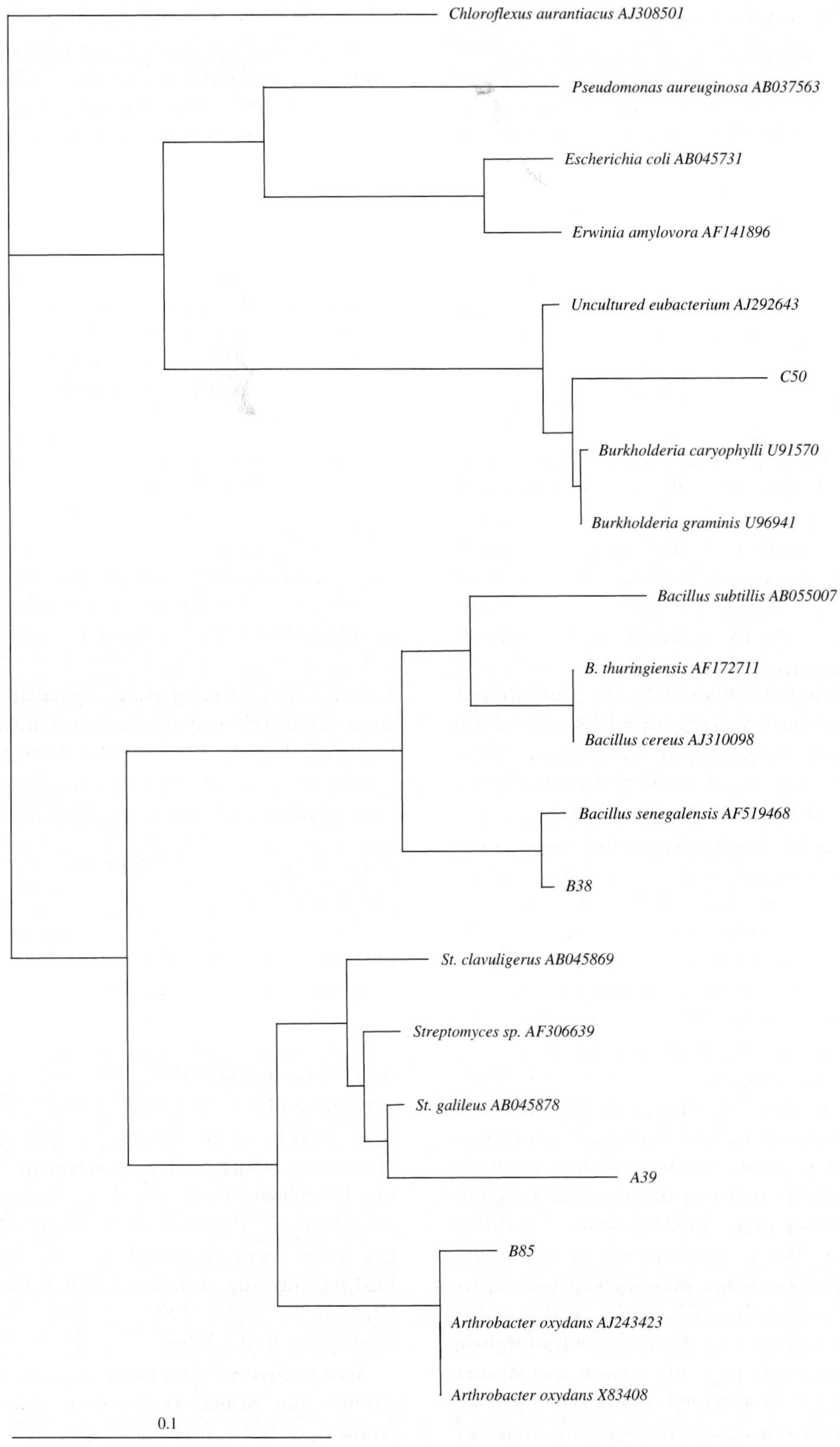

Figure 2. Parsimony tree of strains that showed plant growth promotion on *Cistus* seedlings.

Höflich, 1995). Another requirement for efficacy is that the inoculated bacteria must establish within the rhizobacterial communities without disrupting their equilibrium, since it has been reported that a quick recovery of the system following inoculation provides the best results on plant growth (Ramos et al., 2002).

In addition to the abovementioned factors, the differential effects found in this study may be due to the mechanisms of growth promotion of the assayed strains. It has been shown recently that a number of bacterial genes related to nutrition are induced under stressful conditions in the rhizosphere (Rainey, 1999). This fact together with the characteristics of soils in which the screening was carried out (low in iron and in available phosphate (see ‚Materials and methods' section)), explains the predominance of the CAS-PDYA trait reported in this study. Our results support that the plant selects the most beneficial bacteria for the rhizosphere, since a low nutrient substrate selects for those strains able to provide the scant nutrients in soil.

The differential effect of the seven CAS-PDYA strains on *Cistus* growth may be attributed to the differential induction of genes related to nutrition in the rhizosphere mentioned before. The use of a nutrient-rich soil such as peat may determine that those genes related to siderophore production and phosphate mobilisation are not expressed since they are not necessary. Therefore, those strains that still enhanced *Cistus* seedling growth may have displayed other PGPR traits not related to improvement of nutrition, such as production of plant growth regulators, as it is the case of B38 (auxin production) and B85 (ACC degradation). Consistent with this hypothesis, production of plant growth regulators by rhizosphere microorganisms has been reported (Gonzalez-Lopez et al., 1986; Gutierrez Mañero et al., 1996, 2001). Plant growth regulators of the auxin type and ethylene affect root growth pattern and root system structure, leading to improved nutrient absorption (Selvadurai et al. 1991).

In conclusion, the rhizosphere of *Cistus* has proved to be a good source for PGPR strains that improve seedling growth and enhance their adaptative capacity. Effective PGPR-*Cistus* teams to recover degraded ecosystems must rely on rhizobacteria that are able to improve nutrient acquisition, as indicated by the predominant phenotypic activities in the rhizosphere. Further research should be carried out to confirm the colonisation capacity of the most effective strain C50, identified as *Burkholderia*, as well as its effectiveness in different conditions.

Aknowledgements

This study has been financed by project FEDER 1FD97-1441.We would like to thank Linda Hamalainen for editorial help and Mª Josefa Fernandez for technical support. M.T. Pereyra was formerly a Comunidad Autonoma de Madrid postdoctoral student.

References

Alexander D B and Zuberer D A 1991 Use of chrome azurol S reagents to evaluate siderophore production by rhizosphere bacteria. Biol. Fertil. Soils 12, 39–45.

Benizri E, Courtade A, Picard C and Guckert A 1998 Role of maize root exudates in the production of auxins by *Pseudomonas fluorescens* M.3.1. Soil Biol. Biochem. 30, 1481–1484.

Bowen G D and Rovira A D 1999 The rhizosphere and its management to improve plant growth. Adv. Agron. 66, 1–103.

Burdman S, Jurkevith E and Okon Y 2000 Recent advances in the use of plant growth promoting rhizobacteria (PGPR) in agriculture. Microb. Interact. Agric. Forest. 2, 229–250.

Cattelan A J, Hartel P G and Fuhrmann J J 1999 Screeening for plant growth-promoting rhizobacteria to promote early soybean growth. Soil Sci. Soc. Am. J. 63, 1670–1680.

Clark A G and Lanigan C M S 1993 Prospects for estimating nucleotide divergence with RAPDs. Mol. Biol. Evol. 10, 1096–1111.

de Freitas J R, Banerjee M R and Germida J J 1997 Phosphate-solubilizing rhizobacteria enhance the growth and yield buy not phosphorus uptake of canola (*Brassica napus* L.). Biol. Fertil. Soils 24, 358–364.

di Cello F, Bevivino A, Chiarini L, Fani R, Paffetti D, Tabacchioni S and Dalmastri C 1997 Biodiversity of a *Burkholderia cepacia* population isolated from the maize rhizosphere at different plant growth stages. Appl. Environ. Microbiol. 63, 4485–4493.

Enebak S A, Wei G and Kloepper J W 1997 Effects of plant growth-promoting rhizobacteria on loblolly and slash pine seedlings. Forest Sci. 44, 139–144.

Glick B R, Karaturovic D M and Newell P C 1995 A novel procedure for rapid isolation of plant growth promoting pseudomonads. Can. J. Microbiol. 41, 533–536.

Goddard V J, Bailey M J, Darrah P, Lilley A K and Thompson I P 2001 Monitoring temporal and spatial variation in rhizosphere bacterial population diversity: A community approach for the improved selection of rhizosphere competent bacteria. Plant Soil 232, 181–193.

Gonzalez-Lopez J, Salmeron V, Martinez-Toledo M V, Ballesteros F and Ramos-Cormenzana A 1986 Production of auxins, gibberellins and cytokinins by *Azotobacter vinelandii* ATTCC12837 in chemically-defined media and dialyses soil media. Soil Biol. Biochem. 18, 119–120.

Gutierrez Mañero F J, Acero N, Lucas J A and Probanza A 1996 The influence of native rhizobacteria on European alder [*Alnus glutinosa* (L.) Gaertn.] growth. II. Characterization of growth promoting and growth inhibiting strains. Plant Soil 182, 67–74.

Gutierrez Mañero F J, Ramos B, Lucas García J A, Probanza A and Barrientos M L 2002 Systemic induction of terpenic compounds in *D. lanata*. J. Plant Physiol. 160, 105–113.

Gutierrez Mañero F J, Ramos Solano B, Probanza A, Mehouachi J, Tadeo F R and Talon M 2001 The plant growth-promoting rhizobacteria *Bacillus pumilus* and *Bacillus licheniformis* produce high amounts of physiologically active gibberellins. Physiol. Plant. 111, 1–7.

Harman J H 1967 Modern Factor Analysis. 2Univ. Chicago Press, Chicago.

Kloepper J W, Scrhoth M N and Miller T D 1980 Effects of rhizosphere colonization by plant growth-promoting rhizobacteria on potato plant development and yield. Phytopathology 70, 1078–1082.

Kucey R M N 1983 Phosphate-solubilizing bacteria and fungi in various cultivated and virgin Alberta soils. Can. J. Soil Sci. 63, 671–678.

Louws F J, Rademaker J L W and de Bruijn F J 1999 The three Ds of PCR-based genomic analysis of phytobacteria: Diversity, detection and disease diagnosis. Annu. Rev. Phytpathol. 37, 81–125.

Lucas García J A, Probanza A, Ramos B and Gutierrez Mañero F J 2001 Genetic variability of rhizobacteria from wild populations of four *Lupinus* species based on PCR-RAPDs. J. Plant Nutr. Soil Sci. 164, 1–7.

Lynch J M 1990 The Rhizosphere. Wiley-Interscience, Chichester, England.

Marilley L and Aragno M 1999 Phylogenetic diversity of bacterial communities differing in degree of proximity of *Lolium perenne* and *Trifolium repens* roots. Appl. Soil Ecol. 13, 127–136.

Marten P, Brueckner S and Berg G 2001 Biological plant protection using rhizobacteria – an environmental friendly alternative for biologicl control of soilborne and seedborne phytopathogenic fungi. Gesunde Pflanzen 53, 224–234.

Nei M and Miller J C 1990 A simple method for estimating average number of nucleotide substitutons within and between populations from restriction data. Genetics 125, 873–879.

Probanza A, Mateos J L, Lucas J A, Ramos B, de Felipe M R and Gutierrez Mañero F J 2001 Effects of inoculation with PGPR *Bacillus* and *Pisolitus tinctorius* on *Pinus pinea* L. growth, bacterial rhizosphere colonization and mycorrhizal infection. Microbial Ecol. 41, 140–148.

Rainey P B 1999 Adaptation of *Pseudomonas fluorescens* to the plant rhizosphere. Environ. Microbiol. 1, 243–257.

Ramamoorthy V, Viswanathan R, Raguchander T, Prakasam V and Samiyappan R 2001 Induction of systemic resistance by plant growth promoting rhizobacteria in crop plants against pests and diseases. Crop Protect. 20, 1–11.

Ramos B, Lucas García J A, Probanza A, Barrientos M L and Gutierrez Mañero F J 2002 Alterations in the rhizobacterial community associated with European alder growth when inoculated with PGPR strain *Bacillus licheniformis*. Environ. Exp. Bot. 49, 61–68.

Richardson A E 2001 Prospects for using soil microorganism to improve the acquisition of phosphorus by plants. Aust. J. Plant Physiol. 28, 897–906.

Rodriguez H, Rossolini G M, Gonzalez T, Li J and Glick B R 2000 Isolation of a gene from *Burkholderia cepacia* IS16 encoding a protein that facilitates phosphatase activity. Curr. Microbiol. 40, 362–366.

Selvadurai E L, Brown A E and Hamilton J T G 1991 Production of indole-3-acetic acid analogues by strains of *Bacillus cereus* in relation to their influence on seedling development. Soil Biol. Biochem. 23, 401–403.

Van Loon L C, Bakker P A H M and Pieterse C M J 1998 Systemic resistance induced by rhizosphere bacteria. Annu. Rev. Phytopathol. 36, 453–483.

Wayne L G, Brenner D J, Colwell R R, Grimont P A D, Kandler O, Kirchevsdy M I, Moore L H, Moore W E C, Murray R G E, Stackerbrandt E, Starr M P and Turper H G 1987 Report of the ad hoc committee on reconciliation of approaches to bacterial systematics. Int. J. Syst. Bacteriol. 37, 463–464.

Wiehe W and Höflich G 1995 Establishment of plant growth promoting bacteria in the rhizosphere of subsequent plants after harvest of the inoculated precrops. Microbiol. Res. 150, 331–336.

Phosphate-solubilizing microorganisms isolated from rhizospheric and bulk soils of colonizer plants at an abandoned rock phosphate mine

I. Reyes[1], A. Valery & Z. Valduz
Laboratorio de Biofertilizantes, Decanato de Investigación, Universidad Nacional Experimental del Táchira, Paramillo, San Cristóbal, Táchira, República Bolivariana de Venezuela. [1]*Corresponding author**

Received: 31 May 2006

Key words: rock phosphate, hydroxyapatite solubilization, biodiversity, rhizosphere, *Penicillium* sp., *Azotobacter* sp.

Abstract

The abandoned "Monte-Fresco" rock phosphate mine in Táchira, Venezuela, was sampled to study the biodiversity of phosphate-solubilizing microorganisms (PSM). Rhizosphere and bulk soils were sampled from colonizer plant species growing at a mined site where pH and soluble P were higher than the values found at a near by unmined and shrubby soil. Counting and isolating of PSM choosing strains showing high solubilization halos in a solid minimal medium with hydroxyapatite as phosphate source were evaluated using ammonia or nitrate as nitrogen sources and dextrose, sucrose, and mannitol as carbohydrate sources. A larger number of PSM were found in the rhizospheric than in the bulk soil. Six fungal strains belonging to the genus *Penicillium* and with high hydroxyapatite dissolution capacities were isolated from bulk soil of colonizer plants. Five of these strains had similar phenotypes to *Penicillium rugulosum* IR-94MF1 but they solubilized hydroxyapatite at different degrees with both nitrogen sources. From 15 strains of Gram-negative bacteria isolated from the rhizosphere of colonizer plants, 5 were identified as diazotrophic free-living encapsulated *Azotobacter* species able to use ammonium and/or nitrate to dissolve hydroxyapatite with glucose, sucrose and/or mannitol. Different nitrogen and carbohydrate sources are parameters to be considered to further characterize the diversity of PSM.

Introduction

Interactions between microorganisms that release organic acids and other products onto the surfaces of minerals may liberate ions from their surface layers. In this sense, rock phosphate dissolution by microorganisms directly affects fertility of soils (Reyes et al. 2002). The rhizosphere is a dynamic changing environment that differs from bulk soil both in physical and chemical properties (Bowen and Rovira 1999). In this sense, plant root exudates selectively influence the growth of microorganisms that colonize the rhizosphere when altering the chemistry of soil aggregates and concurrently, rhizospheric microbial populations change the composition and quantity of root exudates through their effect on plant nutrition (Bowen and Rovira 1999; Glick 1995). In mineral soils, bacterial and fungal populations increase in abundance and diversity as minerals are weathered and transformed to soil (Banfied et al. 1999). During the initial stages of weathering, apatite rock phosphate is replaced by chemically- or microbially-precipitated secondary phosphate minerals (e.g., strengite and variscite) and these are completely solubilized in the soil after microbial colonization (Banfied et al. 1999). It has been demonstrated that fungi (Reyes et al. 2001; Vassilev et al. 1996) and bacteria (Goldstein 1995) release organic acids such

* E-mail: isreyes@unet.edu.ve

E. Velázquez and C. Rodríguez-Barrueco (eds.),
First International Meeting on Microbial Phosphate Solubilization, 69–75

as citric, gluconic, and keto-gluconic, to dissolve phosphates and other complexing compounds such as siderophores (Watteau and Berthelin 1994), and reducing mechanisms of cations also liberate phosphates into soil (Altomare et al. 1999).

Colonizer plants growing in soil disturbed by strip mining after exploitation of rock phosphates may reflect in their rhizospheres the processes of mineral transformations of such soils. Spoil banks, left by strip mining, are composed of debris of the rock phosphate, skeletal minerals, and other mixed materials from the disturbed soil horizons. Colonizer plant species could bring some insights about strategies that nature uses for mineral cycling and revegetation of a disturbed land. The purpose of the present work is to determine the phosphate-solubilizing microorganisms (PSM) diversity existing in a rock phosphate mine soil. In the mined site of the mine, diverse strains of *Penicillium*, *Azotobacter* and other unidentified bacteria and fungi were found associated to the colonizer plant species' bulk and rhizosphere soils, respectively. In order to characterize the diversity among their phosphate solubilizing (PS) metabolism *Penicillium* sp. and *Azotobacter* sp. strains were assessed using different carbohydrate and nitrogen sources.

Materials and methods

Site characterization and sampling

Monte Fresco rock phosphate mine is a low-solubility fluorapatite situated in the Andean piedmont at the southwest region of Venezuela (1000 m a.s.l.). Two sites were sampled: an area mined and abandoned since 1994 characterized by a low vegetation stratum distributed in a random-clustered pattern, and a second site, used as a control for the bulk soil, located in an unmined area around 150 m of the first one composed of a shrubby pasture. Sampling of the rhizosphere and bulk soils was done after the end of the rainy season. The unmined and mined sites were sampled for their bulk soils to determine total and solubilizing cultivable fungi and bacteria populations. The mined soil was sampled for the rhizosphere soils of different colonizer plants, which were not present in the shrubby pasture site, in order to isolate Monte Fresco phosphate solubilizers. For the bulk soils composite samples were obtained from the first 10 cm depth and for the rhizosphere soils the whole plants were transported in a cool container with a square of soil around the roots. Some relevant chemical and physical properties of the sampled bulk soils were determined: P, Ca^{2+}, Mg^{2+}, K^{+}, % organic matter, pH and soil texture. Adsorbed forms of phosphate were extracted following the method of Bray and Kurtz (1945), where neutral ammonium fluoride is used as the extraction reagent for soils with a $pH < 7.5$. PO_4-P was analyzed by spectrophotometry with the vanadate–molybdate reagent (Tandom et al. 1968) and results were expressed as ppm of available P. Ca^{2+}, Mg^{2+} and K^{+} were extracted with ammonium acetate (Black 1965) and determined by atomic absorption spectrophotometry, and organic matter in soils by the Walkley and Black (1934) method.

Microbial community counts of bulk and rhizosphere soils

For the rhizosphere soils, the root system of colonizer plants was separated from soil by hand shaking to dislodge loosely adhering soil and 5 long roots containing between 1–2 g d.w. of rhizosphere soil was placed in 100 ml sterile saline solution (Reyes et al. 2002). After shaking for 30 min at 200-rev min^{-1}, 0.1 ml samples from appropriate dilutions of all soils were plated by spreading. All dilutions were plated by triplicate. Total culturable bacteria and fungi were cultivated and counted on nutrient agar (Merk) and PDA (BBL), respectively; and PS bacteria and fungi were grown on a minimal medium containing per litre of distilled water: NH4Cl, 0.4 g; KNO_3, 0.78 g; NaCl, 0.1 g; $MgSO_4{\cdot}7H_2O$, 0.5 g; $CaCl_2{\cdot}2H_2O$, 0.1 g; $FeSO_4{\cdot}7H_2O$, 0.5 mg; $MnSO_4{\cdot}H_2O$, 1.56 mg; $ZnSO_4{\cdot}7H_2O$, 1.40 mg; vitamin B_{12}, 2 µg; glucose, 10 g and agar, 20 g (Reyes et al. 1999). Phosphorus was added to the medium as sparingly soluble phosphate in the form of hydroxyapatite at 2.5 g l^{-1} and the pH was adjusted to 6.5. Plates were incubated at 28°C and total culturable microorganisms were counted between 2 and 4 days, until the highest number of colonies was reached, and PS between 3 and 7 days, allowing colonies to show the

Mps$^+$ phenotype (Reyes et al. 1999). Colonies that produced a clear halo in hydroxyapatite medium indicated PS activity through the production of solubilizing bound phosphate substances released into the agar as described elsewhere (Kucey 1989; Reyes et al. 1999). Counts of culturable bacteria and fungi were expressed as CFU g^{-1} of dry soil.

Physiological characterization of PS strains

Colonies showing the highest PS activities, or bigger haloes, in the isolation hydroxyapatite minimal medium were selected. Several morphological and biochemical characteristics as motility, nitrate reduction, growth in *Azotobacter* medium, citrate growth, and fermentation of sugars (lactose, maltose, sorbitol, inositol, fructose, mannitol, glucose and sucrose) were determined to further characterize hydroxyapatite solubilizing bacteria (Gerhardt et al. 1994). Hydroxyapatite solubilization haloes and colony growth of the PS strains were measured in the minimal solid medium using dextrose, mannitol, and sucrose as carbohydrate sources and ammonium and nitrate as nitrogen sources.

Data analysis

Homogeneity of variance and analysis of variance (ANOVA) at $P < 0.05$ were conducted with the program Statgraphics Plus 4.0. All values are means of three replicates.

Results and discussion

Effect of the rock phosphate mine exploitation on total and PS culturable microbial populations of bulk and rhizospheric soil

Limitations of conventional plating techniques pose a major problem in measuring microbial populations in soils. Between 0.1–1% of the soil bacteria observed in the fluorescence microscope can be isolated and cultured on laboratory media, which means that isolated bacteria may account for a minor proportion of the total bacterial diversity in soil (Lawlor et al. 2000; Torsvik et al. 1990). Because this work attempted not only to determine the PSM diversity in the rock phosphate mine soil, but also to isolate phenotypic diversity, the conventional plating technique was used. However, other techniques such as the fractionated centrifugation technique (Fægri et al. 1977) allow the recuperation of higher numbers of microbial populations. Genetic information about the nonculturable microorganisms in soil can be determined by DNA heterogeneity (Torsvik et al. 1990). Other methods such as fatty acid methyl esters (FAMEs) and phospholipid fatty acids (PLFAs) give insight into the microorganisms' functions and activities (Lawlor et al. 2000).

Populations of total culturable fungi were significantly higher in the unmined site of the mine than in the mined and disturbed one while total culturable bacteria populations were similar in both sites (Fig. 1). Populations of PS bacteria were significantly greater in the mined soil, moreover for the mined site, both PS bacteria and fungi accounted for 19% and 20% of their totals, respectively. This suggests an increase of PS bacteria and fungi populations when compared to the unmined site which showed 5% and 4%, respectively (Fig. 1). The former results could have been related to the type of soil, which was composed mainly by rock phosphate debris. This mineral substrate might have induced a selective microbial colonization by the production of phosphate dissolving substances. The presence of higher PSM in the mined site may account for the higher P soluble, Ca^{2+} and pH

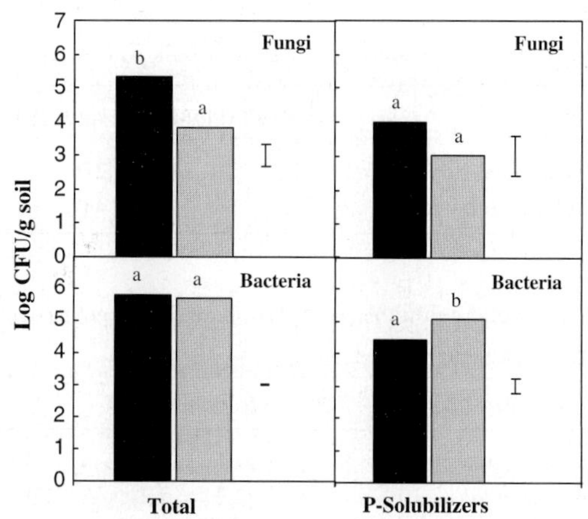

Figure 1. Total and PS culturable fungi and bacteria in the unmined (■) and mined (□) bulk soils. Different letters between soils show values that are significantly different according to ANOVA at a level of $P < 0.05$.

found in this soil compared to the unmined soil which showed higher Mg^{2+}, K^{+} and organic matter contents (Table 1).

Populations of total and PS culturable bacteria and fungi were counted for the rhizosphere soils of four colonizer plants (Table 2) of the mined site. The rhizosphere soil:bulk soil (R:S) ratio was calculated by using the average of culturable bacteria and fungi registered from the rhizospheric soils and the counts obtained from the bulk mined soil. Total culturable bacteria showed the highest R:S value while PS fungi and bacteria were comparable (Table 2). The R:S ratio of different plants range from 23 to 58 for bacteria and from 1.3 to 12 for fungi and varies with the plant, soil, the growth conditions, and the amounts of soil adhering to the roots (Bowen and Rovira 1999). Total culturable bacteria and fungi in the rhizosphere of plants were greater than in the bulk soil, probably due to the high levels of carbon fluxes creating the 'rhizosphere effect' (Bowen and Rovira 1999) used to support bacterial growth (Glick 1994). The colonizer plants *Ageratum conyzoides* (Compositae) and *Mimosa* sp. (Mimosaceae) showed in their rhizospheres the highest culturable PS fungi, 16% and 29%, respectively, while *Iresine herstii* (Amarantaceae) and *Baccharis* sp. (Compositae) showed the highest culturable PS bacteria, 12.6% and 12%, respectively. The culturable microbial rhizospheric counts for the four plants sampled varied from 0–29% for the PS fungi and from 3–13% for the PS bacteria, these results indicate that different plant species from the same type of soil dwell different culturable microbial rhizospheric phosphate populations probably related to their different rhizospheric exudates. It has been shown by Chiarini et al. (1998), that soil type influenced bacterial community structure of maize roots, and that plant development selectively influenced some bacterial groups present in the rhizosphere, such as pseudomonas, fast growing bacteria (r-strategists) and slow growing bacteria (K-strategists). The fungi PS populations found for most of the sampled colonizer plants at the rhizosphere level may indicate benefits that plants obtain from these rhizosphere populations. For instance, besides many biologically active substances produced by microorganisms, PS populations could act as both nutrient releasers from the phosphoric rock and as soil builders. Phosphate dissolving substances such as organic acids released by fungi (Reyes et al. 2002) may chelate metals from the phosphate granules causing its dissolution. Moreover, the fungal hyphae-penetrating action may also, as roots do, disaggregate cracked coarse fragments of rock. Banfield et al. (1999) showed that the fungal hyphae contributed to the physical weathering, by penetrating mineral cleavages and grain boundaries, and the exposure of the crystals' to microbial colonization. Fungi has also the ability to traverse distances more easily than bacteria (Whitelaw 2000) besides the improvement of soil aggregation and quality as reported for the extraradical mycelium of symbiotic fungi (Barea et al. 2002).

Table 1. Physico-chemical properties of bulk soils sampled at Monte Fresco rock phosphate mine

Soil	P avail (ppm)	Ca (ppm)	Mg (ppm)	K (ppm)	Organic matter (%)	pH (1:2)
Unexploited (loam)	7	276	89	92	3.8	4.8
Exploited (clay loam)	27	447	27	54	3.3	5.4

Table 2. Total culturable and phosphate solubilizing bacteria and fungi from the mined soil of the Monte Fresco rock phosphate mine

Microorganism	*Iresine herbstii*	*Ageratum conyzoides*	*Baccharis* sp.	*Mimosa* sp.	R:S[a]
Total fungi ($\times 10^4$)	21.1	10.1	24.6	6.2	15.5/0.4 = 22
Total bacteria ($\times 10^6$)	22.2	50.6	6.6	25.3	26.6/0.6 = 44
Solubilizing fungi ($\times 10^4$)	0.0[b]	1.6	2.4	1.8	1.5/0.1 = 15
Solubilizing bacteria ($\times 10^6$)	2.8	2.0	0.8	0.8	1.6/0.1 = 16

[a]Values for microbial populations of the mined bulk soil are the same than those from Figure 1.
[b]Values from dilutions showed < 100 culturable solubilizing fungi/ml.

Several fungal strains mostly belonging to the common soil genera of *Penicillium* and *Aspergillus* have been reported as plant growth promoters by their P-solubilizing activities (Whitelaw 2000; Reyes et al. 2002).

The original source of most soils and plant P is apatite (Ahn 1993). After Banfield et al. (1999), cells that compose a diverse microbial community are attached to mineral surfaces and polymer-mediated dissolution, transport and/or recrystallization allow rocks to be transformed to soil. Comprehension of the microbial mechanisms of action for apatites solubilization and nutrient-enrichment of P-poor soils may enable the development of more efficient phosphate biofertilizers.

Isolation and characterization of PS strains from the exploited mine soil

From the bulk and rhizosphere soils, several strains of fungi were isolated among them, five isolates of *Penicillium* sp. which all have a close phenotype to *Penicillium rugulosum* IR-94MF1 isolated in 1994 from a different site of the mine (Reyes et al. 1999). The isolated *Pencillium* strains were showing among other isolated fungi, *Aspergillus* sp., *Trichoderma* sp., and other unidentified strains, the highest PS activity. The fact that several *Penicillium* strains have been studied for their PS activity and some have become biofertilizers (Whitelaw 2000), make these isolated strains an important microbial resource. Data in Table 3 shows the highest values for the hydroxyapatite halo when using sucrose rather than dextrose; similarly, ammonium gave the highest values among all fungal strains except ZV-01MF9. The Halo/Colony ratio varied from 1.5 to 3.0 for ammonium and 1.7 to 2.4 for nitrate. The fact to use both ammonium and nitrate renders these strains good competitors with potential as biofertilizers for tropical soils.

Fifteen Gram-negative bacterial PS strains, showing the biggest clarification haloes, were isolated from the rhizosphere of colonizer plants. Evaluations of biochemical and hydroxyapatite solubilization tests showed that all the isolated strains were different but two of them. Results of the strains for the biochemical tests were as follows: 10 citrate (+), 11 nitrate reduction (+), 9 lactose (+), 8 maltose (+), 12 sorbitol (+), 12 inositol (+), 11 fructose (+), 11 mannitol (+), 13 glucose (+) and 12 sucrose (+). Among the bacterial strains, eight were found to be free-living diazotrophs able to grow on the *Azotobacter* medium and five among them were identified as encapsulated *Azotobacter* sp. A further characterization of the *Azotobacter* strains, due to their potential as biofertilizers, showed variations for the hydroxyapatite solubilization metabolism; however, none of the five strains were able to solubilize hydroxyapatite with mannitol and nitrate treatment. Moreover, the strain H01-Scm showed hydroxyapatite solubilization only when using mannitol and sucrose with ammonium (Table 3).

Molecular techniques to discriminate different species and ribotypes such as amplifiedd 16S ribosomal DNA restriction analysis (ARDRA) and random amplified polymorphic DNA (RAPD) have to be used in order to confirm the strains diversity of the PS *Penicillium rugulosum* and *Azotobacter* sp. The ARDRA has been used to demostrate genetic biodiversity of rhizosphere soil isolates (Picard et al. 2000).

Reports, such as the introduction of non-indigenous PS wild and genetically modified strains into a soil–plant system affect differently the microbial colonization of the rhizosphere (Reyes et al. 2002), show that it is a microenvironment subjected to significant microbial variations due to microbial–plant–soil interactions. It has been observed that inoculation of plants with P-solubilizing fungi could encourage the proliferation of other P-solubilizing fungi in the rhizosphere (Whitelaw 2000). In a natural soil environment, where there is a high P-insoluble content as in the mined Monte Fresco soil, it could be expected to find high P-solubilizing microbial populations adapted metabolically through different mechanisms of action to obtain P from sparingly soluble P-sources. The different microbial strains isolated from the rhizosphere of colonizer plants from the Monte Fresco mine showed variations in vitro for their PS activities as a result of the use of different carbohydrate and nitrogen sources, accordingly their expressed mechanisms of action to dissolve phosphates in the rhizosphere would produce a constant P-solubilizing microenvironment, with a concomitant P uptake by plant roots. Moreover, the hydroxyapatite solubilization phenotype diversity that was found demonstrated a range of microbial strategies for PS activities associated to the

Table 3. Hydroxyapatite halo/colony growth ratio of *Penicillium* sp. and *Azotobacter* sp. strains showing the highest solubilizing activity and isolated from Monte Fresco rock phosphate mine soil when using different carbohydrate and nitrogen sources after 7 days of incubation at 28°C

Strain	Carbohydrate	Halo:Colony ratio	
		Ammonium	Nitrate
Penicillium sp.			
IR-94MF1	Dextrose	2.0	1.8
	Sucrose	2.3	2.0
ZV-01MF1	Dextrose	2.1	1.9
	Sucrose	2.6	2.2
ZV-01MF2	Dextrose	2.2	2.1
	Sucrose	3.0	2.4
ZV-01MF4	Dextrose	1.8	1.7
	Sucrose	2.9	1.9
ZV-01MF5	Dextrose	2.0	1.9
	Sucrose	2.8	2.2
ZV-01MF9	Dextrose	1.5	1.8
	Sucrose	2.6	2.2
Azotobacter sp.			
AV01-MF1b	Dextrose	3.2	2.8
	Mannitol	2.5	0.0
	Sucrose	3.7	4.0
AV01-MF5	Dextrose	3.5	4.3
	Mannitol	2.5	0.0
	Sucrose	3.8	4.3
AV01-MF9b	Dextrose	4.9	4.8
	Mannitol	2.5	0.0
	Sucrose	4.5	4.2
H01-Scm	Dextrose	0.0	0.0
	Mannitol	1.1	0.0
	Sucrose	2.2	0.0
H01-Smp	Dextrose	3.1	4.1
	Mannitol	2.7	0.0
	Sucrose	3.5	2.1

availability of nutrients that should be understood in order to develop successful phosphate and nitrogen biofertilizers.

Acknowledgements

All authors are grateful to Douglas Torrelles for his technical assistance in the laboratory and to Dr. Hani Antoun for the critical revision of the manuscript. This research was supported by the Universidad Nacional Experimental del Táchira (UNET) of Venezuela project 02-001-2001.

References

Ahn P M 1993 Tropical soils and fertilizer use. Intermediate tropical agriculture series. Longman Scientific and Technical. Essex, England.

Altomare C, Norvell A W, Björkman T and Harman G E 1999 Solubilization of phosphates and micronutrients by the plant-growth-promoting and biocontrol fungus *Trichoderma harzianum* Rifai 1295–22. Appl Environ Microbiol 65, 2926–2933.

Banfield J F, Barker W W, Welch S A and Taunton A 1999 Biological impact on mineral dissolution: application of the lichen model to understanding mineral weathering in the rhizosphere. Proc Natl Acad Sci USA 38, 3404–3411.

Barea J M, Azcón R and Azcón-Aguilar C 2002 Mycorrhizosphere interactions to improve plant fitness and soil quality. Antonie van Leeuwenkoek 81, 343–351.

Black C A 1965 Methods of soil analysis. Part 1 No. 9 in the series Agronomy. ASA, Inc. Publisher, Madison, Wisconsin, USA.

Bowen G D and Rovira A D 1999 The rhizosphere and its management to improve plant growth. Adv Agron 66, 1–102.

Bray R H and Kurtz L T 1945 Determination of total, organic, and available forms of phosphorus in soils. Soil Sci 59, 439–445.

Chiarini L, Bevivino A, Dalmastri C, Nacamulli C and Tabacchioni S 1998 Influence of plant development, cultivar and soil type on microbial colonization of maize roots. Appl Soil Ecol 8, 11–18.

Fægri A, Torsvik V L and Goksöyr J 1977 Bacterial and fungal activities in soil: separation of bacteria and fungi by a rapid fractionated centrifugation technique. Soil Biol Biochem 9, 105–112.

Gerhardt P, Murray R G E, Wood W A and Krieg N R 1994 Methods for general and molecular bacteriology. ASM, Washington.

Glick B R 1995 The enhancement of plant growth by free-living bacteria. Can J Microbiol 41, 109–117.

Goldstein A H 1995 Recent progress in understanding the molecular genetics and biochemistry of calcium phosphate solubilization by gram negative bacteria. Biol Agric Hort 12, 185–193.

Kucey R M N, Janzen H H and Leggett M E 1989 Microbially mediated increases in plant-available phosphorus. Adv Agron 42, 199–228.

Lawlor K, Knight B P, Barbosa-Jefferson V L, Lane P W, Lilley A K, Paton G I, McGrath S P, O'Flaherty S M and Hirsch P R 2000 Comparison of methods to investigate microbial populations in soils under different agricultural management. FEMS Microbiol Ecol 33, 129–137.

Picard C, Di Cello F, Ventura M, Fani R and Guckert A 2000 Frequency and biodiveersity of 2,4-diacetylphloroglucinol-producing bacteria isolated from the maize rhizosphere at different stages of plant growth. Appl Environ Microbiol 66, 948–955.

Reyes I, Bernier L, Simard R R, Tanguay P and Antoun H 1999 Characteristics of phosphate solubilization by an isolate of a tropical *Pencillium rugulosum* and two UV-induced mutants. FEMS Microbiol Ecol 28, 291–295.

Reyes I, Baziramakenga R, Bernier L and Antoun H 2001 Solubilization of phosphate rocks and minerals by a wild-type strain and two UV-induced mutants of *Penicillium rugulosum*. Soil Biol Biochem 33, 1741–1746.

Reyes I, Bernier L and Antoun H 2002 Rock phosphate solubilization and colonization of maize rhizosphere by wild and genetically modified strains of *Penicillium rugulosum*. Microb. Ecol 44, 39–48.

Tandon H L S, Cescas MP and Tyner E H 1968 An acid-free vanadate–molybdate reagent for the determination of total phosphorus in soils. Soil Sci Soc Am Proc 32, 48–51.

Torsvik V, Goksøyr J and Daae F L 1990 High diversity in DNA of soil bacteria. Appl Environ Microbiol 56, 782–787.

Vassilev N, Fenice M and Federici F 1996 Rock phosphate solubilization with gluconic acid produced by immobilized *Penicillium variabile* P16. Biotechnol Tech 10, 585–588.

Walkley A and Black I A 1934 An examination of degtjareff method for determining soil organic matter and a proposed modification of the chromic acid titration method. Soil Sci 37, 29–37.

Watteau F and Berthelin J 1994 Microbial dissolution of iron and aluminium from soil minerals: efficiency and specificity of hydroxamate siderophores compared to aliphatic acids. Eur J Soil Biol 30, 1–9.

Whitelaw M A 2000 Growth promotion of plants inoculated with phosphate-solubilizing fungi. Adv Agron 69, 99–151.

Microbial solubilization of rock phosphate on media containing agro-industrial wastes and effect of the resulting products on plant growth and P uptake

N. Vassilev[1,]*, A. Medina[2], R. Azcon[2] & M. Vassileva[1]
[1]*Department of Chemical Engineering, Faculty of Sciences, University of Granada, c/Fuentenueva s/n, Granada, 18071, Spain.* [2]*Estacion Experimental del Zaidin, Granada, Spain. Corresponding author**

Received: 31 May 2006

Key words: agro-industrial wastes, fermentations, fungal microorganisms, plant growth, rock phosphate solubilization

Abstract

Four agro-industrial wastes were assayed as substrates for microbial solubilization of rock phosphate (RP). Sugar beet wastes (SB), olive cake (OC) and olive mill wastewaters (OMWW) were treated by *Aspergillus niger*, and dry olive cake (DOC) was treated by *Phanerochaete chrysosporium*. In conditions of solid-state fermentation 46% of SB and 21% of OC were mineralized by *A. niger* while 16% of DOC was mineralized by *P. chrysosporium*. Repeated-batch mode of fermentation was employed for treatment of OMWW by immobilized *A. niger*, which resulted in conversion of 80% of the fermentable sugars. Acidification of all media treated by *A. niger* was registered with a simultaneous solubilization of 59.7% (SB), 42.6% (OC), and 36.4% (OMWW) of the total P present in the RP. The same mechanism of RP solubilization was observed in DOC-based medium inoculated with *P. chrysosporium* but other mechanisms were probably involved during the process. A series of microcosm experiments were then performed in the greenhouse to evaluate the effectiveness of the resulting fermented products. All amendments improved plant growth and P acquisition, which were further enhanced by mycorrhizal inoculation. The level of all studied parameters including the root mycorrhizal colonization depended on the substrate characteristics. The reported biotechnological schemes offer a potential application particularly for degraded soils.

Introduction

Although phosphorus (P) is quite abundant in many soils, it is one of the major plant nutrients limiting plant growth. P is added to soil in the form of phosphate fertilizers, part of, which is utilized by plants but another part rapidly forms insoluble complexes with soil constituents, thus lowering the overall P use efficiency. Therefore, frequent application of soluble forms of inorganic P is needed. However, in practice, as the capacity of soil to bind P is limited, many soils receive P in excess of crop requirements which results in its leaching to the ground water. The runoff from P-loaded soil is accepted as the main factor in eutrophication of natural water reservoirs (Del Campillo et al., 1999). In view of environmental concerns and current developments in sustainability, research efforts are concentrated on elaboration of agro-techniques that involve the use of less expensive, though less bioavailable, sources of plant nutrients such as rock phosphate (RP).

It is accepted that there is no substitute of RP as a source of P. However, and particularly for non-acidic soils, a minimum processing is required before application. Even when the soil

* Fax: +34-958-248992
E-mail: nbvass@yahoo.com

E. Velázquez and C. Rodríguez-Barrueco (eds.),
First International Meeting on Microbial Phosphate Solubilization, 77–84

acidity is below 5.5–6.0 it has been shown that only after 4 years of annual direct application, does RP become as effective as superphosphate (Ghani et al., 1994). It is well established that RP application is not economically feasible, particularly at soil conditions characterized by a high P sorption capacity, low cation exchange capacity, high pH, low rainfall, low organic matter content, and low microbial activity (Simpson et al., 1997). For these reasons, various strategies for RP solubilization have been recently proposed with an increasing emphasis on application of P-solubilizing microorganisms (Rodriguez and Fraga, 1999; Vassilev et al., 2001; Whitelaw, 2000).

A number of *in vitro* studies have shown that bacteria, fungi and actinomycetes are able to liberate phosphate ions from sparingly soluble inorganic P-bearing compounds (Kucey et al., 1989). The P-solubilizing activity is determined by the ability of the microbes to release metabolites such as organic acids, which through their hydroxyl and carboxyl groups chelate the cations bound to phosphate, the latter being converted to soluble forms (Sagoe et al., 1998). In any case, metabolizable C compounds must be applied to the microbes to ensure their growth, organic acid production, and, simultaneously, RP solubilization. On the other hand, the choice of C sources determines the mode of the fermentation processes and the form of application of the microbial cultures. A wide range of carbohydrates, have been tested such as glucose, sucrose, fructose, xylose, starch, etc. (Cerezine et al., 1988). Studies, which include alternative carbohydrate sources such as agro-industrial wastes should also be taken into consideration, bearing in mind the importance of utilizing renewable resources and the simplicity of the cultivation equipment. Processes for RP microbial solubilization based on inexpensive and abundant organic substrates seem to be economically attractive for soils, which are characterized by low contents of organic matter and of soluble P.

The aim of this work was to evaluate solubilization of rock phosphate by fungi utilizing agro-industrial wastes, typical for Southern Spain, such as sugar beet wastes (SB), olive cake (OC), dry olive cake (DOC), and olive mill waste waters (OMWW).

Materials and methods

Microorganisms

The strain of *Aspergillus niger* NB2 used throughout this study was obtained from the Culture Collection of the Institute of Microbiology, Bulgarian Academy of Sciences, and was maintained on potato-dextrose agar slants at 4 °C. It was proved to produce only citric acid on complex substrates (Vassilev et al., 1986) and mineralize lignocellulosic materials (Vassilev et al., 1998). For inoculum preparation, *A. niger* was grown on a slant at 30 °C for 7 days and spores were scraped in sterile distilled water. The spore amount was measured by optical density measurement at 750 nm following calibration between this data and direct haemocytometer counting.

Phanerochaete chrysosporium was obtained from the Culture Collection of the Faculty of Pharmacy, University of Granada, Spain. It was maintained on malt extract plates at 4 °C. *P. chrysosporium* was incubated at 26 °C for 7 days before use as inoculum in fermentation experiments. A spore suspension for inoculation was prepared by dislodging spores from the plate surface in sterile distilled water. The spore number was measured as described for *A. niger*.

Culture media

Sugar beet wastes (SB), olive cake (OC), dry olive cake (DOC), and olive mill wastewaters (OMWW) were used as substrates in the fermentation trials. The characteristics of wastes were determined before fermentation (Table 1). The solid residues were dried in a 60° oven and then ground to pass a 2-mm pore screen. Portions of 15 g of each solid substrate were placed in 250-ml Erlenmeyer flasks. Preliminary studies (data not reported) were carried out to determine the media composition for each waste–microorganism combination using as criteria microbial growth, pH, and acid production. Czapek-Dox mineral salt solution, 40 mL, was added to the treatments with SB and OC, while DOC was mixed with 40 mL distilled water. OMWW was supplemented with $(NH_4)_2SO_4$, 2.5 g L^{-1} and $MgSO_4 \cdot 7H_2O$, 0.2 g L^{-1} and 100 ml of this OMWW-based medium was then

Table 1. Characteristics of sugar beet waste (SB), olive cake (OC[a]), dry olive cake (DOC[a]), and olive mill wastewaters (OMWW[a])

Component	SB	OC	DOC	OMWW
Cellulose (%)	29	24	18	ND
Hemicellulose (%)	23	8	16	ND
Lignin (%)	5	30	26	ND
C_{total} (g kg^{-1} dw)	520	532	464	29[b]
N_{total} (g kg^{-1} dw)	7	9	11	2[b]
P_{total} (g kg^{-1} dw)	0.7	0.8	0.6	0.4[b]

[a]Polyphenol content: 3.3 g kg^{-1}, OC; 2.1 g kg^{-1}, DOC; 6 g L^{-1}, OMWW.
[b](g L^{-1}).

poured into 250-ml Erlenmeyer flasks. Rock phosphate (Morocco fluorapatite, 12.8% P, 1 mm mesh) was added when necessary to all treatments at a rate of 0.75 g per flask. Media were sterilized by autoclaving at 120 °C for 30 min. Spore suspension of *A. niger* (1.2×10^7) and *P. chrysosporium* (2.3×10^6) were carefully spread over the surface of the respective media. OMWW was treated with *A. niger* that had been passively immobilized in polyurethane 0.5-cm^3 cubes. The immobilization system was prepared as described previously (Vassilev et al., 1993) on a medium containing g L^{-1}: glucose, 60; NH_4NO_3, 2.0; KH_2PO_4, 1.0; $MgSO_4 \cdot 7H_2O$, 0.5; yeast extract, 1.0.

Culture conditions

All experiments were carried out in 250-ml Erlenmeyer flasks (in triplicate) with or without RP. Solid-state fermentations were performed with SB, OC (inoculated with *A. niger*), and DOC (inoculated with *P. chrysosporium*) at 30 °C for 20 days. Repeated-batch mode of fermentation was employed in experiments with OMWW treated with immobilized *A. niger* at 30 °C in shaken culture at 200 rpm. In this case the medium was changed every 48 h during five fermentation cycles. Mixtures of waste materials and RP, microbially treated or not, were further used in soil–plant experiments.

Soil–plant experiment

The treatments used in this experiment were as follows: (i) SB/RP: control; SB/RP treated with *A. niger*; (ii) OC/RP: control; OC/RP treated with *A. niger*; (iii) DOC/RP: control; DOC/RP treated with *P. chrysosporium*; (iv) OMWW/RP: control; OMWW/RP treated with immobilized *A. niger*. The fermentation products from treatments (i)–(iv), prepared as described before, were mixed with a steam-sterilized soil–sand mixture (1:1, v/v) and left for equilibration for 4 weeks at room temperature. Topsoil (0–20 cm) from a field of Granada (Spain) province was used. The main soil characteristics were pH 7.5; 8 μg P g^{-1} (Olsen test); organic carbon 0.46%; total N 0.046%. Waste materials and RP mixtures microbially treated or not, were added to the soil at a rate supplying 5 g solid waste in 100 g soil or 5 ml liquid waste in 100 g soil with a corresponding amount of 0.15 g RP in 100 g soil. Three seedlings of *Trifolium repens* were transplanted in each pot (d = 12.2 cm; 500 g capacity; five pots per treatment) inoculated or not with the arbuscular mycorrhizal fungus *Glomus deserticola*. In treatments with DOC treated or not with *P. chrysosporium*, the experimental plant was *Dorycnium pentaphyllum*. All pots received 1 mL (10^8 cells per ml) of *Rhizobium trifoli* suspension. The AM inoculum consisted of 5 g spores, mycelium and mycorrhizal root fragments and was applied to each of the corresponding pots in the bottom of a 5-cm deep hole. The plants were grown in a greenhouse under a day/night cycle of 16/8 h, 21/15 °C, and 50% relative humidity. Water loss was compensated by watering every day after weighing.

Analytical methods

Analyses of fermentation products were carried out by homogenising 4-g sample in 96 mL distilled water. After centrifugation the supernatant was analyzed for citric acid content by the method of Taussky (1949). Medium pH was measured with a glass electrode and titratable acidity (TA) was determined by titrating each sample to pH 7.0 with NaOH. Percentage of citric acid concentration of the total acidity was calculated as described by Tsay and To (1987). Phosphorus content was determined by the molybdo-vanado method described by Lachica et al. (1973). Solubilization productivity was expressed as mg soluble P per kg of waste per hour. Weight loss of lignocellulose during the fermentation processes was calculated on ash content basis according to Kumar and Sign (1990) and presented as a percent of mineralization. Lignin, cellulose, and hemicellulose contents were measured according to the method of Goering and Van Soest (1970). Total carbon was determined by the method of Jackson (1960) and total nitrogen was measured as described by Baethgen and Alley (1989). Total phenolic content in OC, DOC, and OMWW was determined as described by Ribereau-Gayon (1968).

The plants were harvested after 2 months. Shoot dry weight was recorded after drying at 70 °C. Shoot P content was determined by the molybdo-vanado method described by Lachica et al. (1973). The percentage of mycorrhizal root length was determined by microscopic examination of stained root samples (Phillips and Hayman, 1970) using the gridline intersect method of Giovanetti and Mosse (1980).

Results

Fermentation experiments

The results obtained after 20 days of solid-state fermentation with or without RP showed that SB, OC, and DOC proved to be excellent substrates as rapid growth of *A. niger* and *P. chrysosporium* was observed, particularly at the beginning (the first 3–4 days) of the cultivation process (data not shown). The highest final dry biomass was measured in treatments with SB while the lowest mycelia production was registered in flasks with OMWW-based medium treated with immobilized *A. niger*. The slight neutralizing effect should be mentioned when RP was added to the fermentation media without changing significantly the values measured for fungal biomass, pH and titratable acidity. There was a drop in the initial pH of 6.8–7.0, which at the end of the studied period reached its lowest value in treatments with SB (Table 2). The final pH was higher when DOC was used as substrate treated with *P. chrysosporium*. The production of citric acid, independently of the presence of RP, was well pronounced in all treatments with *A. niger* thus accounting for 91, 75, and 68% of the titratable acidity measured at the end of the fermentation processes based on OMWW, SB, and OC, respectively. The percent of substrate mineralization ranged from 16% in experiments with DOC to 80% in OMWW treatments. The concentrations of polyphenols dropped three times in treatments with OC, DOC, and OMWW and reached 1.2 g kg^{-1}, 0.7 g kg^{-1}, and 2 g L^{-1}, respectively.

Under these conditions, *A. niger* and *P. chrysosporium* were able to solubilize the phosphate rock supplemented in the fermentation media. The highest concentration of soluble P was found in SB-based treatments with *A. niger*. However, the solubilization productivity of 1.24 mg P kg^{-1} h^{-1} obtained in this case was lower as compared to 10.6 mg P L^{-1} h^{-1} reached in OMWW-based repeated-batch process with polyurethane foam immobilized *A. niger*. On the other hand, the percentage of soluble P of total P in the RP, 59.7% and 42.6%, was higher in the treatments with SB and OC treated with *A. niger* in conditions of solid-state fermentations as compared to 32.6% and 36.4% in the case of DOC and OMWW treated by *P. chrysosporium* and immobilized *A. niger*, respectively.

The soluble P concentrations measured in the absence of RP in the fermentation media were low ranging from 21 mg kg^{-1} in the solid-state fermentation with DOC to 156 mg L^{-1} in OMWW-based repeated-batch fermentations. These values of soluble P could not be explained by the microbial action and most likely corresponded to the soluble P fraction of each crude waste. The highest total soluble P concentration of 635 mg kg^{-1} was obtained in SB/RP-based solid-state fermentation and the lowest one was 347 mg kg^{-1} in the treatment with DOC/RP mixture.

Soil–plant experiments

Dry matter and P concentration of shoots of non-mycorrhizal and mycorrhizal plants are presented in Table 3. Overall, plant responses to combinations of agroindustrial wastes and RP depended on whether the wastes were previously treated by *A. niger* and *P. chrysosporium*. Shoot dry weight of plants grown in soil amended with pre-treated SB waste and RP was increased more than five times and reached 330 mg per pot, compared with treatment supplemented with untreated SB/RP. Similarly, two-fold and

Table 2. pH, titratable acidity (TA), soluble P, and solubilization productivity of fermentation processes by *A. niger* and *P. chrysosporium* on media containing sugar beet waste (SB), olive cake (OC), dry olive cake (DOC), and olive mill wastewaters (OMWW) supplemented or not with rock phosphate (RP)

Substrate	Microorganism	PH/TA (mmol kg^{-1})	P_{sol}[d] (mg kg^{-1})	Productivity (mg P kg^{-1} h^{-1})
SB	*A. niger*	2.6/64 ± 3	38 ± 0.7	0.08
SB + RP	*A. niger*	2.9/53 ± 1	597 ± 6	1.24
OC	*A. niger*	3.1/20 ± 0.3	79 ± 1.1	0.16
OC + RP	*A. niger*	3.7/13 ± 0.4	426 ± 4.2	0.89
DOC	*P. chrysosporium*	5.6/8 ± 0.1	21 ± 0.2	0.04
DOC + RP	*P. chrysosporium*	5.7/6 ± 0.2	326 ± 2.9	0.68
OMWW[c]	*A. niger*	3.8/91 ± 3.0	156 ± 3.1[a]	3.25[b]
OMWW[c] + RP	*A. niger*	4.1/72 ± 1.8	364 ± 0.8[a]	10.6[b]

[a]mg L^{-1}.
[b]mg P L^{-1} h^{-1}.
[c]Average result after five repeated-batch cycles.
[d]Soluble P concentrations obtained in experiments with and without RP must be sum to calculate the total soluble P.

three-fold plant growth increase was registered in treatments amended with microbially treated OMWW/RP, DOC/RP, and OC/RP, respectively. In treatments supplemented with untreated wastes and RP, the highest shoot P concentrations of 1.75 and 1.2 mg g^{-1} shoot dry weight were found in mycorrhizal and non-mycorrhizal *T. repens*, respectively, grown in presence of OMWW. Increased P plant acquisition was demonstrated in all soil–plant systems amended with previously treated wastes. The highest, eight-fold increase of P concentration, was measured in mycorrhizal plants grown in microbially treated SB/RP-amended soil.

Microscopic observations of plant roots showed that only AM-inoculated plants were root colonized. Taken individually, the presence of *G. deserticola* influenced positively the plant weights and P uptake in shoots in each treatment. The percentage of AM root length colonization was 58% in the treatments with plants grown in soil amended with untreated SB/RP but decreased in presence of untreated OC, DOC, and OMWW. However, this parameter increased

Table 3. Shoot dry weight (DW) and P content of *Trifolium repens* and *Dorycnium pentaphyllum* as affected by fermentation-resulting products and mycorrhizal inoculation

Substrate/ plant	AM	**Mycorrhiz. (%)		*Shoot DW (mg per pot)		*P concen. in shoots (mg g^{-1} shoot DW)	
		C	+TW	C	+TW	C	+TW
SB/	+	58a	42b	60	330	0.50	4.1
T. repens	–			10	34	0.41	1.5
OC/	+	41ab	51ab	44	141	0.53	2.02
T. repens	–			41	110	0.30	1.78
DOC/	+	38b	71a	280	520	0.20	0.8
D. pentaph.	–			70	290	0.07	0.6
OMWW/	+	32b	48b	110	250	1.75	3.68
T. repens	–			80	170	1.20	3.41

C: soil enriched with untreated waste and RP.
+TW: C + microbially treated RP-enriched waste.
*LSD < 0.5.
**Within each column, means followed by the same letters are not significantly different ($P < 0.05$) using Duncan's multiple range tests.

in all treatments supplemented with microbially treated olive oil wastes and RP except in SB/RP-amended soil where an 18% decrease was measured.

Discussion

This work has proved possible RP solubilization in conditions of solid-state fermentation, on media containing solid agro-industrial wastes. In general, lignocellulosic materials do not give good yields of organic acids without some pre-treatment because of the slow rate of hydrolysis and low level of available sugars. However, the microorganisms used in these experiments are known to have lignocellulolytic activity and particularly *P. chrysosporium* is characterized by its high lignolytic enzyme activity (Bastawde, 1992; Kerem and Hadar, 1993). Therefore, SB, OC, and DOC were mineralized successfully in fermentation systems by *A. niger* and *P. chrysosporium*. It is important to note that in all treatments with *A. niger*, a high value of acidity (mainly citric acid) was measured, while the titratable acidity in the treatment with *P. chrysosporium* was only 6 mmol kg^{-1}. Recently, *P. chrysosporium* was reported to produce low-molecular weight organic acids thus lowering pH outside of the fungal hyphae (Makela et al., 2002). In addition, some metal chelating compounds released by white-rot fungi, including *P. chrysosporium*, have been reported to take part in wood mineralization (Milagres et al., 2002).

The overall effect of the extracellular compounds with their properties of chelators is the most likely reason for some RP solubilization by *P. chrysosporium* although other mechanisms could be also involved bearing in mind that the above mentioned processes occur in the early stage of fungal development and the amounts of the released compounds is low. However, this work demonstrated that, although *P. chrysosporium* is not a typical organic acid producer, it can be used efficiently in RP solubilization providing a percentage of soluble P at least comparable to that obtained by *A. niger*.

The microbial solubilization of RP based on the liquid agro-industrial waste, OMWW, demonstrated the highest system productivity of 10.6 mg P L^{-1} h^{-1}. The most likely explanation is the application of immobilized living cells during the repeated-batch fermentation. Immobilization is known to prevent shear stresses, typical for submerged operations, thus ensuring higher metabolic activity, catalytic longevity, and stability (Vassilev and Vassileva, 1992). The main advantage of solubilizing RP by immobilized microbial cells is that for a period of one batch cycle (48 h), the immobilized system provides an amount of soluble P equal or higher that that obtained by free fungal cultures in conditions of solid-state and even conventional submerged process.

Both *A. niger* and *P. chrysosporium* were able to decrease the concentration of total phenols. Filamentous fungi and white-rot fungi are known to significantly lower the concentration of phenolic compounds in residues of olive oil extraction processes (Blanquez et al., 2002; D'Annibale et al., 2003), which was confirmed in this work.

Our results from microcosm experiments confirmed the important role of mycorrhizal fungi in P uptake, particularly in combination with P-solubilizing microorganisms. By the use of isotopic ^{32}P dilution technique, we have recently reported that mycorrhizal plants benefited from P solubilized from RP by *A. niger* (Vassilev et al., 2002).

The mycorrhizal development in this work depended on the nature of the waste material. Root mycorrhizal colonization in control plants, amended with untreated mixtures of OC, DOC, OMWW, and RP was lower than that measured in soil amended with microbially treated substrate/RP mixtures. The addition of *A. niger*-treated SB/RP resulted in lower percentage of AM root length colonization compared with the treatment amended with untreated SB/RP. It is now well established that cellulose stimulates mycorrhizal development in certain conditions (Gryndler et al., 2002). SB (as a cellulose carrier) stimulated the mycorrhizal colonization of the experimental plants when added untreated. However, lowering of this parameter was observed in case of treated SB/RP amendment, which could be caused partly by the substrate mineralization and/or by the presence of soluble P in the fermented product.

A different trend was observed in all experiments based on OC, DOC, and OMWW. Here, the lower root mycorrhization in the control soil–plant systems could be explained by the presence of polyphenol compounds which are well known antimicrobial and phytotoxic agents

(Rodriguez et al., 1988). Information regarding the effect of residues of olive oil extraction processes on AM development is scarce. Recently, Martin et al. (2002) have reported that the application of untreated DOC decreased the percentage of AM colonization of plants. The obtained data showed that AM fungi increased the phytotoxicity of DOC in soybean and lettuce. Our results with olive oil wastes confirm these statements which were more pronounced particularly in the case of plants amended with untreated OMWW where the highest concentration of polyphenols of 6 g L^{-1} was the most likely reason for the lowest value of AM colonization compared with all other treatments amended with untreated wastes. Under these conditions, however, *T. repens* was able to accept high concentrations of P, which was due to the high concentration of plant available soluble P (156 mg L^{-1}) in the crude OMWW. The results reported here indicated that the detoxification of olive oil wastes by *A. niger* and *P. chrysosporium* increased the level of mycorrhiza formation along with all effects related to plant growth and P uptake. The experiment with DOC/RP, previously treated with *P. chrysosporium* should be mentioned as it provided the lowest both total polyphenol and total soluble P concentrations which resulted in a high percentage of root length AM colonization compared with the control treatments. In this case the P concentration of 0.8 mg g^{-1} shoot dry weight was the lowest one among all treatments amended with treated wastes which is, however, most likely caused by the dilution (high plant growth) effect.

Because of the different physical and chemical waste characteristics it would be difficult to strictly compare their effect on RP solubilization and further on plant growth and P uptake. However, some conclusions can be drawn:

1. *Aspergillus niger* and *Phanerochaete chrysosporium* were able to grow and simultaneously solubilize the phosphate rock independently of the mode of fermentation.
2. The process of RP solubilization was related to the release of organic acids, mainly citric acid, particularly in the treatments with *A. niger*. In cultures with *P. chrysosporium* the process of RP dissolution depended on the same mechanism during the first stage of fermentation but other mechanisms could be later involved.
3. The concentration of phenolic compounds in residues of olive oil extraction process decreased in microbially treated OC, DOC, and OMWW.
4. Microbially treated wastes and RP enhanced plant growth and P uptake. Inoculation with *G. deserticola* appeared to have an important role in these processes by increasing the overall positive effect of the introduced amendments.
5. The level of root mycorrhization depended on the substrate characteristics and its microbial treatment.

Further studies should be carried out to determine the effect of untreated and microbially treated wastes and the corresponding microorganisms solubilizers on the soil physical, chemical, microbiological, and biochemical quality in greenhouse and natural conditions.

References

Baethgen W E and Alley M M 1989 A manual colorimetric procedure for measuring ammonium nitrogen in soil and plant Kjeldahl digests. Comm. Soil Sci. Plant Anal. 20, 961–969.

Bastawde K B 1992 Cellulolytic enzymes of a thermotolerant *Aspergillus terreus* strain and their action on cellulosic substrates. W. J. Microbiol. Biotechnol. 8, 45–49.

Blanquez P, Caminal G, Sarra M, Vicent M T and Gabarell X 2002 Olive oil mill waste waters decoloration and detoxification in a bioreactor by white rot fungus *Phanerochaete flavido-alba*. Biotechnol. Prog. 18: 660–662.

Cerezine P C, Nahas E and Banzatto D A 1988 Soluble phosphate accumulation by *Aspergillus niger* from fluorapatite. Appl. Microbiol. Biotechnol. 29, 501–505.

D'Annibale A, Federici F, Fenice M, Sermanni G, Petruccioli M and Vassilev N 2003 Residues of olive oil extraction process: Possible biotechnological approaches. RichMac Magazine March: 1–4.

Del Campillo M C, Van Der Zee S and Torrent J 1999 Modelling long-term phosphorus leaching and changes in phosphorus fertility in excessively fertilized acid sandy soils. Eur. J. Soil Sci. 50, 391–399.

Ghani A, Rajan S S S and Lee A 1994 Enhancement of rock phosphate solubility through biological processes. Soil Biol. Biochem. 26, 127–136.

Giovannetti M and Mosse B 1980 An evaluation of techniques for measuring vesicular–arbuscular mycorrhizal infection in roots. New Phytol. 84, 489–500.

Goering H K and Van Soest P J 1970 Forage Fiber Analysis. Agriculture Handbook 379. Agricultural Research Service, USDA, Washington DC, pp. 1–12.

Gryndler M, Vosatka M, Hrselova H, Chvatalova I and Jansa J 2002 Interaction between arbuscular mycorrhizal fungi and cellulose in growth substrate. Appl. Soil Ecol. 19, 279–288.

Jackson M L 1960 Chemical Analysis. Prentice-Hill, Englewood Cliffs, NJ.

Kerem Z and Hadar Y 1993 Effect of manganese on lignin degradation by *Pleurotus ostreatus* during solid-state fermentation. Appl. Environ. Microbiol. 59, 4115–4120.

Kucey R M N, Jansen H H and Leggett M E 1989 Microbially mediated increases in plant-available phosphorus. Adv. Agron. 42, 199–228.

Kumar N and Sign K 1990 Chemical and microbiological changes during solid substrate fermentation of wheat straw with *Coprinus fimetatus*. Biol. Wastes 33, 231–242.

Lachica M, Aguilar A and Yanez J 1973 Metodos analiticos en la Estacion Experimental del Zaidin. Anal. Edad. Agrobiol. 32, 1033–1047.

Makela M, Galkin S, Hatakka A and Lundell T 2002 Production of organic acids and oxalate decarboxylase in lignin-degrading white-rot fungi. Enz. Microb. Technol. 30, 542–549.

Martin J, Sampedro I, Garcia-Romera I, Garcia-Garrido J M and Ocampo J A 2002 Arbuscular mycorrhizal colonization and growth of soybean (*Glycine max*) and lettuce (*Lactuc sativa*) and phytotoxic effects of olive mill residues. Soil Biol. Biochem. 34, 1769–1775.

Milagres A M F, Arantes V, Medeiros C L and Machuca A 2002 Production of metal chelating compounds by white and brown-rot fungi and their comparative abilities for pulp bleaching. Enz. Microb. Technol. 30, 562–565.

Phillips J M and Hayman D S 1970 Improved procedure for clearing roots and staining parasitic and vesicular–arbuscular mycorrhizal fungi for rapid assessment of infection. Trans. Br. Mycol. Soc. 55, 159–161.

Ribereau-Gayon P 1968 Les composes phenoliques des vegetaux. Dumond, Paris.

Rodriguez H and Fraga R 1999 Phosphate solubilizing bacteria and their role in plant growth promotion. Biotechnol. Adv. 17, 319–339.

Rodriguez M, Perez J, Ramos-Cormenzana A and Martinez J 1988 Effect of extracts obtained from olive oil mill waste waters on *Bacillus megaterium* ATCC 333085. J. Appl. Bacteriol. 64, 219–226.

Sagoe C I, Ando T, Kouno K and Nagaoka T 1998 Relative importance of protons and solution calcium concentration in phosphate rock dissolution by organic acids. Soil Sci. Plant Nutr. 44, 617–625.

Simpson P G, Sale P W G and Tennakoon SB 1997 An economic analysis of the field performance of North Carolina reactive phosphate rock compared with single superphosphate for selected sites from the National Reactive Phosphate Rock Project. Austr. J. Exp. Agric. 37, 1061–1076.

Taussky H H 1949 A microcolorimetric method for the determination of citric acid. J. Biol. Chem. 181, 195–198.

Tsay S S and To K Y 1987 Citric acid production using immobilized conidia of *Aspergillus niger* TMB 2022. Biotechnol. Bioeng. 29, 297–304.

Vassilev N and Vassileva M 1992 Production of organic acids by immobilized filamentous fungi. Mycol. Res 96, 563–570.

Vassilev N, Vassileva M and Ganchev I 1986 Citric acid production by *Aspergillus niger* on starch hydrolysate media. Acta Microbiol. Bulg. 18, 62–67.

Vassilev N, Vassileva M and Spassova D 1993 Production of gluconic acid by *Aspergillus niger* immobilized on polyurethane foam. Appl. Microbiol. Biotechnol. 39, 285–288.

Vassilev N, Vassileva M, Azcon R and Barea JM 2002 The use of ^{32}P dilution techniques to evaluate the effect of mycorrhizal inoculation on plant uptake of P from products of fermentation mixtures including agrowastes, *Aspergillus niger* and rock phosphate. *In* Assessment of Soil Phosphorus Status and Management of Phosphoric Fertilizers to Optimise Crop Production. Ed. F Sikora. International Atomic Energy Agency, TECDOC-1272, Vienna, Austria.

Vassilev N, Vassileva M, Fenice M and Federici F 2001 Immobilized cell technology applied in solubilization of insoluble inorganic (rock) phosphates and P plant acquisition. Biores. Technol. 79, 263–271.

Vassilev N, Baca M T, Vassileva M, Franco I, Azcon R and De Nobili M 1998 Mineralization of three agro-industrial wastes by an acid-producing strain of *Aspergillus niger*. *In* The Science of Composting. Eds. M de Bertoldi, P Sequi, B Lemmes and T Papi. pp. 1375–1379. Chapman and Hall, London.

Whitelaw M A 2000 Growth promotion of plants inoculated with phosphate-solubilizing fungi. Adv. Agron. 69, 99–151.

Making microorganisms mobilize soil phosphorus

A. E. Richardson*
CSIRO Plant Industry, PO Box 1600, Canberra ACT 2601, Australia

Received 21 December 2002. Accepted in revised form 2 January 2003

Key words: microorganisms, soil, phosphorus, mineralization, solubilization, immobilization

Abstract

Microorganisms are involved in a range of processes that affect the transformation of soil phosphorus (P) and are thus an important component of the soil P cycle. In particular, soil microorganisms are effective in releasing P from inorganic and organic pools of total soil P through solubilization and mineralization. The microbial biomass in soil also contains a significant quantity of immobilized P that is potentially available to plants. Microorganisms therefore are critical for the transfer of P from poorly available soil pools to plant available forms and are important for maintaining P in readily available pools. These processes are likely to be most significant in the rhizosphere of plants.

Introduction

Microorganisms are an integral component of the soil phosphorus (P) cycle and are important for the transfer of P between different pools of soil P. Consequently, there has been longstanding interest in the manipulation of soil microorganisms to improve the P nutrition of plants, with the objective of increasing the overall efficiency of P-use in agricultural systems. This interest stems from the fact that P deficiency is widespread on soils throughout the world, that P fertilizer represents a major cost for agricultural production and that the efficiency of P-use by plants from soil and fertilizer sources is poor. Furthermore, P is a finite resource and, based on current rate of use, it is expected that the worlds' known reserves of high-quality rock phosphate will be depleted within the current century (Isherwood, 2000). Beyond this time, the production of phosphate-based fertilizers will require the processing of lower-grade rock phosphates at significantly higher cost. Alternatively, the direct use of rock phosphates as fertilizers will require an effective means for solubilization. These issues are particularly relevant to soils throughout developing countries and on acidic soils in tropical and subtropical regions (Hedley et al., 1995). It is also imperative that management of P fertilizers in agricultural environments is improved (particularly in more highly P fertilized environments) so that any adverse environmental effects due to P losses are minimized (Tunney et al., 1997).

The concept of using soil microorganisms to improve mobilization of poorly available forms of soil P is not new. It is now over 50 years since Gerretsen (1948) first showed that pure cultures of soil bacteria could increase the P nutrition of plants through increased solubility of Ca-phosphates. A large volume of literature has since been published and a great deal has been promised. However, it is fair to say that not much has been delivered. Clearly, microbial-plant interactions in soil environments are complex and, with few exceptions, have proven difficult to manipulate (Richardson, 2001). Therefore, the challenge remains. Indeed, opportunities for exploiting soil microorganisms for P-mobilization are improved as knowledge of the processes and understanding

* FAX No: +61-2-6246-5000.
E-mail: alan.richardson@csiro.au

E. Velázquez and C. Rodríguez-Barrueco (eds.),
First International Meeting on Microbial Phosphate Solubilization, 85–90

of the ecology of microorganisms in soil environments is gained. Such opportunities are further enhanced with the advent of new techniques. These include the possibility for direct manipulation of organisms through gene technology. In this paper, recent issues concerning the mobilization of soil P by microorganisms are summarized and some opportunities for the future are discussed.

Phosphorus mobilization by soil microorganisms

Microorganisms directly affect the ability of plants to acquire P from soil through a number of structural or process-mediated mechanisms. These include, (i) an increase in the surface area of roots by either an extension of existing root systems (e.g., mycorrhizal associations) or by enhancement of root branching and root hair development (i.e., growth stimulation through phytohormones), (ii) by displacement of sorption equilibria that results in increased net transfer of phosphate ions into soil solution or an increase in the mobility of organic forms of P and (iii) through stimulation of metabolic processes that are effective in directly solubilizing and mineralizing P from poorly available forms of inorganic and organic P (Figure 1). These processes include the excretion of hydrogen ions, the release of organic anions, the production of siderophores and the production of phosphatase enzymes that are able to hydrolyse soil organic P. In particular, organic anions and associated protons are effective in solubilizing precipitated forms of soil P (e.g., Fe- and Al-P in acid soils, Ca-P in alkaline soils), chelating metal ions that may be associated with complexed forms of P or may facilitate the release of adsorbed P through ligand exchange reactions (Jones, 1998).

However, distinction between the roles of microbial processes, in comparison to direct effects of plant mechanisms, on P mobilization in soil is poorly understood. It is well established that plant roots effectively increase P acquisition through modified root growth and architecture and similarly produce metabolites that directly influence P availability (Raghothama, 1999). Processes such as rhizosphere acidification, exudation of organic anions and secretion of phosphatases from plant roots occur in response to P deficiency, and are established mechanisms by which plants acquire P (Randall et al., 2001; Hinsinger et al., 2005; Richardson et al., 2005). Furthermore, it has been suggested that microbial-mediated processes on their own may be insignificant in soil environments, and are unlikely to mobilize sufficient P for plant requirements (Tinker, 1980). This argument remains to be resolved. On the other hand, the importance of the microbial biomass for P cycling in soil and the potential of this P to contribute to plant P nutrition is more difficult to deny (Jakobsen et al., 2005).

Soil microbial biomass phosphorus and contribution to plant nutrition

The microbial biomass in soil contains a significant amount of P (typically 10–50 kg P/ha, but as high as 100 kg P/ha) and generally accounts for 2–5% of the total P and around 10–15% of the soil organic P. Importantly, microbial P is a

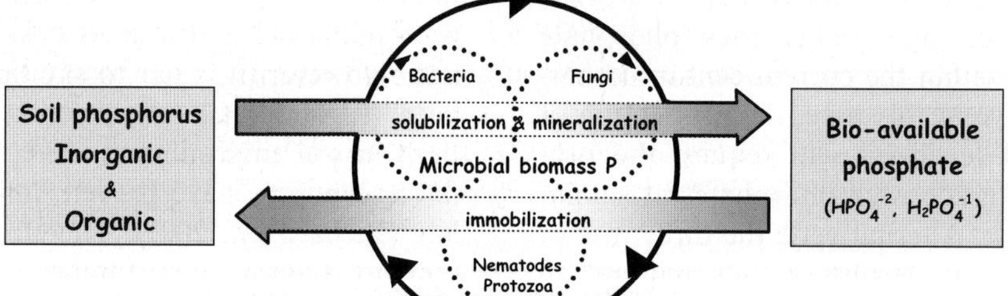

Figure 1. Schematic representation of soil phosphorus mobilization by microorganisms.

dynamic component of the soil P cycle and is responsive to soil fertility, seasonal conditions and management practices (Richardson, 1994; Oberson and Joner, 2005). Whilst the P content of microbial biomass may vary considerably in relation to microbial C, it is evident that significant pools are maintained even in soils considered to be P deficient for plant growth (Oberson et al., 2001). This indicates that microorganisms in soil are highly efficient in acquiring P to meet their own requirements. In addition, it has been shown that soil microorganisms are capable of rapidly assimilating P supplied from fertilizer or as plant residues. For instance, McLaughlin et al. (1988) showed that some 25% of P in labelled crop residues was incorporated into microbial biomass within 7 days.

A number of studies have highlighted the potential importance of microbial P in providing available P to plants. Seasonal dynamics indicate that significant amounts of P are released from the biomass in response to soil moisture deficiency and it is estimated that soil microbial P is completely turned over at least annually (He et al., 1997). More recent studies have found that the rate of P-flux through the microbial biomass is potentially much greater (Oehl et al., 2001; Oberson et al., 2001). Incubation studies using labelled phosphate have shown both a rapid incorporation of P into biomass (within 2–3 days) and concomitant release of the P back to soil solution. Importantly, these transfers of P occurred in the absence of any significant changes in the size of the microbial P pool. Highest rates of P cycling through the biomass were evident in P-deficient soil and in soils that received organic inputs, as distinct from those that were P-fertilized. The capacity of the microbial biomass to immobilize P was also increased by the provision of soluble C, which resulted in an increase in both the size of the microbial P pool and its rate of turnover.

These observations have important implications concerning the contribution of microbial P to plant nutrition. First, the significance of P immobilization within the soil microflora and its effect on the 'short-term' availability of P to plants is not clear. Likewise, processes that affect the release of P from the microbial biomass and its subsequent availability to plants require further investigation. Although P in microorganisms occurs predominantly in organic forms (or as polyphosphates), the P appears to be rapidly mineralized and is readily available for uptake by plant roots (Macklon et al., 1997). However, in soil environments the availability of released P will be influenced by spatial and temporal factors and will also be subject to further immobilization (by both soil micro and macroflora and fauna) and other physico-chemical reactions of P in soil. The actual contribution that P-turnover through microorganisms makes to the mobilization of soil P therefore remains to be fully determined. Nevertheless, it is well known that soil P is significantly depleted in close proximity to roots (Hinsinger et al., 2005; Richardson et al., 2005), that roots release significant quantities of C that is available to soil microorganisms and that microbial populations in the rhizosphere are enhanced by many orders of magnitude (Bowen and Rovira, 1999). Therefore, the potential for turnover of P by rhizosphere microorganisms is substantial, and further work needs to be undertaken to quantify it in terms of enhancing plant P nutrition.

Using microorganisms to improve soil phosphorus availability

Recognition that microorganisms are important for P mobilization in soil has led to research effort directed at improving plant P nutrition. Essentially, there are two major strategies for manipulating soil microorganisms.

Management of existing microbial populations to optimize their capacity to mobilize P

Success with this approach requires detailed knowledge of how soil management practices (e.g., crop rotations, soil amendments, cultivation, etc.) impact on microbial abundance, diversity and presence of various functional groups and how these relate to the magnitude and availability of different soil P fractions. The manipulation of VA mycorrhizas in soil through crop rotation is one example of how populations might be managed to increase the availability of soil P to plants (Thompson, 1994). Increased mineralization of organic P generally occurs in response to soil cultivation and crop rotation has been shown to increase the rate of P cycling through the microbial biomass. For example, incorporation of organic residues through legume rotation resulted in higher biological activity and increased microbial P uptake and release (Oberson et al., 2001). Although the contribution of P released through these processes needs to be evaluated in relation to plant uptake, such observations indicate that management opportunities do exist for increasing the cycling of P and its maintenance in plant-available pools. Elucidation as to whether or not the availability of this P can be synchronized with plant requirements, or be targeted to the rhizosphere, remains a significant challenge.

The use of specific microbial inoculants to increase P mobilization

A range of soil microorganisms able to solubilize precipitated forms of P or mineralize organic P has been characterized. Typically, such organisms have been isolated using cultural procedures, with species of *Pseudomonas* and *Bacillus* bacteria and *Aspergillus* and *Penicillium* fungi being predominant (Jakobsen et al., 2005). These organisms are commonly associated with the rhizosphere and, when inoculated onto plants, often result in improved growth and P nutrition with responses being observed under both glasshouse and field conditions (see reviews by Kucey et al., 1989; Rodríguez and Frago, 1999; Whitelaw, 2000). Despite this, there are few examples of successful application of microbial inoculants. Essentially, a lack of consistent performance under different environmental conditions in the field has precluded their wider use. A number of factors can be identified to explain this variable performance (Richardson, 2001). They include (i) poor understanding of the actual mechanisms involved in plant growth promotion where, in fact, P mobilization may not necessarily be the primary mechanism involved, (ii) selection of microorganisms by laboratory screening may be insufficiently rigorous when organisms are required to mobilize P in soil environments, (iii) the apparent lack of any specific association between phosphate solubilizing microorganisms and host plants, (iv) poor understanding of interactions between physical and chemical characteristics of soil and how these interact with biological P availability, (v) poor knowledge of how to deliver microorganisms into soil environments and of how to establish them as dominant components of complex microbial communities and, in particular, of their capability of colonizing the rhizosphere and (vi) in most instances the benefits of microbial mobilization of P may in fact be indirect. In short, whilst microorganisms may directly solubilize P to meet their own requirements, subsequent benefits to plants may only occur following turnover of the microbial biomass.

It is evident therefore that the proposition for developing inoculants for routine application remains problematic and a number of issues still need to be addressed. Nevertheless, some microorganisms show consistent plant growth promotion under glasshouse and field conditions and have been developed as commercial inoculants (e.g., *Penicillium* spp., Leggett et al., 2001). Although growth promotion in such cases is generally associated with increased plant P nutrition, it is difficult in many cases to ascertain whether increased P-mobilization is either the cause or the consequence of the response (Wakelin et al., 2004). Similarly, many other organisms (including mixed populations of soil bacteria and fungi, that often are only poorly characterized) have been promoted as commercial inoculants with claims that they increase plant growth through P-mobilization. Unfortunately, in most cases detailed evidence to support these claims is rarely provided.

Prospects for enhancing phosphorus mobilization by soil microorganisms

There seems little doubt that soil microorganisms are essential for the cycling of P in terrestrial ecosystems and as such, play an important role either directly or indirectly in mediating phosphate availability to plants. However, attempts to capitalize on microbial processes to increase plant access to P from soil and/or fertilizer sources have generally been met with limited success. It is reasonable to assume that future opportunities will increase as understanding of the processes of P mobilization and the ecology of microorganisms in soil environments improves. Development of novel techniques and access to new technologies will be important. Recent developments in microbial community analysis that do not rely on cultural procedures will provide better understanding of how microorganisms interact in complex environments. For example, procedures are now available for detection and direct visualization of specific microorganisms in the rhizosphere (Watt et al., 2006).

Opportunity also exists for genetic manipulation of soil microorganisms. It is now a reality that gene technologies can be used to enhance specific traits that may increase an organism's capacity to mobilize soil P directly, enhance its ability to colonize the rhizosphere (i.e., rhizosphere competence, Lugtenberg et al., 2001), or perhaps to form specific associations with plant roots (Bowen and Rovira, 1999). Identifying microbial traits that are associated with P mobilization is an important step in this regard and subsequent isolation and manipulation of candidate genes is required. Over-expression of microbial phosphatase genes, mineral phosphate solubilisation (*mps*) genes and genes that are directly associated with organic acid biosynthesis (e.g., citrate synthase, phosphoenol pyruvate carboxylase) are examples (Gyaneshwar et al., 2002). Alternatively, microorganisms may provide a novel source of genes for directly modifying plants. For instance, it has been reported that, when expressed in roots, a bacterial citrate synthase gene increases the exudation of organic anions and significantly improves plant access to soil P (López-Bucio et al., 2000). The reproducibility of this approach has however been questioned (Delhaize et al., 2001). The ability of plants to use P from phytate, which is a predominant form of organic P in most soils, has been shown to be dependent on the presence of soil microorganisms and utilisation of phytate-P was significantly improved when an *Aspergillus* phytase gene is expressed directly in plant roots (Richardson et al., 2001a, b). However, plant responses in soil have been found to be less consistent and largely dependent on soil type (George et al., 2004; George et al., 2005). Nevertheless these examples highlight that new opportunities for direct manipulation of microorganisms and/or plants do exist. However, it is also important to recognize that any approach to increase the mobilization of soil P through genetic modification of soil microorganisms or plants will also need to satisfy a range of community and environmental issues.

The promise of exploiting soil microorganisms to increase mobilization of soil P remains. Whether or not this will be achieved through better management of soil microbial communities, by development of more effective microbial inoculants, through the genetic manipulation of specific organisms, or with a combination of these approaches is not known. What is clear though is that soil microorganisms play an important role in the mobilization of soil P and that detailed understanding of their contribution to the cycling of P in soil-plant systems is required for the development of sustainable agriculture.

References

Bowen G D and Rovira A D 1999 The rhizosphere and its management to improve plant growth. Adv. Agron. 66, 1–102.

Delhaize E, Hebb D M and Ryan P R 2001 Expression of a *Pseudomonas aeruginosa* citrate synthase gene in tobacco is not associated with either citrate accumulation or efflux. Plant Physiol. 125, 2059–2067.

George T S, Simpson R J, Hadobas P A and Richardson A E 2004 Expression of a fungal phytase gene in *Nicotiana tobacum* improves phosphorus nutrition in plants grown in amended soil. Plant Biotech. J. 3, 129–140.

George T S, Richardson A E, Smith J B, Hadobas P A and Simpson R J 2005 Limitations to the potential of transgenic *Trifolium subterraneum* L. plants that exude phytase, when grown in soils with a range of organic P content. Plant Soil 258, 263–274.

Gerretsen F C 1948 The influence of microorganisms on the phosphate intake by the plant. Plant Soil 1, 51–81.

Gyaneshwar P, Naresh Kumar G, Parekh L J and Poole P S 2002 Role of soil microorganisms in improving P nutrition of plants. Plant Soil 245, 83–93.

He Z L, Wu J, O'Donnell A G and Syers J K 1997 Seasonal responses in microbial biomass carbon, phosphorus and sulphur in soils under pasture. Biol. Fert. Soils 24, 421–428.

Hedley M J, Mortvedt J J, Bolan N S and Syers J K 1995 Phosphorus fertility management in agroecosystems. *In* Phosphorus in the Global Environment: Transfers, Cycles and Management. Ed. H Tiessen. pp. 59–92. Wiley, Chichester.

Hinsinger P, Gobran G R, Gregory P J and Wenzel W W 2005 Rhizosphere geometry and heterogeneity arising from root-mediated physical and chemical processes. New Phytol. 168, 293–303.

Isherwood K F 2000 Mineral Fertilizer Use and the Environment. International Fertilizer Industry Association/United Nations Environment Programme, Paris.

Jakobsen I, Leggett M E and Richardson A E 2005 Rhizosphere microorganisms and plant phosphorus uptake. *In* Phosphorus, Agriculture and the Environment. Eds. J T Sims and A N Sharpley. pp. 437–494. American Society for Agronomy, Madison.

Jones D L 1998 Organic acids in the rhizosphere – a critical review. Plant Soil 205, 25–44.

Kucey R M N, Janzen H and Leggett M E 1989 Microbial mediated increases in plant-available phosphorus. Adv. Agron. 42, 199–228.

Leggett M, Glendive S and Holloway G 2001 Phosphate-solubilising microorganisms and their use. *In* Plant Nutrient Acquisition: New Perspectives. Eds. N Ae, J Arihara, K Okada and A Srinivasan. pp. 299–318. Springer-Verlag, Tokyo.

López-Bucio J, de la Vega O M, Guevara-García A and Herrera-Estrella L 2000 Enhanced phosphorus uptake in transgenic tobacco plants that overproduce citrate. Nat. Biotech. 18, 450–453.

Lugtenberg B J J, Dekkers L and Bloemberg G V 2001 Molecular determinants of rhizosphere colonization by *Pseudomonas*. Ann. Rev. Phytopathol. 39, 461–490.

Macklon A E S, Grayston S J, Shand C A, Sim A, Sellars S and Ord B G 1997 Uptake and transport of phosphorus by *Agrostis capillaris* seedlings from rapidly hydrolysed organic sources extracted from ^{32}P-labelled bacterial cultures. Plant Soil 190, 163–167.

McLaughlin M J, Alston A M and Martin J K 1988 Phosphorus cycling in wheat-pasture rotations. II. The role of the microbial biomass in phosphorus cycling. Aust. J. Soil Res. 26, 333–342.

Oberson A, Friesen D K, Rao I M, Bühler S and Frossard E 2001 Phosphorus transformations in an oxisol under contrasting land-use systems: the role of the microbial biomass. Plant Soil 237, 197–210.

Oberson A and Joner E J 2005 Microbial turnover of phosphorus in soil. *In* Organic Phosphorus in the Environment. Eds. B L Turner, E Frossard and D S Baldwin. pp. 133–164. CAB International, Wallingford.

Oehl F, Oberson A, Probst M, Fliessbach A, Roth H R and Frossard E 2001 Kinetics of microbial phosphorus uptake in cultivated soils. Biol. Fert. Soils 34, 32–41.

Raghothama K G 1999 Phosphate acquisition. Ann. Rev. Plant Physiol. Plant Mol. Biol. 50, 665–693.

Randall P J, Hayes J E, Hocking P J and Richardson A E 2001 Root exudates in phosphorus acquisition by plants. *In* Plant Nutrient Acquisition: New Perspectives. Eds. N Ae, J Arihara, K Okada and A Srinivasan. pp. 71–100. Springer-Verlag, Tokyo.

Richardson A E 1994 Soil microorganisms and phosphorus availability. *In* Soil Biota: Management in Sustainable Farming Systems. Eds. C E Pankhurst, B M Doube, V S S R Gupta and P R Grace. pp. 50–62. CSIRO, Australia.

Richardson A E 2001 Prospects for using soil microorganisms to improve the acquisition of phosphorus by plants. Aust. J. Plant Physiol. 28, 897–906.

Richardson A E, Hadobas P A, Hayes J E, O'Hara C P and Simpson R J 2001a Utilization of phosphorus by pasture plants supplied with *myo*-inositol hexaphosphate is enhanced by the presence of soil microorganisms. Plant Soil 229, 47–56.

Richardson A E, Hadobas P A and Hayes J E 2001b Extracellular secretion of *Aspergillus* phytase from *Arabidopsis* roots enables plants to obtain phosphorus from phytate. Plant J. 25, 641–649.

Richardson A E, George T S, Hens M and Simpson RJ (2005) Utilisation of soil organic phosphorus by higher plants. *In* Organic Phosphorus in the Environment. Eds. B L Turner, E Frossard and D S Baldwin. pp. 165–184. CAB International, Wallingford.

Rodríguez H and Frago R 1999 Phosphate solubilizing bacteria and their role in plant growth promotion. Biotechnol. Adv. 17, 319–339.

Thompson J P 1994 What is the potential for management of mycorrhizas in agriculture? *In* Management of Mycorrhizas in Agriculture, Horticulture and Forestry. Eds. A D Robson, L K Abbott and N Malajczuk. pp. 191–200. Kluwer Academic Publishers, Dordrecht.

Tinker P B 1980 The role of rhizosphere microorganisms in mediating phosphorus uptake by plants. *In* The Role of Phosphorus in Agriculture. Eds. F E Kwasenah, E C Sample and E J Kamprath. pp. 617–654. American Society of Agronomy, Madison.

Tunney H, Carton O T, Brookes P C and Johnston A E 1997 Phosphorus Loss From Soil to Water. CAB International, Oxon.

Wakelin S A, Warren R A and Ryder M H 2004 Effect of soil properties on growth promotion of wheat by *Penicillium radicum*. Aust. J. Soil Res. 42, 897–904.

Watt M, Hugenholtz P, White R, Vinall K 2006 Numbers and locations of native bacteria on field grown wheat roots quantified by fluorescence *in situ* hybridization (FISH). Environ. Microbiol. 8, 871–884.

Whitelaw M 2000 Growth promotion of plants inoculated with phosphate-solubilizing fungi. Adv. Agron. 69, 99–151.

Future trends in research on microbial phosphate solubilization: one hundred years of insolubility

Alan H. Goldstein*
Center for Biomaterials Research, Alfred University, 2 Pine Street , Alfred, New York, 14802, USA

Received 14 October 2002. Accepted in revised form 2 January 2003

Key words: glucose dehydrogenase, gluconic acid, mineral phosphate

Abstract

For over one hundred years, agricultural microbiologists and soil ecologists have studied the ability of a wide range of soil microorganisms to dissolve poorly soluble mineral phosphates. In the case of calcium phosphates, a significant body of evidence has been developed to show that Gram negative bacteria exhibiting superior mineral phosphate solubilizing (MPS) capabilities utilize the direct oxidase pathway. This pathway (also called nonphosphorylating oxidation) produces gluconic acid and 2-ketogluconic acid directly in the periplasmic space. These strong organic acids can dissolve poorly soluble calcium phosphates such as hydroxyapatite and rock phosphate ore (e.g. fluroapatite). Therefore, we propose that the conservation of the direct oxidation pathway in rhizobacteria may, at least in part, result from the mutualistic advantage provided by the MPS trait. This article contains a brief literature review, some examples of ongoing work in our lab and, finally, a proposal for a unified terminology for the classification of microorganisms capable of solubilizing or mobilizing P in the soil or other ecosystems.

Introduction

Soils are often high in insoluble mineral phosphates but deficient in the soluble orthophosphate (Pi) essential for the growth of most plants and microorganisms (Goldstein 1986; Sperber 1957; Sundara Rao and Sinha 1963; Tinker 1980). In agricultural crop production, phosphorous is second only to nitrogen in importance as a fertilizer amendment so that phosphorus fertilizers are the world's second largest bulk agricultural chemical and, therefore, the second most widely applied chemical on Earth (Banik and Dey, 1982; Goldstein 2000; Goldstein et al. 1993). There is a broad spectrum of mineral phosphate chemistries; but in arid to semiarid soils the predominant forms are the calcium phosphates (Hausenbuiller 1972). Calcium phosphates are soluble to varying degrees in the presence of the wide array of organic acids produced by microorganisms. Other biosolubilization mechanisms exist as well, so that conversion of mineral phosphates to Pi is generically attributed to microorganisms in most representations of global P cycling. With respect to plant growth, some workers have postulated that associations between plant roots and mineral phosphate solubilizing (MPS) microorganisms could play an important role in phosphorus nutrition in many natural and agroecosystems (Bagyaraj et al., 2000; Gerretsen, 1948; Goldstein, 1986). As a result, an enormous amount of research has been conducted over the last hundred years involving isolation and characterization of mineral phosphate solubilizing microorganisms from many soils with the goal of developing P biofertilizers that would accomplish much the same function

* FAX No: +607-871-2354.
E-mail: fgoldste@alfred.edu

E. Velázquez and C. Rodríguez-Barrueco (eds.),
First International Meeting on Microbial Phosphate Solubilization, 91–96

as biological nitrogen fixation. To date, the results of these efforts have been problematic (c.f. Agnihotri, 1970; Asea et al., 1988; Illmer and Schinner, 1992; Kucey et al., 1989; Richardson, 1994; Sperber, 1957). Our observation that, in Gram-negative rhizobacteria, extracellular oxidation of glucose to gluconic acid and 2-ketogluconic acid via the direct oxidation pathway provides the biochemical basis for highly efficacious calcium phosphate solubilization and may, in fact, be the basis for the evolution of mutualistic plant–bacteria relationships in some phosphate-limited soil ecosystems (Babu-Khan et al., 1995; Goldstein, 1994, 1995; Goldstein and Liu, 1987; Goldstein et al., 1999).

Materials and methods

Our methods for screening for the quinoprotein glucose dehydrogenase are well documented (Babu-Khan et al., 1995; Liu et al., 1992). The basic experimental protocols for 'trapping' either apoglucose dehydrogenase or pqq genes are summarized in Figure 1. The Mineral Phosphate Solubilizing (MPS^+) phenotype is usually identified by the production of clearing zones of solubilization as shown in Figure 2.

Results and discussion

Soils are often high in insoluble mineral and organic phosphates but deficient in available orthophosphate (Pi; Goldstein 1995; Richardson 1994; Sundara Rao and Sinha 1963; Tinker 1980). Phosphorous is second only to nitrogen as an essential nutrient for plant growth and development. Soil amendment with fertilizer P, produced via chemical processing of rock phosphate ore, is an absolute requirement in order to feed the world's population. For over one hundred years workers have recognized the ability of soil microorganisms to solubilize Pi from insoluble (i.e. nutritionally unavailable) organic and mineral phosphates (Gerretsen, 1948; Goldstein 1986; Sundara Rao and Sinha 1963). Molecular characterization of the biochemical/genetic pathways of mirobial P solubilization will enhance our understanding of phosphate cycling in natural ecosystems and make an essential contribution to the development of the renewable agricultural practices necessary for the long term environmental stability of our planet (Goldstein, 1994, 1995).

A wide range of microbial biosolubilization mechanisms exist, so that much of the global cycling of insoluble organic and inorganic soil phosphates is attributed to bacteria and fungi (Banik and Dey, 1982; Sperber, 1957; Sundara Rao and Sinha, 1963; Tinker, 1980). The genetic and biochemical mechanisms for this solubilization are as varied as the spectrum of P-containing soil compounds. High molecular weight organic phosphates are ubiquitous, whereas soil type and pH will control the chemistry of insoluble mineral phosphate precipitates. Large amounts of iron or aluminium phosphates are often found in acidic soils whereas, in arid to semiarid soils, calcium phosphates (CaPs) predominate. The limiting levels of Pi in most soils provides the ecophysiological basis for positing associations between plant roots and mineral phosphate solubilizing (MPS) and/or organic P solubilizing microorganisms. These associations are assumed to play an important role in phosphorus nutrition in many natural and agro-ecosystems. As a result, an enormous amount of research has been conducted involving isolation and characterization of MPS and organic P solubilizing microorganisms from a wide range of soils. In general, the goals have been to understand P cycling and/or to develop P biofertilizers analagous to biological nitrogen fixation. To date the results of these efforts have been problematic.

A general discussion of the full range of microbial mineral and organic phosphate biosolubilization is beyond the scope of this presentation, so the author will focus on calcium phosphates as a model (and widely applicable) system. Calcium phosphates may be dissolved by a wide array of microbe-generated organic acids. By comparison with organic phosphates, the molecular genetic and biochemical bases of bacterial transformations of poorly soluble CaPs has been difficult to resolve into a systematic field of study. The reason may be the simplicity of the solubilization mechanism. Calcium phosphates are dissolved by acidification. Therefore, any bacterium that acidifies its external medium will show some level of MPS activity. In most soils, proton substitution reactions are driven by

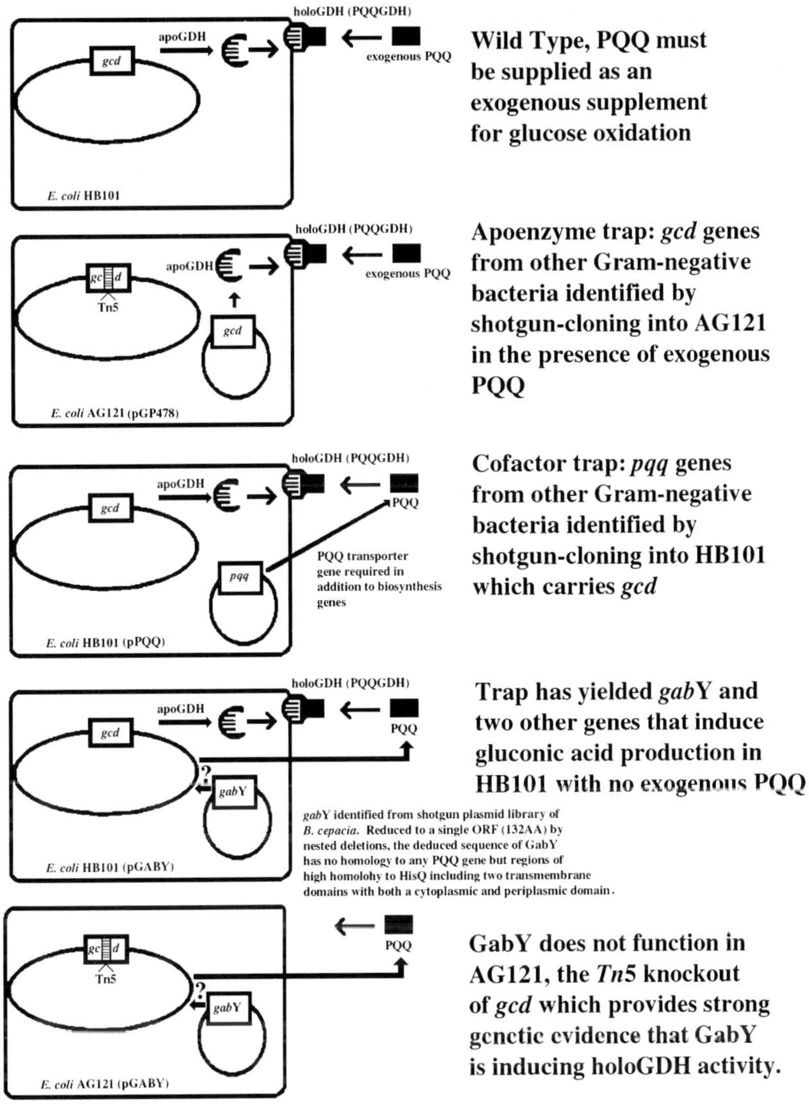

Figure 1. Generalized scheme for 'trapping' apoglucose dehydrogenase or pqq genes. Other genes such as *gab*Y have also been 'trapped'. As previously noted, *E. coli* does not produce sufficient acid to solubilize highly insoluble mineral phosphates such as tri-calcium phosphate (TCP). However, since *E. coli* produces apoGDH constitutively, cloning of *pqq* genes into wild type *E. coli* or cloning apoGDH genes (designated *gcd* in *E. coli*) into an *E. coli gcd* knockout (e.g. AG121,) in the presence of exogenous PQQ will produce the typical clearing zone shown in Figure 2.

microbial production of organic acids; represented generically by the Equation (1) (Goldstein, 2000):

$$[Ca^{2+}]_n[PO_4^{3-}]_m + [HA] = [H^+][PO_4{}^{3-}] + [Ca^{2+}][A^-] \quad (1)$$

There is no stoichiometry in Equation (1) because of the complexity of CaP chemistry and the multiplicity of naturally occurring organic acids (HAs) with differing numbers of dissociable protons. It may be seen that any number of organic acids (HAs) may be substituted into Equation (1) with varying efficacies depending on the solubility of the CaP, the number of acidic protons, and pK_a(s) of the organic acid. Therefore, one is immediately confronted with the question of how to investigate the MPS phenomenon in a way that can determine if some bacteria produce acids as a strategy to make Pi bioavailable. In

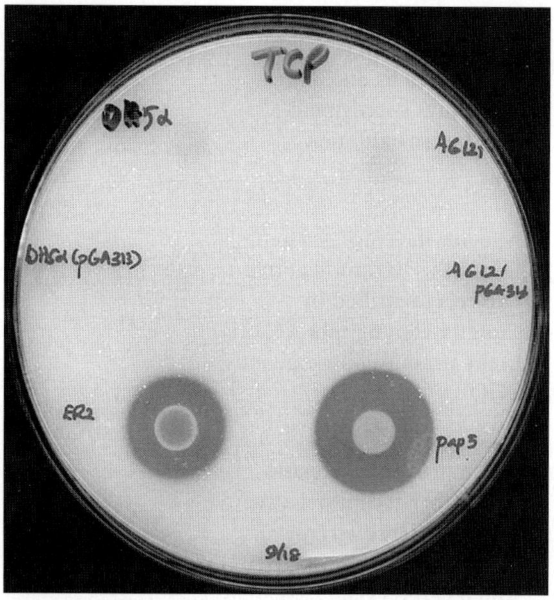

Figure 2. Classic MPS+ phenotype expressed by *Serratia marcescens* ER2 (endorhizosphere isolate) or *Gluconoacetobacter oxydans* PAP3. Note that none of the *E. coli* strains (DH5a, AG121, DH5a[pGA313], AG121 [pGA313]) can create this zone. Cloning of *pqq* genes into DH5a or *gcd* into Ag121 (with exogenous PQQ) would create an equivalent clearing zone.

that case the organic acid production could be considered a *phenotype*, defined as a 'characteristic of the organism'. Alternatively, the proximity of organic acid-producing bacteria and CaPs within the soil will result in some solubilization. This could be called a MPS *effect*, as in 'to cause Pi to come into being'. As previously mentioned, an enormous amount of applied microbiology research has been conducted in this field.

It is generally accepted that the most common mechanism for the MPS phenomenon with respect to calcium phosphates is the acidification of the medium via biosynthesis and release of organic acids. However, the multiplicity of acid-generating metabolic pathways available to bacteria has made it impossible to develop a unified approach to the microbiology of CaP solubilization such as is available for the study of N_2 fixation; where all pathways must converge at some variation of the nitrogenase system. However, literally thousands of Gram-negative rhizobacterial isolates have been screened for the MPS trait on minimal CaP medium with glucose as the carbon source. When using these screening criteria, isolates expressing the direct oxidation pathway (Duine, 1991) far surpass others in their ability to dissolve calcium phosphates. These bacteria have been designated as having a MPS^+ phenotype (Goldstein, 1986, 1995 Goldstein et al., 1993). MPS^+ bacteria can dissolve highly insoluble phosphates such as rock phosphate ore (RPO, flourapatite) because of the extremely low pK_a(s) of the glucose oxidation products; gluconic acid, and 2-ketogluconic acid (pK_a(s) of ~3.4 and ~2.6 respectively). In addition, since these acids are produced in the periplasmic space, protons are efficiently released into the extracellular medium, or rhizosphere space *in vivo*. Superior MPS^+ strains not only have direct oxidation genes but express this metabolic pathway at a high level so that there is a direct correlation between acid production and CaP dissolution. It is also of interest to note that Pi starvation can induce the direct oxidation pathway in some strains. Therefore, a thorough review of the literature and twenty years of directed studies leads this author to the conclusion that, in many ecosystems, a relationship exists between highly efficacious Gram-negative MPS^+ bacteria and the expression of the direct oxidation pathway. The question of rhizosphere mutualism and even symbiosis between efficacious MPS bacteria and higher plants remains to be clarified (c.f. Goldstein et al., 1999).

Direct oxidation (a.k.a. nonphosphorylating oxidation) is one of the four major metabolic pathways for glucose (aldose) utilization by bacteria. For many bacterial species, the direct oxidation pathway is the primary mechanism for aldose sugar utilization (Duine, 1991). The first oxidation is catalyzed by the quinoprotein glucose dehydrogenase; so named because it is a member of the group of bacterial enzymes that utilize the redox cofactor PQQ (2,7,9-tricarboxyl-1H-pyrrolo[2,3-f]quinoline-4,5-dione). PQQGDH is a membrane-bound dehydrogenase whose catalytic domain is located on the outer face of the cytoplasmic membrane. This enzyme transfers electrons from aldose sugars directly to ubiquinone in the plasma membrane via two electron, two proton oxidations mediated by the cofactor PQQ. Direct oxidation of glucose to gluconic acid generates a transmembrane proton motive force (PMF) that may be used for bioenergetic and/or membrane transport functions while the dissociable proton of gluconic acid is available for CaP solubilization. MPS^+ bacteria usually carry out the second periplasmic oxidation of gluconic acid to 2-ketogluconic acid via gluconate dehydrogenase.

We have proposed that the direct oxidation pathway is the metabolic basis for the superior MPS^+ phenotype in Gram-negative bacteria (Goldstein et al., 1993). This provides workers with both a unifying metabolic strategy, and a set of biochemical and genetic probes with which to systematically identify and evaluate the role of a specific subpopulation of rhizosphere bacteria in P cycling. In addition, the demonstrated efficacy of the direct oxidation pathway for the dissolution of fluroapatites has provided a potential strategy for large-scale bioprocessing of rock phosphate ores. In terms of future research, successful molecular cloning of direct oxidation pathway genes provides the tools with which to study relationships between the population dynamics of MPS^+ bacteria in the rhizosphere and Pi in the soil solution. Available probes include several apo-glucose dehydrogenase genes, PQQ biosynthesis genes, and at least one gluconate dehydrogenase gene. We have used such genetic probes to help demonstrate the presence of unique populations of Gram-negative MPS^+ rhizobacteria in two alkaline desert soil environments where the levels of poorly soluble calcium phosphates are extremely high but Pi is undetectable in bulk soil extracts. Both MPS^+ populations were capable of high levels of direct oxidation of glucose, and the presence of the quinoprotein glucose dehydrogenase was confirmed by both enzymatic and molecular biology assays. In one case, a unique rhizobacterial population of *Enterobacter cloacae* expressed the direct oxidation pathway only in the presence of compounds washed from the root of the host plant, *Helianthus* sp. providing preliminary evidence for mutualism in this highly alkaline soil (pH 10; Goldstein et al., 1999).

With respect to agriculture, bioprocessing of rock phosphate ore (RPO) to inorganic phosphate may provide an energy efficient, environmentally desirable alternative to current technology for industrial P fertilizer production (Goldstein et al., 1993, 2000). Bioconversion occurs at a low temperature and is more selective to phosphate extraction than conventional processes. This increased selectivity of attack may reduce the solubilization of undesirable ore contaminants such as radionuclides and toxic metals. The process uses carbohydrate as an energy/proton source as opposed to the 'wet-acid' chemical process that uses concentrated sulfuric acid. The bioprocessing of phosphate ore is not as sensitive to ore quality as are conventional approaches. This may allow lower grade ore deposits and tailings, not presently of any value, to be used. Finally, appropriate formulation may allow the bioprocess to be utilized for *in situ* bioconversion of RPO in the soil or even specifically in the rhizosphere of plant roots. Currently, we are implementing bioprocess engineering strategies that involve both conventional fermentation technology and advanced methods such as directed evolution to produce a system that can compete with the economics of conventional wet-acid phosphate facilities. In our approach, glucose produced via biomass conversion (an intermediate step in ethanol production) is used as feedstock for direct oxidation-mediated biosolubilization of rock phosphate ore. Looking towards the future it is reasonable to propose that, using the tools of biotechnology, biophosphorous fertilization is an achievable goal that lends itself well to the global imperative of sustainable agricultural production.

Acknowledgement

This work has been supported, in part, by the Fierer Endowment to AHG.

References

Agnihotri V P 1970 Solubilization of insoluble phosphates by some soil fungi isolated from nursery seed beds. Can. J. Microbiol. 16, 877–880.

Asea P., Kucy R M N and Stewart J W B 1988 Inorganic phosphate solubilization by two *Penicillium* species in solution culture and soil. Soil. Biol. Biochem. 20, 459–464.

Babu-Khan S, Yeo T C, Martin W L, Duron M D, Rogers R D and Goldstein A H 1995 Cloning of a mineral phosphate solubilizing gene from *Pseudomonas cepacia*. Appl. Env. Microbiol. 61, 972–978.

Bagyaraj D J, Krishnaraj P U and Khanuja S P S 2000 Mineral phosphate solubilization: agronomic implications, mechanism and molecular genetics. Proc Indian Nat. Sci. Acad. (PINSA) 66, 69–82.

Banik S and Dey B K 1982 Available phosphate content of an alluvial soil as influenced by inoculation of some isolated phosphate solubilizing microorganisms. Plant Soil 69, 353–364.

Duine J A 1991 Quinoproteins: enzymes containing the quinoid cofactor pyrroloquinoline quinone, topaquinone or tryptophan-tryptophan quinone. Eur. J. Biochem. 200, 271–284.

Gerretsen F C 1948 The influence of microorganisms on the phosphate intake by the plant. Plant Soil 1, 51–81.

Goldstein A H 1986 Bacterial mineral phosphate solubilization: Historical perspective and future prospects. Am. J. Altern. Agric. 1, 57–65.

Goldstein A H and Liu S T 1987 Molecular cloning and regulation of a mineral phosphate solubilizing gene from *Erwinia herbicola*. Bio/Technology 5, 72–74.

Goldstein A H, Rogers R D and Mead G 1993 Separating phosphate from ores via bioprocessing. Bio/Technology 11, 1250–1254.

Goldstein A H 1994 Involvement of the quinoprotein glucose dehydrogenase in the solubilization of exogenous phosphates by gram negative bacteria. *In* Phosphate in microorganisms: cellular and molecular biology. Eds. A Torriani-Gorini, E Yagil and S Silver. pp. 197–203. ASM Press, Washington, D.C.

Goldstein A H 1995 Recent progress in understanding the molecular genetics and biochemistry of calcium phosphate solubilization by gram negative bacteria. Biol. Agric. Horticult. 12, 185–193.

Goldstein A H, Braverman K E and Osorio N 1999 Evidence for mutualism between a plant growing in a phosphate-limited desert environment and a mineral phosphate solubilizing (MPS) rhizobacteria. FEMS Microbiol. Ecol. 30, 295–300.

Goldstein A H 2000 Bioprocessing of rock phosphate ore: essential technical considerations for the development of a successful commercial technology. Proceedings of the 4th International Fertilizer Association Technical Conference. IFA, Paris.

Hausenbuiller R L 1972 Soil Science, Principles and Practices. Wm. C. Brown Company, Dubuque, Iowa, 269 p.

Illmer P and Schinner F 1992 Solubilization of inorganic phosphates by microorganisms isolated from forest soils. Soil Biol. Biochem. 24, 389–395.

Kucey R W N, Tanzen H H and Leggett M E 1989 Microbially mediated increases in plant available phosphorus. Adv. Agron. 42, 199–228.

Liu S-T, Lee L-Y, Tai C-Y, Horng C-H, Chang Y-S, Wolfram J H, Rogers R D and Goldstein A H 1992 Cloning of an *Erwinia herbicola* gene necessary for gluconic acid production and enhanced mineral phosphate solubilization in *Escherichia coli* HB101. J. Bacteriol. 174, 5814–5819.

Richardson A E 1994 Soil microoganisms and phosphorus availability. Soil Biota 50, 35–39.

Sperber J I 1957 Solution of mineral phosphates by soil bacteria. Nature. 180, 994–995.

Sundara Rao W B and Sinha M K 1963 Phosphate dissolving microorganisms in the soil and rhizosphere. Indian J. Agric. Sci. 33, 272–278.

Tinker P B 1980 The role of rhizosphere microorganisms in phosphorus uptake by plants. *In* The role of rhizosphere microorganism in phosphorus uptake by plants. Eds. Khasawneh F E, et al. pp. 617–647. American Society for Agronomy Press, Madison.

Molecular methods for biodiversity analysis of phosphate solubilizing microorganisms (PSM)

A. Peix[1], E. Velázquez[2] & E. Martínez-Molina[2,3]
[1]*Departamento de Producción Vegetal, IRNA-CSIC, Salamanca, Spain.* [2]*Departamento de Microbiología y Genética, Edificio Departamental de Biología, Universidad de Salamanca, 37007, Salamanca, Spain.* [3]*Corresponding author**

Received 20 November 2002. Accepted in revised form 2 January 2003

Key words: biodiversity analysis, DNA, fingerprint, phosphate solubilizing microorganisms, RNA

Abstract

Although the phosphate solubilizing potential is not a very common characteristic among microorganisms, phosphate solubilizers belonging to diverse groups of microorganisms, especially bacteria, are known. The ecological role of these microorganisms in soil is very important, as they take part in the biogeochemical cycles of the main nutrient elements in the ecosystems. Thus, it is necessary to study the composition and dynamics of these microbial populations to reach a better understanding of soil microbial diversity and nutrient uptake by plants. The study of populations of microorganisms, which share the common characteristic of phosphate solubilization has great complexity, because they belong to very diverse groups sometimes not closely related under a phylogenetic point of view. Therefore good techniques are needed to perform the analysis and identification of PSM populations. The molecular techniques based on nucleic acid composition are excellent tools for this purpose, as they are precise, reproducible and not dependent on culture media composition or growth phase of microorganisms. In this paper main molecular methods based on electrophoresis of nucleic acids as LMW RNA profiling and PCR-based techniques specially DNA involving approaches are reviewed and discussed, highlighting the main advantages and drawbacks of the different methods.

Molecular techniques for microbial population studies

Although the phosphate solubilizing potential is not a very common characteristic among microorganisms, at present phosphate solubilizers belonging to diverse groups of microorganisms, especially bacteria, are known.

From the bacteria, most of the known PS species belong to pseudomonads, bacilli and rhizobia groups, although there are other Genera such as *Acinetobacter* or *Enterobacter*, which have been described as phosphate solubilizers.

With the new phylogenetic classification of bacteria, the outlines of which are recorded in Bergey's Manual (http://www.cme.msu.edu/bergeys/), the pseudomonads group has been divided into diverse Genera able to solubilize phosphate in plates, and these Genera have not only been separated from the original Genus but are also to be found in different subclasses of Proteobacteria. For example, *Burkholderia* (*Pseudomonas*) *cepacia* has been included in β-Proteobacteria. The same has occurred with rhizobia, which have been divided into several families. In this way, for example *Mesorhizobium* (*Rhizobium*) *mediterraneum* has been separated from family *Rhizobiaceae* to family *Phyllobacteriaceae*. Within bacilli, *Paenibacillus* (*Bacillus*) *polymyxa*

* FAX No: +34-923224876.
E-mail: emm@usal.es

E. Velázquez and C. Rodríguez-Barrueco (eds.),
First International Meeting on Microbial Phosphate Solubilization, 97–100

has been included in the new family *Paenibacillaceae*. Within eukaryotes, although they have not been widely studied, the phosphate solubilization by fungi of the Genera *Penicillium* and *Aspergillus* (Hermosa, pers. comm.), as well as by some yeast as *Yarrowia lypolytica* is known. Therefore, it is necessary to have techniques, which allow the analysis of biodiversity of PS populations and the identification of the different isolates.

Nowadays, techniques based on nucleic acids electrophoresis are the most commonly used for the analysis of microbial populations. Of these, the most used is DNA, although certain molecules of RNA have been proposed as molecular fingerprints of microorganisms, which may be applied to the study of microbial populations (Höfle, 1988). These molecules have been named low molecular weight RNA (LMW RNA) and they include the 5S rRNA and tRNA in prokaryotes (both in archaea and bacteria). In eukaryotes, besides these molecules, the 5.8S rRNA forms part of the LMW RNA profiles. This differentiating characteristic between prokaryotes and eukaryotes is very useful when analyzing isolates from complex populations, since in the eukaryotes there is always one molecule more (5.8S rRNA) which, besides, has different sizes for each eukaryotic genus, when a voltage gradient electrophoresis is used for separating these molecules, when they have similar sizes (Cruz-Sánchez et al., 1997).

The results obtained up to now have allowed researchers to establish that LMW RNA molecules separated by staircase electrophoresis are molecular images of each microbial species both in the case of eukaryotes and prokaryotes (Velázquez et al., 2001 a,b). According to the results obtained so far, each prokaryote genus has a characteristic 5S rRNA zone and each eukaryote genus has a different combined zone of 5S and 5.8S rRNA. On the other hand tRNA profiles, both class 1 and class 2, are different in each species of prokaryotes and eukaryotes belonging either to the same or different genus. It has also been demonstrated that LMW RNA profiles of all the strains belonging to the same species are identical (Velázquez et al., 2001b). Hence LMW RNA profiles are an excellent technique for analyzing population diversity as they can be applied to a large number of strains in a quick, easy and precise way without intra-specific variations.

In spite of all this, for the time being, DNA is far more widely used in taxonomy and biodiversity studies, especially due to the fact that PCR is more and more available and affordable and analyses can be carried out in almost every laboratory.

Techniques based on PCR

In spite of all this, for the time being, DNA is far more widely used in taxonomy and biodiversity studies, especially due to the fact that PCR is more and more available and affordable and analyses can be carried out in almost every laboratory.

There are numerous PCR-based techniques that can be applied to the study of biodiversity. Among them, some are based on sequencing of reaction products but techniques of this kind are more useful in taxonomy than in studies of large populations, as sequencing is still too complex a method to be applied to many strains at the same time. For this reason, other techniques have been developed in order to obtain the same results without the need of gene sequencing. Such techniques are known as DGGE and TGGE which are based on amplification of a G + C hyper-variable content zone which allows the separation of bands of the same size (generally in 16S rDNA) based on their G + C content obtained by means of variation of temperature during electrophoresis in polyacrylamide gels through a time or spatial gradient (Muyzer, 1999). The fingerprints obtained through these techniques do not give direct taxonomic information, for which later analysis have to be done to assigne the population fingerprints, and this is the main limit.

Other techniques are based on the amplification of certain molecules followed by digestion with several restriction enzymes, obtaining RFLP profiles which can be mathematically analysed to establish clusters (Marsh, 1999; Osborn et al., 2000). Nevertheless, the usefulness of RFLP in taxonomy has not been well established because are dependent on the type of molecule and restriction enzymes used. While DGGE and TGGE have been applied to prokaryotes, for the time being the RFLP profiling has also been widely applied to eukaryotes and therefore can be a very useful technique in biodiversity studies (Osborn et al., 2000). However, it has the disadvantage that the taxonomic level which this technique can differentiate is not well established since, depending on the zone where restriction is applied it may show differences at genus level in some groups, at species level in others and even at strain level in the case of ITS1 and ITS2 in fungi. ITS1 and ITS2 profiles are used a lot in eukaryotes because they are very variable, especially in different genera allowing the establishment of differences directly according to the size of the bands amplified with the same primers. The amplified ribosomal DNA restriction analysis (ARDRA) has been succesfully applied to ecological population studies, and has the advantage of the taxonomic resolution at species level (Heyndrickx et al., 1996).

Finally, other techniques are based on direct electrophoresis of amplified fragments by PCR. In these cases the difference in the various procedures usually depends on the type of primer used. Sometimes small primers and low annealing temperatures are used; these lead to random amplifications within the microbial genome thus obtaining RAPD profiles (Di Cello et al., 1997). Intra-specific variations have been described in the RAPD both in bacteria and in fungi and therefore this technique is of great use in biodiversity studies, but at an infra-specific level. In other cases one or two primers designed on a basis of repeated sequences in the genome of the microorganisms are used. This is the case of Box-PCR, which uses one single primer and of ERIC-PCR (Hulton et al., 1991; Versalovic et al., 1994) and rep-PCR, which employ two primers in order to obtain DNA profiles (Versalovic et al., 1994). In all these cases intra-specific variations have also been observed, so the usefulness of these techniques is very similar to that of RAPD. Another technique is the rDNA internal-spacer analysis (RISA), based on the amplification region of the spacer region located between the 16S and the 23S rRNA genes. It is a very good tool for population fingerprinting, and allows resolution at species level. It may be useful for strain characterization, as it shows great variability in very closely related taxonomic groups (Nagpal et al., 1998).

Recently a new procedure has been described (Rivas et al., 2001) for obtaining DNA profiles that are effective in biodiversity and taxonomy since the variations observed are established at subspecies level within the same species. This method is based on the use of the two universal primers used to make a complete amplification of the 16S rDNA increasing its concentration about ten times and applying an annealing temperature of 50–55 °C.

Some techniques for analyzing biodiversity are based on extra-chromosomal material, which, in some cases, involves a high percentage of the genetic material of the bacteria, above all in the case of fast-growing rhizobia. The plasmid profiles analysed by horizontal electrophoresis in agarose gels are a useful tool for the analysis of population diversity that is usually applied in the case of rhizobia (Hartmann and Amarger, 1991). The advantages of this technique is that plasmids are the easiest nucleic acid molecule to study, their size differentiate among strains, and plasmid DNA is easily extracted from cells. The main drawbacks are the lack of stability in some strains, the plasmid transfer among strains and the fact that some strains do not contain plasmids.

As well as nucleic acids, proteins are also normally used in microbial biodiversity studies although they have the disadvantage of the fact that the culture conditions always have to be the same in order to avoid the existence of variations in the electrophoretic profiles (SDS-PAGE) (Lowry et al., 1951) which have their origin in differences in culture media composition or in incubation conditions (temperature, time, etc.). The same can be said of the LPS profiles (Tsai and Frasch, 1982), which are useful in biodiversity studies whenever they are analysed under the same culture conditions in all the strains.

Therefore, at present a wide range of molecular tools is available for biodiversity analysis that facilitate the study at different taxonomic levels which can be applied to large populations in order to establish clusters of different microorganisms which is of great help in the later identification of representative strains from the different groups obtained.

References

Cruz-Sánchez J M, Velázquez E, Mateos P and Martínez-Molina E 1997 Enhancement of resolution of low molecular weight RNA profiles by staircase electrophoresis. Electrophoresis 18, 1909–1911.

Di Cello F, Bevivino A, Chiarini L, Fani R, Paffetti D, Tabacchioni S and Dalmastri C 1997 Biodiversity of a *Burkholderia cepacia* population isolated from the maize rhizosphere at different plant growth stages. Appl. Environ. Microbiol. 63, 4485–4493.

Hartmann A and Amarger N 1991 Genotypic diversity of an indigenous *Rhizobium meliloti* field population assessed by plasmid profiles, DNA fingerprinting, and insertion sequence typing. Can. J. Microbiol. 37, 600–608.

Heyndrickx M, Vauterin L, Vandamme P, Kersters K and de Vos P 1996 Applicability of combined amplified ribosomal DNA restriction analysis (ARDRA) patterns in bacterial phylogeny and taxonomy. J. Microbiol. Meth. 26, 247–259.

Höfle M G 1988 Identification of bacteria by low molecular weight RNA profiles: a new chemotaxonomic approach. J. Microbiol. Meth. 8, 235–248.

Hulton C S J, Higgins C P and Sharp P M 1991 ERIC sequences: a novel family of repetitive elements in the genomes of *Escherichia coli, Salmonella typhimurium* and other enterobacteria. Mol. Microbiol. 5, 825–834.

Lowry O H, Rosenbrough N J, Farr A L and Randal R J 1951 Protein measurement with Folin phenol reagent. J. Biol. Chem. 193, 265–275.

Marsh T L 1999 Terminal restriction fragment length ploymorphism (T-RFLP): an emerging method for characterizing diversity among homologous populations of amplification products. Curr. Opin. Microbiol. 2, 323–327.

Muyzer G 1999 DGGE/TGGE a method for identifying genes from natural ecosystems. Curr. Opin. Microbiol. 2, 317–322.

Nagpal M L, Fox K F and Fox A 1998 Utility of 16S–23S rRNA spacer region methodology: how similar are interspace regions within a genome and between strains for closely related organisms?. J. Microbiol. Meth. 33, 211–219.

Osborn A M, Moore E R B and Timmis K N 2000 An evaluation of terminal-restriction fragment length ploymorphism (T-RFLP) analysis for the study of microbial community structure and dynamics. Environ. Microbiol. 2, 39–50.

Rivas R, Velázquez E, Valverde A, Mateos P F and Martínez-Molina E 2001 A two primers random amplified polymorphic DNA procedure to obtain polymerase chain reaction fingerprints of bacterial species. Electrophoresis 22, 1086–1089.

Tsai C M and Frasch C E 1982 A sensitive silver stain for detecting lipopolysaccharides in polyacrylamide gels. Anal. Biochem. 119, 115–119.

Velázquez E, Igual J M, Willems A, Fernández M P, Muñoz E, Mateos P F, Abril A, Toro N, Normand P, Cervantes E, Gillis M and Martínez-Molina E 2001a Description of *Mesorhizobium chacoense* sp. nov. that nodulates *Prosopis alba* in the Chaco Arido region (Argentina). Int. J. Syst. Evol. Microbiol. 51, 1011–1021.

Velázquez E, Trujillo M E, Peix A, Palomo J L, García-Benavides P, Mateos P F, Ventosa A and Martínez-Molina E 2001b Stable low molecular weight RNA analyzed by staircase electrophoresis, a molecular signature for both prokaryotic and eukaryotic microorganisms. Syst. Appl. Microbiol. 24, 490–499.

Versalovic J, Schneider M, De Bruijn F J and Lupski J R 1994 Genomic fingerprinting of bacteria using repetitive sequence-based polymerase chain reaction. Methods Mol. Cell. Biol. 5, 25–40.

Taxonomy of phosphate solublizing bacteria★

P. Kämpfer*

Institut für Angewandte Mikrobiologie, Justus-Liebig Universität Gießen, Heinrich-Buff-Ring 26-32, D-35392, Gießen, Germany

Received 9 December 2002. Accepted in revised form 2 January 2003

Key words: classical methods, diversity, molecular methods, phosphate solubilizing bacteria, species identification, taxonomy

Abstract

Although phosphate solubilizing capabilities seem to be widespread within bacterial taxa, it is surprising, that the description of phosphate solubilizing bacteria is restricted to relatively few bacterial genera. Among the bacteria, strains belonging to the genus *Rhizobium* and related organisms have been investigated most extensively until now. In addition, several other organisms belonging to taxonomically different and phylogenetic largely unrelated genera (e.g. *Bacillus*, *Paenibacillus*, *Escherichia*, *Enterobacter*, *Rahnella*, *Pseudomonas*, *Burkholderia* and some others) have been studied for their potential to solubilize phosphate. This report gives a short overview on the taxonomy of these genera and some reommendations for identification of unknown organisms with phosphate solubilizing potential in the light of microbial diversity in soil and the current species "definition" in bacteriology.

Introduction

Phosphorus is an essential nutrient for plant growth and development, and besides to nitrogen it is one of the most important elements in crop production (Bowen and Rovira, 1999; Hedley et al., 1995). Because the availability of phosphorus to plants is restricted by various factors, it seems reasonable to study microorganisms that are able to solubilize phosphate from soil and promote its uptake by plants (Gerettsen, 1984; Jana et al, 2001; Johri et al., 1999; Kucey et al., 1989; Macklon et al., 1997; McLaughlin et al., 1988). During the last 10 years knowledge on phosphate solubilizing microorganisms increased significantly (Raghothama, 1999; Richarson, 2001; Richardson et al., 2001; Rodríguez and Fraga, 1999). Several strains of bacterial and fungal species have been described and investigated in detail for their phosphate-solubilizing capabilities (Glick, 1995; He et al., 1997; Leggett et al., 2001).

Identification of phophate solubilizing bacteria is only provided in publications either dealing with model studies or those concerned with a restricted number of isolates. In broader studies, the identification of single isolates is often neglected, as this requires time- and material consuming methods connected with a detailed taxonomic identification.

It is widely accepted, that the diversity of bacteria in soil is very poorly known and it has been estimated that less than one percent of the bacteria from natural communities can be subcultured in the laboratory. This certainly leads to an underestimation of the bacterial diversity.

Diversity can be defined as richness, or number of "types" (Hughes et al., 2001). The term

★ as of 2003

* FAX No: +49-641-9937359.
E-mail: peter.kaempfer@agrar.uni-giessen.de

E. Velázquez and C. Rodríguez-Barrueco (eds.),
First International Meeting on Microbial Phosphate Solubilization, 101–106

"type" may be replaced by the term "species", however, several ecologists have doubted that the present species definition in bacteriology really reflects diversity. Species are the basis of the taxonomic scheme. They are the universally accepted lowest taxonomic cluster of living organisms, and are most often considered to be the units to measure biodiversity.

Some microbial ecologists favour a natural species concept for prokaryotes, which is mainly based on the results of direct molecular analyses of natural microbial populations. However, as pointed out elswhere (Rosselló-Mora and Amann, 2001; Rosselló-Mora and Kämpfer, 2003) species should be a pragmatic unit. Pragmatic does not mean that the unit is easy to circumscribe (in terms of applied methodologies), but that the constructed taxonomic scheme should be operational and predictive.

Some general developments in the species definition in bacteriology

There has been a long and controversial discussion on the problem of the definition of a "species" in bacteriology (some aspects on the history are summarized by Rosselló-Mora and Amann (2001) and Rosselló-Mora and Kämpfer (2003). Today's prokaryotic species concept results from empirical improvements of what has been thought to be a unit. The circumscription of the species has been optimized through the development of microbiological methods that reveal both genomic and phenotypic properties of prokaryotes, which cannot be retrieved by simple observation.

The species definition in bacteriology is a practical one. Especially, the conclusions and recommendations of the Ad Hoc Committee on Reconciliation of Approaches to Bacterial Systematics (Wayne et al., 1987) have provided bacteriologists with a uniform definition of prokaryotic species that has been widely used in systematic studies.

Since 1987 the introduction of new methods have provided new opportunities for systematics of prokaryotes, some of which have already been realized. A remarkable breakthrough in microbial systematics occurred when 16S rRNA gene sequence was used in cladistic analyses to draw genealogical (phylogenetic) trees that represent lines of descent. However, as noted by Young (2001), the term phylogeny in this context is rarely defined precisely. Microbes have left no detailed fossil record, and phylogenetic inferences depend almost entirely on the indirect evidence provided by the analysis of sequence data that are considered to express rates of genetic change (nucleotide base mutation) in time. In this context, for microorganisms it has been assumed that changes in molecular sequences can be used to infer reliable historical relationships. This assumption, in connection with the easy use of PCR, to amplify conserved 16S rRNA or 23S rRNA and to sequence it, has lead to an enormous increase in the numbers of so-called phylogenetic studies. Unfortunately, 16S rRNA gene sequence lacks resolving power at the prokaryotic species level, as defined by DNA–DNA similarity, but permits the identification of the phylogenetic position of new organisms. Its use in new species characterization has been routinely used although it has never been implicitly necessary. It should be pointed out, however, that the simultaneous application of a set of techniques is necessary to understand the diversity of the strains under study, following the so called 'polyphasic approach' (Vandamme et al., 1996).

Further developments of particular interest include determination of inter- and intra-species relatedness by rapid DNA typing methods (AFLP, RAPD, Rep-PCR, PFGE), gene clusters (ribotyping of *rrn* operons), individual genes (ARDRA of 16S rDNA), and intergenic 16S-23S rDNA spacer regions (ISR) (Stackebrandt et al., 2002). On the basis of these developments, an Ad Hoc committee for the Re-Evaluation of the Species Definition in Bacteriology met in February 2002 to discuss these new developments. The recommendations of this committee have been published recently (Stackebrandt et al., 2002). It was a major outcome of the discussions, that the advantages of molecular biology will influence our thinking and understanding of bacteral diversity and hence the unit of diversity which may be defined as "species". It was concluded, that a species "is a category that circumscribes a (preferably) genomically coherent group of individual isolates/strains sharing a high degree of similarity in (many) independent features, comparatively tested under highly standardized conditions".

Identification of unknown isolates

Identification is the pragmatic branch of taxonomy. Most microbial ecologists wish to simply apply a set of standardized techniques for identifcation of unknown isolates in order to identify (or classify) them without going deep into the philosophical discussions on taxonomic systems.

However, for both identification and classification, however, it is important to make correct use of the definition in order to avoid confusions, misidentifications and/or misclassifications. Some recommendations for identifying or classifying an unknown are given by Rosselló-Mora and Amman (2001) and can be summarized as follows:

1. The isolation and collection of an adequate number of strains of the taxon is highly recommended; they should be all used for comparison.
2. The recognition of the closest related taxa through 16S rRNA gene analysis and phenotypic characteristics is necessary for genus allocation. At least, the type strains of these related taxa should be included in the taxonomic analyses.
3. The use of 70% DNA-similarity (or 5 °C a ΔT_{m}) as absolute limits for circumscribing a new species (Stackebrandt et al., 2002; Wayne et al., 1987) should be critically reflected. The current concept allows more relaxed DNA–DNA similarity boundaries, and an internal genomic heterogeneity is permitted. A single species can consist of several genomic groups (genomovars) which do not necessarily have to be classified as different species. DNA–DNA reassociation experiments might be substituted by alternative techniques, however, they should be in sufficient congruence with the results retrieved by DNA–DNA hybridization experiments (Stackebrandt et al., 2002).
4. It is essential to characterize the phenotype of the organisms. Although commercially available test-kits based on biochemical and physiological characters may be useful (API, BIOLOG etc.), sometimes the information retrieved from the results might be insufficient. A critical view on the usefulness of these test-kits often designed for medically important organisms is demanded. The phenotype is not only described by metabolism, there are in addition chemotaxonomic markers (fatty acids, polyamines, quinones...), which are in some cases more stable, as they are independent from cultivation conditions. The more accurately and exhaustively the phenotype is described, the better the circumscription.

Some short reports on the taxonomy of bacterial genera involved in phosphate solubilization.

The following small paragraphs give some selected informations on the taxonomy of bacterial genera and species reported in connection with phosphate solublizing properties. Such a compilation can never be exhaustive, and only very few aspects are given from an arbitrarily selected current publication of new species within these genera. The number of new bacterial species is growing rapidly. The current basis is supported by the nearly 7700 prokaryotic species that appear to be correctly classified (Euzéby, 2002; http://www.bacterio.cict.fr/). It is highly recommended to visit this page before describing a new species.

Rhizobium and relatives

Most of the literature on bacterial phosphate solubilization has been published on strains belonging to the genus *Rhizobium* and related genera (Chabot et al., 1996; Goldstein et al., 1999; Ratti et al., 2001; Rodríguez and Fraga, 1999; Rojas et al., 2001). The complex taxonomic situation of this genus and related genera will be provided by A. Willems in this book (Willems, 2006).

A list of "rhizobia" on the basis of the publication of Young et al. (2001) is as follows:

Azorhizobium caulinodans, *Bradyrhizobium elkanii*, *B. japonicum*, *Mesorhizobium amorphae*, *M. ciceri*, *M. huakuii*, *M. loti*, *M. mediterraneum*, *M. plurifarium*, *M. tianshanense*, *Phyllobacterium myrsinacearum*, *P. rubiacearum*, *Rhizobium etli*, *R. galegae*, *R. gallicum*, *R. giardinii*, *R. hainanensis*, *R. huautlense*, *R. leguminosarum*, *R. mongolen se*, *R. radiobacter*, *R. rhizogenes*, *R. rubi*, *R. tropici*, *R. undicola*, *R. vitis*, *Sinorhizobium fredii*, *E. medicae*, *E. meliloti*, *E. sahelense*, *E. terangae* and *E. xinjiangense*. Complete lists of species are given on the pages: http://www.bacterio.cict.fr/p/[genus name].html.

The "rhizobia" are currently classified into the genera: *Rhizobium* (comprising *Rhizobium*, *Allorhizobium* and *Agrobacterium Ensifer*),

Mesorhizobium, *Phyllobacterium*, *Bradyrhizobium* and *Azorhizobium*.

Pseudomonas and *Burkholderia*

A list of *Pseudomonas* species (on the basis of the paper of Dabboussi et al., 2002) is as follows: *P. aeruginosa*, *P. agarici*, *P. alcaligenes*, *P. amygdali*, *P. asplenii*, *P. aureofaciens*, *P. balearica*, *P. caricapapayae*, *P. chlororaphis*, *P. cichorii*, *P. citronellolis*, *P. coronafaciens*, *P. corrugata*, *P. ficuserectae*, *P. flavescens*, *P. fluorescens*, *P. libanensis*, *P. marginalis*, *P. mendocina*, *P. monteilii*, *P. mosselii*, *P. mucidolens*, *P. oleovorans*, *P. pseudoalcaligenes*, *P. putida*, *P. resinovorans*, *P. rhodesiae*, *P. stutzeri*, *P. synxantha*, *P. syringae*, *P. taetrolens*, *P. tolaasii*, *P. veronii*, and *P. viridiflava*.

A complete list is given on the page: http://www.bacterio.cict.fr/p/pseudomonas.html. given below. Since the 1980s, species of the genus *Pseudomonas* sensu lato have been classified into the genera: *Pseudomonas* sensu stricto, *Chryseomonas*, *Flavimonas*, *Burkholderia*, *Comamonas*, *Acidovorax*, *Hydrogenophaga*, *Telluria*, *Stenotroph omonas*, *Brevundimonas*, *Aminobacter*, *Sphinogomonas*, *Delftia*, and others. Within that time the number of authentic *Pseudomonas* species have increased to 40: Several species often reported in connection with phosphate solubilization (Lugtenberg et al., 2001).

A recent publication of Brämer et al. (2001) summarized the current situation of this genus, which contains at present 18 species: *B. andropogonis*, *B. caribensis*, *B. caryophylli*, *B. cepacia*, *B. gladioli*, *B. glathei*, *B. glumae*, *B. graminis*, *B. kururiensis*, *B. multivorans*, *B. phenazinium*, *B. plantarii*, *B. pseudomallei*, *B. pyrrocinia*, *B. sacchari*, *B. stabilis*, *B. thailandensis*, and *B. vietnamiensis*. Also in this case, a complete list is given on the page: http://www.bacterio.cict.fr/p/burkholderia.html. *B. cepacia* is most often reported in connection with phosphate solubilization (Babu-Khan et al., 1995).

Bacillus and *Paenibacillus*

At present the taxomomy of the genus *Bacillus* sensu lato is in a state of transition. Sequence analyses of rRNA have provided (and still provide) a firm basis for the division of *Bacillus* into several phylogenetically distinct genera: *Alicylcobacillus*, *Amphibacillus*, *Bacillus* sensu stricto, *Brevibacillus*, *Geobacillus*, *Gracilibacillus*, *Paenibacillus*, *Salibacillus*, *Jeotgalibacillus*, *Marinibacillus*, *Virgibacillus*, *Ureibacillus*, *Thermobacillus*, *Sporolactobacillus*, and others. This procedure is still under way. For a complete list of validly published species, see: http://www.bacterio.cict.fr/p/[genus name].html.

A summary of the present situation has been given recently (Berkeley et al., 2002).

The species *Bacillus licheniformis* and *Paenibacillus polymyxa* are often reported in connection with phosphate solubilization.

Enterobacteriaceae

Also within the family *E*nterobacteriaceae, within the last years the number of new genera and species increased significantly: In addition to the genera: *Citrobacter*, *Edwardsiella*, *Enterobacter*, *Escherichia*, *Erwinia*, *Hafnia*, *Klebsiella*, *Morganella*, *Proteus*, *Providencia*, *Salmonella*, *Serratia*, *Shigella*, and *Yersinia*, "new" genera have been described: *Arsenophonus*, *Brenneria*, *Buchneria*, *Budvicia*, *Buttiauxella*, *Calymmatobacterium*, *Cedecea*, *Ewingella*, *Kluyvera*, *Leclercia*, *Leminorella*, *Moellerella*, *Obesumbacterium*, *Pantoea*, *Pectobacterium*, *Photorhabdus*, *Pragia*, *Rahnella*, *Sodalis*, *Tatumella*, *Trabulsiella*, *Wigglesworthia*, *Xenorhabdus*, and *Yokenella* . Members of the genera *Escherichia*, *Erwinia*, *Serratia*, *Pantoea*, and *Rahnella* are most often found with the capability of phosphate solubilization (Goldstein and Liu, 1987; Kim et al., 1998; Liu et al., 1992; Rodríguez et al., 2000; Rossolini et al., 1998).

Conclusions

For the discussion on biodiversity of phosphate solubilizing bacteria a dialogue between taxonomists, populations geneticists and microbial ecologists is essential to find a common language in the future.

It is anticipated that in the future a better understanding of the overall processes of phosphate solubilization is connected with a better understanding of the diversity and interactions of these microorganisms (including the main players) and the description of this diversity (preferably in taxonomical terms).

References

Babu-Khan S, Yeo T C, Martin W L, Duron M R, Rogers R D and Goldstein A H 1995 Cloning of a mineral phosphate-solubilizing gene from *Pseudomonas cepacia*. Appl. Environ. Microbiol. 61, 972–978.

Berkeley R, Heyndrickx M, Logan N and De Vos P 2002 Applications and Systematics of *Bacillus* and Relatives. Blackwell Publishing, Malden, USA.

Bowen G D and Rovira A D 1999 The rhizosphere and its management to improve plant growth. Adv. Agron. 66, 1–102.

Brämer C O, Vandamme P, da Silva L F, Gomez J G C and Steinbüchel A 2001 *Burkholderia sacchari* sp. nov., a polyhydroxyalkanoate-accumulating bacterium isolated from soil of a sugar-cane plantation in Brazil. Int. J. Syst. Evol. Microbiol. 51, 1709–1713.

Chabot R, Antoun H, Kloepper J W and Beauchamp C J 1996 Root colonization of maize and lettuce by bioluminescent *Rhizobium leguminosarum* biovar *phaseoli*. Appl. Environ. Microbiol. 62, 2767–2772.

Dabboussi F, Hamze M, Singer E, Geoffroy V, Meyer J M and Izard D 2002 *Pseudomonas mosselii* sp. nov., a novel species isolated from clinical specimens. Int. J. Syst. Evol. Microbiol. 52, 363–376.

Gerretsen F C 1984 The influence of microorganisms on the phosphate intake by the plant. Plant Soil 1, 51–81.

Glick B R 1995 The enhancement of plant growth by free living bacteria. Can. J. Microbiol. 41, 109–117.

Goldstein A and Liu S T 1987 Molecular cloning and regulation of a mineral phosphate solubilizing gene from *Erwinia herbicola*. Biotechnology 5, 72–74.

Goldstein A H, Braverman K and Osorio N 1999 Evidence for mutualism between a plant growing in a phosphate-limited desert environment and a mineral phosphate solubilizing (MPS) rhizobacterium. FEMS Microbiol. Ecol. 30, 295–300.

He Z L, Wu J, O'Donnell A G and Syers J K 1997 Seasonal responses in microbial biomass carbon, phosphorus and sulphur in soils under pasture. Biol. Fertil. Soils 24, 421–428.

Hedley M J, Mortvedt J J, Bolan N S and Syers J K 1995 Phosphorus fertility management in agroecosystems. *In* Phosphorus in the Global Environment: Transfers, Cycles and Management. Ed. H Tiessen. pp. 59–92. Wiley & Sons, Chichester.

Hughes J B, Hellmann J J, Ricketts T H and Bohannan B J M 2001 Counting the uncountable: statistic approaches to estimating microbial diversity. Appl. Environ. Microbiol. 67, 4399–4406.

Jana B B, Chakraborty P, Biswas J K and Ganguly S 2001 Biogeochemical cycling bacteria as indices of pond fertilization: importance of CNP ratios of input fertilizers. J. Appl. Microbiol. 90, 733–740.

Johri J K, Surange S and Nautiyal C S 1999 Occurrence of salt, pH, and temperature-tolerant, phosphate-solubilizing bacteria in alkaline soils. Curr. Microbiol. 39, 89–93.

Kim K Y, Jordan D and Krishnan H B 1998 Expression of genes from *Rahnella aquatilis* that are necessary for mineral phosphate solubilization in *Escherichia coli*. FEMS Microbiol. Lett. 159, 121–127.

Kucey R M N, Janzen H H and Leggett M E 1989 Microbially mediated increases in plant-available phosphorus. Adv. Agron. 42, 199–228.

Leggett M, Gleddie S and Holloway G 2001 Phosphate-solubilising microorganisms and their use. *In* Plant Nutrient Acquisition: New Perspectives. Eds. N Ae, J Arihara, K Okada and A Srinivasan. pp. 299–318. Springer Verlag, Tokyo.

Liu S T, Lee L Y, Tai C Y, Hug C H, Chang Y S, Wolfram J H and Rogers Goldstein A H 1992 Cloning of an *Erwinia herbicola* gene necessary for gluconi acid production and enhanced mineral phosphate solubiliza in *Escherichia coli* HB101: nucleotide sequence and probab involvement in biosynthesis of the coenzyme pyrroloquinoli quinone. J. Bacteriol. 174, 5814–5819.

Lugtenberg B J J, Dekkers L and Bloemberg G V 2001 Molecular determinants of rhizosphere colonization by *Pseudomonas*. Annu. Rev. Phytopathol. 39, 461–490.

Macklon A E S, Grayston S J, Shand C A, Sim A, Sellars S and Ord B G 1997 Uptake and transport of phosphorus by *Agrostis capillaris* seedlings from rapidly hydrolysed organic sources extracted from ^{32}P-labelled bacterial cultures. Plant Soil 190, 163–167.

McLaughlin M J, Alston A M and Martin J K 1988 Phosphorus cycling in wheat-pasture rotations. II. The role of the microbial biomass in phosphorus cycling. Aust. J. Soil Res. 26, 333–342.

Raghothama K G 1999 Phosphate acquisition. Annu. Rev. Plant Physiol. Plant Mol. Biol. 50, 665–693.

Ratti N, Kumar S, Verma H N and Gautam S P 2001 Improvement in bioavailability of tricalcium phosphate to *Cymbopogon martini* var. *motia* by rhizobacteria, AMF an *Azospirillum* inoculation. Microbiol. Res. 156, 145–149.

Richardson A E 2001 Prospects for using soil microorganisms to improve the acquisition of phosphorus by plants. Aust. J. Plant Physiol. 28, 897–906.

Richardson A E, Hadobas P A, Hayes J E, O'Hara C P and Simpson R J 2001 Utilization of phosphorus by pasture plants supplied with *myo*-inositol hexaphosphate is enhanced by the presence of soil microorganisms. Plant Soil 229, 47–56.

Rodríguez H and Fraga R 1999 Phosphate solubilizing bacteria and their role in plant growth promotion. Biotechnol. Adv. 17, 319–339.

Rodriguez H, Gonzalez T and Selman G 2000 Expression of a mineral phosphate solubilizing gene from *Erwinia herbicola* in two rhizobacterial strains. J. Biotechnol. 84, 155–161.

Rojas A, Holguin G, Glick B R and Bashan Y 2001 Synergism between *Phyllobacterium* sp. (N(2)-fixer) and *Bacillus licheniformis* (P-solubilizer), both from a semiarid mangrove rhizosphere. FEMS Microbiol. Ecol. 35, 181–187.

Rosselló-Mora R and Amann R 2001 The Species concept for prokaryotes. FEMS Microbiol. Rev. 25, 39–67.

Rosselló-Mora R and Kämpfer P 2004 Defining microbial diversity – the species concept for prokaryotic and eukaryotic microorgansms. *In* Microbial Diversity and Bioprospectives, Ed. A Bull. ASM Press, Washington, 29–39.

Rossolini G M, Schippa S, Ricccio M L, Berlutti F, Macaskie L E and Thaller M C 1998 Bacterial nonspecific acid phosphohydrolases: physiology, evolution, and use as tools in microbial biotechnology. Cell. Mol. Life Sci. 54, 833–850.

Stackebrandt E, Frederiksen W, Garrity G M, Grimont P A D, Kämpfer P, Maiden M C J, Nesme X, Rosselló-Mora R, Swings J, Trüper H G, Vauterin L, Ward A C and Whitman W B 2002 Report of the ad hoc committee for the re-evaluation of the species definition in bacteriology. Int. J. Syst. Evol. Microbiol. 52, 1043–1047.

Vandamme P, Pot B, Gillis M, De Vos P, Kersters K and Swings J 1996 Polyphasic taxonomy, a consensus approach to bacterial systematics. Microb. Rev. 60, 407–438.

Wayne L G, Brenner D J, Colwell R R, Grimont P A D, Kandler O, Krichevsky M I, Moore L H, Moore W E C, Murray R G E, Stackebrandt E, Starr M P and Trüper H G 1987 Report of the ad hoc committee on reconciliation of approaches to bacterial systematics. Int. J. Syst. Bacteriol. 37, 463–464.

Willems A 2006 Taxonomy of Rhizobia – On Overview. Plant Soil 287, 3–14.

Young J M 2001 Implications of alternative classifications and horizontal gene transfer for bacterial taxonomy. Int. J. Syst. Evol. Microbiol. 51, 945–953.

Young J M, Kuykendall L D, Martínez-Romero E, Kerr A and Sawada H 2001 A revision of *Rhizobium* Frank 1889, with an emended description of the genus, and the inclusion of all species of *Agrobacterium* Conn 1942 and *Allorhizobium undicola* de Lajudie et al. 1998 as new combinations: *Rhizobium radiobacter*, *R. rhizogenes*, *R. rubi*, *R. undicola* and *R. vitis*. Int. J. Syst. Evol. Microbiol. 51, 89–103.

Taxonomy of filamentous fungi and yeasts that solubilizes phosphate

E. Velázquez & M. E. Trujillo[1]
*Departamento de Microbiología y Genética, Lab 209, Edificio Departamental de Biología, Universidad de Salamanca, Pl. Doctores de la Reina s/n, 37007 Salamanca, Spain. [1]Corresponding author**

Received 14 October 2002. Accepted in revised form 2 January 2003

Key words: Phosphate solubilizing fungi, identification

Abstract

The solubilization of phosphates in the soil is one of the possible ways to increase the crop yield. Some rhizospheric microorganisms may mobilize the soluble phosphorous to the plants and therefore they may be used as biofertilizers. From the rhizospheric microorganisms, the bacteria are the most studied. Nevertheless, some fungal genera are well known as P solubilizers. The phosphate solubilizing filamentous fungi are more studied than the yeasts, mainly *Penicillium* and *Aspergillus* strains. The identification of strains of fungal isolates is more complex than in the case of bacteria because the taxonomy of fungi is based on morphological characteristics. Nevertheless, in the last years, the techniques based on molecular characteristics have changed the species concept in this group of eukaryotic microorganisms and they are very useful in the identification of the new isolates.

Introduction

Phosphorous is, after nitrogen, a basic element to plant growth that is added to soils as soluble inorganic phosphate (NPK). The main problem of this practice is that phosphate becomes insoluble in contact with some elements present in the soil such as calcium or aluminium. These salts are solubilized at pH acid or through microbial processes. Many species of microorganisms have been reported as phosphate solubilizers, the most of them within bacteria. Nevertheless, in recent times several studies have shown that many species of fungi can also solubilize phosphate and that the soluble phosphorus is mobilized to the plants. Within fungi, the mycorrhizae (Mahmood et al., 2001; Vassilev et al., 2001; Villegas and Fortin, 2002) are the most wide and important group of P solubilizers, but also other filamentous fungi and yeasts are able to solubilize phosphate.

Among filamentous fungi that solubilize phosphate, the genera *Aspergillus* and *Penicillium* (Fenice et al., 2000; Gharieb, 2000; Khan and Khan, 2002; Narloch et al., 2002; Narsian and Patel, 2000; Reyes et al., 1999, 2002; Vázquez et al., 2000; Wahid and Mehana, 2000; Whitelaw et al., 1999) are the most representative although strains of *Trichoderma* (Altomare et al., 1999) and *Rhizoctonia solani* (Jacobs et al., 2002) have also been reported as P solubilizers. Among the yeasts, only a few studies have been conducted to assess their ability to solubilize phosphate; these include *Yarrowia lipolytica* (Vassilev et al., 2001), *Schizosaccharomyces pombe* and *Pichia fermentans* (Velázquez et al., unpublished results). As more studies are conducted a wider diversity of phosphate-solubilizing filamentous fungi are expected to be described. Therefore, the knowledge of taxonomy of fungi is very important for those that work with these microorganisms.

Fungal systematics is still based mainly on morphological criteria, and most fungi are usually recognized and identified basically on their

* FAX No: +34923224876.
E-mail: mett@usal.es

E. Velázquez and C. Rodríguez-Barrueco (eds.),
First International Meeting on Microbial Phosphate Solubilization, 107–109

phenotypes. Numerous alternative approaches have been developed, including nutritional and physiological studies, serologic tests, secondary metabolites, and fatty acids. Although some of these are very useful for identifying poorly differentiated fungi such as yeasts and black yeasts, they are only complementary tools of morphological data in most cases. Molecular biology techniques, especially the analysis of rDNA sequences, are currently used for reliable phylogenetic studies, which enable a more natural classification system to be established.

The dual modality of fungal propagation (sexual and asexual) has meant that since the last century there has been a dual nomenclature. Since each phase has been described in total ignorance of the existence of the other in many cases, the *International Code of Botanical Nomenclature* maintains that it is legal to give them separate binomials. With the advent of molecular approaches in fungal taxonomy, some mycologists have advocated abandoning the dual system of naming because unified classification of all fungi may be possible on the basis of the rDNA sequences of the anamorphs.

Phylogenetic information gained from sequence analyses, especially of ribosomal DNA, has revolutionized our knowledge of the classification of this group over the last decade. Comparisons of the 18S rDNA sequences have been performed to assess the relationships of the major groups of living organisms (Prescott et al., 1999). For phylogeny of filamentous fungi, the 18S sequence is mostly used completely or in subunits of over 600 bp. In the yeasts, the D1 and D2 variable regions of the 25S rDNA are almost exclusively used (Kurztman and Blanz, 1998).

The correct identification of fungi is of great practical importance in clinical mycology, plant pathology, biodeterioration, biotechnology, and environmental studies. Fungi comprise one of the largest eukaryotic microbial groups where a significant number of species have been described predominately on phenotypic characteristics, metabolic features, and the morphology of the vegetative and sexual states of these organisms. Since the distinguishing morphological characteristics are frequently too limited and difficult to apply by non-experienced researchers in the identification of a fungus, physiological and biochemical techniques are applied, as has been routinely done for the yeasts. However, for poorly differentiated filamentous fungi, these methods are laborious, time-consuming, variable and provide insufficient taxonomic resolution. The development of molecular methods has opened up new paths to study the fungi as these techniques are universally applicable.

Among the electrophoretic methods, restriction length polymorphism (RFLP) and Random primer polymorphic DNA (RAPDS) are particularly useful for taxonomy. RFLP-based typing methods have been used to reveal anamorph–teleomorph connections. Random primed methods are particularly useful to determine relationships below the level of species, but depending on the length of the primers and the recognized taxonomic diversity of the group under study, the method may help to discriminate species.

More recently, a new electrophoretic technique had allows the separation of stable low molecular weight (LMW) RNA that in fungi includes 5.8s rRNA, 5S rRNA and class 1 and class 2 tRNA (Velázquez et al., 2000). The LMW RNA profiles are molecular signatures of each species of both prokaryotes and eukaryotes (Velázquez et al., 2001). In fungi, the 5.8S–5S rRNA zone are different and characteristic for each genus and the tRNA profiles are different and characteristic for each species of the same or different genera.

Therefore, currently there are numerous techniques that can be applied to fungal taxonomy, but it is necessary an exhaustive revision of the classification and nomenclature of filamentous fungi and yeast as well as the criteria for taxa delimitation and description of new species.

References

Altomare C, Norvell W A, Borjkman T and Harman G E 1999 Solubilization of phosphates and micronutrients by the plant growth promoting and biocontrol fungus *Trichodermna harzianum* Rifai 1295-22. Appl. Environ. Microbiol. 65, 2926–2933.

Fenice M, Seblman L, Federici F and Vassilev N 2000 Application of encapsulated *Penicillium variabile* P16 in solubilization of rock phosphate. Bioresour. Technol. 73, 157–162.

Gharieb M M 2000 Nutritional effects on oxalic acid production and solubilization of gypsum by *Aspergillus niger*. Mycol. Res. 104, 550–556.

Jacobs H, Boswell G P, Ritz K, Davidson F A and Gadd G M 2002 Solubilization of calcium phosphate as a consequence of carbon translocation by *Rhizoctonia solani*. FEMS Microbiol. Ecol. 40, 65–71.

Khan M R and Khan S M 2002 Effect of root-dip treatment with certain phosphate solubilizing microorganisms. Bioresour. Technol. 85, 213–215.

Kurtzman C P and Blanz P A 1998 Ribosomal RNA/DNA seqeunce comparisons for assesing phylogenetic relationships. *In* The Yeast: A Taxonomic Study (4th edn). Eds. C P Kurtzman and J W Fell. pp. 69–74. Elsevier, Amsterdam.

Mahmood S, Finlay R D, Erland S and Wallander H 2001 Solubilisation and colonisation of wood ash by ecthomycorrhizal fungi isolated from a wood ash fertilised spruce forest. FEMS Microbiol. Ecol. 35, 151–161.

Narloch C, de Oliveira V L, dos-Anjos T J and Silva G N 2002 Responses of radish culture to phosphate solubilizing bacteria. 2002. Pesquisa Agropecuaria Brasileira 37, 841–845.

Narsian V and Patel H H 2000 *Aspergillus aculeatus* as a rock phosphate solubilizer. Soil Biol. Biochem. 32, 559–565.

Prescott R, Harley J P and Klein D A 1999 Microbiology. *In* The Fungi (Eumycota), Slime molds, and Water Molds. Eds. R Prescott, J P Harley, D A Klein. pp. 522–564. WCB MacGraw-Hill, Boston.

Reyes I, Bernier L, Simard R R, Tanguay P and Antoun H 1999 Characteristics of phosphate solubilization by an isolate of a troopical *Penicillium rugulosum* and two UV-induced mutants. FEMS Microbiol. Ecol. 28, 291–295.

Reyes I, Bernier L and Antoun H 2002 Rock phosphate solubilization and colonization of maize rizosphere by wild and genetically modified strains of *Penicillium rugulosum*. Microbial Ecol. 44, 39–48.

Vassilev N, Vassileva M, Azcón R and Medina A 2001 Preparation of gel-entrapped mycorrhizal inoculum in the presence or absence of *Yarrowia lypolytica*. Biotechnol. Lett. 23, 907–909.

Vázquez P, Holguin G, Puente M E, López-Cortés A and Bashan Y 2000 Phosphate-solubilizing microorganisms associated with the rhizosphere of mangroves in a semiarid coastal lagoon. Biol. Fertil. Soils 30, 460–468.

Velázquez E, Calvo O, Cervantes E, Mateos P F, Tamame M and Martínez-Molina E 2000 Staircase electrophoresis profiles of stable Low Molecular Weight RNA as yeast fingerprinting. Int. J. Syst. Evol. Microbiol. 50, 917–923.

Velázquez E, Trujillo M E, Peix A, Palomo J L, García-Benavides P, Mateos P, Ventosa A and Martínez-Molina E 2001 Stable low molecular weight RNA analyzed by staircase electrophoresis, a molecular signature for both prokaryotic and eukaryotic microorganisms. Syst. Appl. Microbiol. 24, 490–499.

Villegas J and Fortin J A 2002 Phosphorous solubilization and pH changes as a result of the interactions between soil bacteria and arbuscular mycorrhizal fungi in a medium containing NO_3^- as nitrogen source. Can. J. Bot. 80, 571–576.

Wahid O A A and Mehana T A 2000 Impact of phosphate solubilizing fungi on the yield and phosphorous-uptake by wheat and faba bean plants. Microbiol. Res. 155, 221–227.

Whitelaw M A, Harden T J and Helyar K R 1999 Phosphate solubilisation in the solution culture by the soil fungus *Penicillium radici*. Soil Biol. Biochem. 31, 655–665.

Phosphate solubilizing microorganisms: Effect of carbon, nitrogen, and phosphorus sources

Ely Nahas*
Departamento de Produção Vegetal-FCAV, Universidade Estadual Paulista-UNESP, Rodovia Prof. Paulo Donato Castellane km 5, 14884-900, Jaboticabal, SP, Brazil.

Received 26 November 2002. Accepted in revised form 2 January 2003

Key words: bacteria, fungi, P-fertilizer, P-solubilization control

Abstract

Although most soils contain large amounts of total phosphorus, they are deficient in phosphates available to plants. However significant populations of soil microorganisms present the ability to dissolve poorly soluble mineral phosphates. Most of these microorganisms are heterotrophs and depend on carbon and energy sources that can be found in the rhizosphere or by recycling crop residues. Besides, nitrogen and phosphorus sources may be considered as control factors in soil, because as carbon sources, they influence microorganisms growth and consequently their solubilization capacity. The principal mechanism for mineral phosphate solubilization is the production of organic acids. Some studies indicate that the physiology and biochemistry of C, N, and P play a role in the phosphate solubilization process. Is was suggested that increased numbers of phosphate solubilizing microorganisms in soil may be significant in relation to the phosphorus economy of the plant.

Introduction

Many agricultural soils are generally low in available phosphorus and some of them sorb large amounts of applied phosphate fertilizer (Oberson et al., 2001). Consequently, efforts have been made to study the role of soil microorganisms in the solubilization of inorganic phosphates to improve phosphorus availability for plant growth (Asea et al., 1988; Nahas, 1996).

Phosphate-solubilizing microorganisms dissolve insoluble phosphates by the production of inorganic or organic acids and/or by the decrease of the pH, producing available phosphate that can be up taken by the plants (Cerezine et al., 1988; Sperber, 1958; Rodríguez and Fraga, 1999). It was found another mechanism in *Penicillium aurantiogriseum* and *Pseudomonas* sp. in which calcium phosphates were solubilized by the excretion of the proton H^+ accompanying respiration or NH_4^+ assimilation (Illmer and Schinner, 1995). The solubilizing ability of a microorganism is related to its organic acid production that binds cations such as Al^{3+}, Ca^{2+}, and Fe^{3+} from mineral phosphates and releases soluble P for plant nutrition. Although the final pH of the culture medium or acid production are important to minerals solubilization, these factors were not always correlated to the soluble phosphate produced (Nahas, 1999).

However, microorganisms have to scavenge the soil nutrients that they need for their growth. The availability and the nature of these nutrients are important factors in the mechanism of mineral phosphate solubilization. The aim of this review is to summarize the research on the behavior of C, N, and P sources on bacteria and fungi and their effect on the production of organic acids, pH decrease and solubilization

* FAX No: 55 +16 322-4275.
E-mail: enahas@fcav.unesp.br

E. Velázquez and C. Rodríguez-Barrueco (eds.),
First International Meeting on Microbial Phosphate Solubilization, 111–115

of hardly soluble phosphates. The relationship among organic acids, soluble phosphate production and pH decreased was discussed.

Effect of carbon sources

As heterotrophic organisms, phosphate solubilizers need a carbon source and energy for both the synthesis of new cell material and the oxidation of carbon compounds (Moat and Foster, 1988). Rhizosphere soils present water-soluble C and N compounds mainly as carbohydrates and organic acids and a small portion of amino acids (Merbach et al., 1999; Ryan et al., 2001). It is well known that a considerable number of microorganisms was associated with the plant rhizosphere due to its carbon concentration (Lynch and Whipps, 1990). Therefore, the proportion of phosphate-dissolving bacteria in the rhizosphere of several crop plants was enhanced in relation to the root-free soil (Katznelson et al., 1962). It was also observed that the percentage of bacteria P-solubilizing was greater in the rhizoplane of pigeon pea than in the adhering soil (Finger et al., 2002, unpublished results).

Decomposition of plant residues is a mechanism by which carbon sources are recycled in the soil. Organic constituents of plants as cellulose, hemicelluloses, and lignin are the most abundant polymers that need be degraded by the microorganisms to provide the carbon sources in the soil (Vinopal and Romano, 2000). Then, the level of microbial activity for phosphate solubilization depends of the organic matter degradation and the production of suitable carbon sources for P-solubilizers growth. The solubilization of two types of rock phosphates increased significantly during decomposition of wheat straw and cattle urine (Singh and Amberger, 1991). Alternative possibilities have been based on the use of easily utilizable C sources, which markedly increased the activity of solubilization by the use of microorganisms able to excrete organic acids (Nahas et al., 1990; Vassilev et al., 1996).

There are several considerations by which microorganisms can metabolize carbohydrates. When sugars are available in the medium they are taken up into the cell, however their utilization is dependent on an enzyme system that is constitutive in some organisms and inducible in others (Brock et al., 1994). Catabolic pathways convert sugars into a group of metabolic intermediates as organic acids and others by a set of pathways recognized as a central metabolism (Vinopal and Romano, 2000). Some of these acids are found in living cells as intermediates of tricarboxylic acid (TCA) cycle. However, the production of organic acids is dependent of both the metabolic routes generated by the microorganism and the regulation of enzyme activity involved in these pathways.

These aspects need to be borne in mind when one is extrapolating from knowledge of the synthesis of organic acids by microorganisms in soil. The considerable amount of information on acid production by *A. niger* indicated the complexities of the metabolic processes involved (Kubicek and Röhr, 1986). In addition to the central biochemical pathways, many bacterial species appear to have the capacity to use glucose by the direct oxidation path. This is supported by experiments showing that certain strains of Gram-negative bacteria dissolve calcium phosphates by the production of gluconic and 2-ketogluconic acids by enzymes located on periplasmic space (Goldstein et al., 1999).

Many works have reported the effect of type and concentration of carbon sources on the production of acids (Gupta et al., 1976; Xu et al., 1989). Among the carbohydrates tested in the culture media with *Aspergillus niger*, fructose, glucose, xylose, sucrose, and starch enhanced fluorapatite solubilization more than galactose and maltose, although total phosphorus accumulation was 78% with fructose as opposed to 59–69% with the other carbon sources (Cerezine et al., 1988). The solubilization of hydroxyapatite and $FePO_4$ by the fungus *P. rugulosum* was smaller in presence of glucose or maltose than of sucrose (Reyes et al. 1999b). A significant relationship was detected between solubilization measured as soluble phosphate and mycelium dry weight, final pH and titratable acidity in the culture medium. The amount of soluble phosphate was negatively correlated with pH and titratable acidity and positively correlated with mycelium dry weight. However, a significant correlation ($P < 0.05$) was obtained between titratable acidity and soluble phosphate up to the third day of fungal growth ($r = 0.58$) then point determinations may not reflect a behavior as a whole in the environment. Thus, other factors should also be considered, i.e., the time and conditions of growth of the microorganism. Fluorapatite solubilization was not enhanced when increasing amounts of fructose (0.25–4.00%, w/v) or vinasse were added to the culture medium as carbon sources. The maximum contents of soluble phosphate produced corresponded to 1.2% fructose and this was probably due to soluble phosphate uptake by the *A. niger* fungus (Cerezine et al., 1988; Nahas et al., 1990). Higher P solubilization was found when *P. radicum* was grown in culture medium supplemented with 30 g sucrose L^{-1} than with 10 g L^{-1} (Whitelaw et al., 1999).

Effect of nitrogen sources

Increased amounts of soluble P as a result of the ammonium addition to the culture media have been documented (Asea et al., 1988; Whitelaw et al., 1999). Phosphate solubilization was related to the proton (H^+) excretion accompanying NH_4^+ assimilation (Roos and Luckner, 1984). In *Pseudomonas fluorescens*, solubilization of zinc phosphate was due to both an excretion of H^+ and the production of gluconic acid (Di Simine et al., 1998). Wenzel et al. (1994) reported that the secretion of H^+ could solubilize phosphate by exchange with Ca^{2+}. According to this mechanism, Villegas and Fortin (2001) found that P soluble was related with variations in the pH of the medium in NH_4^+ cultures of *Glomus intraradices*. When ammonium nitrogen salts were added to culture medium inoculated with *A. niger*, fluorapatite solubilization was enhanced more than that of organic or nitrate sources (Cerezine et al., 1988). However, the addition of ammonium nitrate to vinasse used as culture medium enhanced *A. niger* growth but rock phosphate solubilization was severely reduced. This effect was due to greater fungal growth, as well as to lower acid production and lower pH values compared to control (Nahas et al., 1990). Smaller growth on an ammonium medium than on nitrate medium was observed, however this was due to a low PEP carboxykinase activity in *Hebeloma cylindrosporum*, enzyme that catalyzes carbon dioxide fixation assuring that TCA cycle intermediates can be replenished (Scheromm et al., 1990).

Additionally to the type of N source, the concentration of ammonium also had an important effect. In the fungus *P. rugulosum*, lower concentration of ammonium in the growth media decreased phosphate solubilization and citric acid production (Reyes et al., 1999a). The highest growth and gluconic acid production were observed at 1.5 g ammonium L^{-1} for *A. foetidus* (Kara and Bozdemir, 1998). The effect of ammonium, nitrate, and phosphate on the growth in *P. simplicissimum* and secretion of TCA cycle intermediates was studied by Gallmetzer and Burgstaller (2002) who observed that the highest productivity was found during phosphate-limited growth. Although we did not assess the specific mechanism concerning the effect of N sources, it appears feasible to believe that the effect of N sources on P solubilization would be a consequence of multiple paths.

Effect of phosphorus sources

Along with the effects of carbon and nitrogen sources on the solubilization of insoluble phosphates, different levels of the soluble phosphates can also affect this process in two additional ways.

First, the process of solubilization by the bacteria *Escherichia coli* and *Erwinia herbicola* was found to be regulated by the external phosphate levels, as also observed for phosphatases (Goldstein, 1986). A similar mechanism was found in *A. niger* (Nahas and Assis, 1992). Acid phosphatase production and fluorapatite solubilization were decreased when the concentration of soluble phosphate was enhanced at 2 m*M* Pi. According to Chabot et al. (1998), P-solubilization activity of rhizobial strains was repressed in the presence of 0.5% soluble P.

Second, soil fertilization with phosphorus is important to both plant and microorganism growth. Increasing active exudation from roots provides a substantial amount of organic acids that enhance solubilization (Dinkelaker et al., 1989). Then, when soluble phosphate was exhausted, the solubilization process was triggered, with a consequent increase in soluble phosphate in the soil.

Numerous studies have been done in relation to the process of solubilization by the fungus *A. niger* (Bojinova et al., 1997; Cerezine et al., 1998; Illmer et al., 1995; Vassilev et al., 1996) but it appears that the data for confident generalization are still lacking. Then, considering the proposed mechanisms for the phosphate solubilization, it must be taken into consideration the fact that the production of citric acid by *A. niger* need some requirements related to the levels of C, N, and P sources. According to Mattey (1992), high levels of the C source (sucrose, molasses, etc.) and low levels of N (ammonium salts, urea, etc.) and P (potassium phosphate) were the general requirements for citric acid production. Citrate excretion is an energy-spilling process because it does not need the synthesis of much NADH (Gallmetzer and Burgstaller, 2002).

Different factors related to the microbial isolate and to the type of insoluble phosphate (Gafsa, Araxá and Patos rock phosphates and calcium phosphate) affected the final pH values, the levels of titratable acidity and, consequently the amounts of solubilized phosphate (Nahas, 1996). The lowest pH values were observed with the phosphates from Araxá and Patos. The levels of acids produced and the solubilization levels were in the following sequence: calcium phosphate > Gafsa > Araxá > Patos, i.e., they were in the same order of agronomic efficiency as these phosphates, indicating that the agronomic potential of rock phosphate may be related to its solubilizing ability or vice-versa. The ability of phosphate rocks and minerals solubilization by *P. rugulosum* was related to the production of citric and gluconic acids or both, respectively (Reyes et al., 2001).

Sixteen soil isolates were studied for their ability to solubilize rock phosphate and calcium phosphate in culture medium (Nahas, 1996). There was a correlation between final pH value and titratable acidity ($r = -0.29$ to -0.87) and between titratable acidity and soluble phosphate ($r = 0.22$ to 0.99). The correlation values ranged as a function of insoluble phosphate and of the group of microorganisms considered. When each isolate was analyzed separately, a correlation between the variables studied was observed in only some isolates. However, significant responses were obtained between titratable acidity and soluble phosphate for all phosphates used. It was observed that in soil, the microbial interaction is more effective in the solubilization process. If there were a dependence of solubilization on the type of acid secreted (Ryan, 2001), thus, the microorganism community would permit a better response because of a higher diversity among the acids secreted.

In conclusion, it seems clear that some factors related to the nature or the contents of C, N and P sources affected the solubilization of insoluble phosphates by the P-solubilizing microbial communities. Some carbon sources found in soil enhanced P dissolution. Ammonium salts were more effective than nitrate or organic N sources. Phosphorus sources influenced both the growth of P-solubilizer microorganisms and the repression of the enzyme system related to the solubilization mechanism. Available phosphate was only 2.2% of total P, and the supply of phosphate to plants depended on microbial solubilization of insoluble phosphates.

Acknowledgments

The author thank Mrs. Clemencia M. Söndahl who kindly revised the English version of this manuscript. I also wish to thank CNPq for a fellowship.

References

Asea P E A, Kucey R M N and Stewart J W B 1988 Inorganic phosphate solubilization by two *Penicillium* species in solution culture and soil. Soil Biol. Biochem. 20, 459–464.

Bojinova D, Velkova R, Grancharov I and Zhelev S 1997 The bioconversion of tunisian phosphorite using *Aspergillus niger*. Nutrient Cycl. Agroecosyst. 47, 227–232.

Brock T D, Madigan M T, Martinko J M and Parker J 1994 Biology of Microorganisms. Prentice, New Jersey. 909 pp.

Cerezine P C, Nahas E and Banzatto D A 1988 Soluble phosphate accumulation by *Aspergillus niger* from fluorapatite. Appl. Microbiol. Biotech. 29, 501–505.

Chabot R, Beauchamp C J, Kloepper J W and Antoun H 1998 Effect of phosphorus on root colonization and growth promotion of maize by bioluminescent mutants of phosphate-solubilizing *Rhizobium leguminosarum* biovar *phaseoli*. Soil. Biol. Biochem. 30, 1615–1618.

Dinkelaker B, Romheld V and Marschner H 1989 Citric acid excretion and precipitation of calcium citrate in the rhizospehre of white lupin (*Lupinus albus* L.). Plant Cell Environ. 12, 285–292.

Di Simine C D, Sayer J A and Gadd G M 1998 Solubilization of zinc phosphate by a strain of *Pseudomonas fluorescens* isolated from a forest soil. Biol. Fert. Soils 28(1), 87–94.

Gallmetzer M and Burgstaller W 2002 Efflux of organic acids in *Penicillium simplicissimum* is an energy-spilling process, adjusting the catabolic carbon flow to the nutrient supply and the activity of catabolic pathways. Microbiol. Reading 148, 1143–1149.

Goldstein A H 1986 Bacterial solubilization of mineral phosphates: historical perspective and future prospects. Amer. J. Alter. Agric. 1, 51–57.

Goldstein A H, Braverman K and Osorio N 1999 Evidence for mutualism between a plant growing in a phosphate-limited desert environment and a mineral phosphate solubilizing (MPS) rhizobacterium. FEMS Microbiol. Ecol. 30, 295–300.

Gupta J K, Heding L G and Jorgensen O B 1976 Effect of sugars, hydrogen ion concentration and ammonium nitrate on the formation of citric acid by *Aspergillus niger*. Acta Microbiol. Acad. Sci. Hung. 23, 63–67.

Illmer P and Schinner F 1995 Solubilization of inorganic calcium phosphates-solubilization mechanisms. Soil Biol. Biochem. 27, 257–263.

Illmer P, Barbato A and Schinner F 1995 Solubilization of hardly-soluble $AlPO_4$ with P-solubilizing microorganisms. Soil Biol. Biochem. 27, 265–270.

Kara A and Bozdemir T O 1998 Optimization of the growth parameters of *Aspergillus foetidus*. Acta Biotech. 18, 327–338.

Katznelson H, Peterson E A and Rouatt J W 1962 Phosphate-dissolving microorganisms on seed and in the root zone of plants. Can. J. Bot. 40, 1181–1186.

Kubicek C P and Röhr M 1986 Citric acid fermentation. CRC Critical Rev. Biotech. 3, 331–373.

Lynch J M and Whipps J M 1990 Substrate flow in the rhizosphere. Plant Soil 128, 1–10.

Mattey M 1992 The production of organic acids. Rev. Biotech. 12, 87–122.

Merbach W, Mirus E, Knof G, Remus R, Ruppel S, Russow R, Gransee A and Schulze J 1999 Release of carbon and nitrogen compounds by plant roots and their possible ecological importance. J. Plant Nutr. Soil Sci. 162, 373–383.

Moat A G and Foster J W 1988 Microbial Physiology, 2nd edn. Wiley, New York, 597 pp.

Nahas E, Banzatto D A and Assis L C 1990 Fluorapatite solubilization by *Aspergillus niger* in vinasse medium. Soil. Biol. Biochem. 22, 1097–1101.

Nahas E and Assis L C 1992 Effect of phosphate on the solubilization of fluorapatite by *Aspergillus niger*. Rev. Microbiol. 23, 37–42.

Nahas E 1996 Factors determining rock phosphate solubilization by microorganisms isolated from soil. World J. Microbiol. Biotech. 12, 567–572.

Nahas E 1999 *In* Inter-relação fertilidade, biologia do solo e nutrição de plantas. Ed. Siqueira J O et al. pp. 467–486. Viçosa. SBCS Lavras. UFLA/DCS.

Oberson A, Friesen D K, Rao I M, Bühler S and Frossard E 2001 Phosphorus transformations in an oxisol under contrasting land-use systems: The role of the soil microbial biomass. Plant Soil 237, 197–210.

Reyes I, Bernier L, Simard R R and Antoun H 1999a Effect of nitrogen source on the solubilization of different inorganic phosphates by Na isolate of *Penicillium rugulosum* and two UV-induced mutants. FEMS Microbiol. Ecol. 28, 281–290.

Reyes I, Bernier L, Simard R R, Tanguay P and Antoun H 1999b Characteristics of phosphate solubilization by an isolate of a tropical *Penicillium rugulosum* and two UV-induced mutants. FEMS Microbiol. Ecol. 28, 291–295.

Reyes I, Baziramakenga R, Bernier L and Antoun H 2001 Solubilization of phosphate rocks and minerals by a wild-type strain and two UV-induced mutants of *Penicillium rugulosum*. Soil Biol. Biochem. 33, 1741–1747.

Rodríguez H and Fraga R 1999 Phosphate solubilizing bacteria and their role in plant growth promotion. Biotech. Adv. 17, 319–339.

Roos W and Luckner M 1984 Relationships between proton extrusion and fluxes of ammonium ions and organic acids in *Penicillium cyclopium*. J. Gen. Microbiol. 130, 1007–1014.

Ryan P R, Delhaize E and Jones D L 2001 Function and mechanism of organic anion exudation from plant roots. Annu. Rev. Plant Physiol. 52, 527–560.

Scheromm P, Plassard C and Salsac L 1990 Effect of nitrate and ammonium nutrition on the metabolism of the ectomycorrhizal basidiomycete, *Hebeloma cylindrosporum* Romagn. New Phytol. 114, 227–234.

Singh C P and Amberger A 1991 Solubilization and availability of phosphorus during decomposition of rock phosphate enriched straw and urine. Biol. Agric. Hort. 7, 261–269.

Sperber J I 1958 Solution of apatite by soil microorganisms producing organic acids. Aust. J. Agron. Res. 9, 782–787.

Vassilev N, Franco I, Vassileva M and Azcon R 1996 Improved plant growth with rock phosphate solubilized by *Aspergillus niger* grown on sugar-beet waste. Bioresource Technol. 55, 237–241.

Villegas J and Fortin J A 2001 Phosphorus solubilization and pH changes as a result of the interactions between soil bacteria and arbuscular mycorrhizal fungi on a medium containing NH_4^+ as nitrogen source. Can. J. Bot. 79, 865–870.

Vinopal R T and Romano A H 2000 Carbohydrate synthesis and metabolism. *In* Encyclopedia of Microbiology, Vol. 1, 2nd edn. Ed. J. Lederberg. pp. 647–668. Academic, San Diego.

Wenzel C L, Ashford A E and Summerell B A 1994 Phosphate-solubilizing bacteria associated with proteoid roots of seedlings of waratah [*Telopea speciosissima* (Sm.) R.Br.]. New Phytol. 128, 487–496.

Whitelaw M A, Harden T J and Helyar K R 1999 Phosphate solubilisation in solution culture by the soil fungus *Penicillium radicum*. Soil Biol. Biochem. 31, 655–665.

Xu D B, Madrid C P, Röhr M and Kubicek C P 1989 The influence of type and concentration of the carbon source on production of citric acid by *Aspergillus niger*. Appl. Microbiol. Biotech. 30, 553–558.

Efficacy of organic acid secreting bacteria in solubilization of rock phosphate in acidic alfisols

S. Srivastava[1], M. T. Kausalya[1], G. Archana[2], O. P. Rupela[3] & G. Naresh-Kumar[1,4]
[1]*Department of Biochemistry, Faculty of Science, M. S. University of Baroda, Vadodara, 390 002, India.* [2]*Department of Microbiology, Faculty of Science, M. S. University of Baroda, Vadodara, India.* [3]*Plant Physiology, Natural Resource Management Program, International Crops Research Institute for the Semi-Arid Tropics (ICRISAT), Patancheru, Andhra Pradesh, India.* [4]*Corresponding author**

Received 27 December 2002. Accepted in revised form 2 January 2003

Key words: Alfisol, *Enterobacter asburiae* PSI3, P-solubilisation, organic acid, rock phosphate, phosphate solubilizing microorganisms

Abstract

Many phosphate solubilizing microorganisms (PSMs) solubilize calcium phosphate (CaP) complexes by the secretion of organic acids. Alfisols, however, are rich in ferric phosphate (FeP) and aluminium phosphate (AlP) complexes. The efficacy of organic acid secreting PSMs at releasing P in alfisols is not well understood. The potential of naturally existing PSMs at solubilizing mineral phosphates of alfisols was investigated. Native soil bacteria grew well on alfisol and alfisol amended with rock phosphate (RP) when supplemented with 100 m*M* glucose and 10 m*M* KNO_3; but did not release P in soil solution. Two CaP solubilizing bacteria, *Enterobacter asburiae* PSI3 and *Bacillus coagulans*, were tested for their ability to grow on minimal media containing AlP or FeP as sole P sources. *E. asburiae* PSI3 could grow on and acidify the medium with either of the P sources, whereas *B. coagulans* was unable to grow. *E. asburiae* PSI3: *B. coagulans* and another RP solubilizing bacterium, *Citrobacter koseri*, failed to release P from alfisol as well as alfisol amended with RP. Addition of citric, oxalic, tartaric, gluconic, succinic, lactic and acetic acids at concentrations sufficient to decrease the pH to less than 4.0 failed to release P from alfisol. However, addition of organic acids to alfisol supplemented with RP brought about release of P to different extents, oxalic acid being the most effective. After the addition of 2 m*M* KH_2PO_4 to alfisol, a rapid decrease in soluble P to about 60 μM was observed within 30 min indicating very high P fixing ability. In order to extrapolate the difference of P solubilising ability of *E. asburiae* PSI3 in alfisol and vertisol, mung bean plants were seed inoculated with *E. asburiae* PSI3. Significant decrease in the root length was found in vertisols but not in alfisol.

Introduction

Acid soils such as alfisols, aridisols, entisols and ultisols occupy about 30% of world's icefree land area (von Uexull and Mutert, 1995). Historically acid soils have resisted permanent settlement and agricultural use. Deficiency of nutrients such as P and toxicity of aluminium (Al) chiefly mark the crop growth in acid soils. Of the total land area of the world, about 24.2% is considered potentially arable land and most of this belongs to acid soils. Thus, acid soils represents largest potential land area for future agricultural development. With the increase in food demand, there is

* E-mail: gnaresh_k@yahoo.co.in

E. Velázquez and C. Rodríguez-Barrueco (eds.),
First International Meeting on Microbial Phosphate Solubilization, 117–124

greater pressure to develop acid soils for agricultural use.

In alfisols most of the inorganic phosphate is in the iron bound form and minor quantity is complexed with aluminium (Ae et al., 1990, 1991). To enhance P nutrition to plants grown in acidic soils, direct application of rock phosphate (RP) as P fertilizer has been studied (Rajan et al., 1996). RP are rich in calcium phosphate (CaP) complexes and hence are suggested to be soluble in acidic soils The advantages being cheaper than inorganic P-fertilizer; creates less environmental pollution as it requires minimum processing and its dissolution results in slow release of P in the soil. Various reports show the usage of partially acidulated rock phosphate (PARP) in acid soil to provide P nutrition (Abekoe, 1998; Bolland, 1992; Butegwa, 1996; Marwah, 1983). However, the use of RP is still limited due to a number of factors controlling its dissolution in soil and availability to plants. The effectiveness of RP differs with soil properties, climatic conditions and nature of the crops (Nahas, 1996; Rajan et al., 1996). In one study, 228 soil samples were collected from South Western Australia and incubated with North Carolina RP (Huges, 1998). It was found that amounts of RP dissolved were generally low and only 29 of the soils dissolved more than 40% of added RP.

One of the attractive approaches to solubilize RP would be to use microorganisms that are able to secrete organic acids (Gadd, 1999; Vassilev et al., 1996). Application of RP along with P-solubilising bacteria and vesicular arbuscular mycorrhiza (VAM) had shown to increase the P nutrition of plants in alkaline soils (Bolan et al., 1987; Omar, 1998; Piccini and Azcon, 1987; Vassileva et al., 1998). It is known that low molecular weight organic acids effectively chelate Fe and Al from ferric phosphate (FeP) and aluminium phosphate (AlP), thereby solubilising P (Bolan et al., 1994; Gilroy and Jones, 2000; Jones and Darrah, 1994; Kpomblekou and Tabatabai, 1994). Thus, alfisols having low buffering capacity and acidic pH seems to be conducive for direct applications of RP as P fertilizer along with organic acid secreting bacteria.

Citrobacter koseri and *Bacillus coagulans* have been shown to solubilize RP by secretion of a mixture of several organic acids at low (less than 5 m*M*) concentrations (Gyaneshwar et al., 1998). *E. asburiae* PSI3 solubilizes RP by secretion of 50 m*M* gluconic acid (Gyaneshwar et al., 1999). In this paper, we investigated the effect of these PSMs and organic acids in solubilising P from alfisol as well as RP in alfisol.

Material and methods

Bacterial strains and media

Culture conditions for *C. koseri*, *B. coagulans* and *E. asburiae* PSI3 are as described (Gyaneshwar et al., 1998, 1999). 'Senegal' RP was used in all experiments. It has 121.6 mg P/g of RP as determined by total P estimation method (Bowman, 1988).

Solid-medium experiments to assess AlP or FeP solubilisation

The P solubilizing microorganisms were grown on 1.5% agar plates containing 100 m*M* glucose, 10 m*M* NH_4Cl and FeP or AlP at 1 mg/mL as the sole P source. The pH of the media was adjusted to 8.0 by adding 10 m*M* or 100 m*M* Tris.Cl buffer. Methyl red indicator dye was used at 0.1% in plates.

Effect of nutrient supplementation on growth and P release by native microorganisms from alfisol

Alfisol was obtained from agricultural soil near Bangalore, India. The pH of the soil solution was 6.5 and total P content was 2.68 mg P/g of alfisol. Alfisol was suspended at 0.5 g/mL in sterile medium containing 100 m*M* glucose and 10 m*M* KNO_3. RP was added wherever specified. The organisms were allowed to grow aerobically in the soil suspension at 30 °C on rotary shaker at 200 rev/min. Samples were collected at 0 h and 72 h. Total number of microorganisms was determined by culturable count on Luria agar. The supernatant of the soil solution collected by centrifugation of the soil sample at 5000 rev/min for 10 min was used to determine pH and P.

P solubilisation from alfisol in the absence and presence of RP by C. koseri, B. coagulans and E. asburiae PSI3

Alfisol was taken at 0.5 g/mL of medium containing 100 m*M* glucose, 10 m*M* KNO_3 and native organisms eliminated by autoclaving. For inoculation, an overnight grown culture of the bacteria was washed twice with saline and subsequently diluted 1:100 times. 100 μL of this dilution was used to inoculate 60 mL of alfisol soil suspension. The inoculated soil suspension was kept for aerobic growth on rotary shaker at 200 rev/min at 30 °C. Samples were collected at 0 h and 72 h for estimating viable count, pH and solution P as above. Wherever mentioned RP at 40 mg/g of soil was added.

Effect of organic acids on P release from alfisol in the absence and presence of RP

Alfisol was taken at 0.5 g/mL in water with RP supplementation as specified. Concentrated stocks of different organic acids were added to achieve different final concentrations. Aliquots of the suspension was taken at different time intervals and centrifuged at 5000 rev/ min for 10 min. The supernatant thus obtained was used to measure pH and P.

Determination of P fixing ability of alfisol

Alfisol was taken at 0.5 g/mL in water to which a final concentration of 2 m*M* KH_2PO_4 was added and shaken vigorously at room temperature (RT). Aliquots of this suspension were taken at different time intervals, centrifuged at 5000 rev/ min and supernatant thus obtained was used to measure pH and P.

Plant Inoculation studies

Characteristics of the alkaline vertisol and alfisol soil samples are shown in Table 1. Seven inch pots were used, each pot containing 3 kg of soil. Six seeds were sown which were later thinned to 3/pot. Each set of variant was taken in quadruplets. Seeds were allowed to germinate and grow till 30 days, at the end of which root length was measured.

Analytical methods

Estimation of total phosphorus in soil and RP was done as described by Bowman (1988). Phosphate estimations were done by the ascorbate method (Ames, 1964).

Results

Effect of nutrient supplementation on growth and P release by native microorganisms present in alfisols

Native microorganisms were assessed for their ability to solubilize P from alfisol when sufficient amount of carbon (C) and nitrogen (N) sources were added in the form of 100 m*M* glucose and 10 m*M* KNO_3. Native microorganisms showed distinct growth over a period of 72 h but failed to lower the pH of the soil suspension and did not release P in solution (Table 2). Similar results were obtained when the alfisol soil suspension was supplemented with 40 mg RP/g of soil.

P-solubilisation abilities of E. asburiae PSI3 and B. coagulans

E. asburiae PSI3 could grow utilizing AlP, FeP or RP as sole P sources when the minimal

Table 1. Soil Characteristics

Soil type	pH	EC (m.mha/ cm)	Organic C (%)	Exchangeable (ppm)				Mineral N (ppm) (NH_4 + NO_3)	CEC	Total N (ppm)	Micronutrients DTPA Extractable (ppm)				
				K	Ca	Mg	Na				Available P	Fe	Zn	Mn	Cu
Red soil	6.62	0.12	0.55	167	1904	264	74	11	10.95	742	1.7	22.6	0.62	32.5	1.36
Vertisol	8.18	0.28	0.68	285	11730	556	104	21.1	40.15	805	7.1	11.1	0.56	30.2	1.3

Table 2. Efficacy of native microflora in solubilizing P from alfisol and from RP in alfisol when supplemented with glucose (100 m*M*) and KNO_3 (10 m*M*)

Time (h)	Alfisol			Alfisol with RP[a]		
	Viable count (cfu/mL)	pH of soil suspension	Solution P (μM)	Viable count (cfu/mL)	pH of soil suspension	Solution P (μM)
0	$(3.06 \pm 1.2) \times 10^5$	6.7 ± 0.02	U.D.[b]	$(2.76 \pm 0.83) \times 10^5$	6.58 ± 0.34	U.D.
72	$(7.33 \pm 0.8) \times 10^8$	6.41 ± 0.07	U.D.	$(6.80 \pm 0.98) \times 10^7$	6.84 ± 0.05	U.D.

[a]RP was added at 40 mg /g of soil.
[b]U.D.: Undetectable.
Values expressed as mean ± S.D. for minimum of three independent experiments.

medium plates are buffered at 10 m*M* and 100 m*M* Tris.Cl pH-8.0. At 10 m*M* Tris buffer, *E. asburiae* PSI3 showed acid secretion (as indicated by change in color of methyl red change from yellow to red) whereas *B. coagulans* could neither grow nor secrete organic acid in plates containing FeP, AlP or RP (Figure 1).

Upon inoculation of *E. asburiae* PSI3, *C. koseri* and *B. coagulans* separately in alfisol soil solution containing 100 m*M* glucose and 10 m*M* KNO_3 each of the bacteria showed significant increase in their viable count with concomitant lowering of the soil pH (Table 3). However, there was no detectable P released. When RP was supplemented, a similar pattern was observed in growth and solution pH but no discernable P release was seen upto 72 h. RP solubilisation was also monitored in alfisol at 1 h time intervals for the first 8 h with *E. asburiae* PSI3. The pH of the medium dropped from 6.4 to 4.2 within 7 h but P release could not be detected at any time point (data not shown).

Efficacy of organic acids at releasing P from alfisol soil suspension with or without RP supplementation

In order to understand the nature and concentration of organic acids required to solubilize alfisol, soluble P was determined after the addition of acetic, lactic, succinic, tartaric, citric, oxalic and gluconic acids to alfisol in the range of 10–50 m*M*. Soluble P was not detectable in all cases even though pH of the soil suspension dropped below 4.0 (data not given). When alfisol was supplemented with 40 mg RP per g of soil, addition of organic acids resulted in drop in pH to less than 4.0 as well as release of P in the supernatant (Table 4). The effectiveness in solubilising P

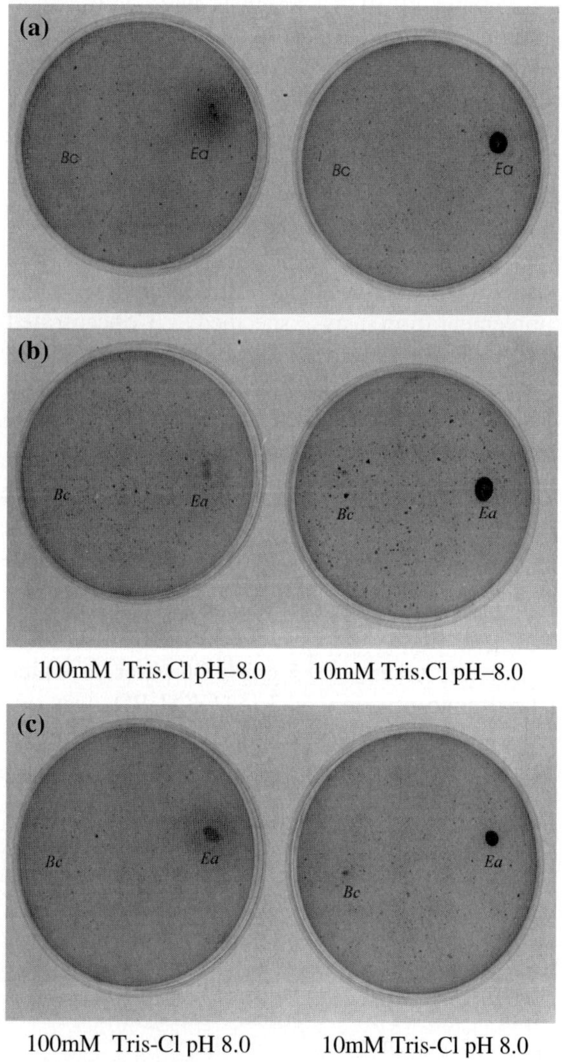

Figure 1. Solubilization of AlP, FeP and rock phosphates by *E. asburiae* PSI3 and B. *coagulans*. A – AlP MM agar plate; B – FeP MM agar plate; C – RP MM agar plate; *Bc* – *B. coagulans*; *Ea* – *E. asburiae* PSI3.

Table 3. Efficacy of PSMs in P-solubilization from alfisol and RP in alfisol when supplemented with 100 m*M* glucose and 10 m*M* KNO_3

Microorganism	0 h			72 h		
	Viable count (cfu/mL)	pH	Solution P (μM)	Viable count (cfu/mL)	pH	Solution P (μM)
Without RP						
E. asburiae PSI3	$(2.1 \pm 1.2) \times 10^4$	5.91 ± 0.4	U.D. [b]	$(1.6 \pm 2.4) \times 10^7$	4.43 ± 0.21	U.D.
C. koseri	$(5.6 \pm 0.9) \times 10^4$	6.12 ± 0.32	U.D.	$(8.6 \pm 3.1) \times 10^8$	4.31 ± 0.08	U.D.
B. coagulans	$(3.4 \pm 1.2) \times 10^4$	6.59 ± 0.08	U.D.	$(5.3 + 3.1) \times 10^7$	4.63 ± 0.11	U.D.
With RP[a]						
E. asburiae PSI3	$(1.0 \pm 0.8) \times 10^5$	6.32 ± 0.41	U.D.	$(5.8 \pm 0.12) \times 10^8$	4.6 ± 0.36	U.D.
C. koseri	$(3.3 \pm 1.2) \times 10^5$	5.98 ± 0.37	U.D.	$(9.8 \pm 1.8) \times 10^7$	4.64 ± 0.17	U.D.
B. coagulans	$(1.8 + 2.3) \times 10^4$	6.59 ± 0.08	U.D.	$(4.3 \pm 1.4) \times 10^7$	4.63 ± 0.11	U.D.

[a]RP was added at 40 mg/g of soil.
[b]U.D.: Undetectable.
Values expressed as mean ± S. D. for minimum of three independent experiments.

Table 4. Efficacy of different organic acids at P-solubilization from alfisol supplemented with RP (40 mg/g soil)

Organic acid	Concentration (m*M*)	0 h		24 h	
		pH	Solution P (μM)	PH	Solution P (μM)
None	–	6.73 ± 0.21	U.D. [a]	6.69 ± 0.02	U.D.
Oxalic	10	3.42 ± 0.05	1560 ± 4.0	3.14 ± 0.01	1092 ± 2.4
Citric	10	3.45 ± 0.04	409 ± 8.0	3.22 ± 0.02	387 ± 3.0
Citric	20	2.80 ± 0.034	850 ± 6.5	3.18 ± 0.431	646 ± 2.9
Tartaric	20	3.50 ± 0.02	550 ± 6.0	3.60 ± 0.23	502 ± 2.1
Gluconic	50	3.40 ± 0.05	800 ± 2.0	3.71 ± 0.01	300 ± 5.0
Lactic	50	3.25 ± 0.015	920 ± 7.0	3.36 ± 0.02	200 ± 3.0
Succinic	50	3.64 ± 0.04	332 ± 7.0	$3.66 + 0.07$	359 ± 5.3
Acetic	50	3.80 ± 0.23	102 ± 2.0	4.45 ± 0.3	80 ± 2.0

[a]U. D. : Undetectable. Values expressed as mean ± S.D. for minimum of three independent experiments.

from RP amended alfisol is in the order oxalic > citric > tartaric > gluconic > lactic > succinic > acetic acid.

Since oxalic acid was found to be most efficient, it was of interest to determine the minimum amounts of oxalic acid and RP needed to release sufficient P in alfisols. 5 m*M* oxalic acid could solubilize P in the range of 217–440 μM when RP was from 10–30 mg per g of soil (Table 5). Increasing oxalic acid to 10 m*M* could

Table 5. Solubilization of RP by oxalic acid in alfisol

Amount of RP (mg/g soil)	Before adding the oxalic acid		30 min after adding the oxalic acid.	
	pH	Solution P (μM)	pH	Solution P (μM)
10[a]	6.59 ± 0.4	U.D.[c]	3.79 ± 0.34	217 ± 2.4
20[a]	6.76 ± 0.5	U.D.	3.70 ± 0.56	264 ± 1.8
30[a]	6.75 ± 1.0	U.D.	3.69 ± 0.43	443 ± 2.7
2[b]	$6.65 + 1.0$	U.D.	$3.50 + 0.50$	$120 + 1.2$

[a]5 m*M* oxalic acid.
[b]10 m*M* oxalic acid.
[c]U. D.: Undetectable.
Values expressed as mean ± S.D. for minimum of three independent experiments.

bring about solubilization of 120 μM P when 2 mg RP was added /g of alfisol.

In order to study the P fixation property of the soil sample, 2 mM KH_2PO_4 was added to the alfisol soil and P was estimated in the supernatant at different time intervals. P levels in the supernatant rapidly decreased to 200 ± 7.0 μM immediately after addition and the P levels were 62 ± 2.6 μM and 22 ± 2.3 μM after 30 min and 24 h, respectively.

Effect of inoculation of mung bean with E. asburiae PSI3 in alfisol and vertisol

Effectiveness of P-solubilising *E. asburiae* PSI3 on growth of mung bean was monitored in pot conditions. Compared to the uninoculated controls, the root length of mung bean significantly decreased when *E. asburiae* PSI3 was inoculated in alkaline vertisol whereas such decrease was absent in alfisols (Figure 2).

Discussion

Present study demonstrates that native bacteria of alfisol could grow well in the presence of sufficient amounts of C and N but did not bring about drop in pH or release P in soil solution. Native microorganisms also failed to solubilize P from RP supplemented alfisols. This result is

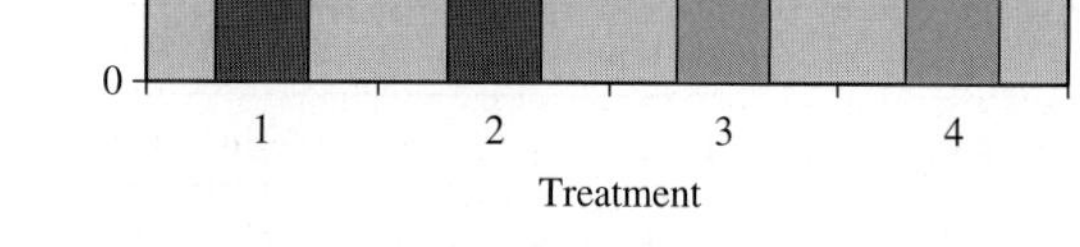

(1) Uninoculated plants in alkaline vertisol.

(2) Plants inoculated with *E. asburiae* PSI3 in alkaline vertisol.

(3) Uninoculated plants in alfisol.

(4) Plants inoculated with *E. asburiae* PSI3 in alfisol.

Figure 2. Effect of *E. asburiae* PSI3 inoculation on length of roots of mung bean when grown in alfisol and vertisol.

surprising since alfisols have less buffering capacity and the abundance of PSMs has been found at 10^3–10^5 cfu/gm in many soils (Kucey et al., 1989).

B. coagulans and *E. asburiae* PSI3 are known to solubilize RP and *E. asburiae* PSI3 is also capable of releasing P from alkaline vertisol (Gyaneshwar et al., 1999). When these bacteria were checked for solubilisation of FeP and AlP, *E. asburiae* PSI3 could grow well on the medium buffered with 10 mM as well as 100 mM Tris–Cl pH 8.0 but showed acidification only on medium buffered with 10 mM Tris pH 8.0. *B. coagulans* could not grow even in plates containing 10 mM Tris pH 8.0. This is in agreement with the organic acid secretion abilities of these two bacteria. *B. coagulans* secretes 1.3 mM succinic, 1.4 mM lactic, 1.4 mM citric and 4.7 mM acetic acids (Gyaneshwar et al., 1998) whereas *E. asburiae* PSI3 secretes 50 mM gluconic acid when grown under similar conditions (Gyaneshwar et al., 1999). Similar results of solubilization of FeP and AlP were reported for *Penicillium rugulosum*, which secretes approximately 45 mM gluconic acid (Reyes et al., 1999a, b).

C. koseri, B. coagulans and *E. asburiae PSI3* could grow and acidify the alfisol soil solution to pH 4.0 but no soluble P was found. There was no release in P even after the supplementation of RP to alfisol. Since RP gets solubilized in aqueous solutions when the pH is less than 5.0 absence of soluble P when the alfisol pH is less than 4.0 could be attributed to the refixation of released P with iron and aluminium oxides and hydroxides present in alfisols (Marwah, 1983). Instant decrease of P from the solution after alfisol was supplemented with 2 mM KH_2PO_4 clearly demonstrated efficient refixation of soluble P by alfisol soil used in this study.

Since known PSMs were found to be ineffective in alfisols, it was necessary to know the nature and amount of organic acids required for P release from alfisols and alfisols supplemented with RP. Addition of organic acids to alfisol did not result in release of detectable amounts of P although the pH of the suspension was reduced to less than 3.0. When RP was added to alfisols, all organic acids used in this study brought about a drop in pH to below 3.8 but the amount of P released ranged from 102 μM with 50 mM acetic acid to 1.5 mM with 10 mM oxalic acid.

As yet the only PSM known to secrete oxalic acid (10 mM) is *Penicillium billaii* (Cunningham and Kuiack 1992). It would be interesting to know phosphate solubilisation ability of this organism with alfisols amended with RP.

Pot experiments where *E. asburiae* PSI3 was inoculated in the root rhizosphere of mung bean plants showed that root elongation was low in case of plants grown in vertisol whereas in case of alfisol there was no significant effect on root length as compared to the uninoculated controls. It is known that under P deficiency, plants develop longer roots to increase their surface area for enhanced nutrient absorption (Gilroy and Jones, 2000). Thus the pot experiments suggest that *E. asburiae* PSI3 seems to be effective in vertisol but not in alfisol.

As there is increasing need of harnessing acidic soils for increasing agricultural produce, it is necessary to screen bacteria which can solubilize P in acidic soils. Our results indicate that mere supplementing screening media with Al–P or Fe–P would not be most ideal to select effective PSMs. The screening medium should also take into account of P fixing properties of alfisols. Further, PSMs secreting at least 5 m*M* oxalic acid would prove to be effective when used in alfisols amended with RP.

Acknowledgements

One of the authors (S.S.) is supported by fellowship from University Grants Commission, India.

References

Ae N, Arihara J, Okada K, Yoshihra T and Johansen C 1990 Phosphorus uptake by pigeon pea and its role in cropping systems of the Indian subcontinent. Science 248, 477–490.

Ae N, Arihara J, Okada K 1991 Phosphorus response of chick pea and evaluation of phosphorus availability in Indian alfisols and vertisols Phosphorus Nutrition of Grain Legumes. India: International Crop Research Institute for the Semi- Arid Tropics. ISBN 92–9066–200. 33P.

Abekoe M K 1998 Fertilizer P transformation and P availability in hillslope soils of Northern Ghana. Nutr. Cycl. Agroecosyst. 52, 45–54.

Ames BN 1964 Assay of inorganic phosphate, total phosphate and phosphatases. Meth. Enzymol. 8, 115–118.

Bolan N S, Robson A D and Barrow N I 1987 Effect of vesicular arbuscular mycorrhizae on availability of iron phosphates in plants. Plant Soil. 99, 401–410.

Bolan N S, Naidu R, Mahimairaja S and Baskaran S 1994 Influence of low-molecular-weight organic acids on the solubilization of phosphates. Biol. Fertil. Soils. 18, 311–319.

Bolland M D A 1992 Agronomic effectiveness of Partially Acidulated Rock Phosphate (PARP) and fused calcium-magnesium phosphate compared with superphosphate. Fertil. Res. 32, 169–183.

Bowman R A 1988 A rapid method to determine total phosphorus in soils. Soil Sci. Soc. Am. J. 52, 1301–1304.

Butegwa C N 1996 Agronomic evaluation of fertilizer products derived from Sukulu hills phosphate rock. Fertil. Res. 44, 113–122.

Cunningham J E and Kuiack C 1992 Production of citric and oxalic acids and solubilization of calcium phosphate by *Penicillium bilaii*. Appl. Environ. Microbiol. 58, 1451–1458.

Gadd G M 1999 Fungal production of citric and oxalic acid: Importance in metal speciation, physiology and biogeochemical processes. Adv. Microb. Physiol. 41, 48–92.

Gilroy S and Jones D L 2000 Through form to function: root hair development and nutrient uptake. Trends Plant Sci. 5, 56–60.

Gyaneshwar P, Naresh kumar G and Parekh L J 1998 Effect of buffering on the phosphate solubilising ability of microorganisms. World J. Microbiol. Biotechnol. 14, 669–673.

Gyaneshwar P, Parekh L J, Archana G, Poole P S, Collins M D, Hutson R A and Naresh kumar G 1999 Involvement of a phosphate starvation inducible glucose dehydrogenase in soil phosphate solubilisation by *Enterobacter asburiae*. FEMS Microbiol. Lett. 171, 223–229.

Huges J C 1998 The dissolution of North Carolina phosphate rock in some South-Western Australian soils. Fertil. Res. 38, 249–253.

Jones D L and Darrah P R 1994 Role of root derived organic acids in the mobilization of nutrients from the rhizosphere. Plant Soil. 166, 247–257.

Kpomblekou A K and Tabatabai M A 1994 Effect of organic acids on the release of phosphorus from phosphate rocks. Soil Sci. 158, 442–448.

Kucey R M N, Janzen H H and Leggett M E 1989 Microbially mediated increases in plant-available phosphorus. Adv. Agron. 42, 198–228.

Marwah B C 1983 Partially acidulated rock phosphate as a source of fertilizer phosphorus with special reference to high P-fixing acid soil - A review. Proc. Indian Nat. Sci. Acad. B49, 436–446.

Nahas E 1996 Factors determining rock phosphate solubilization by microorganisms. World J. Microbiol. Biotechnol. 12, 562–572.

Omar S A 1998 The role of rock-phosphate-solubilizing fungi and vesicular-arbuscular-mycorrhiza (VAM) in the growth of wheat plants fertilized with rock phosphate. World J. Microbiol. Biotechnol. 14, 211–218.

Piccini D and Azcon R 1987 Effect of phosphate-solubilising bacteria and vesicular-arbuscular mycorrhizal fungi on the utilization of Bayovar rock phosphate by alfalfa plant using a sand -vermiculite medium. Plant Soil 101, 45–50.

Rajan S S S, Watkinson J H and Sinclair A G 1996 Phosphate rocks for direct application to soils. Adv. Agron. 77, 57–159.

Reyes I, Bernier L, Simard R R and Antoun H 1999a Effect of nitrogen source on solubilization of different phosphates by an isolate of *Penicillium rugulosum* and two UV-induced mutants. FEMS Microbiol. Lett. 28, 281–290.

Reyes I, Bernier L., Simard R R, Tanguay P and Antoun H 1999b Characteristics of phosphate solubilization by an isolate of a tropical *Penicillium rugulosum* and two UV-induced mutants. FEMS Microbiol. Lett. 28, 291–295.

Vassilev N, Irena F, Maria V and Azcon R 1996 Improved plant growth with rock phosphate solubilized by *Aspergillus niger* grown on sugar-beet waste. Biores. Technol. 55, 237–241.

Vassileva M, Vassilev N and Azcon R 1998 Rock phosphate solubilisation by *Aspergillus niger* on olive cake-based medium and its further application in a soil-plant system. World J. Microbiol. Biotechnol. 14, 281–284.

Von Uexull H R and Mutert E 1995 Global extent, development and economic impact of acid soils. Plant Soil 171, 1–15.

Effect of phosphorous solubilizing bacteria on the rhizobia–legume simbiosis

Susana B. Rosas[1], M. Rovera, J.A. Andrés & N.S. Correa
Facultad de Ciencias Exactas, Laboratorio de Fisiología Vegetal, Físico-Químicas y Naturales, Universidad Nacional de Río Cuarto, Campus Universitario, X5804ZAB, Río Cuarto, Córdoba, República Argentina.
[1]*Corresponding author**

Received 2 December 2002. Accepted in revised form 2 January 2003

Key words: co-inoculation, phosphorous solubilizers, *Pseudomonas*

Abstract

Alfalfa and soybean are the most important leguminous plants in the agricultural system of the semi-arid region of Argentina. The possible action of phosphorous solubilizing bacteria on the leguminous-rhizobia simbiosis was studied since in this region the available phosphorous distribution (Pd) is not uniform. The strains used were *Sinorhizobium meliloti 3DOh13*, excellent solubilizer of Fe and P for alfalfa, *Bradyrhizobium japonicum TIIIB* for soybean and two strains of *Pseudomonas putida* (Sp21 and Sp22) solubilizers of P for the treatments on growth promotion through an improvement in the nodulation and biological nitrogen fixing activity. The rhizobia strains assayed are capable of coexisting with *Pseudomonas*, with meaningful differences being observed in percentage of nodulation and air portion in co-inoculated soybean.

Introduction

Alfalfa is the most important forage species in the agricultural system, by the diffusion of its culture as well as by the quality of nutrients that the forage provides (Viglizzo, 1995). Furthermore the effect that this leguminous has on soil fertility is very important as well as the contribution of its radical system to the improvement and the conservation of the soil structure (Vance, 1997).

Also soybean is an important grain, being Argentina one of the main leading exporting countries in the world. The distribution of phosphorous is very variable in the central region of Argentina and low yields have been associated with lack of this mineral.

Numerous microorganisms, especially those associated with roots, have the ability to increase plant growth and productivity. In some cases, this effect has been suggested to involve solubilization of otherwise unavailable mineral nutrients (Goldstein, 1995).

There is evidence of the fact that some kinds of Pseudomonas increase absorption of N, P and K, in addition to serving as biocontrol of phytopathogenic fungi and to produce phytohormones in the rizosphere, which promotes greater growth of the plants (O'Sullivan and O'Gara, 1992).

In general, Pseudomonas fluorescent can promote plant growth by production of siderophores that capture ferric oxides to convert them into forms available for roots. This increases radical volume (Peter et al., 1987). The insoluble inorganic compounds of phosphorus [$Ca_3(PO_4)_2$] are not totally available for the plants, but these can be converted by bacteria into phosphates available for the plant roots. The principal active strains in this conversion belong to the genus *Pseudomonas, Mycobacterium, Micrococcus, Bacillus* y *Flavobacterium* (Alexander, 1981; Asea et al., 1988; Salih et al., 1989).

* FAX No: + 54-358-4680280/4676232.
E-mail: srosas@exa.unrc.edu.ar

E. Velázquez and C. Rodríguez-Barrueco (eds.),
First International Meeting on Microbial Phosphate Solubilization, 125–128

The techniques implementation that employ microorganisms of the rhizosphere capable of mobilizing iron and phosphorus is the objective of the present work, in order to ensure the sustainability of the agricultural production systems, reducing the environmental pollution risks that are originated by the use of chemical fertilizers.

Materials and methods

Bacterial strains

Sinorhizobium meliloti 3DOh13 and *Bradyrhizobium japonicum TIIIB* was maintained on YEM (yeast extract mannitol) agarized medium (Vincent 1970). *Pseudomonas putida SP21* and *SP22* was maintained on TS (tripticase soya) agarized medium (Britania Laboratory, Argentina). All strains were obtained and identified by our laboratories.

Phosphate solubilization

The medium used containing 2 g yeast extract, 20 g glucose, 2 g phosphate tricalcium, 0.06 g actidione, 15 g agar made up to 1000 mL with water, at pH 7 was used. The medium was inoculated with the relevant strains and incubated at 28 °C for 5 days. Bacterial colonies forming clarification halos were considered to be phosphate solubilizers.

Production of siderophores

The Chrome azurol S method described by Rosas and Schröder (1992) was used. Plates were incubated at 28 °C for 5 days, and microorganisms exhibiting an orange halo were considered to be producers of siderophores.

Plant material

Soybean (*Glycine max* L.) seeds were disinfected for 20 min with 0.4% calcium hypochlorite solution and alfalfa (*Medicago sativa* L.) seeds were scarified by shaking for 15 min in concentrated sulphuric acid, then disinfected with 70% ethanol for 3 min. Seeds were then washed will several changes of sterile distilled water.

Inoculation assays

Seeds were transferred under aseptic conditions onto the surface of perlite/sand (2:1) bed and allowed to germinate. After 48 h from sowing the seeds were inoculated with the corresponding rhizobia strain and with the *Pseudomonas* strains. Bacterial cultures were obtained in the media described following standard procedures cointaining 10^9 CFU mL^{-1} for rhizobia and 10^6 CFU mL^{-1} for Pseudomonas, adjusted by optical density. Two milliliters of each inoculum were applied to the root system of each seedling in the planting hole.

Plants were watered alternately with sterile distilled water and a modification of nitrogen-free Jensen's solution (Vincent, 1970) where the source of phosphorous was changed to phosphate tricalcium.

Uninoculated controls were watered in the same manner, but with the addition of 0.5% KNO_3 L^{-1} to the original Jensen's solution (control with soluble phosphorous) and to the Jensen's modified solution with phosphate tricalcium (control with insoluble phosphorous).

The plants were grown in a chamber under controlled conditions: a 16 h day at 28 °C, and 8 h night at 16 °C, and a light intensity of 220 $\mu E\ m^{-2}\ s^{-1}$. Forty days after sowing the plants were harvested in order to evaluate nodulation and fresh and dry weight.

Results and discussion

S. meliloti 3DOh13 presents solubilization of Fe and P while *B. japonicum TIIIB* is a poor phosphate solubilizer and siderophore producer. *P. putida SP21* and *SP22* mobilize Fe and P with greater efficiency than *B. japonicum TIIIB* and *S. meliloti 3DOh13*, with mobilization halos greater to 15 mm (Table 1).

Tables 2 and 3 show the results of the experiences of soybean and alfalfa inoculation with the rhizobia and *Pseudomonas* strains. The co-inoculated bacteria are capable to coexist without provoking deleterious effects on the plants. Growth of seedlings of alfalfa and soybean inoculated with *P. putida* were not affected.

It was observed that the plants inoculated with *S. meliloti 3DOh13* did not show meaningful differences with respect to the two controls

Table 1. Production of siderophores and phosphate solubilization of bacterial strains

Strain	Production of siderophores (diameter of halo in mm)	Phosphate solubilization (diameter of halo in mm)
S. meliloti 3DOh13	6.0 ± 0.7	3.2 ± 0.4
B. japonicum TIIIB	2.2 ± 0.5	1.0 ± 1.2
P. putida SP21	16.1 ± 0.9	18.1 ± 2.7
P. putida SP22	17.5 ± 1.3	21.0 ± 3.1

Results represent the mean of three replications per strain.

Table 2. Inoculation in soybean with phosphate solubilizer and nitrogen fixation bacteria

Microbial treatment	Root fresh weight (g)	Root dry weight (mg)	Shoot fresh weight (g)	Shoot dry weight (g)	Number of nodules (no. plant^{-1})	Dry weight of nodules (mg plant^{-1})
Control N + soluble P	1.32 ± 0.11	0.14 ± 0.05	3.95 ± 0.41	0.43 ± 0.05	–	–
Control N + insoluble P	0.81 ± 0.14	0.09 ± 0.03	2.83 ± 0.32	0.38 ± 0.07	–	–
B. japonicum TIIIB	0.93 ± 0.07	0.11 ± 0.02	3.04 ± 0.22	0.40 ± 0.04	35.0 ± 9.1	69.6 ± 15.3
B. japonicum TIIIB + *P. putida SP 21*	1.42 ± 0.09	0.19 ± 0.02	3.79 ± 0.27	0.45 ± 0.07	58.2 ± 11.6	148.4 ± 17.1
B. japonicum TIIIB + *P. putida SP 22*	1.27 ± 0.12	0.07 ± 0.03	3.68 ± 0.33	0.40 ± 0.06	61.6 ± 10.5	169.8 ± 12.3

Results represent the mean of three replications per treatment of 25 plants each.

Table 3. Inoculation in alfalfa with phosphate solubilizer and nitrogen fixation bacteria

Microbial treatment	Root fresh weight (mg)	Root dry weight (mg)	Shoot fresh weight (mg)	Shoot dry weight (mg)	Number of nodules (no. plant^{-1})	Dry weigtht of nodules (mg plant^{-1})
Control N + soluble P	32.4 ± 1.6	1.6 ± 0.2	72.3 ± 3.8	15.1 ± 1.5	–	–
Control N + insoluble P	27.3 ± 1.2	1.4 ± 0.1	68.6 ± 4.1	12.3 ± 1.5	–	–
S. meliloti 3DOh13	29.4 ± 1.3	1.5 ± 0.3	67.4 ± 3.5	11.6 ± 1.7	7.6 ± 2.3	12.3 ± 2.4
S. meliloti 3DOh13 + *P. putida SP 21*	29.8 ± 1.4	1.6 ± 0.2	68.9 ± 2.9	12.4 ± 1.2	6.8 ± 1.5	10.2 ± 1.8
S. meliloti 3DOh13 + *P. putida SP 22*	28.1 ± 1.3	1.5 ± 0.1	66.4 ± 4.2	10.3 ± 1.1	7.1 ± 1.1	10.5 ± 2.0

Results represent the mean of three replication per treatment of 25 plants each.

supplied with mineral nitrogen. This did not occur in soybean inoculated with *B. japonicum TIIIB*. When alfalfa was co-inoculated with *S. meliloti* and *Pseudomonas* differences were not observed with respect to the inoculation with *S. meliloti* alone, while for soybean a greater number of nodules and fresh and dry weight was registered when the co-inoculation with *B. japonicum TIIIB* and *Pseudomonas* was completed.

This result can be due to the fact that *Pseudomonas* provides P supplying the deficiencies of *B. japonicum TIIIB* with respect to *S. meliloti*.

Acknowledgements

The authors are gratefully indebted to Secretaría de Ciencia y Técnica of Universidad Nacional de Río Cuarto.

References

Alexander M 1981 Introducción a la microbiología del suelo. AGT Editor, México DF, México.

Asea P E A, Kucey R M N and Stewart J W B 1988 Inorganic phosphate solubilization by two *Penicillium* species in solution culture. Soil Biol. Biochem. 20, 459–464.

Goldstein A H 1995 Recent progress in understanding the molecular genetics and biochemistry of calcium phosphate solubilization of gram negative bacteria. Biol. Agric. Hortic. 12, 185–193.

O'Sullivan D J and O'Gara D 1992 Trails of fluorescent Pseudomonas spp. involved in suppression of plant root pathogens. Microbiol. Rev. 56, 662–676.

Peter A H M B, Barker A W, Marugg J D, Weisbeek P J and Schippers B 1987 Bioassay for studying the role of siderophores in potato growth stimulation by *Pseudomonas* spp. in short potato rotations. Soil Biol. Biochem. 19, 443–449.

Rosas S B and Schröder E 1992 Informe UNESCO. Short Time Fellowships in Biotechnology.

Salih H M, Yonka A I, Abdul-Rahem A M and Munam B H 1989 Availability of phosphorous in calcareous soil treated with rock phosphate or superphosphate as affected by phosphate dissolving fungi. Plant Soil 120, 181–185.

Vance C 1997 Enhanced agricultural sustainability through biological nitrogen fixation. *In* Biological Fixation of Nitrogen for Ecology and Sustainable Agriculture. Eds. A Legocki, A Bothe and A Pühler. pp. 179–186. Springer-Verlag, Germany.

Viglizzo E F 1995 El rol de la alfalfa en los sistemas de producción. *In* La Alfalfa en la Argentina. Eds. E H Hijano and A Navarro. pp. 260–272. Instituto Nacional de Tecnología Agropecuaria (INTA), Buenos Aires, Argentina.

Vincent J M 1970 A manual for the practical study of the root nodule bacteria. IBP Handbook N 15. Blackwell Scientific Publications, Oxford, UK.

Defense response in bean roots is not affected by low phosphate nutrition

L. Alvarez-Manrique, A. Richards & E. Soriano[1]
Instituto de Investigaciones Químico-Biológicas, UMSNH, A.P. 50-3, Morelia, Mich., México. [1]*Corresponding author**

Received 9 December 2002. Accepted in revised form 2 January 2003

Key words: bean, phosphate nutrition, phytoalexins

Abstract

Mycorrhizal symbiosis in plants changes the relationship between plant roots and soil pathogens, leading sometimes to an increase in disease resistance. Since phosphate uptake is the main effect of mycorrhizal colonization, we measured the effect of the levels of phosphate in plant nutrition on the defense response in bean roots. Bean plants growing in a nutrient solution with either 9 or 85 mg/L were elicited to compare phytoalexin accumulation to measure the effect at short and long term of phosphate supply and defense response. Adequate phosphate nutrition did not increase phytoalexin levels in the roots, meanwhile roots lacking phosphate did respond accumulating a phytoalexin. This suggests that the increased resistance observed in mycorrhizal roots is not due to increased accessibility of phosphate in colonized roots.

Introduction

To obtain material resources, humanity has transformed the environment along its evolution. It is well known that agricultural practices have become harmful to the ecosystem leading to more ecologically conscious methods of cultivation for food production (Toledo et al., 1985).

In Latin America in general, soil dedicated to farming are deficient in phosphate content. In low-phosphate soil, plants rely in different mechanisms to acquire such valuable mineral. Symbiosis with microorganisms, particularly mycorrhizal, is a very common solution to phosphate acquisition for plants (Francis and Read, 1994). Besides, mycorrhizal colonization in agriculture is a recommended method of fertilization that is benign to ecosystems (Cook and Baker, 1989).

Moreover, changes in plant metabolism after mycorrhizal symbiosis is reported (Gianinazzi-Person and Gianinazzi, 1983; Graham, 1988) as affecting the relationship of plant roots and soil phytopathogens. Often this change give an increase in disease resistance to the plant where the mechanism is not well understood as yet. Since phosphate uptake is the main effect of mycorrhizal colonization, it is possible that increment of the nutrient in the plant tissue is related to increase in disease resistance. Antimicrobial compounds such as phytoalexins are part of the active defense in plants, which contribute to an increase in disease resistance (Soriano and Heredia, 1990). Ability to accumulate phytoalexins as a measure of increased resistance status, was compared in bean plants that were exposed to adequate phosphate nutrition with bean plants grown under a lack of adequate phosphate conditions.

* FAX No: +52-443-3167436.
E-mail: esoriano@zeus.umich.mx

Material and methods

Plant material

Bean seeds (*Phaseolus vulgaris* L.) cv flor de mayo were surface sterilized by washing with a

E. Velázquez and C. Rodríguez-Barrueco (eds.),
First International Meeting on Microbial Phosphate Solubilization, 129–133

neutral detergent and treating them with a chloride solution (1% Cl) during 10 min. After rinsing the seeds with distilled water, they were allowed to germinate at 30 °C (2–3 days).

The seedlings were placed in pots containing vermiculite (6 seed/pots) and were watered daily with a nutrient solution (150 mL each). The solution contained 60 mg/L Mg $SO_4 \cdot 7H_2O$, 195 mg/L $CaNO_3 \cdot 4H_2O$, 1.5 mg/L $Fe_2(SO_4)_3 \cdot 7H_2O$, 1.8 mg/L EDTA-$Na^+ \cdot 2H_2O$, 0.72 mg/L Boric acid, 4.45 mg/L Mn $Cl_2 \cdot 4H_2O$, 0.055 $ZnSO_4 \cdot 7H_2O$, 0.020 $CnSO_4 \cdot 2H_2O$, 126 mg/L KNO_3. From this, low phosphate solution was prepared which contained 9 mg/L K_2HPO_4 and high phosphate solution, which contained 85 mg/L K_2HPO_4.

Elicitor preparation

For the preparation of endogenous elicitor, the method of Dixon et al. (1989) was used with some modifications. Bean cell walls were extracted from 7 days old plants grown under a 14 h photoperiod. For release of elicitors, enzymic hydrolysis of the cell walls was employed rather than of sodium polypectate. After hydrolysis, the mixture was fractionated on columns of Biogel P-6 (1.0 × 1.32 cm) and Biogel P-4 (1.0 × 140 cm) eluted with distilled water. An oligogalacturonide (PD = 10) with high elicitor activity was obtained (Soriano and García, 1993). This elicitor has shown effectiveness in bean tissue treatment (Cano et al., 1994).

Root inoculacion

Bean plants 7 days old grown under either phosphate solution were treated with the elicitor solution by applying 10 mL (0.5 mg/mL galacturonic acid eq.) around the root in the pots. The nutrient solution was applied by the bottom of the pots to avoid diluting the elicitor solution. Control plants did not received elicitor solution. Three plants samples were used for each determination and this was made twice.

Phytoalexin extraction

According to the experiment, at 10 h or 7 days intervals, plants were harvested and the main root length was measured. After weighing the entire roots, these were homogenized with 95% ethanol (10 mL/g fresh tissue). The filtered homogenate was dried under vacuum and the residue was recovered in 30 mL ethyl acetate: water (v/v). The organic phase was collected after three times extraction with the ethylacetate: water (v/v) mixture and dried with sodium sulphite and then under vacuum. The residue recovered with 500μL ethyl acetate was separated by thin layer chromatography. Solvent system was chloroform: methanol (20:1)(v/v) and analytical gel plates with fluorescent indicator (Sigma-Chemical) were used.

Phaseollin was eluted with ethanol and absorbance at 289 nm was determined. To calculate phaseollin concentration, the known extinction coefficient was used (Theodorou et al., 1982).

Results and discussion

To measure phytoalexin response in roots, first the effect in root growth was determined. The increase in fresh weight of the bean roots grown under adequate mineral nutrition in either presence or absence of the phytoalexin elicitor showed that growth of the roots was slightly inhibited by the elicitor in the first 24 h of treatment. Afterwards, however, root growth resumed its rate (Figure 1). Phaseollin accumulation after

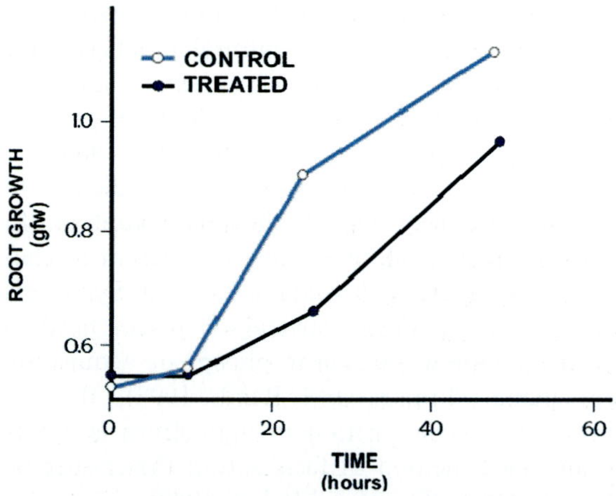

Figure 1. Root growth in elicited bean plants. Bean plants 7 days old growing in vermiculite were watered daily with 150 mL nutrient solution that contained 85 mg/L phosphate. Around the roots 10 mL elicitor solution (0.5 mg/mL galacturonic acid equivalents) was applied and the plants were harvested at 12, 24 and 48 h after treatment. The roots were removed to be weighted. Control plants did not received elicitor solution. Experiment made twice $n = 3$.

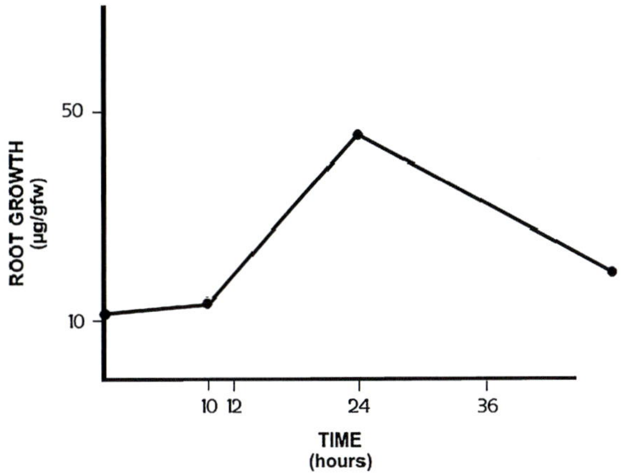

Figure 2. Phytoalexin response in roots. Bean roots elicited with oligogalacturonide when growing in nutrient solution (85 mg/L phosphate) were harvested to extract phytoalexin. Increase in phaseollin above control levels is plotted against time.

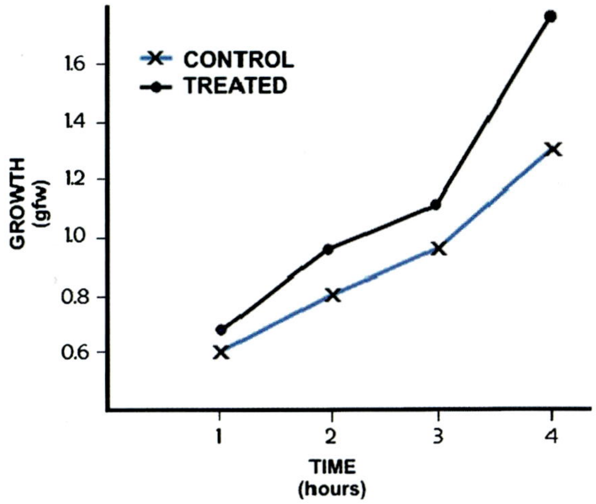

Figure 3. Root growth in elicited bean plants during a long period of time. The plants growing by hydroponics during 4 weeks watered with a nutrient solution (85 mg/L phosphate) were treated at the beginning of culture with the oligogalacturonide elicitor. At the end of each week the plants were harvested, and the fresh weight of the root system was weighted. Control plants did not received elicitor.

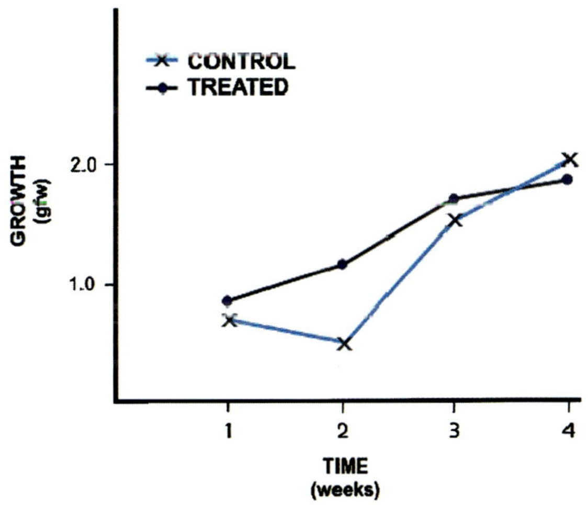

Figure 4. Root growth in elicited bean plants under low phosphate nutrition. Plants growing for 4 weeks under 2 nutrient solution with a low phosphate content (9 mg/L) were elicited with oligogalacturonide and the plants were harvested each week to determine root system fresh weight.

elicitation in bean roots was high and transient, with levels up to three times higher in elicited than in control roots at 24 h. Later, phaseollin levels decreased to basal levels (Figure 2). Most studies on phytoalexin formation in leguminosae have been made in aerial tissues (Bailey and Mansfield, 1982), but in different families of plants, the presence of phytoalexins on roots, especially as glucosides, has been reported (Higgins and Bates, 1994). Flavonoid accumulation in roots infected with vesicular-arbuscular mycorrhizal fungi has been reported that appears in those cells, which contain fungal structures (García-Garrido and Ocampo, 2002; Morandi et al., 1984). In this study, it was found that bean root tissue has thc ability to rccognizc and respond to the defense-elicitor. Moreover it is interesting to note, that the kinetics of phaseollin accumulation in bean roots is similar to that occurring in bean hypocotyls, leaves or pods (Zavala et al., 1989). These results suggest that all tissues in the plant posses a similar biochemical mechanism of defense response.

The effect of phosphate on root growth for plants elicited at the beginning of development, bean plants were grown up to 4 weeks under nutrition with either low or adequate content of phosphate. These last plants showed a constant increase in fresh weight along the 4 weeks but in presence of the elicitor the roots showed a stimulatory effect after the first week of treatment (Figure 3). On the other hand lowering the phosphate content of the nutrient solution caused a delay of growth in the first 2 weeks; afterwards, the roots resumed growth; such effect of low phosphate on growth however, was prevented by elicitor addition; the reason for such an effect is unknown (Figure 4). At the end of the experi-

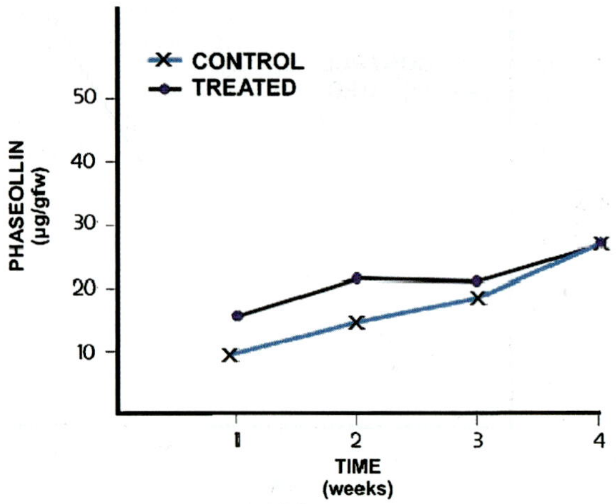

Figure 5. Phytoalexin response in root during development. Bean plants growing for 4 weeks under adequate phosphate nutrition (85 mg/L) were treated with the defense elicitor at the beginning of growth (0.5 mg/mL galacturonic acid eq.) The roots were removed each week to extract phaseollin content. Control plants did not received elicitor.

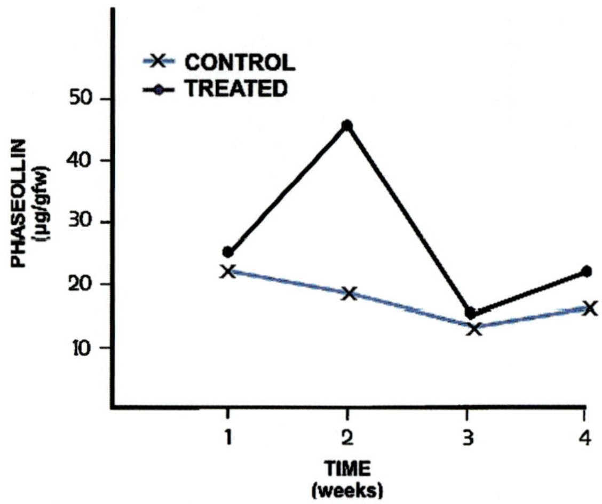

Figure 6. Defense response in roots under low phosphate nutrition. Bean plants growing for 4 weeks in the presence or absence of defense elicitor were irrigated with the nutrient solution that contained only 9 mg/L phosphate. The roots were removed each week and phaseollin content was determined.

ment (4 weeks) both control and low-phosphate roots with or without elicitor reached the same weight. It seems that during early development, the roots require enough phosphate in the nutrient solution to develop but after 4 weeks the amount of phosphate is not growth limiting.

The defense response of roots under low and adequate phosphate levels seemed to be completely different. Bean roots growing with 85 mg/mL phosphate in the nutrient solution when exposed to oligouronide elicitor did not show phytoalexin accumulation along the 4 weeks of development (Figure 5). On the other hand, when the bean plants were growing under low-phosphate, the presence of the oligouronide elicitor caused an increase in phytoalexin levels during the first 2 weeks of age of the plants. Later on, basal levels of phaseollin were observed (Figure 6). The defense response in these roots correlates with the period when plant growth was slowed down; it is possible that the stress caused by lack of phosphate accounts for the defense response in these roots. Several defense responses in plants are associated with stress conditions and the phenylpropanoid pathway seems to be switched on in such cases (Dixon et al., 1989). Therefore, it is possible that in mycorrhizal roots, the observed increase in disease resistance is not related to disponibility phosphate that is supplied by the fungus.

Acknowledgements

The authors thank Universidad Michoacana de San Nicolás de Hidalgo for financial support to the work and Consejo Nacional de Ciencia y Tecnología for fellowship to Alvarez-Manriquez, Alan Richards and Eva Soriano.

References

Bailey J A, Mansfield J W 1982 "Phytoalexins" Ed. Blackie Glasgow and London p. 40.

Cano H, Zavala G, Soriano E and López-Romero E 1994 Oligosaccharides plant and fungal elicit phaseollin accumulation with similar kinetics in bean (*Phaseolus vulgaris* L.) Plants. Rev. Mex. Fitopatol. 12, 163–173.

Cook R J and Baker K F 1989 The Nature and Practice of Biological Control of Pathogens. APS Press, 131 p.

Dixon R A, Jennings A C, Davies L A, Gerrish C and Murthy D L 1989 Elicitor active components from french bean hypocotyls. Physiol. Mol. Plant Pathol. 34, 99–115.

Francis R and Read D J 1994 The contribution of mycorrhizal fungi to the determination of plant community structure. Plant Soil 159, 11–25.

Gianinazzi-Pearson V and Gianinazzi S 1983 The physiology of vesicular-arbuscular mycorrhizal roots. Plant Soil 71, 197–209.

Graham J H 1988 Interactions of mycorrhizal roots with soil-borne pathogens and other organisms: An introduction. Phytopathology 78, 465–466.

García-Garrido J M and Ocampo J A 2002 Regulation of the plant defense response in arbuscular mycorrhizal symbiosis. J. Exp. Bot. 53, 1377–1386.

Higgins V J and Bates D K 1994 Phytoalexins in forage legumes. *In* Handbook of Phytoalexin Metabolism. Eds. M Daniel and R P Purkayastha. pp. 391–404. Marcel Dekker, Inc.

Morandi D, Bailey J A and Soriano E and García E 1993 Microassay to measure phytoalexin response in protoplast of bean. Phytochem. Anal. 4, 82–85.

Soriano E and Heredia B 1990 Phytoalexin production in cross protection of bean. Biol. Control Tests 5, 9.

Theodorou M K, Scanlon J C M and Smith I M 1982 Infection and phytoalexin accumulation in French bean leaves injected with spores of *Colletotrichum lindemuthianum*. Phytopathology 103, 189–197.

Toledo V M, Carabias J Mapes C y Toledo C 1985 Ecología y autosuficiencia alimentaria. Siglo Veintiuno Eds. México.

Zavala G, Cano H and Soriano E 1989 Phytoalexins accumulated in bean tissues in response to treatment. Biol. Plant. 31, 221–226.

Solubilization of phosphate by a strain of *Rhizobium leguminosarum* bv. trifolii isolated from *Phaseolus vulgaris* in El Chaco Arido soil (Argentina)

A. Abril[1], J. L. Zurdo-Piñeiro[2], A. Peix[3,4], R. Rivas[2] & E. Velázquez[2]
[1]*Cátedra de Microbiología Agrícola, Facultad de Agronomía, Universidad de Córdoba, Córdoba, Argentina.* [2]*Departamento de Microbiología y Genética, Universidad de Salamanca, Salamanca, Spain.* [3]*Instituto de Recursos Naturales y Agrobiología IRNA, CSIC, Salamanca, C/ Cordel de Merinas, 40-52, 37008, Spain.* [4]*Corresponding author**

Received 7 December 2002. Accepted in revised form 2 January 2003

Key words: Phaseolus vulgaris, phosphate solubilizing bacteria, phosphorous mobilization, *Rhizobium*

Abstract

Several strains were isolated from *Phaseolus vulgaris* plants growing in a soil from El Chaco Arido (Argentina). Although most of strains nodulating *Phaseolus* are not P-solubilizers in plates containing bicalcium phosphate, we tested the isolates and among them, one strain (ARPV02) was able to solubilize phosphate in plates. This strain showed a high ability to nodulate and to fix nitrogen in common bean. Sequencing of 16S rRNA was performed in the strain ARPV02, showing a 100% similarity with the former type strain of *Rhizobium trifolii* ATCC14480. This strain is currently considered as a biovar of species *Rhizobium leguminosarum* together biovars viceae and phaseoli. These biovars have been defined basing on their ability to nodulate a concrete group of legumes. In this way, the strains belonging to biovar trifolii nodulate *Trifolium*. However, in previous studies we have shown that in Spanish soils strains from this biovar nodulate *Phaseolus*. Besides the solubilization of phosphate, the strain ARPV02 isolated in this study is able to movilize phosphorous to common bean plants.

Introduction

The group of rhizobia is considered one of the most powerful P-solubilizers and some of them, such as *Rhizobium leguminosarum* are able to mobilize phosphorous to plants (Halder et al., 1990; Rodríguez and Fraga, 1999). Species from genus *Phaseolus*, indigenous from American continent, are one of the most important legumes for human nutrition. Currently, six rhizobial species have been identified in common bean nodules (Velázquez et al., 2001). These species are present in several geographical locations, although some species have been mainly found in European soils. For example, in North Spain most of the isolates from nodules of *P. vulgaris* belong to *R. leguminosarum* biovar trifolii, whereas in American soils *R. etli* and *R. tropici* are the most frequent species. Although strains nodulating *Phaseolus* have been reported as phosphate solubilizers (Antoun et al., 1998; Chabot et al., 1998), according to our previous results, in general, the strains nodulating *Phaseolus* showed a low ability to solubilize phosphate (Peix et al., 2001), including strains from *R. leguminosarum* biovar trifolii. However, some strains of this biovar that are associated with roots of rice in Egypt are described as P-solubilizers (Yanni et al., 1997, 2001). At the present it is difficult to establish the factors affecting phosphate solubilization, but an important step to clarify them it is the identification of the strains solubilizing phosphate from diverse origins. Therefore, the aim of

* FAX No: +34-923-224876.
E-mail: alvarp@usal.es

E. Velázquez and C. Rodríguez-Barrueco (eds.),
First International Meeting on Microbial Phosphate Solubilization, 135–138

this work was to identify a strain isolated in El Chaco Arido (Argentina) from *P. vulgaris* by using molecular techniques and to study the interaction of this strain with its host from the point of view of the symbiosis and phosphorous mobilization.

Material and methods

Bacterial strains and evaluation of bicalcium phosphate solubilization of rhizobial strains

Isolation was made according to Vincent (1970) using yeast manitol agar – YMA (Bergersen, 1961) from young effective nodules. The ability to solubilize bicalcium-phosphate of the strains isolated in this study was tested in Petri dishes containing YED (yeast extract 0.5%; glucose 1% and agar 2%) supplemented with a 0.2% of bicalcium phosphate (YED-P). A suspension of each strain was inoculated in this medium and the plates were incubated for 7 days until the solubilization zone surrounding the colonies was observed (de Freitas et al., 1997).

16S sequencing and analysis

DNA extraction was carried out as previously described (Rivas et al., 2001). The amplification of 16S rDNA and its sequencing was performed according to the method already described (Rivas et al., 2002). The sequence obtained was compared with those from the GenBank using the FASTA program (Pearson and Lipman, 1988).

Mobilization of phosphorous in plants

Experiments to study the P mobilization by strain ARPV02 to common bean plants were performed with common bean and were conducted in pots containing vermiculite as sterile support. The pots were placed in a plant growth chamber with mixed incandescent and fluorescent lighting (400 microeinsteins $m^{-2} s^{-1}$; 400–700 nm), programmed for a 16 h photoperiod, day–night cycle, with a temperature varying from 15 °C to 27 °C (night–day), and 50–60% relative humidity. Fifteen pots were used for each treatment. The seeds were placed in each pot at a depth of 2 cm. The experimental design was as follows. Treatment 1: seeds inoculated with the strain ARPV02 and adding 0.2% bicalcium phosphate to the vermiculite. Treatment 2: seeds inoculated with the strain *R. etli* CFN42^{T} and adding 0.2% bicalcium phosphate to the vermiculite. The strain *R. etli* CFN42^{T} was used as reference because forms effective nodules in common bean plants and it is not able to solubilize phosphate *in vitro*.

For inoculation, the strains were grown in Petri dishes with YMB (Bergersen, 1961) for 5 days. After that, sterile water was added to the plates to obtain a suspension with ca. 10^{8} cells mL^{-1}. For inoculation we added 1 mL of the suspension of strain to each seed placed in Petri dishes. The seeds were dried overnight at room temperature. At harvest (30 days) the dry weight of the aerial part of the plants of common bean was determined. Plant N, P, K, Ca and Mg content was measured according to the A.O.A.C. methods (Johnson, 1990). The data obtained were analyzed by one-way analysis of variance, with the mean values compared using the Fisher's Protected LSD (Least Significant Differences) ($P = 0.05$).

Results and discussion

Evaluation of bicalcium phosphate solubilization

Several strains were isolated from nodules of *P. vulgaris* growing in El Chaco Arido (Argentina). Only a strain, ARPV02, was found to solubilize actively phosphate *in vitro*. No solubilization was observed in the case of *R. etli* CFN42^{T}.

16S rDNA sequence analysis

The strain ARPV02 was identified at genus level using 16S rDNA complete sequence. This sequence (accession number in GenBank AY196964) showed a 100% similarity with that of *R. leguminosarum* bv. trifolii ATCC18840 that was the type strain of the former species *R. trifolii*.

Mobilization of phosphorous in plants

The results of the inoculation assays are shown in Table 1. No significant differences were observed

Table 1. Symbiotic characteristics of strain *Rhizobium leguminosarum* bv. trifolii ARPV02 compared to those of *R. etli*

Strain	Nodules per plant	Dry weight per plant (mg)	Total N per plant (mg)	Total P per plant (mg)	Total Ca per plant ((g)	Total Mg per plant ((g)	Total K per plant (mg)
Rhizobium etli CFN42^T	170^a	1163^a	27.7^a	4.9^a	10.6^a	31.2^a	14.3^a
R. leguminosarum bv. trifolii ARPV02	190^a	1010^a	29.0^a	6.3^b	12.0^a	34.0^a	7.6^b

Values followed by the same letter are no significantly different from each other at $P = 0.05$ according to Fisher's Protected LSD (Least Significant Differences).

in the most of parameters between the plants inoculated with the strain ARPV02 compared to the plants inoculated with strain *R. etli* CFN42^T. Nevertheless, although these values are at the limit of statistical significance, the number of nodules and the fixed nitrogen were higher in plants inoculated with the strain ARPV02. A significant decrease was found only in K content per plant. A significant increase in the total P was observed in plants inoculated with strain ARPV02 compared to the plants inoculated with strain *R. etli* CFN42^T. This result is in agreement with those found in chickpea plants when they were inoculated with a phosphate solubilizing strain of *Mesorhizobium mediterraneum* (Peix et al., 2001).

The results of the present work also showed that *R. leguminosarum* bv. trifolii is present in American soils as well as in European countries (Velázquez et al., 2001). Nevertheless, many strains must be analyzed to establish the prevalence of this species in American soils because in Argentina *Phaseolus vulgaris* varieties imported from Europe are used whereas in other American countries local varieties are commonly used.

Acknowledgements

This work was supported by the Junta de Castilla y León. AP is grateful to the TLINKS European Research Project for a position as Postdoctoral Research Fellow. We also wish to thank the sample analysis staff of the IRNA for their collaboration.

References

Antoun H, Beauchamp C J, Goussard N, Chabot R and Lalande R 1998 Potential of *Rhizobium* and *Bradyrhizobium* species as growth promoting rhizobacteria on non-legumes: effect on radishes (*Raphanus sativus* L.). Plant Soil 204, 57–67.

Bergersen F J 1961 The growth of *Rhizobium* in synthetic media. Aust. J. Biol. Sci. 14, 349–360.

Chabot R, Beauchamp C J, Kloepper J W and Antoun H 1998 Effect of phosphorous on root colonization and growth promotion of maize by bioluminiscent mutants of phosphate-solubilizing *Rhizobium leguminosarum* biovar. *phaseoli*. Soil Biol. Biochem. 30, 1615–1618.

de Freitas J R, Banerjee M R and Germida J J 1997 Phosphate-solubilizing rhizobacteria enhance the growth and yield but not phosphorous uptake in canola (*Brassica napus* L.). Biol. Fertil. Soils 24, 358–364.

Halder A K, Mishra A K and Chakrabartty P K 1990 Solubilization of phosphatic compounds by *Rhizobium*. Indian J. Microbiol. 30, 311–314.

Jonhson F J 1990 Detection method of nitrogen (total) in fertilizers. *In* Methods of Analysis of the Association of Official Analytical Chemists. Ed. K Elrich. pp. 17–19. Association of Official Analytical Chemists, USA.

Pearson W R and Lipman D J 1988 Improved tools for biological sequence comparison. Proc. Nat. Acad. Sci. USA 85, 2444–2448.

Peix Rivas-Boyero Mateos A. A A. P F, Rodríguez-Barrueco C, Martínez-Molina E and Velázquez E 2001 Growth promotion of chickpea and barley by a phosphate solubilizing strain of *Mesorhizobium mediterraneum* under growth chamber conditions. Soil Biol. Biochem. 33, 103–110.

Rivas R, Velázquez E, Valverde A, Mateos P F and Martínez-Molina E 2001 A two primers random amplified polymorphic DNA procedure to obtain polymerase chain reaction fingerprints of bacterial species. Electrophoresis 22, 1086–1089.

Rivas R, Velázquez E, Willems A, Vizcaíno N, Subba-Rao N S, Mateos P F, Gillis M, Dazzo F B and Martínez-Molina E 2002 A new species of *Devosia* that forms a nitrogen-fixing root-nodule symbiosis with the aquatic legume *Neptunia natans* (L. f.) Druce. Appl. Environ. Microbiol. 68, 5217–5222.

Rodríguez H and Fraga R 1999 Phosphate solubilizing bacteria and their role in plant growth promotion. Biotech. Adv. 17, 319–339.

Velázquez E, Martínez-Romero E, Rodríguez-Navarro D N, Trujillo M E, Daza A, Mateos P F, Martinez-Molina E and Van Berkum P 2001 Characterization of rhizobial isolates of *Phaseolus vulgaris* by staircase electrophoresis of low molecular weight RNA. Appl. Environ. Microbiol. 67, 1008–1010.

Vincent 1970 The cultivation, isolation and maintenance of rhizobia. *In* A Manual for the Practical Study of Root-Nodule. Ed. J M Vincent. pp. 1–13. Blackwell Scientific Publications, Oxford.

Yanni Y, Rizk R, Corich V, Squartini A, Ninke K, Philip-Hollingsworth S, Orgambide G, de Bruijn F, Stoltzfus J, Buckley D, Schmidt T, Mateos P, Ladha J and Dazzo F 1997 Natural endophytic association between *Rhizobium leguminosarum* bv. *trifolii* and rice roots and assesment of its potential to promote rice growth. Plant Soil 194, 99–114.

Yanni Y, Rizk R, Abd-El Fattah F, Squartini A, Corich V, de Bruijn F, Rademaker J, Maya-Flores J, Ostrom P, Vega-Hernández M, Hollingsworth R, Martínez-Molina E, Mateos P, Velázquez E, Wopereis J, Triplett E, Umali-García U, Rolfe B, Ladha J K, Hill J and Dazzo F B 2001 The beneficial plant growth-promoting association of *Rhizobium leguminosarum* bv. *trifolii* with rice roots. Aust. J. Plant Physiol. 28, 845–870.

Effect of phosphate solubilizing bacteria on role of *Rhizobium* on nodulation by soybean

D.L. Wasule[1,2,3], S.R. Wadyalkar[1,3] & A.N. Buldeo[1,3]
[1]*Department of Plant Pathology, College of Agriculture, Nagpur, India.* [2]*Central Institute for Cotton Research, P.B.No.2 Sankarnager P.O., Nagpur, 440010, India.* [3]*Corresponding author**

Received 23 December 2002. Accepted in revised form 2 January 2003

Key words: *Bradyrhizobium japonicum*, Phosphate solubilizing microorganisms, soybean

Abstract

Studies were conducted in laboratory to find out the most effective phosphate solubilizer. The phosphate solubilizing microorganisms were isolated from rhizosphere on Pikovskaya's solid medium by serial dilution. Most efficient phosphate solubilizers were identified on Pikovskaya's solid medium by measuring clear zone around the colony and measurement of pH. The result indicates that *Aspergillus awamori* among fungi and *Pseudomonas striata* among bacteria produce large sized clear zones around the colony i.e. (0.5 cm) and change the pH of medium from initial 5.8 to 2.5 and 4.5, respectively. To determine the effect of phosphate solubilizing bacteria on role of *Rhizobium* on nodulation, nodule dry weight, dry matter of plant, 1000 seed weight and yield a field experiment was conducted with eight treatments i.e. *Rhizobium* + PSB, *Rhizobium*, PSB, Full fertilizer dose, Half fertilizer dose, Full fertilizer dose + *Rhizobium* + PSB, Half fertilizer dose + *Rhizobium* + PSB and Control. *Rhizobium* + PSB yielded maximum number of nodules (67.13) and nodule dry weight (107.73 mg) *Rhizobium* alone showed maximum production of dry matter (3.63 gm). Full fertilizer dose + *Rhizobium* + PSB gave highest 1000 seed weight (109.92 gm). Half fertilizer dose + *Rhizobium* + PSB gave highest yield (10.67 q/ha) which was equivalent to yield recorded with Full fertilizer dose + *Rhizobium* + PSB (10.66 q/ha) and *Rhizobium* + PSB (10.63 q/ha).

Introduction

Phosphate is a non-renewable and important major plant nutrient. Bacteria and fungi have been reported to be active in solubilizing insoluble inorganic phosphate with high efficiency (Gaur, 1990). An attempt was therefore made to find out the most effective strain of phosphorus solubilizing micro-organism in soil by evaluating their phosphate solubilizing capacity.

The efficiency of phosphate fertilizer is very low due to chemical fixation within a short period of its application in soil complex besides poor solubility of native soil phosphorus sometimes there is a built up of insoluble phosphorus due to phosphatic fertilizers applied over a long period. In this situation seed or soil inoculation of phosphate solubilizing microorganism may benefit the crops by increasing phosphorus availability from insoluble source (Gaur, 1990). The investigation was undertaken to study the effect of Phosphate solubilizing microorganisms on role of *Rhizobium* on nodulation by soybean.

* E-mail: DhirajWasule@rediff.com

Materials and methods

Characterization of bacteria

Bradyrhizobium japonicum culture was isolated from root nodules on yeast extract mannitol agar medium. Phosphate solubilizing microorganisms were isolated from rhizosphere from the field of

E. Velázquez and C. Rodríguez-Barrueco (eds.),
First International Meeting on Microbial Phosphate Solubilization, 139–142

Department of Plant Pathology College of Agriculture, Nagpur, Maharashtra. On Pikovaskya's solid medium (Pikovskaya, 1948) by serial dilution method described by Dhingra and Sinclair (1993) the colonies of isolates showed clear zones around indicating the dissolution of tricalcium phosphate into monocalcium phosphate due to the secretion of acids by isolates. The isolates showing phosphate solubilizing ability were 8 fungi and 4 bacteria identified as *Aspergillus awamori, Aspergillus niger, Aspergillus flavous, Aspergillus fumigatus, Penicilium spp., Curvularia lunata, Trichoderma viride, Fusarium spp. Pseudomonas striata, Bacillus polymyxa, PSB-1, PSB-2.* The identification of bacteria was confirmed by I.A.R.I .New Delhi and that of fungi by Department of Plant Pathology College of Agriculture, Nagpur. These isolates were tested for their efficiency by inoculating them independently on Pikovaskaya's solid medium by pin point inoculation and incubated at 28 ±1 °C under aseptic conditions. The clear zone around the colony was measured after four days. The most effective Phosphate solubilizing microorganisms change the pH of medium to acidic. Pikovskya's broth medium was prepared in a 500 mL conical flask that was inoculated and incubated at 28 ± 1 °C for 15 days. The pH of filtrate was measured after separating the bacterial and fungal biomass.

Field experiment

Field experiment was conducted during 1999–2000 at Department of Plant Pathology College of Agriculture, Nagpur, Maharashtra, under rain fed condition with eight treatments and three replications in a randomized block design .The soil of experimental field contained total Nitrogen 0.07%, available Phosphorus 22–40 kg/ha having pH 7.8 .The treatments are, T_1Rhizobium + PSB, T_2 Rhizobium,T_3 PSB, T_4 Full fertilizer dose, T_5 Half fertilizer dose, T_6 Full fertilizer dose + Rhizobium + PSB, T_7 Half fertilizer dose + Rhizobium + PSB and T_8 Control. For co-inoculation *B. japonicum* and *P. striata* were applied in equal amounts (5 g/kg seed) as a seed treatments. Basic dose of 30 kgN/ha in the form of urea, 60 kgP/ha in the form of single super phosphate and K_2O 20 kg/ha in the form of murate of potash were uniformly applied . Soybean (*Glycine max* (L.)Merrill) cv.JS-335 was sown in 3 M X2.7 M plots having 45 × 5 cm spacing. 75 kg/ha seed rate was been used. Observation on nodule number, nodule dry weight and plant dry weight were taken 60 days after sowing and 1000 seed test weight and yield was recorded after harvest of crops.

Result and discussion

Characterization of bacteria

The results in Table 1 show that the fungus *A. awamori* and the bacterium *P. striata* form large size clear zones around (50 and 45 mm) i.e. they found to be more effective than the rest of the isolates. The pH of the cultural filtrate turned acidic with all cultures indicating production of organic acids. The maximum decrease in pH inoculated broth filtrate was recorded with *A. awamori* in fungi and *P. striata* in bacteria from initial 5.8 to 2.5 and 4.5, respectively. Fungi were proved to be better solubilizers as compared to bacteria. *A. awamori* was found to be most superior over all fungal isolates (Singate et al., 1987; Singh et al. 1984) and among the bacteria *P. striata* solubilized more amount of tricalcium phosphate (Arora and Gaur, 1979).

Effect of treatments on number of nodules per plant 60 DAS

All the treatments were found significantly superior over control. Treatments T_1 (*Rhizobium* + PSB) produced highest number of nodules (67.13) per plant but it was similar with treatments T_3 (PSB) i.e. 61.33 and T_2 (*Rhizobium*) i.e. 60.33 (Table 2).

Effect of treatments on dry weight of nodules

All the treatments were found significantly superior over control. The treatments T_1 (*Rhizobium* + PSB) gave highest dry weight of nodules (107.73 mg), but it was found similar with treatments T_3(PSB) i.e. 104.33 mg and T_2 (*Rhizobium*) i.e.103.67 mg (Table 2).

Effect of treatments on dry matter of plant

All the treatments were found significantly superior over control. The treatments T_2 (*Rhizobium*) produced highest dry matter (Table 2).

Table 1. Measurement of clear zone formed and the change in pH of Pikovskaya's medium by the microorganisms from this study

No.	Microorganism	Clear zone (mm)	pH of medium
1.	*Aspergillus awamori*	50	2.5
2.	*Aspergillus niger*	40	3.1
3.	*Aspergillus flavus*	35	3.3
4.	*Aspergillus fumigatus*	35	3.5
5.	*Penicillium spp.*	40	3.8
6.	*Curvularia lunata*	40	5.2
7.	*Trichoderma viride*	20	4.5
8.	*Fusarium spp.*	20	5.2
9.	*Pseudomonas striata*	50	4.5
10.	*Bacillus polymyxa*	45	4.8
11.	*PSB-1*	30	5.1
12.	*PSB-2*	30	5.1
13.	*Control (uninoculated)*	–	5.8

Table 2. Effect of *Bradyrhizobium japonicum* and *Pseudomonas striata* as co-inoculation on symbiotic traits, number of nodules, dry weight of nodules, dry matter of plant, 1000 seed weight and yield

Sl no	Treatment	Number of nodules/ plant 60DAS	Dry weight of nodules/plant (mg)	Dry matter of plant (gm)	1000 seed weight (gm)	Yield Q/ha
1.	T_1	67.13	107.73	3.06	106.28	10.63
2.	T_2	60.33	103.67	3.64	105.05	10.39
3.	T_3	61.33	104.33	3.27	103.17	10.35
4.	T_4	50.13	81.6	3.4	104.23	10.35
5.	T_5	53.33	79.2	3.36	104.03	10.32
6.	T_6	56.47	93.13	3.55	109.92	10.66
7.	T_7	58.33	94.8	3.46	109.89	10.67
8.	T_8 (control)	43.33	70.8	1.94	98.79	9.97
C.D. at 5%		6.89	4.14	0.0384	0.388	0.078

Effect of treatments on 1000 seed weight

All the treatments were found significantly superior over control. The treatments (Full fertilizer dose + *Rhizobium* + PSB) gave highest 1000 seed weight (109.92 g), but the treatment was found similar with Half fertilizer dose + *Rhizobium* +PSB, 109.89 g (Table 2).

Effect of treatments on yield

All the treatments were found significantly superior over control. Treatment T_7 (Half fertilizer dose + *Rhizobium* + PSB) gave highest yield 10.67 q/ha but it was found similar with the treatment T_6 (Full fertilizer dose + *Rhizobium* + PSB) 10.66 q/ha and T_1 (*Rhizobium* + PSB) 10.63 q/ha. (Table 2).

These results indicate that the co-inoculation of *B. japonicum* and *P. striata* gives better results in all traits and it is possible to achieve soybean yields equivalent to the Half fertilizer dose + *Rhizobium* + PSB, Full fertilizer dose + *Rhizobium* + PSB and Rhizobium + PSB. In other words, it is possible to save and replace entire quantity of chemical fertilizer with cheaper co-inoculation of *B. japonicum* and *P. striata* without affecting the soybean yield and other aspects.

References

Arora D and Gaur A C 1979 Microbial solubilization of different inorganic phosphates. Indian J. Expt. Biol. 17, 1258–1261.

Dhingra O D and Sinclair J B 1993 Basic Plant Pathology Methods. CBS publisher Delhi, India. 179 pp.

Gaur A C 1990 Phosphate Solubilizing Micro-Organisms as Bio Fertilizers. Omega scientific publishers, New Delhi, India 11 pp.

Pikovskaya R I 1948 Mobilization of phosphorus in soil in connection with vital activity of some microbial species. Microbiologiya 17, 362–370.

Singate V V, Rasal P H and Patil P L 1987 Screening of organisms for P solubilizing ability. Maharashtra .agric. Univ. 12, 121–122.

Singh H P, Pareek R P and Singh T A 1984 Solubilization of rock phosphate by phosphate solubilizer in broth. Curr. Sci. 53, 12–13.

Phaseolus lunatus is nodulated by a phosphate solubilizing strain of *Sinorhizobium meliloti* in a Peruvian soil

E. Ormeño[1], R. Torres[1], J. Mayo[1], R. Rivas[2], A. Peix[3], E. Velázquez[2] & D. Zúñiga[1]
[1]*Laboratorio de Ecología Microbiana Marino Tabusso, Departamento de Biología, Universidad Agraria de la Molina, Lima, Perú.* [2]*Departamento de Microbiología y Genética, Facultad de Farmacia, Universidad de Salamanca, Salamanca, Spain.* [3]*Instituto de Recursos Naturales y Agrobiología (IRNA, CSIC), C/Cordel de Merinas, 40-52, 37008, Salamanca, Spain.* [3]*Corresponding author**

Received 16 December 2002. Accepted in revised form 2 January 2003

Key words: phosphate solubilizing bacteria, *Sinorhizobium*, *Phaseolus lunatus*

Abstract

The genus *Phaseolus* includes several species indigenous to American continent that belong to family *Leguminosae*. This genus includes several species, some of them only cultivated in American countries. This is the case of *Phaseolus lunatus*. This plant can be nodulated by fast and slow growing rhizobia. At the moment the fast growing species nodulating *Phaseolus* commonly belong to genus *Rhizobium* and more rarely to *Sinorhizobium fredii*. A strain, LMTR32, isolated from *Phaseolus lunatus* growing in Peru soils showed a high ability to solubilize bicalcium phosphate from YED-P plates. The 16S rRNA sequence of this strain showed a 100% similarity with the type strain of *Sinorhizobium meliloti*. The LMW RNA profile of this strain is identical to that of type strain of *Sinorhizobium meliloti* and confirms that the strain LMTR32 belongs to this species. More studies are necessary in order to establish the prevalence of this species in nodules of *Phaseolus lunatus* in Perú, and, in the future, it will be very interesting to perform wider taxonomic studies of rhizobia nodulating *Phaseolus* in different American countries.

Introduction

The genus *Phaseolus* is indigenous to American soils and was spread in the world after America discovery. This fact has supported the hypothesis that American endosymbionts of *Phaseolus* belong to different species than those isolated in other geographical locations. Moreover, during decades, the rhizobiologists had classified the rhizobia according to cross-inoculation groups of legumes. Currently, this classification has not been completely forgotten and certain cross-inoculation groups have been maintained and even has been used to classify the species of genus *Rhizobium* (Jordan, 1984). In this way, the species *Sinorhizobium meliloti* has been considered an exclusive endosymbiont of *Medicago*, *Melilotus* and *Trigonella*, whereas the endosymbionts of *Phaseolus* were classified in species from genus *Rhizobium*. *R. tropici* (Martínez-Romero et al., 1991) and *R. etli* (Segovia et al., 1993) nodulate *Phaseolus* in American soils. *R. gallicum* and *R. giardinii* (Amarger et al., 1997) have been identified from nodules of *Phaseolus* in France. *R. leguminosarum* biovar trifolii has been isolated from nodules of *Phaseolus vulgaris* in Spain (Velázquez et al., 2001a). For the moment, only a species of genus *Sinorhizobium* (*S. fredii*), has been identified in *Phaseolus* nodules (Herrera-Cervera et al., 1999; Sadowsky et al., 1988; Velázquez et al., 2001a) and only one species of

* FAX No: +34-923-224876.
E-mail: alvarp@usal.es

E. Velázquez and C. Rodríguez-Barrueco (eds.),
First International Meeting on Microbial Phosphate Solubilization, 143–147

genus *Rhizobium*, *Rhizobium mongolense*, can nodulate *Phaseolus vulgaris* and *Medicago ruthenica* (van Berkum et al., 1998) that in theory belong to different cross-inoculation groups. All these studies are carried out on *P. vulgaris* because this plant has been spread in many countries and is used in human nutrition. However, the genus *Phaseolus* comprises several species that only are present in South American countries as occurs with *Phaseolus lunatus* that is used as forage in Perú. The endosymbionts of lima bean (*P. lunatus*) have been few studied and commonly belong to genus *Rhizobium* although slow-growing strains have been isolated (Matos and Zúñiga, 2002).

During a study of strains nodulating *Phaseolus lunatus* in Perú we isolated a strain able to induce nodules in this plant and to solubilize phosphate *in vitro*. In this work we have identify this strain using 16S rDNA sequence and LMW RNA profiles and we have analyzed the ability of this strain to mobilize phosphorous to *P. lunatus*.

Materials and methods

Bacterial strains and evaluation of its ability to solubilize phosphate

The isolation of strain LMTR32 was made according to Vincent (1970) using yeast manitol agar – YMA – (Bergersen, 1961) from young effective nodules. The ability to solubilize bicalcium phosphate was tested in Petri dishes containing YED (yeast extract 0.5%; glucose 1% and agar 2%) supplemented with a 0.2% of bicalcium phosphate (YED-P). A suspension of the strain was inoculated in this medium and the plates were incubated for 7 days until the solubilization zone surrounding the colonies was observed (de Freitas et al., 1997).

16S sequencing and analysis

DNA extraction was carried out as previously described (Rivas et al., 2001). The amplification of 16S rDNA and its sequencing was performed according to the method already described (Rivas et al., 2002). The sequence obtained was compared with those from the GenBank using the FASTA program (Pearson and Lipman, 1988).

LMW RNA extraction and SCE LMW RNA profiling

LMW RNA extraction was accomplished following the phenol/chloroform method described by Höfle (1988), using cells grew in tryptone-yeast agar, TY (Beringer, 1974). The following commercial molecules from Boehringer Manheim (Manheim, Germany) and Sigma (St. Louis, MO, USA) were used as reference: 5S rRNA from *Escherichia coli* MRE 600 (120 and 115 nucleotides) (Bidle and Fletcher, 1995), tRNA specific for tyrosine from *E. coli* (85 nucleotides) and tRNA specific for valine from *E. coli* (77 nucleotides) (Sprinzl et al., 1985). Samples containing 3 μg were added to 5 μg of loading solution (300 mg/mL of sucrose, 460 mg/mL of urea, 10 μL/mL 20% SDS, 1 mg/mL xylene cyanol) and, after 10 min of heating at 70 °C, applied to each well. LMW RNA profiles were obtained using staircase electrophoresis (SCE) which was performed in 400 × 360 × 0.4 mm gels in a vertical slab unit (Poker Face SE 1500 Sequencer, Hoeffer Scientific Instruments, San Francisco, CA, USA). The separating gel contained 14% acrylamide/Bis (acrylamide: *N,N*-methylene bisacrylamide 29:1 (w/w), 7 *M* urea in TBE buffer: 100 mM Tris, 83 mM boric acid, 1 mM EDTA, pH: 8.5) in TBE buffer, pH: 8.5. Before running the pre-electrophoresis (30 min at 100 V), the system was stabilized at 50 °C. The running buffer (TBE, ×1.2) was recycled at a flow rate of 300 mL/min with a peristaltic pump (MasterFlex, Cole Parmer Instruments, Chicago, Illinois, USA) (Cruz-Sánchez et al., 1997). After electrophoresis, gels were silver-stained according to Haas et al. (1994).

Mobilization of phosphorous in plants

Experiments to study the P mobilization in plants were performed with common bean and were conducted in pots containing vermiculite as sterile support. The pots were placed in a plant growth chamber with mixed incandescent and fluorescent lighting (400 microeinsteins $m^{-2}\,s^{-1}$; 400–700 nm), programmed for a 16 h photoperiod, day–night cycle, with a temperature varying from 15 to 27 °C (night-day), and 50–60% relative humidity. The experimental design was as follows. Treatment 1: seeds inoculated with the strain LMTR32 and adding 0.2% bicalcium phosphate to the vermiculte. Treatment 2: Control treatment with insoluble phosphate and uninoculated seeds. Fifteen pots were used for each treatment. The seeds were placed in each pot at a depth of 2 cm.

For inoculation, strain LMTR32 was grown in Petri dishes with YMB (Bergersen, 1961) for 5 days. After that, sterile water was added to the plates to obtain a suspension with ca. 10^8 cells mL^{-1}. For inoculation we added 1 mL of the suspension of strain LMTR32 to each seed placed in Petri dishes. The seeds were dried overnight at room temperature.

At harvest (30 days) the dry weight of the aerial part of the plants of common bean was determined. Plant N, P, K, Ca and Mg content was measured according to the A.O.A.C. methods (Johnson, 1990). The data obtained were analyzed by one-way analysis of variance, with the mean values compared using the Fisher's Protected LSD (Least Significant Differences) ($P = 0.05$).

Results and discussion

Ability to solubilize phosphates in vitro

The strain LMTR32 isolated from effective nodules of *P. lunatus* in Perú was able to solubilize phosphate in plates containing bicalcium hydrogen phosphate as P source. The diameter of the halo surrounding the colonies was 5 mm. This diameter is lower than that obtained in *Mesorhizobium* strains, but it is higher than those obtained for strains of genus *Rhizobium* (Peix et al., 2001). The type strains of the species from genus *Rhizobium* nodulating *P. vulgaris* do not show phosphate solubilization in plates containing bicalcium phosphate although some strains nodulating this legume have been reported as P solubilizers (Halder et al., 1990, 1993). There are no data about the phosphate solubilization of strains nodulating *P. lunatus* because no studies have been carried out with isolates from this species. For this reason we have identified the strain LMTR32 and we analyzed the possibility that this strain mobilize phosphorous to the plant.

16S rDNA sequence analysis

The strain LMTR32 was identified in first place using 16S rDNA complete sequence (Accession number AY196963). This sequence showed a 100% similarity with that of *Sinorhizobium meliloti*. At the moment this species has been not identified in nodules of *Phaseolus*, although some authors have reported the existence of strains related with this species that were isolated from *P. vulgaris*. Nevertheless, in these works the strains isolated do not were completely identified and in some cases the identification was based on symbiotic genes and not in ribosomal genes. For example, based on *nodC* and *nifH* sequences and restriction patterns (Mhamdi et al., 2002) have found strains related to *S. meliloti* isolated from *P. vulgaris* in Tunisian soils. Previously, other authors have reported the nodulation of *Phaseolus vulgaris* by strains of *S. meliloti* (Laguerre et al., 2001). In the present study we have used the LMW RNA profiles to identify at species level the strain LMTR32.

LMW RNA profiling

LMW RNA profiles include three zones in prokaryotes: 5S rRNA zone, class 1 and class 2 tRNA. The 5S rRNA zone is characteristic of each genus and the tRNA profile is characteristic of each species from the same or different genus. Therefore, these profiles are molecular signatures for both prokaryotes and eukaryotes microorganisms (Velázquez et al., 2001b). In a previous study we demonstrated that the species that nodulate *Phaseolus vulgaris* can be distinguishable using LMW RNA profiles (Velázquez et al., 2001a). In this study we identified a strain isolated from nodules in South Spain as *Sinorhizobium fredii*. This species had been reported in common bean nodules by other authors (Herrera-Cervera et al., 1999; Sadowsky et al., 1988), but in our work we unambiguously identified this strain using 16S rDNA sequence and LMW RNA profile. For this reason, in the present study we have analyzed the LMW RNA profile of strain LMTR32 to confirm the identification obtained using 16S rDNA sequencing. Figure 1

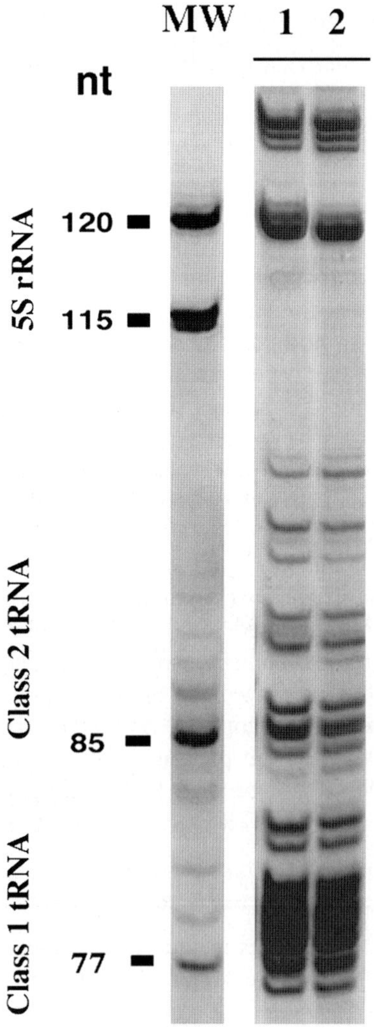

Figure 1. LMW RNA profiles of strain *S. meliloti* ATCC 9930^T (lane 1) and strain LMTR32 (lane 2).

shows the LMW RNA profiles of the type strains of *S. meliloti* (lane 1), the strain LTMR32 (lane 2). The comparison among these strains shows that the LMW RNA profiles are identical in the type strain of *S. meliloti* and in the strain LTMR32 and therefore this strain was identified as *S. meliloti*. These results open a new way in the study of symbiotic relatedness because the promiscuity of strains nodulating legumes seems to be very extended affecting to many rhizobial species. Moreover, unlike the type strain of *S. meliloti*, the strain LMTR32 was able to solubilize phosphate *in vitro*.

Mobilization of phosphorous in plants

The solubilization of phosphate *in vitro* by rhizobia does not involve the mobilization of the P to their hosts. Because of that, we have analyzed if the strain LMTR32 is able to mobilize phosphorous to common bean plants. The results showed that this strain was able to nodulate *P. lunatus* but the number of nodules induced by this strain is low. Moreover the nodules were low effective and therefore dry weight and nitrogen content per plant were also low. Nevertheless, the strain LMTR32 was able to solubilize phosphorous to plants. The increase of the P in plants inoculated with this strain was significantly higher than in the uninoculated plants. Therefore, the strain *S. meliloti* LMTR32 is able to solubilize phosphate *in vitro* and also to mobilize phosphorous to common beans. This result is in agreement with those obtained in the case of other rhizobia (Peix et al., 2001). Concerning to the symbiotic characteristics of this strain, such as nodulation and nitrogen fixation are similar to those presented by strains of *S. fredii* nodulating common bean and they are lower than to those presented by species of genus *Rhizobium* (Rodríguez-Navarro et al., 2000). Nevertheless, the results of this work indicate the great interest of the analysis of bacterial population nodulating legumes in different geographical regions to know the biodiversity of rhizobia that establish relationship with different species and genus of these plants (Table 1).

Table 1. Symbiotic characteristics of strain *S. meliloti* LMTR32 compared to those of *R. etli*

Strain	Number of nodules	Dry weight per plant (mg)	Total N (mg)	Total P (mg)	Total Ca (μg)	Total Mg (μg)	Total K (mg)
Control with insoluble P	0[a]	470[a]	9.4[a]	1.4[a]	3.1[a]	6.0[a]	6.9[a]
Sinorhizobium meliloti LMTR32	*18*[b]	700[b]	17.4[b]	3.1[b]	6.7[b]	18.9[b]	8.4[b]

Values followed by the same letter are no significantly different from each other at $P = 0.05$ according to Fisher's Protected LSD (Least Significant Differences).

Acknowledgements

This work was supported by the Junta de Castilla y León. AP is grateful to the TLINKS European Research Project for a position as Postdoctoral Research Fellow. We also wish to thank the sample analysis staff of the IRNA for their collaboration.

References

Amarger N, Macheret V and Laguerre G 1997 *Rhizobium gallicum* sp. nov. and *Rhizobium giardinii* sp. nov. from *Phaseolus vulgaris* nodules. Int. J. Syst. Bacteriol. 47, 996–1006.

Bergersen F J 1961 The growth of *Rhizobium* in synthetic media. Aust. J. Biol. Sci. 14, 349–360.

Beringer J E 1974 R factors transfer in *Rhizobium leguminosarum*. J. Gen. Microbiol. 84, 188–198.

Bidle K D and Fletcher M 1995 Comparison of free-living and particle-associated bacterial communities in the Chesapake Bay by stable low-molecular-weight RNA analysis. Appl. Environ. Microbiol. 61, 944–952.

Cruz-Sánchez J M, Velázquez E, Mateos P and Martínez-Molina E 1997 Enhancement of resolution of low molecular weight RNA profiles by staircase electrophoresis. Electrophoresis 18, 1909–1911.

de Freitas J R, Banerjee M R and Germida J J 1997 Phosphate-solubilizing rhizobacteria enhance the growth and yield but not phosphorous uptake in canola (*Brassica napus* L.). Biol. Fertil. Soils 24, 358–364.

Haas H, Budowle B and Weiler G 1994 Horizontal polyacrylamide gel electrophoresis for the separation of DNA fragments. Electrophoresis 15, 153–158.

Halder A K, Mishra A K and Chakrabartty P K 1990 Solubilization of phosphatic compounds by *Rhizobium*. Indian J. Microbiol. 30, 311–314.

Halder A K, Mishra A K, Bhattacharyya P and Chakrabartty P K 1993 Solubilization of rock phosphate by *Rhizobium* and *Bradyrhizobium*. J. Gen. Appl. Microbiol. 36, 81–92.

Herrera-Cervera J A, Caballero-Mellado J, Laguerre G, Tichy H V, Requena N, Amarger N, Martínez-Romero E, Olivares J and Sanjuán J 1999 At least five rhizobial species nodulate *Phaseolus vulgaris* in a Spanish soil. FEMS Microbiol. Ecol. 30, 87–97.

Höfle M G 1988 Identification of bacteria by low molecular weight RNA profiles: A new chemotaxonomic approach. J. Microbiol. Meth. 8, 235–248.

Jonhson F J 1990 Detection method of nitrogen (total) in fertilizers. *In* Methods of Analysis of the Association of Official Analytical Chemists. Ed. K Elrich. pp. 17–19. Association of Official Analytical Chemists, USA.

Jordan D C 1984 Family III, *Rhizobiaceae* Conn 1938 321^{AL}. *In* Bergey's Manual of Systematic Bacteriology Vol. 1. Eds. N R Krieg and J G Holt. pp. 234–244. Williams & Wilkins, Baltimore.

Martínez-Romero E, Segovia L, Mercante F M, Franco A A, Graham P and Pardo M A 1991 *Rhizobium tropici*, a novel species nodulating *Phaseolus vulgaris* L. beans and *Leucaena* sp. trees. Int. J. Syst. Bacteriol. 41, 417–426.

Laguerre G, Nour S M, Macheret V, Sanjuán J, Drouin P and Amarguer N 2001 Classification of rhizobia based on *nodC* and *nifH* gene analysis reveals a close phylogenetic relationship among *Phaseolus vulgaris* symbionts. Microbiology 147, 981–993.

Matos G and Zúñiga D 2002 Comportamiento de cepas nativas de rhizobios aisladas de la costa del Perú en dos cultivares de pallar (*Phaseolus lunatus*). Ecol. Appl. 1, 19–24.

Mhamdi R, Laguerre G, Elarbi-Aouani M, Mars M and Amarger N 2002 Different species and symbiotic genotypes of field rhizobia can nodulate *Phaseolus vulgaris* in Tunisian soils. FEMS Microbiol. Ecol. 41, 77–84.

Pearson W R and Lipman D J 1988 Improved tools for biological sequence comparison. Proc. Nat. Acad. Sci. USA 85, 2444–2448.

Peix A., Rivas-Boyero A.A., Mateos P.F., Rodríguez-Barrueco C., Martínez-Molina E. and Velázquez E 2001 Growth promotion of chickpea and barley by a phosphate solubilizing strain of *Mesorhizobium mediterraneum* under growth chamber conditions. Soil Biol. Biochem. 33, 103–110.

Rivas R, Velázquez E, Valverde A, Mateos P F and Martínez-Molina E 2001 A two primers random amplified polymorphic DNA procedure to obtain polymerase chain reaction fingerprints of bacterial species. Electrophoresis 22, 1086–1089.

Rivas R, Velázquez E, Willems A, Vizcaíno N, Subba-Rao N S, Mateos P F, Gillis M, Dazzo F B and Martínez-Molina E 2002 A new species of *Devosia* that forms a nitrogen-fixing root-nodule symbiosis with the aquatic legume *Neptunia natans* (L. f.) Druce. Appl. Environ. Microbiol. 68, 5217–5222.

Rodríguez-Navarro D N, Buendía A M, Camacho M, Lucas M and Santamaría C 2000 Characterization of *Rhizobium* spp. bean isolates from southwest of Spain. Soil. Biol. Biochem. 32, 1601–1613.

Sadowsky M J, Cregan P and Keyser H H 1988 Nodulation and nitrogen fixation efficacy of *Rhizobium fredii* with *Phaseolus vulgaris* genotypes. Appl. Environ. Microbiol. 54, 1907–1910.

Segovia L, Young J P W and Martínez-Romero E 1993 Reclassification of American *Rhizobium leguminosarum* biovar phaseoli type I strains as *Rhizobium etli* sp. nov. Int. J. Syst. Bacteriol. 43, 374–377.

Sprinzl M, Moll J, Meissner F and Hatmann T 1985 Compilation of tRNA sequences. Nucleic Acid Res. 13, 1–49.

van Berkum P, Beyene D, Bao G, Campbell T A and Eardly B D 1998 *Rhizobium mongolense* sp. nov. is one of three rhizobial genotypes identified which nodulate and form nitrogen-fixing symbioses with *Medicago ruthenica* [(L.) Ledebour]. Int. J. Syst. Bacteriol. 48, 13–22.

Velázquez E, Martínez-Romero E, Rodríguez-Navarro D N, Trujillo M E, Daza A, Mateos P F, Martinez-Molina E and Van Berkum P 2001a Characterization of rhizobial isolates of *Phaseolus vulgaris* by staircase electrophoresis of low molecular weight RNA. Appl. Environ. Microbiol. 67, 1008–1010.

Velázquez E, Trujillo M E, Peix A, Palomo J L, García-Benavides P, Mateos P, Ventosa A and Martínez-Molina E 2001b Stable low molecular weight RNA analyzed by staircase electrophoresis, a molecular signature for both prokaryotic and eukaryotic microorganisms. Syst. Appl. Microbiol. 24, 490–499.

Vincent J M 1970 The cultivation, isolation and maintenance of rhizobia. *In* A Manual for the Practical Study of Root-Nodule. Ed. J M Vincent. pp. 1–13. Blackwell Scientific Publications, Oxford.

Phosphate solubilizing rhizobia originating from *Medicago*, *Melilotus* and *Trigonella* grown in a Spanish soil

M. Villar-Igea[1], E. Velázquez[1], R. Rivas[1], A. Willems[2], P. van Berkum[3], M. E. Trujillo[1,4], P. F. Mateos[1], M. Gillis[2] & E. Martínez-Molina[1]
[1]*Departamento de Microbiología y Genética, Lab 209. Edificio Departamental, Campus Miguel de Unamuno, Universidad de Salamanca, Pl. Doctores de la Reina s/n, 37007, Salamanca, Spain.* [2]*Laboratorium voor Microbiologie, Vakgroep Biochemie, Fysiologie en Microbiologie, K.L. Ledeganckstraat 35, B-9000, Gent, Belgium.* [3]*U.S. Department of Agriculture, ARS, Soybean Genomics and Improvement Laboratory, Beltsville, MD, 20705, USA.* [4]*Corresponding author**

Received 15 December 2002. Accepted in revised form 2 January 2003

Key words: alfalfa, identification, phosphate solubilizing bacteria, *Sinorhizobium meliloti*

Abstract

Although phosphate solubilization is a character known to be present in species of *Mesorhizobium*, this property has not been described before in species of *Sinorhizobium*. The type strains of the three species that nodulate *Medicago* species, *Sinorhizobium meliloti*, *S. medicae* and *Rhizobium mongolense*, do not solubilize phosphate from bicalcium phosphate in plate culture. We observed phosphate solubilization among isolates we obtained from nodules of *Medicago sativa*, *Melilotus* and *Trigonella* growing in a Spanish soil. Phenotypic and genetic analyses of these isolates led to the conclusion that they were placed within the genus *Sinorhizobium* with characteristics in common with *S. meliloti* and *S. medicae*. The group of strains solubilizing phosphate is distinguishable to strains from *S. meliloti* and *S. medicae* basing on LMW RNA profiles, TP-RAPD patterns and SDS-PAGE profiles.

Introduction

Phosphate (P) solubilization as a character is widely distributed in rhizobia (Halder et al., 1990, Peix et al., 2001, Rodríguez and Fraga, 1999). Species within the genus *Mesorhizobium* most actively solubilize P *in vitro* (Peix et al., 2001) but this character also has been described as present in some strains that nodulate *Medicago* (Halder et al., 1990), but they were not identified using taxonomic criteria. Generally, rhizobia originating from *Medicago* also nodulate *Melilotus* and *Trigonella*, but isolates originating from these legume hosts are less well studied than those of alfalfa (*M. sativa*).

Alfalfa is important in agriculture as a forage legume, and species of annual medics have value as cover or companion crops. *Melilotus* is important in medicine as an anticoagulant and *Trigonella foenum-graecum* is used as an ingredient in the cosmetics industry and is both used as a spice and a vegetable. Two species of *Sinorhizobium*, *S. meliloti* (de Lajudie et al., 1994) and *S. medicae* (Rome et al., 1996) and one of *Rhizobium*, *R. mongolense* (van Berkum et al., 1998) have been proposed to represent the rhizobia that nodulate alfalfa and related legume species. From our results we conclude that the type strains of these three proposed species do not solubilize phosphate *in vitro*. However, a more comprehensive investigation

* FAX No: Fax no: 34923224876.
E-mail: mett@usal.es

E. Velázquez and C. Rodríguez-Barrueco (eds.),
First International Meeting on Microbial Phosphate Solubilization, 149–156

for P solubilization among different isolates, including those from *Melilotus* and *Trigonella*, has not been done. In this work, we have identified P solubilization among several rhizobial isolates originating from nodules of diverse species of *Medicago*, *Melilotus* and *Trigonella* plants growing in a Spanish soil.

Methods

Bacterial strains

Strains and isolates used in this study are listed in Table 1. The rhizobial isolates were obtained from young Medicago sativa, Medicago. lupulina, Medicago spaherocarpa, Melilotus parviflora, Melilotus alba, Trigonella foenum-graecum and Trigonella monspelliaca plants. Isolations were made according to Vincent (1970) using yeast mannitol agar -YMA- (Bergersen, 1961). The cultures used in further studies were purified from single colonies after 5 days incubation at 28 °C.

Nodulation tests

Surface-sterilized seeds of *Medicago sativa* were used to test the infectivity of the isolates. Seedlings were transferred to pots with sterile vermiculite and watered with nitrogen free Rigaud and Puppo (1975) nutrient solution. Each plant was inoculated with 1 mL of a suspension of each culture containing 8×10^8 cells/mL. The inoculated plants were placed in a plant growth chamber for 20 days and were grown with mixed incandescent and fluorescent lighting (400 microeinsteins m^{-2} s^{-1}; 400 to 700 nm), a 16 h

Table 1. Characteristics of strains used in this study

Strain	Host plant	Geographic origin	Reference	TP-RAPD pattern	Phosphate solubilization
RTM17	*Trigonella monspelliaca*	León (Spain)	This study	I	Positive
SAP11	*Medicago sativa*	Salamanca (Spain)	This study	I	Positive
RTM02	*Trigonella monspelliaca*	León (Spain)	This study	I	Positive
RTM08	*Trigonella monspelliaca*	León (Spain)	This study	I	Positive
RTM11	*Trigonella monspelliaca*	León (Spain)	This study	I	Positive
RMA02	*Melilotus alba*	León (Spain)	This study	I	Positive
RMA05	*Melilotus alba*	León (Spain)	This study	I	Positive
RMA30	*Melilotus alba*	León (Spain)	This study	I	Positive
RMA31	*Melilotus alba*	León (Spain)	This study	I	Positive
RMA32	*Melilotus alba*	León (Spain)	This study	I	Positive
RMO17	*Medicago orbicularis*	León (Spain)	This study	II	Negative
SAF22	*Medicago sativa*	Salamanca (Spain)	This study	II	Negative
RTM18	*Trigonella monspelliaca*	León (Spain)	This study	II	Negative
RMP01	*Melilotux parviflora*	León (Spain)	This study	III	Negative
RMP01	*Melilotux parviflora*	León (Spain)	This study	III	Negative
Sinorhizobium meliloti USDA1002^T (LMG6133^T)	*Medicago sativa*	USA	de Lajudie et al. (1994)	II	Negative
Sinorhizobium medicae USDA1037 ^{T}	*Medicago truncatula*	France	Rome et al. (1996)	III	Negative
Rhizobium mongolense USDA 1844^T	*Medicago ruthenica*	China	van Berkum et al. (1998)	ND	Negative

LMG, Collection of bacteria of the Laboratory voor Microbiologie, Gent, Belgium.
USDA, US Department of Agriculture, Belstville, MD, USA.
ND: No data

photoperiod, day-night cycle, with a constant temperature varying from 25–27 °C, and 50–60% relative humidity. Root nodules appeared 5–10 days after inoculation.

Evaluation of tricalcium phosphate solubilization

The ability of the type strains and the isolates to solubilize bicalcium-phosphate was tested in Petri dishes containing YED-P (0.5% yeast extract, 0.7% glucose, 0.2% bicalcium phosphate and 2% agar). Cultures were plated and were incubated for 7 days until a solubilization zone surrounding the colonies was observed.

DNA extraction

Strains were grown in TY medium (0.4%, tryptone 0.3% yeast extract and 0.09%Ca_2Cl) for 24 h. Cells (1.5 mL of each culture) were collected by centrifugation at room temperature in a microspin centrifuge at 5000×*g* and then washed with 200 μL of a solution of 0.1% sarkosyl in water. The DNA was extracted with 100 μL of 0.05 *M* NaOH (DNA-free) heating at 100 °C for 4 min. Samples were then placed in an ice bath and 900 μL of water was added to each microtube and mixed thoroughly. After an additional centrifugation at 4000×*g* for 3 min, 700 μL of the supernatants were harvested and frozen at −20 °C (28). For sequencing analysis DNA samples were prepared from 10 mL MAG (Modified Arabinose–Gluconate, van Berkum 1990) broth cultures using a Tissue and Blood DNA Extraction kit (Qiagen Inc., Chatsworth, CA).

TP-RAPD patterns

Crude DNA (2 μL) was used as template for obtaining TP-RAPD patterns. PCR was performed using an AmpliTaq Gold reagent kit (Perkin–Elmer Biosystems, California, USA) following the manufacturer's instructions (1.5 m*M* $MgCl_2$, 200 μM of each dNTP and 2 U of Taq polymerase for 25 μl of final volume of reaction). The PCR primers used for amplification were the forward primer 8F (5′-AGAGTTTGATCCTGGCTCAG-3′, *Escherichia coli* positions 8–27) and the reverse primer 1522R 5′-AAGGAGGTGATCCANCCRCA-3′, *Escherichia coli* positions 1502–1522) at a final concentration of 2 μM. We also used another set with the forward primer 879F (5′-GCCTGGGGAGTACGGCCGCA-3′ *Escherichia coli* positions 859–879) and the same reverse primer 1522R (28). PCR conditions were as follows: pre-heating at 95 °C for 9 min; 35 cycles of denaturing at 95 °C for 1 min; annealing at 45 °C for 1 min and extension at 72 °C for 2 min, and a final extension at 72 °C for 7 min. The PCR products were stored at 4 °C.

Eight microliters of PCR product were electrophoresed on 1.5% agarose gel in TBE buffer (100 m*M* Tris, 83 m*M* boric acid, 1 m*M* EDTA, pH: 8.5) at 6 V cm^{-1}, stained in a solution containing 0.5 μL ethidium bromide mL^{-1} and photographed under UV light. Standard VI (Boehringer-Roche, USA) was used as a size marker. Three μL of 6× loading solution (40% glycerol and 0.25% bromophenol blue) were added to each sample.

Determination of nucleotide sequence of the 16S rRNA genes and analysis of the sequence data.

Primers 16Sa and 16Sb (van Berkum and Fuhrmann, 2000) were used for amplification of the 16S rRNA gene locus. The 16S rRNA genes were amplified in 120 μl volumes as described before (van Berkum et al., 1996; Rivas et al., 2002b) with the exception of the primers and the PCR buffer, which was 60 m*M* Tris–HCl, 15 m*M* $(NH_4)_2SO_4$, and 3.5 m*M* $MgCl_2$ at pH 9.0. The PCR products were purified using QIAquick Spin columns (Qiagen Inc., Chatsworth, CA). A Perkin–Elmer 377 DNA Sequencer in combination with a Dye Deoxy Terminator Cycle Sequencing Kit (Perkin–Elmer, Foster City, CA) was used for sequencing the purified PCR products. The sequences were aligned using the PILEUP program in the Wisconsin package of the Genetics Computer Group (Madison, Wis.). Aligned sequences were checked manually and were edited with Genedoc (Nicholas and Nicholas, 1997). Neighbor-joining trees were constructed from Jukes–Cantor distances using the Molecular Evolutionary Genetics Analysis (MEGA) package version 2.1 (Kumar et al., 2001).

Analysis of proteins by SDS-PAGE

Whole-cell protein extracts were prepared and separated by electrophoresis using small modifications of the procedure of Laemmli (1970) as described previously (de Lajudie et al., 1994).

Phenotypic characterization

Phenotypic characterization of isolates and the type strains included pH changes during growth with different carbon sources (acid, basic or neutral), antibiotic resistance and extracellular enzyme production. The carbon sources tested were sucrose, galactose, lactose, L-arabinose, L-rhamnose, trehalose, maltose, adonitol, melibiose and rafinose. The basal medium used contained K_2HPO_4 0.2 g/L, $MgSO_4$ 0.2 g/L, NH_4NO_3 1 g/L, with a solution of vitamins and trace elements according to Bergersen (1961) (1 mL/L), bromothymol blue 0.05 g/L. The pH was adjusted to 7.0 with KH_2PO_4. The discs impregnated with carbon sources (BBL, Beckton Dickinson) were added aseptically to 5 mL of medium. The results obtained were based on the pH changes for the different carbon sources (acid, basic or neutral) and were recorded after five days for fast-growing species, ten days for species of *Mesorhizobium* and 15 days for slow and extra-slow-growing species. Resistances for the antibiotics ampicillin, erythromycin, ciprofloxacin, penicillin, polymyxin, cloxacillin, oxitetracyclin, gentamicin, cefuroxime and neomycin were determined. The basal medium was YMB supplemented with 0.5% of Yeast Extract. Each disc impregnated with an antibiotic was added aseptically to 5 mL of basal medium. Variability in 10 extracellular glucosidases was tested using the chromogenic substrates paranitrophenylsubstrates (PNP): PNP-α-D-arabinopyranoside, PNP-β-D-arabinopyranoside, PNP-α-D-fucopyranoside, PNP-β-D-fucopyranoside, PNP-α-D-galactopyranoside, PNP-β-D-galactopyranoside, PNP-α-D-xylopyranoside, PNP-β-D-xylopyranoside, PNP-α-D-maltopyranoside, and PNP-*N*-acetyl-thio-β-D-glucosaminide at a concentration of 0.4% in 50 m*M* phosphate buffer, pH7. The reactions were done in multiwell plates mixing 50 μL of substrate with 50 μL of the bacterial suspension in sterile water. The suspensions contained 6×10^9 CFU/mL. The suspensions were prepared from bacteria grown on plate culture with Bergersen (1961) minimal medium incubated at 28 °C over 4 days. The multiwell plates with the mixtures were incubated at 28 °C during 4 days and subsequently were developed with the addition of 100 μL 4% sodium carbonate to each well. Development of a yellow colour was considered as a positive result. Abilities to grow at 37 °C and 40 °C or at pH 5 and pH 8 were determined on YMA medium.

Results

Isolation and nodulation

All isolates nodulated and were effective for symbiotic nitrogen fixation with *M. sativa*. Effectiveness for nitrogen fixation was inferred from the presence of nodules and by growth of plants in nitrogen-free medium.

Evaluation of bicalcium phosphate solubilization

The ability of our isolates to solubilize phosphates was determined in Petri dishes containing YED-P (Table 1). Results were considered as positive when the diameter of the clear zones exceeded 5 mm. By this criterion isolates RMA02, RMA05, RTM08, RTM11, RMA30, RMA31, RMA32, RTM02 were considered to solubilize phosphate. Of the isolates tested, SAP11 and RTM17 had little or no ability to solubilize phosphate.

TP-RAPD fingerprinting

TP-RAPD fingerprinting is a new procedure that uses the two universal primers that amplify the 16S rDNA molecule of bacteria (Rivas et al., 2001). Under specific conditions, these primers allow the amplification of a specific set of DNA fragments, producing a specific DNA fingerprint.

Figure 1 shows the TP-RAPD patterns of the new isolates included in this study. As can be seen, the TP-RAPD pattern of these strains (Figure 1a, lanes 1–9) was different from those of reference strains from *S. meliloti* (Figure 1b, lane 10) and *S. medicae* (Figure 1b, lane 11).

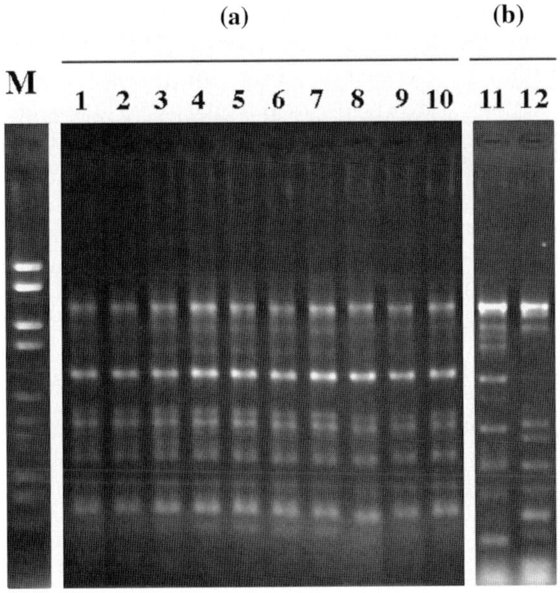

Figure 1. TP-RAPD patterns of strains used in this study. (a) RMA02 (lane 1), RMA05 (lane 2), RMA31 (lane 3), RMA32 (lane 4), RTM08 (lane 5), RTM11 (lane 6), RMA30 (lane 7), RTM02 (lane 8), SAP11 (lane 9) and RTM17 (lane 10). (b) *S. meliloti* USDA1002^T (lane 11) and *S. medicae* USDA1037^T.

The phosphate solubilizing strains from this study dispay same TP-RAPD pattern which is different from that of strains belong to *S. meliloti* and *S. medicae*. According to our previous results, strains that showed the same TP-RAPD pattern belong to the same subspecies (Rivas et al., 2002a). Therefore this PCR-based procedure is very useful to be applied to wide populations of bacteria to select representative strains for sequencing 16S rRNA gene.

16S rDNA sequence analysis

Taking into account the results from TP-RAPD patterns we selected two strains, RTM18 and RMA32, to analyze their 16S rDNA sequence. We also include other strains isolated from plants of cross-inoculation of alfalfa that were not able to solubilize phosphate SAF22, RMO17, RTM18 RMP01 and RMP05. These sequences were aligned with additional 36 sequences of rhizobia and related α-Proteobacteria retrieved from GenBank producing a file with 1470 sites. Misalignments were discovered especially in the variable region starting at site 947 and were corrected using GenDoc. All the isolates for which 16S rRNA gene sequences were determined were placed within the genus *Sinorhizobium* (Figure 2). The sequences of SAF22, RMO17 and RTM18 were identical with that reported for *S. meliloti*, while those of RMP01 and RMP05 were identical with the sequence of *S. medicae*. The isolates RMA32 and RTM17 had identical 16S rRNA gene sequences. Overall 5 nucleotides was the highest number of differences observed among the 16S rRNA gene sequences of these isolates. The two phosphate solubilizing strains belong to the same phylogenetic group which is closely related (more than 99.5% similarity) with the type strain of *S. meliloti*.

Analysis of proteins by SDS-PAGE

Figure 3 shows the protein profiles of the new isolates and of reference strains of *S. medicae* and *S. meliloti*. It is evident that the new strains have a virtually identical and unique profile, different from the reference strains of the *S. medicae* and *S. meliloti*. These results are in agreement with those obtained by TP-RAPD fingerprinting and 16S rRNA sequences and confirm that the phosphate solubilizing strains form a separate group from *S. meliloti* and *S. medicae*. To establish the taxonomic status of strains from this new group more studies must be performed.

Phenotypic characterization

Utilization of different carbon sources was tested in a minimal medium with ammonium nitrate as nitrogen source. Four responses are expected in this medium, acidification, alkalization, no pH change with growth, or no change of pH and no growth. The isolates were placed in two groups but overall had a very similar response in presence of the carbon sources, however, they were distinguishable by pH change with sucrose, trehalose, maltose and raffinose. The pattern of resistance to 10 antibiotics was characteristic of that for species of the genus *Sinorhizobium*. The results of extracellular enzyme production also were characteristic of species within the genus *Sinorhizobium* (Table 2).

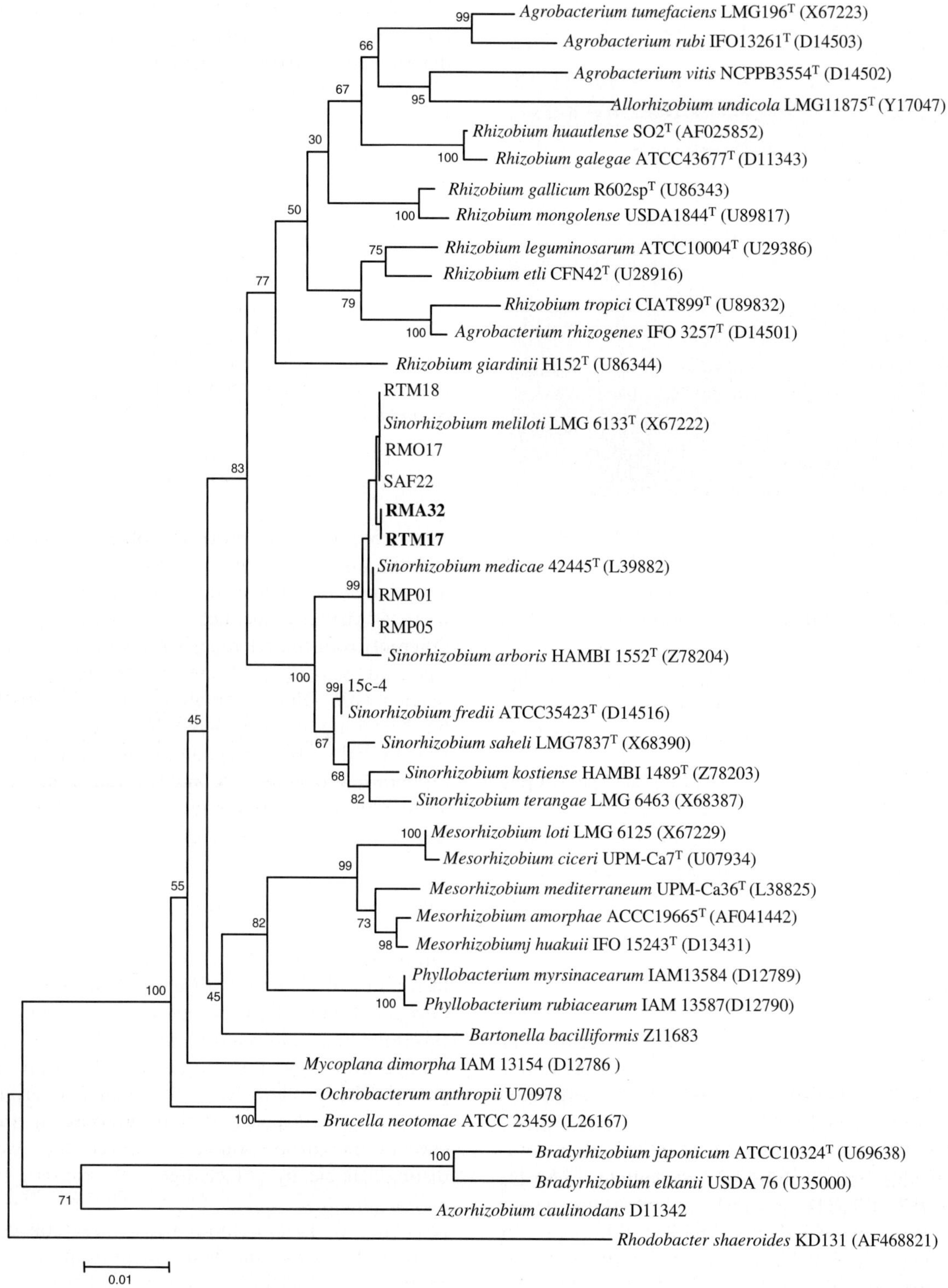

Figure 2. Comparative sequence analysis of 16S rDNA from strains SAP11 and RTM17 and representative strains from the GenBank. The significance of each branch is indicated by a bootstrap value calculated for 1000 subsets. Bar, 1 nt substitutions per 100 nt.

Table 2. Phenotypic characteristics of strains from this study comparing with those of type strains of *S. meliloti* and *S. medicae*

Phenotypic character	*S. meliloti*	*S. medicae*	Phosphate-solubilizing strains group
Carbon source utilization			
Sucrose	A	B	N
Galactose	A	B	A
Lactose	A	B	A
L-Arabinose	A	A	A
Rhamnose	A	B	A
Trehalose	A	N	N
Maltose	A	B	A
Adonitol	A	B	A
Melibiose	A	B	A
Raffinose	A	N	A
Resistance to antibiotics			
Ampicillin	+	+	+
Erythromycin	+	+	+
Ciprofloxacin	−	−	−
Penicillin	+	+	+
Polymyxin	−	−	−
Cloxacillin	+	+	+
Oxitetracyclin	−	−	−
Gentamicin	−	−	−
Cefuroxime	+	+	+
Neomicin	w	−	−
Enzyme activity			
PNP-α-Dara	+	+	+
PNP-β-Dara	w	+	+
PNP-α-Lfuco	−	−	−
PNP-β-Dfuco	+	+	+
PNP-α-Dgal	w	+	+
PNP-β-Dgal	+	+	+
PNP-α-Dxyl	−	−	−
PNP-β-Dxyl	+	+	+
PNP-α-Dmal	−	+	−
PNP-N-ac-thioglc	w	−	−

We have discovered phosphate solubilization as a character among rhizobia that were recovered from nodules of *Medicago*, *Melilotus* and *Trigonella* growing in a field in Spain. From phenotypic and genetic analyses of the isolates we concluded that they were placed in genus *Sinorhizobium*. The isolates are closely related to *S. meliloti* and *S. medicae*, but may be differentiate of both species using several techniques. To our knowledge, this is the first report of phosphate solubilization by strains from genus *Sinorhizobium*.

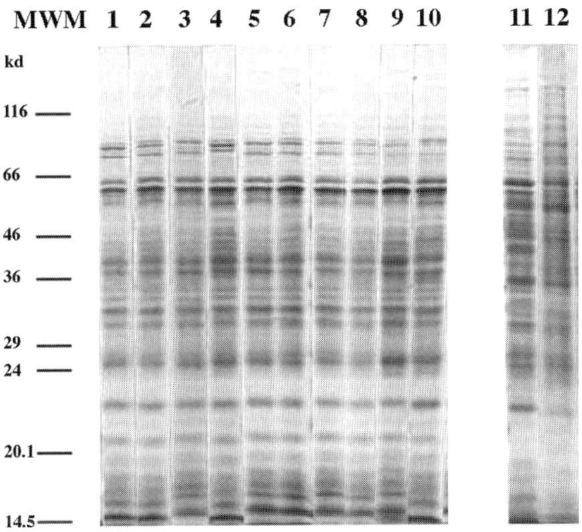

Figure 3. SDS-PAGE patterns of strains from this study. (A) RMA02 (lane 1), RMA05 (lane 2), RMA31 (lane 3), RMA32 (lane 4), RTM08 (lane 5), RTM11 (lane 6), RMA30 (lane 7), RTM02 (lane 8), SAP11 (lane 9) and RTM17 (lane 10). (B) *S. meliloti* USDA1002^T (lane 11) and *S. medicae* USDA1037^T.

References

Bergersen F J 1961 The growth of *Rhizobium* in synthetic media. Aust. J. Biol. Sci. 14, 349–360.

de Lajudie P, Willems A, Pot B, Dewettinck D, Maestrojuan G, Neyra M, Collins MD, Dreyfus B, Kersters K and Gillis M 1994 Polyphasic taxonomy of *Rhizobia*: Enmendation of the genus *Sinorhizobium* and description of *Sinorhizobium meliloti* comb. nov., *Sinorhizobium saheli* sp. nov., and *Sinorhizobium teranga* sp. nov. Int. J. Syst. Bacteriol. 44, 715–733.

Halder A K, Mishra A K and Chakrabartty P K 1990 Solubilization of phosphatic compounds by *Rhizobium*. Indian J. Microbiol. 30, 311–314.

Kumar S, Tamura K, Jakobsen I B and Nei M 2001 Molecular Evolutionary Genetics Analysis Software. Arizona State University, Tempe, Arizona. USA.

Laemmli U K 1970 Cleavage of structural proteins during assembly of the head of bacteriophage T4. Nature 227, 680–685.

Nicholas K B and Nicholas H B 1997. Genedoc: A tool for editing and annotating multiple sequence alignments. Multiple Sequence Alignment Editor and Shading Utility, 2.6.001. Distributed by the authors.

Peix A, Rivas-Boyero A A, Mateos P F, Rodríguez-Barrueco C, Martínez-Molina E and Velázquez E 2001 Growth promotion of chickpea and barley by a phosphate solubilizing strain of *Mesorhizobium mediterraneum* under growth chamber conditions. Soil Biol. Biochem. 33, 103–110.

Rigaud J and Puppo 1975 Indole-3-acetic catabolism by soybean bacteroids. J. Gen. Microbiol. 88, 223–228.

Rivas R, Velázquez E, Valverde A, Mateos P F and Martínez-Molina E 2001 A two primers random amplified polymorphic DNA procedure to obtain polymerase chain reaction fingerprints of bacterial species. Electrophoresis 22, 1086–1089.

Rivas R, Velázquez E, Palomo J L, Mateos P, García-Benavides P and Martínez-Molina E 2002a Rapid identification of *Clavibacter michiganensis* subspecies *sepedonicus* using two primers random amplified polymorphic DNA (TP-RAPD) fingerprints. Eur. J. Plant Pathol. 108, 179–184.

Rivas R, Velázquez E, Willems A, Vizcaíno N, Subba-Rao N S, Mateos P F, Gillis M, Dazzo F B and Martínez-Molina E 2002b A new species of *Devosia* that forms a nitrogen-fixing root-nodule symbiosis with the aquatic legume *Neptunia natans* (L. f.) Druce. Appl. Environ. Microbiol. 68, 5217–5222.

Rome S, Fernández Brunel M P B, Normand P and Cleyet-Marel J C 1996 *Sinorhizobium medicae* sp. nov., isolated from annual *Medicago* spp. Int. J. Syst. Bacteriol. 46, 972–980.

Rodríguez H and Fraga R 1999 Phosphate solubilizing bacteria and their role in plant growth promotion. Biotech. Adv. 17, 319–339.

van Berkum P 1990 Evidence for a third uptake hydrogenase phenotype among the soybean bradyrhizobia. Appl. Environ. Microbiol. 56, 3835–3841.

van Berkum P, Beyene D and Eardly B D 1996 Phylogenetic relationships among *Rhizobium* species nodulating the common bean *Phaseolus vulgaris* L. Int. J. Syst. Bacteriol. 46, 240–244.

van Berkum P, Beyene D, Bao G, Campbell T A and Eardly B D 1998 *Rhizobium mongolense* sp. nov. is one of three rhizobial genotypes identified which nodulate and form nitrogen-fixing symbioses with *Medicago ruthenica* [(L.) Ledebour]. Int. J. Syst. Bacteriol. 48, 13–22.

van Berkum P and Fuhrmann J J 2000 Evolutionary relationships among the soybean bradyrhizobia reconstructed from 16S rRNA gene and Internally Transcribed Spacer region sequence divergence. Int. J. Syst. Evol. Microbiol. 50, 2165–2172.

Vincent J M 1970 The cultivation, isolation and maintenance of rhizobia. *In* A Manual for the Practical Study of Root-Nodule. Ed. J M Vincent. pp. 1–13. Blackwell Scientific Publications, Oxford.

Effect of phosphorous on nodulation and nitrogen fixation by *Phaseolus vulgaris*

M. Olivera, N. Tejera, C. Iribarne, A. Ocaña & C. Lluch[1]
Departamento de Fisiología Vegetal, Facultad de Ciencias, Universidad de Granada, Campus de Fuentenueva s/n, 18071, Granada, Spain. [1]*Corresponding author**

Received 26 November 2002. Accepted in revised form 2 January 2003

Key words: acid phosphatase, amino-acids, N_2 fixation, nodule, *Phaseolus vulgaris*, total soluble sugar

Abstract

The impact of phosphorous on plant growth and symbiotic N_2 fixation in common bean (*P. vulgaris*) plants was investigated. Plants inoculated with *R. tropici* CIAT899 were grown with six P dosage. The P increased plant growth, nodule mass, nitrogenase activity (ARA) and P content, and decreased amino acids and total soluble sugars in the vegetative organs (root, shoot and nodule). The root growth proved less sensitive to P deficiency than did shoot growth, and the leaf area was inhibited at low P. The optimal amount for this symbiosis was 1.5 mM P, this treatment augmented nodule ARA some 20-fold and plant ARA some 70-fold with respect to control.

Introduction

The common bean is the most important food legume especially in Latin-America and Africa and their cultivation is extends into marginal areas. Symbiotic nitrogen fixation (SNF) potential in common bean is considered to be low (Pereira and Bliss, 1987) in comparison with other legumes, as soybean (Israel, 1987). In studies on mineral requirements of symbiosis (Robson, 1983), phosphorous (P) has received considerable attention due to the dramatic effect observed when P fertilizer is applied to nodulated legume in low P soils. This element is one of the most frequently limiting plant nutrients in the tropics and it is estimated that over 50% of soils are limited by P deficiency (CIAT, 1992). Increased nodule number and nitrogenase activity by P addition implies more efficient SNF (Israel, 1987). The mechanism that accounts for the increased SNF has not been elucidates, given that some studies report that P deficiency decreased nitrogenase activity (Vadez et al., 1999), while others do not (Ribet and Drevon, 1996).

High or low P levels are know to induce numerous changes in plant metabolism as carbohydrate content (Rychter and Randall, 1994), total respiration rate (Wanke et al., 1998), amino-acids concentration (Almeida et al., 2000) and enzyme activities associates to P stress like acid phosphatase activity (APA). Bieleski (1973) has noted that APA of nodules may have a significant role in making P more available for plant use.

The objective of this study was to examine common bean plants responses to different P levels in terms of growth parameters and nitrogen fixation metabolism.

Material and methods

Seeds of *P. vulgaris* var. Contender inoculated with *R. tropici* CIAT899 were grown in a solution (Rigaud and Puppo, 1975), in a controlled environmental chamber. Plants were harvested at

* FAX No: 34-958-248995.
E-mail: clluch@ugr.es

E. Velázquez and C. Rodríguez-Barrueco (eds.),
First International Meeting on Microbial Phosphate Solubilization, 157–160

28 days. Leaf area was measured using a photoelectronic planimeter. The treatments consisted in different P levels applied to the nutrient solution (0, 0.1, 0.5, 1.0, 1.5 and 2 mM) as KH_2PO_4.

Nitrogenase was determined by the acetylene reduction activity test (ARA) following the method of Herdina and Silsburry (1990). The nodulated root sample (1 g plus nodules) of each plant was incubated at room temperature in vials containing C_2H_2 (10%, v/v) in air and sealed with serum caps. Aliquots of 0.2 mL were taken after 5 and 10 min incubation and analyzed for ethylene in a Konik Instrument Gas Chromatograph fitted with a Porapak-R column and a flame ionization detector.

The total soluble sugar (TSS) and of free amino-acids content in nodules, were extracted according to Irigoyen et al. (1992), using 1 g of nodules and 12 mL of extraction medium. TSS was determined following a colorimetric method with anthrone reagent, and amino-acids content was assayed using ninhydrine reagent (Yemm and Cocking, 1955). Total nitrogen (N) was analyzed using Kjeldahl method, and the P content was determined with autoanalyzer using the reaction with amidol (Lachica et al., 1973).

Acid phosphatase activity was measured using a modification of the method of Tabatabai and Bremner (1969). Samples 0.5 g in 5 mL extraction buffer were incubated with 1 mM p-nitrofenyl phosphate and 0.2 *M* sodium acetate buffer pH 5.2 for 30 min at 30 °C in water bath. The reaction was stopped by the addition of 4 mL of 0.2 *M* NaOH, and the optical density was measured at 505 nm in spectrophotometer. Controls were assayed by adding NaOH immediately before the extract addition.

Results

The plant dry weight (PDW) increased with the P application. The root:shoot ratio (RSR) decreased with the dosage of P, indicating that the P deficiency reduced the shoot dry weight (SDW), while the root dry weight (RDW) was not affected. Thus, the highest P concentration doubled the SDW, while the RDW increased only 10% (data not shown), resulting in a RSR 50% lower than control. Area leaf response was similar to the PDW, with 0.1 and 0.5 mM P leaf area (LA) increased 25%, and 50–100 % with other treatments (Table 1).

The nodule number (NN) and nodule dry weight (NDW) increased with the P dosage (Table 1). Both parameters showed significant positive correlation with PDW (r = 0.90* and r = 0.83*, respectively). Nodule ARA, expressed in μmol C_2H_4 h^{-1} g $nodule^{-1}$, varied with P-treatments (Table 1). This activity was greatest at 1.5 and 2 mM, reaching 10–20-fold higher than control. Nodule ARA correlated with root P content (r = 0.81*).

Phosphorous content increased in root and shoot. This increase was more intensive in root than in shoot, since with the highest P treatment the P content in this organ was 6-fold that detected in control, whereas it was 4-fold greater in the shoot. The N content (%) in root and shoot was not significantly affected by P application. The APA was 30-fold higher in leaf than in root. In shoot, the APA decreased with the P, dosage, except with 1.0 mM P where this activity increased. On the contrary, in roots increased this activity with the amount of P applied (Table 2).

Table 1. Plant dry weight (PDW) in g $plant^{-1}$, root:shoot ratio (RSR), leaf area (LA) in cm^2, nodule number (NN), nodule dry weight (NDW) in g $plant^{-1}$, acetylene-reduction activity (ARA) in μmol C_2H_4 $(g\ nodule)^{-1}$ h^{-1} in *P. vulgaris* inoculated with the *R. tropici* CIAT899 strain and grown with different P doses

P dosage (mM)	PDW	RSR	LA	NN	NDW	ARA
0.0	1.57^a	0.44^f	286.4^a	83^a	0.054^a	26.23^a
0.1	2.06^b	0.40^e	361.2^b	152^b	0.124^b	68.58^b
0.5	2.49^c	0.38^d	366.6^b	166^c	0.228^e	77.28^b
1.0	2.55^c	0.30^c	436.8^c	199^d	0.190^d	73.72^b
1.5	2.58^c	0.25^a	450.7^c	233^f	0.159^c	556.80^d
2.0	2.63^c	0.22^a	534.5^d	193^e	0.191^d	209.88^c
LSD (0.05)	0.32	0.04	28.5	6.34	0.043	31.83

a–f: Means followed by the same letter within a column do not differ at the $P < 0.05$ probability level using the LSD test.

Table 2. P and N (%) in shoot and root, and acid phosphatase activity (APA, μmol g^{-1} FW h^{-1}) in leaves and roots of *P. vulgaris* inoculated with the *R. tropici* CIAT899 strain and grown with different P doses

P dosage (mM)	P		N		APA	
	Shoot	Roots	Shoot	Roots	Leaves	Roots
0.0	0.23^{a}	0.25^{a}	2.59^{a}	1.19^{ab}	390.81^{e}	9.33^{b}
0.1	0.38^{b}	0.44^{b}	2.59^{a}	1.19^{ab}	393.21^{e}	9.34^{b}
0.5	0.71^{c}	0.82^{c}	2.64^{a}	1.00^{a}	324.30^{c}	9.45^{b}
1.0	0.74^{c}	0.99^{d}	2.47^{a}	1.12^{ab}	374.39^{d}	8.16^{a}
1.5	0.93^{d}	1.62^{f}	2.47^{a}	1.29^{b}	304.63^{b}	11.68^{c}
2.0	1.06^{e}	1.45^{e}	2.40^{a}	1.32^{b}	245.97^{a}	12.26^{d}
LSD (0.05)	0.09	0.14	0.26	0.23	8.64	0.37

a–f: Means followed by the same letter within a column do not differ at the $P < 0.05$ probability level using the LSD test.

Table 3. Amino acid (AA) and total soluble sugar (TSS) content in leaves, roots and nodules (mg g^{-1} FW) of *P. vulgaris* inoculated with the *R. tropici* CIAT899 strain and grown with different P doses

P dosage (mM)	AA			TSS		
	Leaves	Roots	Nodules	Leaves	Roots	Nodules
0.0	7.47^{c}	14.94^{a}	16.17^{d}	8.22^{e}	9.96^{e}	11.45^{f}
0.1	7.06^{c}	19.99^{c}	15.75^{c}	5.17^{a}	8.22^{d}	10.30^{e}
0.5	5.31^{a}	17.58^{b}	15.81^{c}	5.90^{d}	6.54^{c}	6.87^{d}
1.0	6.19^{b}	14.73^{a}	15.52^{c}	5.41^{c}	3.88^{a}	6.15^{c}
1.5	6.43^{b}	14.03^{a}	13.31^{b}	5.40^{b}	5.54^{b}	5.84^{b}
2.0	6.40^{b}	14.59^{a}	12.91^{a}	4.83^{a}	5.39^{b}	5.63^{a}
LSD (0.05)	0.51	1.79	0.38	0.36	0.75	0.28

a–f: Means followed by the same letter within a column do not differ at the $P < 0.05$ probability level using the LSD test.

Phosphorous deficient plant (0 and 0.1 mM) exhibited increased TSS in vegetative organs (nodules, leaves and roots). These TSS accumulated primarily in nodules and roots. In general, increased P appeared not to favour the TSS accumulation in vegetative organs (Table 3). The amino-acids content in root, leaves and nodules, showed a similar trend in the three organs, decreasing slightly as the P increased. The root was the organ that accumulated the greatest amount of amino acids, while the leaf accumulated only 40–48% of the amount in the root.

Discussion

The phosphorous absorption ability has been reported to be strongly correlated with dry matter production (Lynch et al., 1991). In our results PDW was significantly correlated with the P on shoot and roots (in both $r = 0.83^*$). Shoot and roots showed a different behaviour, reflected on RSR. This ratio is a determining factor for the nutrient-uptake capacity of the roots (Larson, 1994). Plants that exhibit reduced shoot growth and increased RSR can be considered P deficient, like P-control plants. We also observed that root growth was less sensitive to P deficiency than shoot growth. The decrease in SDW showed on P-deficient plants can be a direct consequence of a reduction of leaf expansion (Lynch et al., 1991).

The phosphorous treatments did not significantly alter the N content in shoot or root. Different results indicating that the P application increased N content, has been reported by Kolawole and Kang (1997) and Ribet and Drevon (1996). In common bean plants dependent on N_2 fixation, the absence of the relationship between shoot N content and increasing P levels could indicate that nodule functioning requires more P than does plant growth in general.

The phosphorous supply increased the NN, NDW and ARA nodule (Table 1). This nodule response confirms the greater responsiveness of traits associated with biological N_2 fixation than of host plant growth (Israel, 1987). However, our results are not consistent with the finding of Pereira and Bliss (1987) which showed NDW to be correlated with N_2 fixation. The correlation between ARA-nodule and phosphorous content appear to result from efficiency in phosphorous use as well as in efficiency of nodules (Vadez et al., 1999).

According to Burauel et al. (1990) moderate P stress in soybean plants enhanced the assimilate import to the roots. However, in common bean plants our results suggest a significant import to the nodules which seem to constitute the dominant sink under these conditions. TSS accumulation was greater in nodules that in roots and shoots, at least at the low phosphorous treatments (0.1 and 0.5 mM). We confirm that amino-acids content feel with the increased of P. It was found that P limitation was associated with a substantial accumulation of amino-acids (Almeida et al., 2000). Associated with the decline in amino-acids in nodules and roots (Table 3), the phosphorous treatment lowered nitrogenase activity increased (Table 1). Under salt stress Khadri et al. (2001) also found an increased amino-acid concentration in common bean nodules with the depression of nitrogenase activity.

In our results, the optimized amount of P for the *P. vulgaris–R. tropici* CIAT899 symbiosis was 1.5 mM because it increased whole plant dry weight as well as the ARA nodule, that was 20-fold higher than in control.

Acknowledgements

Financial support was obtained through the Andalusian Research Program (AGR–137) and the DGICYT, PB98-1276 project.

References

Almeida J P, Hartwig U A, Freshner M, Nösberger J and Lüsher A 2000 Evidence that P deficiency induces N feedback regulation of symbiotic N_2 fixation in white clover (*Trifolium repens* L.). J. Exp. Bot. 51, 1289–1297.

Bieleski R L 1973 Phosphate pools, phosphate transport, and phosphate availability. Annu. Rev. Plant Physiol. 24, 225–252.

Burauel P, Wieneke J and Führ F 1990 Carbohydrate status in roots of two soybean cultivars: a possible parameter to explain different efficiencies concerning phosphate uptake. *In* Genetics Aspects of Plant Mineral Nutrition. Ed. El Bassan. pp. 111–116. Kluwer Academic Publishers, Dordrecht.

CIAT, 1992. Constraints to and opportunities for improving bean production. A planning document 1993–98. An achievement document 1987–92.CIAT, Cali Colombia.

Herdina J A and Silsbury J H 1990 Estimating nitrogenase activity of faba bean (*Vicia faba* L.) by acetylene reduction assay. Aust. J. Plant Physiol. 17, 489–502.

Irigoyen J J, Emerich D W and Sánchez-Díaz M 1992 Water stress induced changes in concentrations of proline and total soluble sugar in nodulated alfalfa (*Medicago sativa*) plants. Physiol. Plant. 84, 55–60.

Israel D W 1987 Investigation of the role of phosphorus in symbiotic dinitrogen fixation. Plant Physiol. 84, 835–840.

Khadri M, Pliego L, Soussi M, Lluch C and Ocaña A 2001 Ammonium assimilation and ureide metabolism in common bean nodules under salt stress. Agronomie 21, 635–643.

Kolawole G O and Kang B T 1997 Effect of seed size and phosphorus fertilization on growth of select legumes. Commun. Soil Sci. Plant Anal. 28, 1223–1235.

Lachica M, Aguilar A and Yañez J 1973 Análisis foliar. Métodos utilizados en la Estación Experimental del Zaidín C.S.I.C. (II). Anal. Edaf. Agrobiol. XXXII, 1033–1047.

Larson C M 1994 Responses of the nitrate uptake system to external nitrate availability: a whole plant perspective. *In* A Whole Plant Perspective on Carbon–Nitrogen Interactions. Eds. J Roy and E Garner. pp. 31–45. SBB Academic Publish, The Hague Netherlands.

Lynch J, Läuchli A and Epstein E 1991 Vegetative growth of the common bean in response to phosphorus nutrition. Crop Sci. 31, 380–387.

Pereira P A A and Bliss F A 1987 Nitrogen fixation and plant growth of common bean (*Phaseolus vulgaris* L.) at different levels of phosphorus availability. Plant Soil 104, 79–84.

Ribet J and Drevon J J 1996 The phosphorus requirement of N_2 fixation and urea-fed *Acacia mangium*. New Phytol. 132, 383–396.

Rigaud J and Puppo A 1975 Indole-3-acetic acid catabolism by soybean bacteroids. J. Gen. Microbiol. 88, 223–228.

Robson A D 1983 Mineral nutrition. *In* Nitrogen Fixation. Ed. W J Broughton. pp. 36–55. Oxford University Press, Great Britain.

Rychter A M and Randall D D 1994 The effect of phosphate deficiency on carbohydrate metabolism in bean roots. Physiol. Plant 91, 383–388.

Tabatabai M A and Bremner J M 1969 Use of p-nitrophenyl phosphate for assay of soil phosphatase activity. Soil Biol. Biochem. 1, 301–307.

Vadez V, Lasso J H, Beck D P and Drevon J J 1999 Variability of N_2-fixation in common bean under P deficiency is related to P use efficiency. Euphytica 106, 231–242.

Wanke M, Ciereszko I, Podbiekowska M and Rychter A M 1998 Response to phosphate deficiency in bean (*Phaseolus vulgaris* L.). Respiratory metabolism, sugar localization and changes in ultrastructure of bean root cells. Ann. Bot. 82, 809–818.

Yemm E W and Cocking E C 1955 The determination of amino acids with ninhydrin. Analystic 80, 209–213.

Role of arbuscular mycorrhizal fungi in the uptake of phosphorus by micropropagated blackberry (*Rubus fruticosus* var. brazos) plants

Y. Carreón-Abud[1,3], E. Soriano-Bello[2] & M. Martínez-Trujillo[1]

[1]*Laboratorio de Microbiología, Facultad de Biología, Universidad Michoacana de San Nicolás de Hidalgo, Campus Ciudad Universitaria, Morelia, Michoacán, 58066, México.* [2]*Laboratorio de Bioquímica Vegetal, Instituto de Investigaciones Químico Biológicas, Universidad Michoacana de San Nicolás de Hidalgo, Campus Ciudad Universitaria, Morelia, Michoacán, 58066, México.* [3]*Corresponding author**

Received 23 December 2002. Accepted in revised form 2 January 2003

Key words: blackberry, *in vitro*, micropropagated, mycorrhizal arbuscular fungi, phosphorus

Abstract

The beneficial effects of arbuscular mycorrhizae on plant growth have been often related to the increase in the uptake of no mobile nutrients from the soil such as phosphorus. In this work, the data obtained about the increase of phosphorus by plants extracted from *in vitro* cultures with mycorrhizae in relation with non-mycorrhizal plants is shown. The mycorrhizal fungi were isolated and propagated from maize cultures. Micro propagation systems for blackberry were used in the Murashige and Skoog medium, varying the hormone concentration according to the growth. Once the plants were developed, they were transplanted into sterile soil and were inoculated with the previously propagated mycorrhizal fungi and finally transferred to a greenhouse. Harvest was made periodically with the objective of making the evaluation of the different agronomic variables and the amount of phosphorus in the aerial parts of the plant was determined by the colorimetric method of blue molybdate. The percent of phosphorus in the aerial part of *Rubus fruticosus* var. brazos, started to increase since the 30 days of treatment, until it increased an 80% at the end of the assay. This brought more efficiency in the mycorrhized plants to raise the photosynthetic rates in a shorter period of time and to be under a lower stress due to the transplanting process. One of the major effects of mycorrhizal fungi inoculation in plants is the increase of phosphorus absorption ability, by the direct activity of the extramatricial mycelium that allows the exploration of the soil volume. In this way, the mycorrhizal arbuscular fungi make up another chance in the process of getting nutrients for the plants, particularly phosphorus.

Introduction

In natural ecosystems, more than 80% of vascular flowering plants live in symbiotic associations with arbuscular mycorrhizal (AM) fungi (Smith and Read, 1997; Carreón, 2002; Carreón et al., 2000). AM symbioses can be found in ecosystems throughout the world, where they affect plant biodiversity and ecosystem functioning (Van der Heijden et al., 1998).

AM fungi are obligate biotrophs that colonize plant roots to obtain carbon from the plant. In addition to growth within the root cortex, they develop an extensive extraradical mycelium in the surrounding soil. The fungus is able to translocate phosphate from the soil to the interior of the root system, where it is released to the plant (Smith et al., 2001). Phosphorus is one of the mineral nutrients essential

* E-mail: ycabud@yahoo.com.mx

E. Velázquez and C. Rodríguez-Barrueco (eds.),
First International Meeting on Microbial Phosphate Solubilization, 161–165

for plant growth and development. Although the total phosphorus content of soils may be high, phosphorus exists largely as sparingly soluble complexes that are not directly accessible to plants, and it is one of mineral nutrients that limits crop production through the world (Bieleski, 1973).

The beneficial effects of AM fungi on plant growth have been often related to the increase in the uptake of no mobile nutrients from the soil such as phosphorus (Bolan, 1991) This contribution is dependent of external mycelia of AM fungi, which absorb and transport P and other mineral nutrients from beyond the nutrient depletion zone into the associated plant roots (Villegas and Fortin, 2001). It has been proposed that AM fungi increase soil P uptake by increasing the mycorrhizal root absorptive area, by improving nutrient transfer efficiency and utilization of P within the host plant, and by enhancing the solubility of P in the rhizosphere through pH alteration of surrounding soil (Ortas et al., 1996).

Various mechanisms have been suggested from the increase in the uptake of P by mycorrhizal plants. These include: exploration of larger soil volume: faster movement of P into mycorrhizal hyphae, and solubilization of soil phosphorus (Bolan, 1991).

In AM associations, the fungi release phosphate from differentiated hyphae called arbuscules, that develop within the cortical cells, and the plant transports the phosphate across a symbiotic membrane called periarbuscular membrane, into the cortical cell. In *Medicago truncata*, a model legume used widely for studies of root symbioses, it is apparent that the phosphate transporters operate in the root. Soil interface does not participate in symbiotic phosphate transport (Harrison et al., 2002). They cloned and characterized phosphate transporter genes from the AM fungi *Glomus versiforme* and *Glomus intraradices*, and from the roots of a host plant, *Medicago truncata*. Expression analyses and localization studies indicate that each of these transporters has a role in phosphate uptake from the soil solution.

In this work, the data obtained about the increase of phosphorus by plants extracted from *in vitro* cultures with mycorrhiza in relation with non-mycorrhizal plants is shown.

Materials and methods

Blackberry micropropagation

Micropropagation systems were used for blackberries (*Rubus fruticosus*) in Murashige and Skoog (MS) cultivation medium (Murashige and Skoog, 1962), variating the concentration of nutrients and the hormone concentration in the medium, depending on the cultivation stage. The systems *in vitro* were stablished starting from the apex and buds. They were cultivated in the MS medium, with 2 mg L^{-1} of Benciladenin (BA) in a 25 °C temperature and a photoperiod of 16 light hours. After 30 days of the buds growth, these were dissected and incubated in the MS medium with 1 mg L^{-1} of indolbutyric acid (AIB) for the taking roots process.

Plants post in vitro

Plantlets were taken out from the cultivation flask and homogeneous in sizes were chosen. Plants were inoculated with the previously propagated inoculum, which was isolated from the corn rhizosphere that is cultivated in Tiripetio, Mich.

Inoculation was processed with inocule and soil, that contained spores, mycorrhizal propagules and roots with more than 50% of colonization; 10–15 g of inocule per plant.

Treatments

For the *R. fruticosus* plants treated with AM fungi, only two treatments were established, distributed in blocks with 6 repetitions for each one of them. Treatment No. 1, had the micropropagated plants inoculated with mixed inocule (isolated from the studied soil and propagated in "trap plants" in a greenhouse) and Treatment No. 2 corresponded to control plants.

Agronomic variables evaluation

The samplings were done in the same way every 15 days in a 120-day treatment, with destructive harvest and analysing the different agronomic variables. A *Long Ashton* solution with 22 μg of phosphorus per litre was used. The harvest was carried out at 15, 30, 45, 60, 75 days until completing a 120 days period.

The evaluated agronomic variables were the following ones: aerial part size, root size, dry weight, survival percentage and mycorrhizic colonization percentage. From the nutritional point of view the phosphorus contents were determined.

The phosphates were determinated by the procedure of blue molybdate, described by Chapman and Pratt (1973).

Statistical analysis

The obtained results were subjected to statistical treatment. The variance analysis was applied and comparation means by Tukey test was determined.

Results

In the inoculum isolated from maize crops, the next species of fungi were found: *Glomus invermaium*, *Glomus mosseae*, *Glomus agregatum*, and also spores from the genus *Sclerocystis* and *Acaulospora*, which are in process of identification (Figure 1).

Agronomic variables evaluation

In the plants treated with AM fungi, the survival percentage was 100% for each one of the times in which the evaluations were carried out; on the other side, in the control plants the survival percentage ranged from 75% at 15 days to 98% at 120 days (Figure 2).

The mycorrhization percentages that were presented in micropropagated plants inoculated whith mixed inoculum, increased during the course of the experiment. The highest rise was when the plants were 60 days old, kept high until the 105 days old, with a light decrease at 120 days (Figure 3).

The micropropagated plants height results are presented in Figure 4. We can appreciate that from the 45 days until the end of the experiment

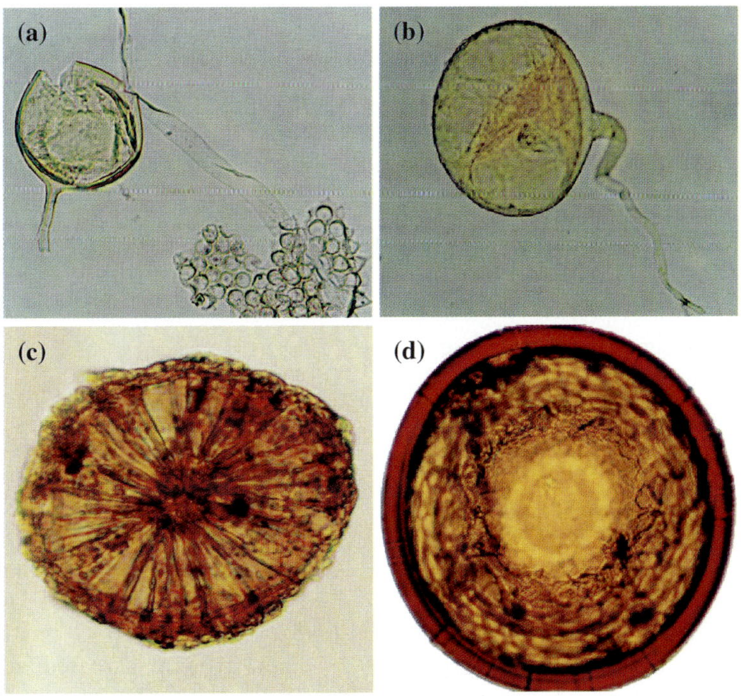

Figure 1. (a) *Glomus* sp. with sporocarp, 40×; (b) *Glomus* sp. with support hyphae, 40×; (c) *Sclerocystis* sp. 40× and (d) *Glomus* sp. 100×.

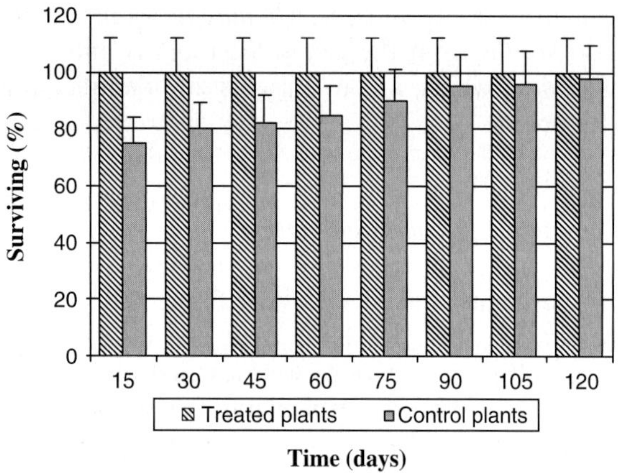

Figure 2. Survival percentage of *R. fruticosus* var. brazos micropropagated plants inoculated with AM fungi under greenhouse conditions.

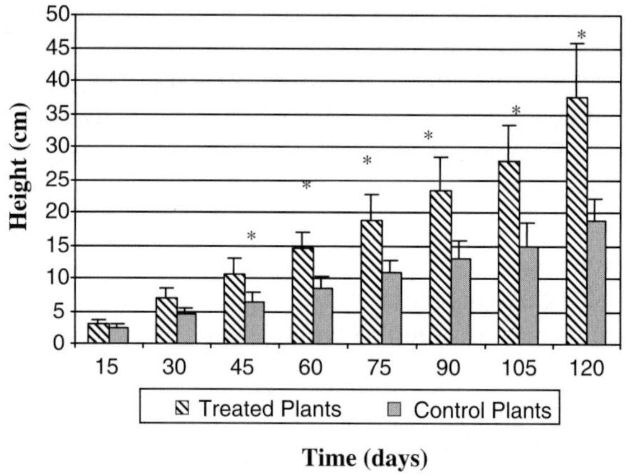

Figure 4. Effect of AM fungi on the height of *R. fruticosus* var. brazos plants in greenhouse conditions. Significative differences (*).

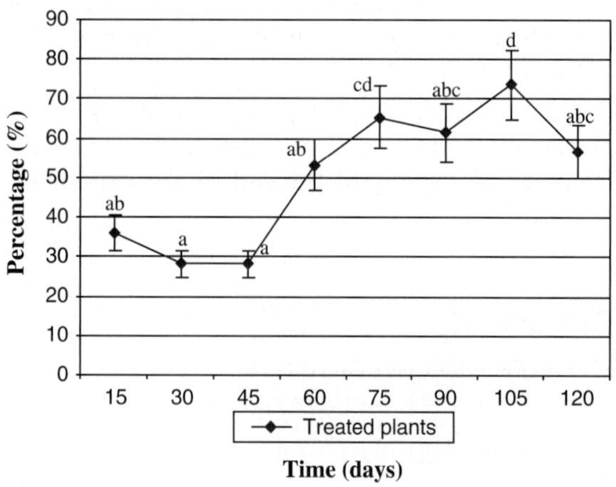

Figure 3. Colonization percentage of *R. fruticosus* var. brazos inoculated with AM fungi in greenhouse conditions.

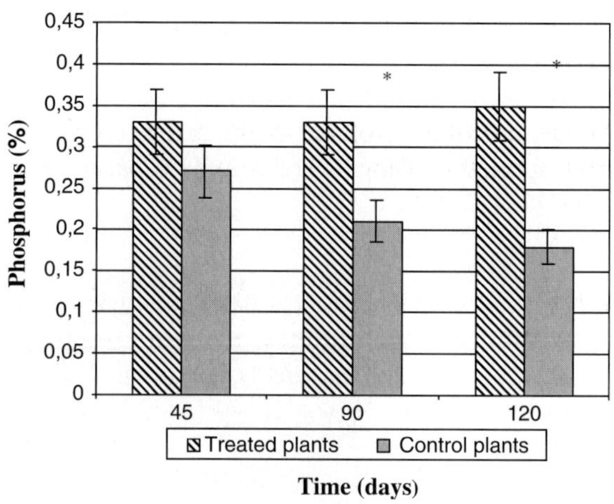

Figure 5. Phosphorus concentration (%) in aerial parts of *R. fruticosus* var. brazos micropropagated plants under greenhouse conditions. Significative differences (*).

(120 days), there was a significative increase in the height of inoculated plants respect to the control ones.

The phosphorus concentration in the treated plants was of 33% at 45 and 90 days old and 0.35% at 120 days old. In control plants the phosphorus concentration was 0.27, 0.21 and 0.04%, respectively, at the same days of treatment (Figure 5).

These results showed that the phosphorus concentration in the aerial part in the beginning is not statistically different in the inoculated plants and the controls, but it becomes significative in the plants with 90–120 days of treatment.

Discussion

Plants treated with AM fungi had surviving percentage of 100% while the non-treated controls had only 75%. This difference could be due to the effect of fungi in improving the root system, since it has been reported that the non-mycorrhized plants have weak root system (Subhan et al., 1998).

Plant growth increased in a 100% compared to plants at the 75 days from the transplanting. This showed that the mycorrhizic association helps *in vitro* plants to establish in field conditions. AM fungi benefits have been demonstrated for micropropagated plants with horticole importance (Sbrana et al., 1994; Wang et al., 1993).

One of the main effects of the mycorrhizic fungi inoculation in plants is the increase in phosphorus absorption by direct activity of the extramatricial mycelium, that allows a greater soil exploration (Bago et al., 2000; Tinker, 1975). This way, the arbuscular mycorrhizic fungi make up another possibility in the plant nutrition process, particularly by enhancing phosphorus uptake.

The phosphorus percentage in *R. fruticosus* had an increase since the 30 days of treatment until it increased up to an 80% compared to the controls at the end of the experiment. This brought a better efficiency in the mycorrhized plants to increase the photosynthetic rates in a shorter period of time and be under less stress due to the transplanting (Nylund and Wallander, 1989; Paul and Kucey, 1981). Phosphorus is required in a primordial way, because building phosphorylized compounds is needed for the photosynthetic metabolism to be carried out (Elmeskaoui et al., 1996).

Acknowledgements

Thanks to Dr. Rafael Salgado Garciglia, for giving us the blackberry plantlets. To M.C. Lourdes Ballesteros Almanza, for her contribution in the experiments and to Miss Yasmín Martínez Carreón for helping in the text elaboration.

References

Bago B, Azcón-Aguilar C, Sachar-Hill Y and Pfeffer P E 2000 El micelio externo de la micorriza arbuscular como puente simbiótico entre la raíz y su entorno. En: Ecología, Fisiología y Biotecnología de la Micorriza Arbuscular. Eds. Alarcón A and Ferrera-Cerrato R. Colegio de Postgraduados. Mundi Prensa. S. A. de C. V. México. 251 p.

Bieleski R L 1973 Phosphate pools, phosphate transport and phosphate availability. Annu. Rev. Plant Physiol. 24, 114–117.

Bolan N S 1991 A critical review on the role of mycorrhizal fungi in the uptake of phosphorus by plants. Plant Soil 134, 189–207.

Carreón Y, Ballesteros L, Salgado R and Alarcón A 2000 Inoculación de hongos micorrízicos arbusculares en plantas de zarzamora (*Rubus sp.)* micropropagadas. En: Ecología, Fisiología y Biotecnología de la Micorriza Arbuscular. Eds. A. Alarcón and R. Ferrera-Cerrato. Colegio de Postgraduados. Mundi Prensa. S. A. de C. V. México. 251 p.

Carreón A Y 2002 Estructura y función de la simbiosis micorrízica arbuscular. Ciencia Nicolaita 31, 65–74.

Chapman H D and Pratt P F 1973 Métodos de análisis para suelos, plantas y aguas. 1ª. Edición. Edit. Trillas. México. 195 p.

Elmeskaoui A, Damont J P, Poulin M J, Piché Y and Desjardins Y 1996 A tripartite culture system for endomycorrhizal inoculation of micropropagated strawberry plantlets in vitro. Mycorrhiza 5, 313–319.

Harrison J M, Dewbre G R and Liu J 2002 A phosphate transporter from *Medicago truncata* involved in the acquisition of phosphate released by arbuscular mycorrhizal fungi. Plant Cell 14, 2413–2429.

Murashige T and Skoog F 1962 A revised medium for rapid growth and bioassays with tobacco tissue cultures. Physiol. Plant. 15, 473–487.

Nylund J E and Wallander H 1989 Effects of ectomycorrhizas on host growth and carbon balance in a semi-hydroponic cultivation system. New Phytol. 112, 389–398.

Ortas I, Harris P J and Rowell D I 1996 Enhanced uptake of phosphorus by mycorrhizal Sorghum plants as influenced by forms of nitrogen. Plant Soil 184, 255–264.

Paul E A and Kucey R M 1981 Carbon flow in microbial associations. Science 213, 415–431.

Sbrana C, Giovannetti M. and Vitagliano C 1994 The effect of mycorrhizal infection on survival anf growth renewal of micropropagated fruit rootstocks. Mycorrhiza 5, 153–156.

Smith S E and Read D J 1997 Mycorrhizal Symbiosis. Academic Press, San Diego, CA 605 pp.

Smith S E, Dickson S and Smith F A 2001 Nutrient transfer in arbuscular mycorrhizas: How are fungal ant plant processes integrated?. Aust. J. Plant Physiol. 28, 683–694.

Subhan S, Sharmila P and Pardha-Saradhi P 1998 *Glomus fasciculatum* alleviates transplation shock of micropropagated *Sesbani sesban*. Plant Cell Rep. 17, 268–272.

Tinker P B 1975 Soil chemistry of phosphorus and mycorrhizal effects on plant growth. *In* Endomycorrhizas. Eds. F E Sanders, B Mosse and P B Tinker. pp. 353–371. Academic Press, London, UK.

Van der Heijden M G A, Klironomos J N, Ursic M, Moutoglis P, Streitwolf-Engel R, Boller Wiemken T A and Sanders I R 1998 Mycorrhizal fungal diversity determine plant biodiversity, ecosystem variability and productivity. Nature 396, 69–72.

Villegas J and Fortin J A 2001 Phosphorus solubilization and pH changes as a result of the interactions between soil bacteria and arbuscular mycorrhizal fungi on a medium containing NH_4^+ as nitrogen source. Can. J. Bot. 79, 865–870.

Wang H, Parent S, Gosselin A and Desjardins D 1993 Vesicular-arbuscular mycorrhizal peat-based substrates enhance symbiosis establishment and growth of tree micropropagated species. J. Am. Soc. Hort. Sci. 118, 896–901.

Effect of plant species and mycorrhizal inoculation on soil phosphate-solubilizing microorganisms in semi-arid Brazil: Growth promotion effect of rhizospheric phosphate-solubilizing microorganisms on *Eucalyptus camaldulensis*

M. R. Scotti[1,2,3], N. Sá[1], I. Marriel[2], L. C. Carvalhais[1], S. R. Matias[1], E. J. Corrêa[1], N. Freitas[1], M. A. Sugai[1] & M. C. Pagano[1]

[1]*Department of Botany, Institute of Biological Sciences, Federal University of Minas Gerais, Avenida Antonio Carlos, 6627, Pampulha, Cep: 31.270-901 Belo Horizonte, Minas Gerais, Brazil.* [2]*EMBRAPA-Maize and Sorghum, Sete Lagoas, MG, Brazil.* [3]*Corresponding author**

Received 19 December 2002. Accepted in revised form 2 January 2003

Key words: *Eucalyptus camaldulensis*, Mycorrhizal fungi, revegetation, *Rhizobium* sp., semiarid, soil phosphate-solubilizing microorganisms, woody caatinga

Abstract

The Jaíba Project is an irrigation enterprise in the north of the state of Minas Gerais and its native vegetation is a dry deciduous forest called woody Caatinga. Two experimental areas (1.5 ha/site) were established in a degraded area using native species intercropped with *Eucalyptus camaldulensis* in three blocks at random. In each experimental area six plots, randomly distributed in each of the three blocks were cultivated as follows: In area A: (1) *Platymenia reticulata* Benth (2) *P. reticulata* inoculated with Rhizobia and spores of Arbuscular Mycorrhizal Fungi (AMF) (3) *Eucalyptus camaldulensis* Dehnh, (4) *Eucalyptus camaldulensis* + AMF (5) *P. reticulata* + *Eucalyptus camaldulensis* + *Tabebuia* sp. (6) *P. reticulata.* (Rhizobia + AMF) + *Eucalyptus camaldulensis* (AMF) + *Tabebuia* sp. In the other area plots were cultivated as follows: (1) *Schinopsis brasiliensis* Engl (2) *Schinopis brasiliensis* + AMF (3) *Eucalyptus camaldulensis* (4) *Eucalyptus camaldulensis* + AMF (5) *Schinopis brasiliensis* + *Myracrodruon urundeuva* Fr. Allen + *Eucalyptus camaldulensis* (6) *Schinopsis brasiliensis* (AMF) + *Eucalyptus camaldulensis* (AMF) + *Myracroduon urundeuva*. Soil samples were taken in the root zone of each cultivated plant and analyzed in relation to the number of phosphate solubilizing microorganisms (PSM) and AMF spores. The results showed that the number of PSM and MF spores was significantly higher in the inoculated *Eucalyptus* rhizosphere, when compared to the native species and also to the non-inoculated *Eucalyptus* plants. The treatment where PSM and AMF populations were increased the plants also showed greatest height and diameter growth and it was not related to soil phosphatase activity. The growth promotion effect of PSM and AMF was confirmed under greenhouse conditions where the double inoculation improved the dry matter production and phosphorus content. Double inoculation of PSM and MF was recommended to *Eucalyptus* plants cultivated in semiarid land.

Introduction

The Jaíba Project is an irrigation enterprise located in semi-arid land of the north of the state of Minas Gerais State and the natural vegetation in the ecological reserve is a "Dry Deciduous Forest" called "Woody Caatinga". The Jaíba's reserve is one of the largest protected areas of woody Caatinga which is under intense anthropic activities and subject to the destructive effects of deforestation and fire, thus showing a very slow rate of regeneration of valuable

* FAX No: +031-4992673/225-6627.
E-mail: scottimuzzi@ig.com.br

E. Velázquez and C. Rodríguez-Barrueco (eds.),
First International Meeting on Microbial Phosphate Solubilization, 167–172

tropical species (Del Rey, 1991). According to Bóo et al. (1997) the sprouting capacity determines the persistence and abundance of woody species. In the reserve subjected to this impact the sprouting of woody species has been restrained by the invasion of pioneer species composed by low load trees and shrub mesh, called "Carrasco vegetation". Thus, a very slow rate of regeneration of valuable tropical species of "Woody Caatinga" has been observed. Therefore, the recovery of the "Woody Caatinga" should have priority. On the other hand, the demand for wood by the farmers becomes a continuous threat for the preserved area, needing a project of wood and energy provisioning for the local populations. With an objective of minimizing the exploratory actions and protection of forest reserve, an afforestation programme was proposed as a model to allow renewable sources of fuel and provide other wood products, supplying the demand for wood in Jaíba's project. In semi-arid ecosystems, phosphorus and nitrogen are limiting elements for plant growth but trees may transfer P from the depths in the soil profile (Salcedo et al., 1997). Different rhizosphere community members may exert effects on plant growth (Westover and Bever, 2001) specially related to phosphorus metabolism (Paul and Rao, 1971) such as mycorrhizal fungi and rock-phosphate-solubilizing microorganisms. It is well known that Mycorrhizal fungi (MF) can improve the phosphorus availability and plant growth (Manjunath et al.,1984; Marques et al, 2001) specially when associated with rock-phosphate-solubilizing microorganisms (Omar, 1998). In addition, AMF can exist saprotrophically, enhancing the uptake of soil organic phosphate (Hodge et al., 2001; Koide and Kabir, 2000). Therefore, for the afforestation programme, *Eucalyptus* plants were intercropped with native woody species inoculated with mycorrhizal fungi and/or rhizobia in an irrigated experimental site subdivided into plots of 1.5 ha. The aim of the study was to evaluate the role of mycorrhizal inoculation over soil microbial activity and phosphate metabolism over plant growth.

Materials and methods

Study site

The study area is situated in semiarid (15°09′03″S 43°49′26″W). The proeminent vegetation was composed by a dense community of interlaced shrubs, characteristic of "Carrasco". The effect of fire or deforestation in the Caatinga ecosystem results in the invasion of "Carrasco vegetation" species. Predominant soil types are yellow podsols with a high infiltration rate and low levels of organic matter (Del Rey, 1991).

Microorganism inoculation and plant cultivation

The slow growing rhizobia strain BHCB-PL19, previously isolated from *Platymenia reticulata* nodules and screened for the effectiveness was provided at 1 ml per plant of *Platymenia reticulata* (10^8 cfu/ml), according to Somasegaran and Hoben (1985). The MF used were *Gigaspora margarita* and *Glomus etunicatum* and was performed by placing 100 spores per fungi specie to inoculate *Platymenia reticulata* and *E. camaldulensis* in site A and *S. brasiliensis* and *E. camaldulensis* in site B. The following treatments were used for each group of 160 plants: (I) Complete fertilization for all species (Somasegaran and Hoben, 1985). (II) Fertilization without nitrogen plus inoculation with the rhizobia strain BHCB-PL19 plus mycorrhizal fungi (MF) for *Platymenia reticulata* (III) Fertilization plus inoculation with arbuscular mycorrhizal fungi(MF) for *Shinopsis brasiliensis* and *Eucalyptus camaldulensis.*

Experimental design

Two experimental sites (1.5 ha/site) were cleared of Carrasco plants and cultivated with native species intercropped with *Eucalyptus camaldulensis* using a completely randomized block. Six treatments with 42 or 48 plants in single and mixed cultivation respectively, were randomly distributed in each of the three blocks. Six plots of 378 m^2 (21×18 m) or 432 m^2 (24×18 m) were distributed in an aleatory way in each block under irrigation. In site A, these six plots were cultivated as follows: (1) *Platymenia reticulata* Benth (2) *P. reticulata* inoculated with rhizobia and spores of arbuscular mycorrhizal fungi (AMF) (3) *Eucalyptus camaldulensis* Dehnh (4) *Eucalyptus camaldulensis* + AMF 5-*P. reticulata* + *Eucalyptus camaldulensis* + *Tabebuia* sp. (6) *P. reticulata* (Rhizobia + AMF) + *Eucalyptus camaldulensis* (AMF) + *Tabebuia* sp. In site B plots were cultivated as follows: (1) *Schinopsis brasiliensis* Engl, (2) *Schinopis brasiliensis* + AMF (3) *Eucalyptus camaldulensis* (4) *Eucalyptus camaldulensis* + AMF (5) *Schinopis brasiliensis* + *Myracrodruon urundeuva* Fr. Allen + *Eucalyptus camaldulensis* (6) *Schinopis brasiliensis* (AMF) + *Eucalyptus camaldulensis* (AMF) + *M. urundeuva.*

Soil microorganisms evaluation and enzyme activity

Soil samples (0–10 cm in depth) were collected from the rhizophere of each cultivated plant, as well as in the Forest Reserve and in Carrasco. Samples were analyzed for the number of Phosphate Solubilizing Microorganisms (PSM) using Pikovskaya's agar (Pikovskaya, 1948), and MF spores were recovered from the field by wet sieving, decanting and sucrose centrifugation (Walker, 1983), and analyzed data were expressed as number of spores/gram of dry soil. Estimation of acid phosphatase rate was determinated by g^{-1} of soil $hour^{-1}$ (Tabatabai, 1982).

Effect of phosphate solubilizing microorganisms (PSM) inoculation

To test the inoculation effect of the PSM isolate (*Aspergillus* sp.), *E. camaldulensis* seedlings were transferred to pots with sand and fertilized according to Somasegaran and Hoben (1985) with diferent P sources as follows: (1) Complete fertilization with 70 ppm of soluble P (2) Complete fertilization with $Ca_3(PO_4)_2$ (70 ppm) (3) Complete fertilization with $Ca_3(PO_4)_2$ + PSM (4) Complete fertilization with $Ca_3(PO_4)_2$ + MF (5) Complete fertilization with $Ca_3(PO_4)_2$ + PSM + MF for 3 months under greenhouse condition.

Sampling and analyses

Growth parameters, diameter of 30 cm above the ground and the height of all plants were recorded after 3 and 4 months in greenhouse and field experiments respectively. The following determinations were then made: (i) shoot dry matter (after drying at 70 °C for 48 h) (ii) phosphorus content (Linderman, 1958). The data was statistically treated by analysis of variance (ANOVA) and means were compared by the Tuckey test.

Results and discussion

Phosphate-solubilizing microorganisms (PSM) population and mycorrhizal spores were increased in *Eucalyptus* rhizosphere, specially when inoculated with MF, contrasting with the other intercropped native species (Figure 1). This effect was higher in experiment A. Mycorrhizal fungi may increase the root exudation and through rhizosphere effect stimulate the PSM population (Barea et al., 2002). This selective rhizosphere effect (Whitelaw, 2000) may be explained by the known high demand of *Eucalyptus* species to phosphorus. If this hypothesis is true its growth would be favored. The data presented in Table 1 confirm that *Eucalyptus* growth was increased specially in experiment A after 8 months of transplanting probably due to the PSM and MF effect .The results of Table 2 also confirm the growth promotion ability of double inoculation with PSM and MF on *Eucalyptus* plants, fertilized with rock phosphate. Double inoculation increased not only the dry matter production but also phosphorus content of *Eucalyptus* leaves contrasting with the single inoculation procedures. Similar results were described by Omar (1998) with wheat plants. Many investigators observed that a high proportion of P-solubilizing microorganisms are concentrated in the rhizosphere of plants (Khan and Bhatnagar, 1977).

Rhizosphere may be considered as the domain in which the plant communicates with soil microorganisms, the place where microorganisms compete or are able to show a synergistic or antagonist function for the benefit or detriment of plant growth. This effect occurs through the flavonoids; in the root exudate which are different compounds for each plant family thus, these compounds may have a selective stimulatory or inhibitory effect over specific microbial groups such as rhizobia and bradyrhizobia (Fisher and Long, 1992) or mycorrhizal fungi (Ishii et al., 1997). Rengel (1997) showed that wheat genotypes tolerant to Zn or Mn deficiency have a capacity to alter biological properties of the rhizosphere, thus increasing the availability of micronutrients. Paul and Rao (1971) also observed a stimulatory effect of PSM in fast growth herbaceous species when compared to woody species like *Sesbania aculeata.*

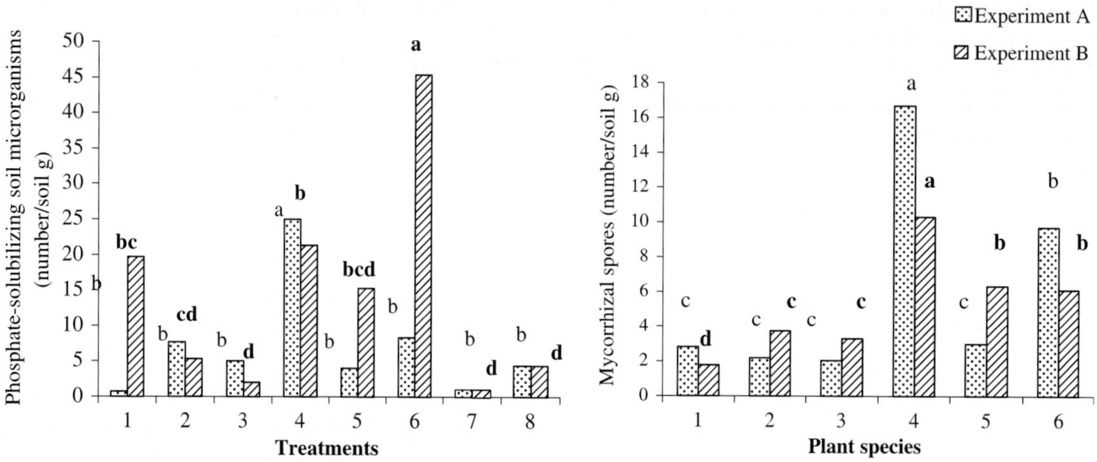

Figure 1. Phosphate-solubilizing Microorganisms (10^{-5})(PSM) number and Mycorrhizal fungi (MF) spore numbers: Experiment A 1 – (1A) *Platymenia reticulata* 2 – (2A) *Platymenia reticulata**, 3 – (3A) *Eucalyptus camaldulensis* 4 – (4A) *Eucalyptus camaldulensis**, 5 – (5A) *Platymenia reticulata* + *Eucalyptus camaldulensis* + *Tabebuia* sp, 6 – (6A) *Platymenia reticulata** + *Eucalyptus camaldulensis** + *Tabebuia* sp) and B (1 – (1B) *Schinopsis brasiliensis*, 2 – (2B) *Schinopsis brasiliensis**,3 – (3B) *Eucalyptus camaldulensis*,4 – (4B) *Eucalyptus camaldulensis* *5 – (5B) *Schinopsis brasiliensis* + *Eucalyptus camaldulensis* + *Myracrodruon urundeuva*, 6 – (6B) *Schinopsis brasiliensis** + *Eucalyptus camaldulensis** + *Myracrodruon urundeuva*, 7 – Carrasco Vegetation, 8 – Forest Reserve, 4 months after transplanting. *: Inoculated with MF. *Means with different letters compared inside of each experiment are significantly different as determined by Tuckey multiple-range test at the 5% confidence level ($P \leq 0.05$). NS: significantly different.

Therefore, eucalyptus as a fast growing species, will demand high phosphorus content for their growth and would be expected that the specific PSM and MF population related to P metabolism will be stimulated. The main mechanism of soil phosphate solubilization has often been due to excretion of organic acids. Oxalic acid is considered the major organic acid produced by *Aspergillus* species (Vassilev et al., 1995). Inorganic phosphate solubilization through organic acid excretion appears to be the most important metabolic mechanism to P availability in experiment A. In contrast, the enzymatic activity may be the preferential source of phosphorus in both the Forest area and site B soils based on the high phosphatase activity (Figure 2). Our data was confirmed by George et al. (2002) who showed that forestry species enhances acid phosphatase

Table 1. Height (H) and Diameter (D) growth of *Eucalyptus camaldulensis* in Experiment A(1 – *Eucalyptus camaldulensis* 2 – *Eucalyptus camaldulensis**, 3 – *Platymenia reticulata* + *Eucalyptus camaldulensis* + *Tabebuia* sp, 4 – *Platymenia reticulata** + *Eucalyptus camaldulensis** + *Tabebuia* sp) and B (3 – *Eucalyptus camaldulensis*, 4 – *Eucalyptus camaldulensis** 5 – *Schinopsis brasiliensis* + *Eucalyptus camaldulensis* + *Myracrodruon urundeuva* ,6 – *Schinopsis brasiliensis** + *Eucalyptus camaldulensis*), 4 months after transplanting. (*Inoculated with MF)

Treatments	Experiment A				Experiment B			
	4 M		8 M		4 M		8 M	
	H	D	H	D	H	D	H	D
1	47.73^{a}	3.4^{NS}	137.5^{ab}	15.45^{a}	43.37^{NS}	3.25^{NS}	123.97^{NS}	13.65^{NS}
2	47.19^{a}	3.5	148.37^{a}	16.2^{a}	47.15	3.42	128.22	14.8
3	42.9^{ab}	3.04	116.13^{b}	12.76^{b}	49.21	3.31	131.00	14.06
4	48.88^{a}	3.05	110.7^{b}	12.47^{b}	47.10	3.04	124.6	12.9

*Means with different letters on each column are significantly different as determined by Tuckey multiple-range test at the 5% confidence level ($P \leq 0.05$).
NS: Not significantly different.

Table 2. Effect of phosphate-solubilizing microorganism (PSM) and Mycorrhizal fungi (MF) inoculation on *Eucalyptus camaldulensis* growth after 3 months in greenhouse conditions

P source + microorganisms inoculation	Height (cm)	Dry matter (g)	P (%)
Control with soluble P	23.8^{NS}	0.373^{ab}	0.303^{ab}
$Ca_3(Po_4)_2$	25.3	0.2391^{b}	0.196^{b}
$Ca_3(Po_4)_2$ + PSM	28.76	0.503^{a}	0.251^{ab}
$Ca_3(Po_4)_2$ + MF	27.10	0.413^{a}	0.278^{ab}
$Ca_3(Po_4)_2$ + PSM + MF	27.0	0.448^{a}	0.406^{a}

*Means with different letters on each column are significantly different as determined by Tuckey multiple-range test at the 5% confidence level ($P \leq 0.05$).
NS: Not significantly different.

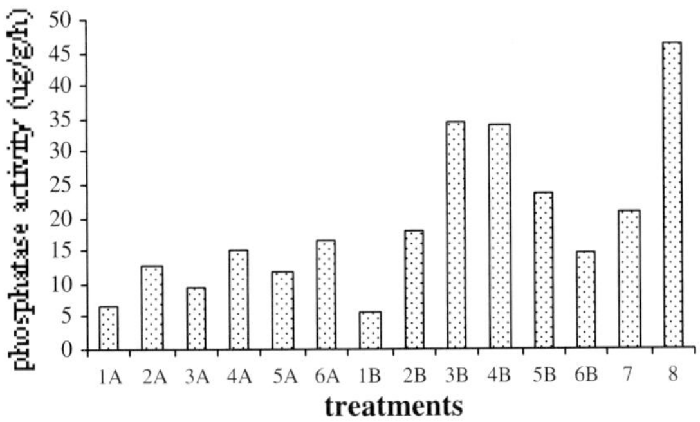

Figure 2. Phosphatase activity (μg/g/h): Experiment A 1 – *Platymenia reticulata* 2 – *Platymenia reticulata**, 3 – *Eucalyptus camaldulensis* 4 – *Eucalyptus camaldulensis* *, 5 – *Platymenia reticulata* + *Eucalyptus camaldulensis* + *Tabebuia* sp, 6 – *Platymenia reticulata** + *Eucalyptus camaldulensis** + *Tabebuia* sp) and B :1 – *Schinopsis brasiliensis*, 2 – *Schinopsis brasiliensis**, 3 – *Eucalyptus camaldulensis*, 4 – *Eucalyptus camaldulensis* *5 – *Schinopsis brasiliensis*. + *Eucalyptus camaldulensis* + *Myracrodruon urundeuva*, 6 – *Schinopsis brasiliensis** + *Eucalyptus camaldulensis** + *Myracrodruon urundeuva*, 7 – Carrasco Vegetation, 8 – Forest Reserve, 4 months after transplanting. *Means with different letters compared inside of each experiment are significantly different as determined by Tuckey multiple-range test at the 5% confidence level ($P \leq 0.05$). NS: significantly different.

activity in their rhizosphere in detriment of organic acid. Soil phosphatase activity may be partially explained by mycorrhizal fungal activity in site B, confirming the results of Rao and Tak (2001) who observed an increase of phosphatase activity in rhizosphere of gypsum mine spoil.

Conclusions

(1) Phosphate-solubilizing microorganisms (PSM) population and Mycorrhizal fungi (MF) were favored in the *Eucalyptus* rhizosphere, specially when inoculated with mycorrhizal fungi as a result of selective rhizosphere effect. (2) The plants which presented an increase in PSM and MF rhizosphere population showed the greatest height growth under field conditions. (3) The growth promoting effect of the PSM and MF was confirmed under greenhouse conditions. (4) Double inoculation of PSM and MF is recommended to *Eucalyptus* plants cultivated in semi-arid land.

Acknowledgements

This study was supported by Ministério do Meio Ambiente/Fundo Nacional do Meio Ambiente. Scholarships from CAPES: Patricia Pinto, Neimar Freitas, Eduardo J. A Correa, are gratefully acknowledged.

References

Barea J M, Azcón R and Azcón-Aguilar C 2002 Mycorrhizosphere interactions to improve plant fitness and soil quality. Antonie van Leeuwenhoek 81, 343–351.

Boó R M, Peláez D V, Bunting S C, Mayor M D and Elía O R 1997 Effect of fire on woody species in central semiarid Argentina. J. Arid Environ. 35, 87–94.

Del Rey 1991 Distrito de Irrigação do Jaíba- Vegetação e manejo de solos. Relatório no: 5, Belo Horizonte, 77 p.

Fisher R F and Long S R 1992 Rhizobium-plant signal exchange. Nature. 357, 655–660.

George T S, Gregory P J, Wood M, Read D and Buresh R J 2002 Phosphatase activity and organic acids in the rhizosphere of potential agroforestry species and maize. Soil Biol. Biochem. 34, 1487–1494.

Hodge A, Campbell C D and Fitter A H 2001 An arbuscular mycorrhizal fungus accelerates decomposition and acquires nitrogen directly from organic material. Nature 413, 297–299.

Ishii T, Narutaki A, Sawada K, Aikawa J, Matsumoto I and Kadoya K 1997 Growth stimulatory substances for vesicular-arbuscular mycorrhizal fungi in Bahia grass (*Paspalum notatum* Flügge.) roots. Plant Soil 196, 301–307.

Khan J A and Bhatnagar R M 1977 Studies on solubilization of insoluble phosphates by microorganisms. Fertil. Technol. 14, 329–333.

Koide R T and Kabir Z 2000 Extraradical hyphae of the mycorrhizal fungus *Glomus intraradices* can hydrolyse organic phosphate. New Phytol. 148, 511–517.

Linderman W 1958 Observation on the behaviour of phosphate compounds in *Chlorella* at the transition from dark to light. United Nations International Conference on the Peaceful. Uses of Atomic Energy. Geneva. 24, 8–15.

Manjunath A, Bragyaraj D J and Gopala Gowada H S 1984 Dual inoculation with V.A. mycorrhiza and *Rhizobium* is beneficial to *Leucaena*. Plant Soil. 78, 445–448.

Marques M S, Pagano M and Scotti M R M 2001 Dual inoculation of a woody legume (*Centrolobium tomentosum*) with rhizobia and mycorrhizal fungi in south-eastern Brazil. Agrofor. Syst. 52, 107–117.

Omar S A 1998 The role of rock-phosphate-solubilizing fungi and vesicular-arbuscular-mycorrhiza (VAM) in growth of wheat plants fertilized with rock phosphate. World J. Microbiol. Biotechnol. 14, 211–218.

Paul M B. and Rao W V B S 1971 Phosphate-dissolving bacteria in the rhizosphere of some cultivated legumes. Plant Soil 35, 127–132.

Pikovskaia R L 1948 Mobilization of phosphorus in soil in connection with vital activities by some microbial species. Mikrobiologia 17, 362–370.

Rao A V and Tak R 2001 Influence of mycorrhizal fungi on the growth of different tree species and their nutrient uptake in gypsum mine spoil in India. Appl. Soil Ecol. 17, 279–284.

Rengel Z 1997 Root exudation and microflora populations in rhizosphere of crop genotypes differing in tolerance to micronutrient deficiency. Plant Soil 196, 255–260.

Salcedo I H, Tiessen H and Sampaio E V S B 1997 Nutrient availability in soil samples of cultivation sites in the semi-arid Caatinga of NE Brazil. Agric. Ecosyst. Environ. 65, 177–186.

Somasegaran P and Hoben H J 1985 Methods in legume-Rhizobium Technology, Ed. Niftal USAID, Hawaii, 367 p.

Tabatabai M A 1982 Soil enzymes. *In* Methods of soil analysis, Part 2. Chemical and microbiological properties 2nd edn. American society for Madison, WI. Eds. Page A L, Miller R H, Kenney D R. pp. 903–947 Iowa State University.

Vassilev N, Barea M T, Vassileva M F L and Azcon R 1995 Rock phosphate solubilization by *Aspergillus niger* grown on sugar-beet waste medium. Appl. Microbiol. Biotechnol. 44, 546–549.

Walker C 1983 Taxonomic concepts in the Endogonaceae: Spore wall characteristics in species descriptions. Mycotaxon 18, 443–455.

Westover K M and Bever J D 2001 Mechanisms of plant species coexistence: Roles of rhizosphere bacteria and root fungal pathogens. Ecology 82, 3285–3294.

Whitelaw M A 2000 Growth promotion of plants inoculated with Phosphate-solubilizing fungi. Adv. Agron. 69, 99–151.

The interactive effects of arbuscular mycorrhizal fungi and rhizobacteria on the growth and nutrients uptake of sorghum in acid soil

J. Widada[1,3], D. I. Damarjaya[2] & S. Kabirun[1]

[1]*Laboratory of Soil and Environmental Microbiology, Department of Soil Science, Gadjah Mada University, Yogyakarta, Indonesia.* [2]*Department of Soil Science, Sriwijaya University, Palembang, Indonesia.* [3]*Corresponding author**

Received 22 December 2002. Accepted in revised form 2 January 2003

Key words: arbuscular mycorrhizal fungi, phosphate-solubilizing bacteria, N-fixing bacteria, siderophore-producing bacteria, sorghum, acid mineral soil

Abstract

The inoculation effects of arbuscular mycorrhizal fungi (AMF) or/and rhizobacteria, (phosphate-solubilizing bacteria, PSB; N_2-fixing bacteria, NFB; and siderophore-producing bacteria, SPB) on the growth and nutrients uptake of sorghum *(Sorghum bicolor)* were studied in acid and low availability phosphate soil. The microbial inocula consisted of the AMFs *Glomus manihotis* and *Entrophospora colombiana*, PSB *Pseudomonas* sp., NFB *Azospirillum lipoferum*, and SPB fluorescent pseudomonad. The inoculation of either AMF or each rhizobacterium improved the plant dry weight and nutrients uptake such as N, P, Fe, and Zn. Dual inoculation of AMF and each rhizobacterium yielded the higher of plant dry weight and nutrients uptake compared to the single inoculation. Dual inoculation of AMF and PSB, AMF and NFB, AMF and SPB increased plant dry weight by 112, 64, and 60 times higher compared to the uninoculated plant, respectively. The rhizobacteria also improved plant colonization by AMF. These results indicated that the interaction of AMF and the selected rhizobacteria has a potential to be developed as biofertilizers in acid soil.

Introduction

Soil acidification and aluminium toxicity are probably the major limiting factors to plant growth and crop production in many agricultural areas of the world (Baligar and Fageria, 1997; Kamprath, 1984). Acidic soils, mainly Ultisols and Oxisols, are very common and abundant in Indonesia where they cover more than 25% of Indonesia's land area (Anonymous, 1987). These soils, having high P-fixing capacity, need intensive P fertilisation rates for obtaining economic yields and have more than half of their total P as organic P (Sanchez, 1976). Much research has been directed towards correcting the problem, including the application of mineral lime, organic residues or other pH-raising materials to replenish plant nutrients and reduce the toxicity of Al and Mn. Because of excessive costs, especially in developing countries, beside application of organic residues, the application of biofertilizers such as plant growth promoting bacteria (PGPR) and arbuscular mycorrhizal fungi (AMF), are considered a key component in low-input-based agro-technologies (Jeffries and Barea, 2001).

Mycorrhizal symbioses are known to play a critical role in plant nutrition, based on the ability of the external mycorrhizal mycelium developing around the host plant roots to efficiently explore a larger volume of soil, thereby enhancing mineral acquisition by the plant (Smith and Read, 1997).

The study about the association of AMF and rhizobacteria had been done extensively. Some

* E-mail: jwidada@lycos.com

E. Velázquez and C. Rodríguez-Barrueco (eds.),
First International Meeting on Microbial Phosphate Solubilization, 173–177

experiment showed that association between AM and rhizobacteria, such as *Azospirillum*, *Azotobacter*, *Pseudomonas*, and phosphate solubilizing bacteria (PSB) had synergic effect on the plant growth (Bagyaraj, 1990; Gaur and Rana, 1990; Negi and Tilak, 1990; Pacovsky, 1988; Tilak, 1990;). However, this kind of research is scare in acid mineral soil, such as Ultisol.

It had been well known that AMF had high adaptability and ability to increase the plant growth and plant production in acid mineral soils (Kabirun and Widada, 1995; Sieverding, 1991). Based on this reason, it was suggested that the association of AMF and rhizobacteria provide better performance compared to the single inoculation in acid mineral soils.

The aim of this research was to evaluate the interactive effects of AMF and rhizobacteria on the growth and nutrient absorption of sorghum in Ultisol.

Table 1. Chemical characteristic of soil used in this study

Chemical characteristics		Criteria*
PH H_2O (1:2,5)	4.9	Low
PH KCl (1:2,5)	3.7	Low
Organic-C (%)	1.6	Very low
Total-N (%)	1.2	High
C/N	1.3	
Available-P Bray 1 (mg kg^{-1})	2.2	Low
Total-P (mg kg^{-1})	232.3	
Exchangeable-Al (cmol(+) kg^{-1})	5.5	Intermediate
Exchangeable-H (cmol(+) kg^{-1})	9.2	Intermediate
Exchangeable Cation (NH_4OAc)		
Exch.-Ca (cmol(+) kg^{-1})	4.7	Intermediate
Exch.-Mg (cmol(+) kg^{-1})	0.07	Low
Exch.-K (cmol(+) kg^{-1})	1.8	Intermediate
Exch.-Na (cmol(+) kg^{-1})	0.3	Low

*Based on Landon (1984).

Materials and methods

Pot experiment using Completely Randomized Design with three replications and 12 treatments as follows: (1) control, without inoculation, (2) inoculated with *Glomus manihotis*, (3) *Entrophospora colombiana*, (4) N-fixing bacterium (NFB), *Azospirillum lipoferum*, (5) siderophore-producing bacterium (SPB), fluorescent pseudomonad, (6) phosphate-solubilizing bacterium (PSB), *Pseudomonas* sp. strain G8, (7) *G. manihotis* + NFB, (8) *G. manihotis* + SPB, (9) *G. manihotis* + PSB, (10) *E. colombiana* + NFB, (11) *E. colombiana* + SPB, and (12) *E. colombiana* + PSB.

Ultisol was taken from West Java, Indonesia, and was sterilized through fumigation using methyl bromide. The following chemical characteristic of soil (Table 1) were analyzed as follows: pH H_2O and pH KCl (1:2.5); organic C (Walkley and Black method), total-N (Kjeldhal method); available P (Bray-1 method), exchangeable H and Al using 1 N KCl extraction, and exchangeable Ca, Mg, K, Na using NH_4OAc extractants (Anderson and Ingram, 1989).

Every pot received 4 kg sterilized soil and was added with plant residue (*Colopogonium cerelium* 5 ton ha^{-1}) as a carbon source, and basal fertilizer. Nitrogen fertilizer was given by the 1/3 amount of recommended level. Phosphate fertilizer was applied as rock phosphate at the level 30 kg P ha^{-1}.

Sorghum bicolor var. UPCA-S_2 was surface sterilized using $HgCl_2$ and aseptically germinated for 2 days, and then removed into the pot experiment. Water content was maintained at 80–85% of water field capacity.

Plant height was observed every 10 days. The plant was harvested at 41 days after planting for determining of the dry weight of shoots and roots. For nutrient analysis, plant material was ground to pass through a 0.5-mm screen and digested in a H_2SO_4–H_2O_2 mixture. P content was quantified by spectrophotometry (Murphy and Riley, 1962), and ferrum, copper, zinc and manganese determined by atomic absorption spectroscopy. The percentage of AMF colonization was determined using the gridline intersect method (Giovanneti and Mosse, 1980).

The data were subjected to analysis of variance using the ANOVA procedures of the SAS Institute, SAS/STAT version 6 (1990). Statistical significance was determined at $P < 0.05$.

Results and discussion

The growth of sorghum was significantly improved by AMF or/and rhizobacteria (N-fixing, P-solubilizing, and siderophore-producing bacteria) inoculation (Table 2). However, the inoculation of

Table 2. Plant growth and AMF infection, 41 days after planting

Treatments	Plant height (cm)	Total dry weight (g)	Shoot-dry weight (g)	Root dry weight (g)	Shoot/ root ratio	AMF coloni-zation (%)
Uninoculated	8.6 g	0.05 d	0.04 d	0.02 e	2.26	0
G. manihot (AMF1)	54.4 cd	1.51 c	0.99 c	0.51 bcde	1.97	54
E. colombiana (AMF2)	34.7 e	0.91 cd	0.60 cd	0.21 bcde	1.96	53
A. lipoferum (NFB)	12.4 g	0.13 d	0.07 d	0.06 de	1.20	0
Fluorescent pseudomonad (SPB)	13.8 g	0.09 d	0.06 d	0.03 e	2.15	0
Pseudomonas sp. G8 (PSB)	34.2 f	1.67 c	0.49 cd	1.18 ab	1.09	0
G. manihotis + NFB	68.1 ab	3.19 b	2.12 b	1.08 abc	1.97	65
G. manihotis + SPB.	61.5 cb	3.02 b	2.06 b	0.96 abcd	2.38	68
G. manihotis + PSB	70.3 a	5.62 a	3.80 a	1.82 a	2.26	2
E. colombiana + NFB	43.7 e	0.83 cd	0.57 b	0.25 cde	2.34	65
E. colombiana + SPB	48.5 d	1.11 cd	0.74 cd	0.38 bcde	2.00	60
E. colombiana + PSB	59.2 c	2.94 b	1.98 c	0.96 abcd	2.05	68

Data in a column followed by the same letter are not significantly different ($P = 0.05$).

AMFs improved sorghum growth better than inoculation of rhizobacteria. Dual inoculation AMF and rhizobacteria resulted in significantly higher of total plant dry weight than these microorganisms were used alone (single inoculation). For all the inoculations effect, dual inoculation with the AMF (*G. manihotis*) and PSB resulted in the highest plant growth response. This dual inoculation able to increase the total dry weight 112 times compared to the uninoculated treatment. Surprisingly, the contribution of all of rhizobacteria used in this study was significant higher when inoculated together with *G. manihotis* than were used alone. Dual inoculation of PSB and *G. manihotis* improved the total dry weight of sorghum 372%, while NFB and SPB improved 211% and 200%, respectively, higher than single inoculation with *G. manihotis* alone. In AMF treatments, inoculation of rhizobacteria also increased the colonization of AMF. These results showed that both AMF and rhizobacteria have significant role on plant growth promotion, and their roles were higher when applied together. Seemly, all rhizobacteria used in this study behaved as mycorrhizal helper bacteria that promote the colonization of AMF. As has been reported previously, the PSB behaved as mycorrhizal-helper bacteria, which improve the colonization of both the indigenous and introduced AMF (Toro et al., 1997). However, the mechanisms by which these bacteria stimulated AMF colonization are still poorly understood (Toro et al., 1997). The production of plant growth promoting substances by rhizobacteria, such as vitamins, amino acids, and hormones may be involved in this interaction (Barea et al., 1997). In other hand, on acid mineral soil, the AMF probably has significant role to stimulate the favorable condition for the survival and growth of rhizobacteria.

The effect of inoculation of AMF or/and rhizobacteria on nutrient concentration in sorghum shoots and roots were showed in Tables 3 and 4, respectively. The concentration of N in sorghum shoots was significantly improved by inoculation of AMF and NFB, *A. lipoferum*. However, inoculation of AMF and rhizobacteria tended to decrease the concentration of N in sorghum roots. The concentration of N in sorghum shoots was lower when inoculated by AMF and rhizobacteria (dual inoculation) than when inoculated by AMF only. The concentration of P in sorghum shoots was improved by inoculation AMF but not by inoculation of rhizobacteria. The concentration of P in sorghum roots was significantly decreased by inoculation of AMF. However, inoculation of PSB was significantly increased the concentration of P in sorghum roots. In general, inoculation of AMF or /and rhizobacteria tended to decrease the concentration of Zn, Cu, Fe, and Mn, both in shoot and root of sorghum.

In the present study root dry weights were improved by inoculation both AMF and

Table 3. The concentration of N, P, Zn, Cu, Fe and Mn of plant shoot, 41 days after planting

Treatments	N %	P %	Zn (mg g^{-1})	Cu (mg g^{-1})	Fe (mg g^{-1})	Mn (mg g^{-1})
Uninoculated	2.261 c	0.056 e	76.176 ab	9.016	256.128 bc	918.712 a
G. manihot (AMF1)	3.143 b	0.214 ab	65.995 abc	11.04	192.035 bc	611.371 bcd
E. colombiana (AMF2)	4.378 a	0.181 bc	52.869 bc	5.060	190.164 bc	525.596 def
A. lipoferum (NFB)	2.903 b	0.058 e	51.888 bc	-	548.688 a	682.640 b
Fluorescent pseudomonad (SPB)	2.415 c	0.025 e	59.800 abc	-	361.008 ab	570.768 cde
Pseudomonas sp. G8 (PSB)	2.453 c	0.041 e	77.403 ab	-	369.840 ab	438.963 f
G. manihotis + NFB	2.877 b	0.225 a	82.616 a	6.992	190.072 bc	654.917 bc
G. manihotis + SPB.	2.195 b	0.168 c	68.448 abc	-	157.749 bc	524.400 def
G. manihotis + PSB	2.406 c	0.101 d	57.899 abc	-	108.867 c	482.816 ef
E. colombiana + NFB	3.044 b	0.160 c	43.976 c	-	300.043 bc	535.259 def
E. colombiana + SPB	3.007 b	0.149 c	51.336 bc	-	218.899 bc	499.069 ef
E. colombiana + PSB	2.523 c	0.096 d	49.741 bc	-	103.960 c	473.003 f

Data in a column followed by the same letter are not significantly different (P = 0.05).

rhizobacteria (Table 2). AMF and rhizobacteria enhanced root growth that, in turn, is better equipment to make use potentially available quantities of other nutrients. This resulted in total nutrients taken up by plant increased (data not shown). Inoculated treatments by AMF and rhizobacteria did not have such a high nutrient concentration because of a dilution effect associated with growth (Table 2).

The dual inoculation *G. manihotis* and PSB provided the best stimulation effect on plant growth in acid mineral soil used in this study. It is possible that the available and transportation of P are main key for plant growth in this soil. The PSB released the fixed phosphorus and subsequently this released P was transported by the external mycorrhizal mycelium to the plant root system, thereby enhancing mineral P acquisition by the plant (Smith and Read, 1997).

In summary, it appears that the described interaction between AMF and rhizobacteria contributed to the plant growth promotion in acid mineral soils due to the plant growth promotion and improvement of nutrient uptake. The potential of dual inoculation with AMF and rhizobacteria needs to be further evaluated under different crop and agroclimatic conditions, particularly in the field.

Table 4. The concentration of N, P, Zn, Cu, Fe, and Mn of plant root, 41 days after planting

Treatments	N %	P %	Zn (mg g^{-1})	Cu (mg g^{-1})	Fe (mg g^{-1})	Mn (mg g^{-1})
Uninoculated	2.836a	0.223efg	198.900 a	0.190 de	680.520 a	539.900 bc
G. manihot (AMF1)	2.358bcd	0.048g	92.867 d	0.717 bc	248.500 cdef	572.300 abc
E. colombiana (AMF2)	2.466bc	0.039g	97.667 d	0.307 cd	283.220 cde	622.900 abc
A. lipoferum (NFB)	2.133de	0.179fg	128.800 b	0.000 e	409.760 b	513.600 c
Fluorescent pseudomonad (SPB)	1.819f	0.262def	99.900 d	0.000 e	222.700 ef	231.800 d
Pseudomonas sp. G8 (PSB)	1.975ef	0.399cde	112.267 bcd	0.151 de	204.887 ef	633.333 abc
G. manihotis + NFB	2.233cde	0.555abc	98.900 d	1.061 a	175.133 f	626.967 abc
G. manihotis + SPB.	2.349bcd	0.557abc	94.500 d	0.880 a	325.833 c	691.467 ab
G. manihotis + PSB	2.165de	0.441bcd	93.267 d	1.081 a	318.100 cd	639.333 abc
E. colombiana + NFB	2.555b	0.706a	126.533 bc	0.119 de	304.260 cd	510.100 c
E. colombiana + SPB	2.412bcd	0.642b	108.567 cd	0.266 de	272.370 cde	531.500 c
E. colombiana + PSB	2.281bcd	0.597abc	100.767 d	0.222 de	242.933 def	698.333 a

Data in a column followed by the same letter are not significantly different (P = 0.05).

References

Anderson J M 1989 Tropical Soil Biology and Fertility. A Handbook of Methods. C.A.B. International, Wallingford, Oxon, U.K p. 171.

Anonymous1987 Statistik Indonesia. Jakarta p 472.

Bagyaraj D J 1990 Biological interaction between VA mycorrhizal fungi and other beneficial soil organism. *In* Current Trend in Mycorrhizal Research. Eds. Jalali and H Chand. pp. 76–77. Tata Energy Research Institute, India.

Baligar V C and Fageria N K 1997 Nutrient use efficiency in acid soils: nutrient management and plant use efficiency. *In* Plant Soil Interactions at Low pH. Ed. A C Moniz, pp. 75–95. Brazilian Soil Science Society, Vicosa.

Barea J M, Azcon-Aguilar C and Azcon R 1997 Interaction between mycorrizal fungi and rhizosphere microorganisms within the context of sustainable soil-plant system. *In* Multitrophic Interaction in Terrestrial System. Eds. A C Gange and V K Brown. pp. 65–77. Blackwell Science, Cambridge England.

Gaur A C and Rana J P S 1990 Role of VA mycorrhizae, phosphate solubilising bacteria and their interaction on growth and uptake of nutrients by wheat crops. *In* Current Trend in Mycorrhizal Research. Eds. Jalali and H Chand. pp. 105–106. Tata Energy Research Institute, India.

Giovannetti M and Mosse B 1980 An evaluation of techniques for measuring vesicular-arbuscular mycorrhizal infection in root. New Phytol. 84, 489–500.

Jeffries P and Barea J M 2001 Arbuscular mycorrhiza - a key component of sustainable plant-soil ecosystems. *In* The Mycota, vol IX Fungal associations. Ed. B Hock. pp. 95–113. Springer, Berlin Heidelberg New York.

Kabirun S and Widada J 1995 Response of soybean grown on acid soils to inoculation of vesicular-arbuscular mycorrhizal fungi. Biotrop Spec. Publ. 56, 131–137.

Kamprath E J 1984 Crop response to lime in soils in the tropics. *In* Soil Acidity and Liming, 2nd edn (Agronomy monograph 9). Ed. F Adams. pp. 643–698. American Society for Agronomy and Soil Science Society of America, Madison, Wis.

Landon J R 1984 Booker Tropical Soil Manual. Booker Agriculture International Limited, London p. 450.

Murphy J and Riley J P 1962 A modified single solution method for the determination of phosphate in natural waters. Anal. Chim. Acta. 27, 31–36.

Negi M and Tilak K V B R 1990 Interaction between *Glomus versiforme* and *Azospirillum brasilense* in barley. *In* Current Trend in Mycorrhizal Research. Eds. Jalali and H Chand. pp. 109–110. Tata Energy Research Institute, India.

Pacovsky R S 1988 Interaction of inoculation of *Azospirillum brasilense* and *Glomus fasciculatum* on sorghum nutrition. Plant and Soil 110: 283–287.

Sanchez P A 1976 Properties and Managenent of Soil in the Tropics. John Willey and Sons, New York p. 618.

SAS Institute 1990 SAS/STAT Users Guide: Statistic, Version 6, 4th edition SAS Institute, Cary, N.C.

Sieverding E 1991 Vesicular-arbuscular Mycorrhiza Management in Tropical Agrosystems. Deutshe GTZ, Eshborn p. 371.

Smith S E and Read D J 1997 Mycorrhizal Symbiosis. Academic Press, London.

Tilak K V B R 1990 Interaction of VA-mycorrhizae with beneficial soil microorganism. 1990 Interaction between *Glomus versiforme* and *Azospirillum brasilense* in barley. *In* Current Trend in Mycorrhizal Research. Eds. Jalali and H Chand. pp. 87–90. Tata Energy Research Institute, India.

Toro M, Azcon R and Barea J M 1997 Improvement of arbuscular mycorrhiza development by inoculation of soil with phosphate-solubilizing rhizobacteria to improve rock phosphate bioavailability (^{32}P) and nutrient cycling. Appl. Environ. Microbiol. 63, 4408–4412.

Fertilizer potential of phosphorus recovered from wastewater treatments

L. E. de-Bashan[1] & Y. Bashan[1,2]
[1]*Environmental Microbiology Group, The Center for Biological Research of the Northwest (CIB), P.O. Box 128, La Paz, BCS 23000, Mexico.* [2]*Corresponding author**

Received 9 December 2002. Accepted in revised form 2 January 2003

Key words: fertilizer, hydroxyapatite, phosphate-solubilizing organisms, phosphate precipitation, phosphorus removal, wastewater treatment

Abstract

Large quantities of phosphate present in wastewater is one of the main causes of eutrophication that negatively affect natural water bodies, both fresh water and marine. It is desirable that water treatment facilities remove phosphorus from the wastewater before it is returned to the environment. In most countries, total removal or at least a significant reduction of phosphorus is obligatory, if not always fulfilled. This mini-review summarizes the options of recovering phosphorus from wastewater as struvite (ammonium-magnesium-phosphate) and hydroxyapatite formation and other feasible options, using the now largely regarded contaminant, phosphorus in wastewater, as a raw material for the fertilizer industry. The future use of phosphate solubilizing microorganisms, applied together with the recovered phosphorus, is proposed.

Abbreviations: EBPR – Enhanced biological phosphorus removal; PSB – Phosphate solubilizing bacteria; PSF – Phosphate solubilizing fungi

Introduction

Large-scale wastewater production is an inevitable consequence of contemporary societies. Wastewater is usually hazardous to human populations and the environment and must be treated prior to disposal into streams, lakes, seas, and on land surfaces. Obligatory anaerobic treatment of domestic and agro-industrial wastewater releases large amounts of phosphorus and nitrogen into wastewater. These nutrients are directly responsible for eutrophication (extraordinary growth of algae as a result of excess nutrients in the water) of water bodies worldwide (Lau et al., 1997; Trépanier et al., 2002). Consequently, disposal of wastewaters produces a constant threat to dwindling fresh water on a global scale (Montaigne and Essick, 2002).

Before discharging wastewater into water bodies, removing phosphate is usually obligatory, even though, in many cases it is not performed, and leads to major contamination on a worldwide level. The wastewater treatment industry presently uses several methods to remove phosphorus. Some are used in large-scale facilities and some are only experimental, and therefore, still used on a small-scale basis (for a process-engineering point of view, see: Stratful et al., 1999; Van Loosdrecht et al., 1997). In all cases, phosphorus is removed by converting the phosphorus ions in wastewater into a solid fraction. This fraction can be insoluble salt precipitates, microbial mass in activated sludge, or plant biomass in constructed wetlands. These approaches do not recycle phosphorus as a truly sustainable product because it is removed with various other

*FAX No: +52-612-125-4710.
E-mail: bashan@cibnor.mx

E. Velázquez and C. Rodríguez-Barrueco (eds.),
First International Meeting on Microbial Phosphate Solubilization, 179–184

waste products, some of which are toxic. The non-solubilized phosphates are either buried at landfills after incineration of the organic matter or used as sludge fertilizer, providing the treatment facility eliminates human pathogens and toxic compounds.

This mini-review analyzes recycling of phosphorus from wastewater as a potential raw material for the phosphate industry.

Use of recovered phosphate as fertilizer

Mined rock phosphate is an abundant and relatively cheap source of phosphate for fertilizer production. At the current rate of exploitation, the high quality portion of the resource will be largely depleted in <100 years, if another source of high-quality phosphate is not identified (Isherwood, 2000). This will create a problem for agriculture production, because lower-grade phosphates will have to be used, significantly increasing production costs. Phosphates recovered from wastewater plants might be a viable source of industrial raw material for manufacture of phosphate fertilizers. The activated sludge (also known as biosolids) commonly discarded in fields from wastewater treatment plants contain considerable phosphorus. Recovery of this phosphorus would help to alleviate the stringent legal restrictions on sludge disposal with low levels of phosphorus in discarded sludge. Currently, this phosphorus is regarded more as a contaminant than a resource. This perspective has started to change.

The most common approach for removing phosphate from wastewater is metal salt precipitation, which makes the precipitate unrecoverable for possible industrial processing into fertilizer (Donnert and Salecker, 1999a, b). Phosphate recovery from municipal wastewater is possible without metal salt precipitation, using existing technologies (Driver et al., 1999; Durrant et al., 1999; Stratful et al., 1999; Strickland, 1999; Woods et al., 1999). It is economically feasible to recover 10–80% of the phosphorus flowing into wastewater treatment facilities. Recovered phosphorus product can be superior in quality to currently available phosphate rock. Today, the recovery approach is attractive only for wastewater treatment plants that use biological nutrient removal (Gaterell et al., 2000; Jeanmaire and Evans, 2001).

The most promising compound for recovery from wastewater plants is magnesium ammonium phosphate hexahydrate ($MgNH_4PO_4 \cdot 6H_2O$), commonly known as struvite, which precipitates spontaneously in some wastewater processes (Booker et al., 1999; Stratful et al., 2001; Williams, 1999). If formation and collection are controlled and cost-effective, struvite might have potential in the fertilizer market. Miniscule quantities of recovered struvite are currently being tested as fertilizer, mainly in Japan. The theoretical potential is much larger, and approaches 67,000 tons of P_2O_5 fertilizer per year for the UK alone, as well as 270,000 tons for Western Europe (Gaterell et al., 2000). Because the compound is receiving increasing attention, this mini-review will elaborate on its formation, potential, and limitations.

Struvite precipitates spontaneously in wastewater treatment environments where high concentrations of soluble phosphorus and ammonium are present. Additional essential conditions are low concentration of suspended solids and pH above 7.5. Precipitation of struvite requires that its components are available simultaneously in the wastewater in the molecular ratio $1(Mg^{+2}):1(NH_4^{+}):1(PO_4^{-3})$. Normally, municipal wastewater is rich in ammonium, but deficient in magnesium, so supplementation of magnesium is required, and this helps to increase solution pH (Munch and Barr, 2001). The pH can also be elevated by adding NaOH, an expensive process, (Stratful et al., 2001) or by air stripping, where aeration of wastewater removes CO_2 and increases pH in the process (Battistoni et al., 1997). Struvite crystals are orthorhombic, ranging from X-shaped to flat trapezoidal, depending on degree of solution saturation. Crystals can grow significantly in 3 h. A natural aging with phosphate precipitation can be obtained in a few days with a different percentage of struvite and hydroxyapatite formed, depending on the sludge liquor composition; supersaturation of the solution with different ions play a major role in dictating the outcomes (Battistoni et al., 2000).

The implication of theoretical knowledge on struvite formation is that, if these conditions can be duplicated and exploited in a practical engineering process, there is potential to economically extract struvite from wastewater in commercial quantities. Precipitating struvite might be done in a dedicated reactor, instead of allowing spontaneous formation (Munch and Barr, 2001; Stratful et al., 2001). Spontaneous formation fouls pipes and other inner surfaces of the treatment process, making operation of the plant inefficient and costly because the struvite must be dissolved with sulfuric acid or broken down manually with hammer and chisel (Durrant et al., 1999; Stratful et al., 2001; Williams 1999). Another advantage of a dedicated reactor is that excess ammonium, which is a normal product of anaerobic wastewater digestion, might yield relatively pure struvite. Today however, many wastewater processes do not fulfill the basic requirements for struvite formation, and the best medium for formation and recovery of struvite is the supernatant of sludge obtained from EBPR. Phosphorus can be obtained from sludge only with biological processes because the commonly used chemical precipitation of phosphorus, which involves addition of iron and aluminum salts, produces a product that cannot be recycled for industrial recovery of phosphorus (Donnert and Salecker, 1999a, b).

There is no common method for recovery of struvite from biological processes. Equipment ranges from complex, patented reactors employing ion exchangers, to simple stirring tanks (Gaterell et al., 2000; Williams, 1999). Recently, a struvite crystallization reactor was developed in Japan. Struvite was produced from the filtrate of anaerobic sludge digestion by adding magnesium hydroxide, and adjusting pH to 8.2–8.8 with sodium hydroxide. A retention time of 10 days allowed the growth of pellets 0.5–1.0 mm in diameter. Recovered struvite was sold to fertilizer companies (Ueno and Fujii, 2001). Another experimental industrial process provided simultaneous removal of phosphate and ammonium ions by selective ion exchange and recovery of the product by chemical precipitation in the form of struvite (Liberti et al., 2001).

As struvite contains both phosphorus and nitrogen, its removal will affect the content of both elements in the leftover sludge, used by farmers as a soil improvement agent and fertilizer. Struvite recovery from wastewater might have a marginal effect on the net content of nitrogen, but greater impact on the concentration of phosphorus in the sludge. Gaterell et al. (2000) calculated that, because sewage has a typical N:P ratio of 8:1 and struvite 1:1: a theoretical maximum of 12.5% of the nitrogen load could be removed as struvite. The practical limit is lower because not all phosphorus can be recovered as struvite with today's technologies. Struvite removal leaves less phosphorus in the sludge. This is beneficial from two perspectives: (i) Sludge applied to fields usually has phosphorus in excess of that needed by plants. The resulting eutrophication of water bodies from leached phosphorus compounds would be reduced, and, (ii) Struvite recovery would help to meet legal requirements imposed on sludge disposal and reduce the area needed for disposal.

Struvite has many uses. The most obvious is as a raw material for the fertilizer industry (Gaterell et al., 2000). It can be used as a material in fire-resistant panels and in cement (Sarkar, 1990; Schuilling and Andrade, 1999). If cheap production methods are developed, it could be used in detergents, cosmetics, and animal feed, all of which use phosphates (Gaterell et al., 2000).

While struvite has many potential uses as fertilizer, as yet, none have been proven commercially profitable. The most promising application is as a slow-release fertilizer (Munch and Barr, 2001) that can be applied in a single high dose without damaging growing plants. The suggested plants are ornamentals, forest out-plantings, turf, orchard trees, and potted plants. This fertilizer might have low leach rates and slowly release nutrients during the growing season.

Highly soluble orthophosphate, serving as the initial phosphorus supply for establishing container plants, could be used together with struvite in a mixed fertilizer product. Struvite could also replace the major fertilizer diammonium phosphate, which is produced by neutralizing phosphoric acid with ammonia. Mixing struvite with phosphoric acid might even yield a superior fertilizer; part slow-release $MgHPO_4$; and part fast-release, highly soluble ammonium phosphate $(NH_4)_2HPO_4$. This might be considerably more cost-effective than commonly used diammonium phosphate fertilizers. Another possible product is untreated granular struvite that can be mixed with peat to serve as a lightweight potting mix (Gaterell et al., 2000). High solubility is not an asset in many fertilizer applications, as in grasslands and forests, where fertilizer is applied once in several years. Slow-release fertilizers are preferred for these cases, and struvite is even more solubilized than the currently used, mostly insoluble rock phosphate. The presence of magnesium in struvite makes it attractive as an alternative to contemporary fertilizers for a few crops, like sugar beets, that require magnesium (J. Driver in: Gaterell et al., 2000).

The phosphate industry is using rock phosphate for detergents, food, and cosmetics by applying a high-energy, high-temperature industrial process to purify the phosphate. It is not likely that struvite will be used in the short term because struvite purification technology is unknown, small amounts of phosphate is needed by these industries, and the currently well-known rock phosphate purification technology. However, as high-quality rock phosphate increases in price, and the industry is forced to use low-quality rock phosphate, pressure will increase to develop purification methods for struvite, which is sometimes purer than rock phosphate.

In summary, present consumption of phosphorus raw material is over one million tons per year. The most likely use of struvite is as a substitute for rock phosphate in the fertilizer industry, since no additional industrial treatment is required, and the material can be used "as is" or with very simple further processing, like granulation and mixing. Furthermore, routine simple treatments used by the fertilizer industry, like application of phosphoric acid, could be used to alter the solubility of struvite, making it more desirable as a fertilizer. Future use of struvite depends mainly on industrial operations that are simple and cost effective and positive field trials that have not been undertaken. Without knowing how much struvite can be recovered efficiently, the cost effectiveness of the industrial operation, and the desire of consumers to use struvite as a fertilizer, it is nearly impossible to assess how soon struvite-based products will enter the market. Perhaps legislative pressure to keep the environment free of phosphorus contamination from sludge disposal and exhaustion of the high-quality rock phosphate supplies, with attendant increase in price, will increase the economic attractiveness of struvite for industrial development.

Another alternative for leftover sludge after phosphorus removal is as a source of ash, which can be recovered, if iron is not used for phosphate precipitation in the treatment plant. Applying Ca, Al, or EBPR increases the recycling potential for ash. Currently, Cu and Zn are still highly concentrated in the ash, prohibiting its use as fertilizer (Schipper et al., 2001). An alternative technology for phosphate removal by conventional precipitation produced phosphate pellets having high-purity and extremely low water content. The process was patented in The Netherlands. The main advantage, apart from removing phosphorus from the water, is that the pellets can be reused in industry (Giesen, 1999).

A simpler option for phosphate recovery as a beneficial raw material is to heat the sludge. Heating activated sludge at 70° C for one hour released most of the phosphorus from polyphosphate. Precipitation of the phosphorus by $CaCl_2$ yielded 75% of the phosphorus without the need of pH adjustment, as is the case for struvite. The precipitate contained more P and less Ca than typical natural rock phosphate. Hence, this has potential as a simple recovery process for usable phosphorus from wastewater material (Kuroda et al., 2002).

Another option is to use a strongly adsorbing filter material that retains phosphorus efficiently, and after phosphorus saturation, using the discarded filter as a fertilizer. Blast furnace slag showed high phosphorus sorption capacity and was used for abiotic sorption of wastewater phosphorus. During this process, calcium concentration also decreased, forming the compound hydroxyapatite. This implies Ca–P precipitation as the predominant phosphorus removal mechanism in the slag. The low solubility of hydroxyapatite might have important implications for its possible production for fertilizer by direct formation (Johansson and Gustafsson, 2000). Seed crystal material made of calcium silicate hydrate (tobermorite crystals), applied for phosphorus removal by crystallization, yielded precipitated hydroxyapatite that can be incorporated into soil and had characteristics of a good plant fertilizer (Moriyama et al., 2001). The direct use of non-soluble phosphate, such as hydroxyapatite, as a fertilizer, will require an effective and cheap means of solubilization. However, this problem might be solved with the use of phosphate-solubilizing bacteria (PSB) and fungi (PSF) (Richardson, 2001; Whitelaw, 2000).

PSB and PSF are common types of soil microorganisms and some are considered plant growth-promoting microorganisms (Bashan et al. 2000; Rojas et al., 2001; Vazquez et al., 2000). They inhabit almost all agricultural and forest soil, belong to numerous genera, and are employed worldwide, mainly in developing countries, as inoculants together with rock phosphate as substrate (Leggett et al., 2001; Rodriguez and Fraga, 1999). Although several methods for mobilization of non-soluble phosphorus (Fe- and Al–P in acid soils, Ca–P in alkaline soils) are documented, the most common is by production of organic acids and microbial release of protons (Jones, 1998).

We propose that precipitated phosphate, recovered from wastewater together with common PSB and PSF, be used as a fertilizer. The organic acids produced by these microorganisms are strong enough to dissolve the struvite and hydroxyapatite obtained from wastewater. There are significant economic and environmental advantages in developing these materials in a wide range of fertilizers. These include very slow-release fertilizers for pasture and forest (phosphates as recovered), a mix of slow- and fast-release fertilizers for most agricultural practices (recovered phosphate mixed with phosphate that is partly dissolved by PSB and/or PSF), and fast-release fertilizers combining recovered phosphate with strong phosphate-solubilizing microorganisms for crops needing rapid start up.

Concluding remarks

Eutrophication of water bodies is a major, global environmental problem. Its main cause is disposal of nutrients (N and P) directly from wastewater plants or indirectly from agriculture runoff and leaching from sludge deposited in landfills and fields. As stringent laws now require that the level of these nutrients be significantly reduced (Gaterell et al., 2000; Stratful et al., 1999), a boost to the economy of existing wastewater industry was given.

The contemporary issue is not eliminating phosphorus *per se* (that has been done efficiently using metal and polymer precipitation), but recycling it. Recycling of phosphorus will convert an acute disposable problem into a raw material that will benefit industry and society, as phosphorus makes a significant contribution, as fertilizer, to the well being of human societies. All tertiary wastewater facilities eliminate phosphorus (either by chemical or biological removal) as a non-recyclable material (metal-phosphate precipitates or precipitates together with numerous other waste materials as sludge). What is needed is a process of phosphorus removal that separates it from other waste components, so that it can be recycled as a fertilizer or an ingredient in other valuable phosphorus products.

The emerging technology of struvite crystallization (magnesium ammonium phosphate) and hydroxyapatite (calcium phosphate), as an alternative phosphorus removal technology may serve as a catalyst for removing phosphorus as a recyclable product benefiting the fertilizer industry. The best way is to combine EBPR with struvite and hydroxyapatite crystallization. It will save chemicals for precipitation, reduce the size of the treatment facility, reduce the volume of effluent to be treated (more concentrated phosphorus in present effluent), and consequently reduce costs (Jeanmaire and Evans, 2001; Stratful et al., 1999). The use of phosphate solubilizing microorganisms together with these future, slow-release fertilizer products may boost the prospects that waste material products will find their way into agricultural practices.

Acknowledgements

This assay was written in memory of the late Mr. Avner Bashan of Israel. We thank the editorial staff at CIB for polishing the English text. This assay was partially supported by a travel grant to the First International Meeting on Phosphate Solubilizing Bacteria by the University of Salamanca, Spain and by the Bashan Foundation.

References

Bashan Y, Moreno M and Troyo E 2000 Growth promotion of the seawater-irrigated oilseed halophyte *Salicornia bigelovii* inoculated with mangrove rhizosphere bacteria and halotolerant *Azospirillum* spp. Biol. Fertil. Soils 32, 265–272.

Battistoni P, Fava G, Pavan P, Musacco A and Cecchi F 1997 Phosphate removal in anaerobic liquors by struvite crystallization without addition of chemicals: preliminary results. Water Res. 31, 2925–2929.

Battistoni P, Pavan P, Prisciandaro M and Cecchi F 2000 Struvite crystallization: A feasible and reliable way to fix phosphorus in anaerobic supernatants. Water Res. 34, 3033–3041.

Booker N A, Priestley A J and Fraser I H 1999 Struvite formation in wastewater treatment plants: Opportunities for nutrient recovery. Environ. Technol. 20, 777–782.

Donnert D and Salecker M 1999a Elimination of phosphorus from municipal and industrial waste water. Water Sci. Technol. 40, 195–202.

Donnert D and Salecker M 1999b Elimination of phosphorus from waste water by crystallization. Environ. Technol. 20, 735–742.

Driver J, Lijmbach D and Steen I 1999 Why recover phosphorus for recycling and how?. Environ Technol. 20, 651–662.

Durrant A E, Scrimshaw M D, Stratful I and Lester J N 1999 Review of the feasibility of recovering phosphate from wastewater for use as a raw material by the phosphate industry. Environ. Technol. 20, 749–758.

Gaterell M R, Gay R, Wilson R, Gochin R J and Lester J N 2000 An economic and environmental evaluation of the opportunities for substituting phosphorus recovered from wastewater treatment works in existing UK fertiliser markets. Environ. Technol. 21, 1067–1084.

Giesen A 1999 Crystallisation process enables environmental friendly phosphate removal at low costs. Environ. Technol. 20, 769–775.

Isherwood K F 2000 Mineral fertilizer use and the environment. International Fertilizer Industry Association / United Nations Environment Programme, Paris. pp. 106.

Jeanmaire N and Evans T 2001 Technico-economic feasibility of P-recovery from municipal wastewaters. Environ. Technol. 22, 1355–1361.

Johansson L and Gustafsson J P 2000 Phosphate removal using blast furnace slags and opoka-mechanisms. Water Res. 34, 259–265.

Jones D L 1998 Organic acids in the rhizosphere – a critical review. Plant Soil 205, 25–44.

Kuroda A, Takiguchi N, Gotanda T, Nomura K, Kato J, Ikeda T and Ohtake H 2002 A simple method to release polyphosphate from activated sludge for phosphorus reuse and recycling. Biotechnol. Bioeng. 78, 333–338.

Lau P S, Tam N F Y and Wong Y S 1997 Wastewater nutrients (N and P) removal by carrageenan and alginate immobilized *Chlorella vulgaris*. Environ. Technol. 18, 945–951.

Leggett M, Gleddie S and Holloway G 2001 Phosphate-solubilising microorganisms and their use. *In* Plant Nutrient Acquisition: New Perspectives. Eds. N Ae, J Arihara, K Okada and A Srinivasan. pp. 299–318. Springer-Verlag, Tokyo.

Liberti L, Petruzzelli D and De Florio L 2001 REM NUT ion exchange plus struvite precipitation process. Environ. Technol. 22, 1313–1324.

Montaigne F and Essick P 2002 Water Pressure. National Geographic 202, 2–33.

Moriyama K, Kojima T, Minawa Y, Matsumoto S and Nakamachi K 2001 Development of artificial seed crystal for crystallization of calcium phosphate. Environ. Technol. 22, 1245–1252.

Munch E V and Barr K 2001 Controlled struvite crystallization for removing phosphorus from anaerobic digester side-streams. Water Res. 35, 151–159.

Richardson A E 2001 Prospects for using soil microorganisms to improve the acquisition of phosphorus by plants. Australian J. Plant Physiol. 28, 897–906.

Rodriguez H and Fraga R 1999 Phosphate solubilizing bacteria and their role in plant growth promotion. Biotechnol. Adv. 17, 319–339.

Rojas A, Holguin G, Glick B R and Bashan Y 2001 Synergism between *Phyllobacterium* sp. (N_2–fixer) and *Bacillus licheniformis* (P-solubilizer), both from a semiarid mangrove rhizosphere. FEMS Microbiol. Ecol. 35, 181–187.

Sarkar A K 1990 Phosphate cement-based fast setting binders. J. Am. Ceram. Soc. Bull. 69, 234–238.

Schipper W J, Klapwijk A, Potjer B, Rulkens W H, Temmink B G, Kiestra F D G and Lijmbach A C M 2001 Phosphate recycling in the phosphorus industry. Environ. Technol. 22, 1337–1345.

Schuiling R D and Andrade A 1999 Recovery of struvite from calf manure. Environ. Technol. 20, 765–768.

Stratful I, Brett S, Scrimshaw M B and Lester J N 1999 Biological phosphorus removal, its role in phosphorus recycling. Environ. Technol. 20, 681–695.

Stratful I, Scrimshaw M D and Lester J N 2001 Conditions influencing the precipitation of magnesium ammonium phosphate. Water Res. 35, 4191–4199.

Strickland J 1999 Perspectives for phosphorus recovery offered by enhanced biological P removal. Environ. Technol. 20, 721–725.

Trépanier C, Parent S, Comeau Y and Bouvrette J 2002 Phosphorus budget as a water quality management tool for closed aquatic mesocosms. Water Res. 36, 1007–1017.

Ueno Y and Fujii M 2001 Three years experience of operating and selling recovered struvite from full-scale plant. Environ. Technol. 22, 1373–1381.

Van Loosdrecht M C M, Hooijmans C M, Brdjanovic D and Heijnen J J 1997 Biological phosphate removal processes. Appl. Microbiol. Biot. 48, 289–296.

Vazquez P, Holguin G, Puente ME, Lopez-Cortes A and Bashan Y 2000 Phosphate-solubilizing microorganisms associated with the rhizosphere of mangroves in a semiarid coastal lagoon. Biol. Fertil. Soils 30, 460–468.

Whitelaw M 2000 Growth promotion of plants inoculated with phosphate-solubilizing fungi. Adv. Agron. 69, 99–151.

Williams S 1999 Struvite precipitation in the sludge stream at slough wastewater treatment plant and opportunities for phosphorus recovery. Environ. Technol. 20, 743–747.

Woods N C, Sock S M and Daigger G T 1999 Phosphorus recovery technology modeling and feasibility evaluation for municipal wastewater treatment plants. Environ. Technol. 20, 663–679.

Microalgae growth-promoting bacteria as "helpers" for microalgae: A novel approach for removing ammonium and phosphorus from municipal wastewater

L. E. de-Bashan, J. P. Hernandez & Y. Bashan[1]
The Environmental Microbiology Group, The Center for Biological Research of the Northwest (CIB), P.O. Box 128, La Paz, BCS, 23000, Mexico. [1]*Corresponding author**

Received 9 December 2002. Accepted in revised form 2 January 2003

Key words: phosphate, nitrate, microalgae, growth promoting bacteria, wastewater

Abstract

A combination of microalgae (*Chlorella vulgaris* or *C. sorokiniana*) and a microalgae growth-promoting bacterium (MGPB, *Azospirillum brasilense* strain Cd), co-immobilized in small alginate beads, was developed to remove nutrients (P and N) from municipal wastewater. This paper describes the most recent technical details necessary for successful co-immobilization of the two microorganisms, and the usefulness of the approach in cleaning the municipal wastewater of the city of La Paz, Mexico. *A. brasilense* Cd significantly enhanced the growth of both *Chlorella* species when the co-immobilized microorganisms were grown in wastewater. *A. brasilense* is incapable of significant removal of nutrients from the wastewater, whereas both microalgae can. Co-immobilization of the two microorganisms was superior to removal by the microalgae alone, reaching removal of up to 100% ammonium, 94% nitrate, and 92% phosphorus within 6 days (varied with the source of the wastewater), compared to 75% ammonium, 84% nitrate, and 89% phosphorus by the microalgae alone. This study shows the potential of co-immobilization of microorganisms in small beads to serve as a treatment for wastewater.

Introduction

Plant growth-promoting bacteria (PGPB) used as inoculants in agricultural experiments are commonplace, both for control of phytopathogens and plant growth promotion (Bashan, 1998). Bacteria of the genus *Azospirillum* are well known as PGPB for numerous crop plants (Bashan and Holguin, 1997). Recently, it was observed that one common strain, the type strain *A. brasilense* Cd, is also capable of promoting many growth parameters of the unicellular microalgae *Chlorella vulgaris* (Gonzalez and Bashan, 2000), and change the cytology, lipids, and pigment production by the microalgae (de-Bashan et al., 2002a; Gonzalez-Bashan et al., 2000; Lebsky et al., 2001). Therefore, it may be considered as a microalga growth-promoting bacterium (MGPB). *C. vulgaris* is commonly used for tertiary wastewater treatment (De la Noüe and De Pauw, 1988; Gonzalez et al., 1997; Tam and Wong, 2000; Valderrama et al., 2002), yet it had not been demonstrated that the observed growth promotion might also yield improve capabilities of microalgae to remove nutrients from natural wastewater. The microbial carrier chosen in this study were alginate beads. Immobilization of microalgae in polysaccharide gels is an experimental way to use these microorganisms for

* FAX No: +52 (612) 125 4710.
E-mail: bashan@cibnor.mx

E. Velázquez and C. Rodríguez-Barrueco (eds.),
First International Meeting on Microbial Phosphate Solubilization, 185–192

PRIOR TO IMMOBILIZATION IN BEADS

Axenic *Chlorella* cultures were cultivated in C30, a sterile mineral medium composed of g/L: KNO_3, 25; $MgSO_4·7H_2O$, 10; KH_2PO_4, 4; K_2HPO_4 1; $FeSO_4·7H_2O$, 1; micronutrients (μg/L H_3BO_3, 2.86; $MnCl_2·4H_2O$, 1.81; $ZnSO_4·7H_2O$, 0.11; $CuSO_4·5H_2O$, 0.09; $NaMoO_4$, 0.021 for 5-6 days, under continuous agitation (150 r.p.m.), 60-80 μmol/m^2/s light intensity, and at 27-30°C.

Azospirillum brasilense Cd DSM 1843 was grown in liquid nutrient broth (from Sigma) or M9 nitrogen-free medium composed of g/L: Na_2HPO_4, 6; KH_2PO_4, 3; NaCl, 1.5, and 1 M $MgSO_4·7H_2O$, 0.01 M $CaCL_2$, 20% malic acid) at 30 ± 2°C and agitated (120 r.p.m.) for 16 to 18 hours.

100 mL

100 mL

CENTRIFUGATION at 4000 *g*

IMMOBILIZATION

100 mL of axenic microalgal culture, containing aprox. 6 x 10^6 cells/mL, was washed twice in sterile saline solution (0.85% NaCl) and re-suspended in 20 mL of the same sterile saline solution.

50 mL of *A. brasilense*, culture were washed twice in saline solution, and resuspended in 20 mL of saline solution. The optical density was adjusted to 1.0 to have aprox. 10^9 CFU/mL.

To co-immobilize the two microorganisms in the same bead, after the washing procedure, each of the cultures was resuspended in 10 mL saline solution, and then mixed to obtain 20 mL total volume.

The culture (either one species or a mix of two) was mixed with 80 mL of a sterile 6,000 cP 2% alginate solution (a solution made of alginate mixed at 14,000 and 3,500 cP), and mixed slowly with a stirrer for 15 minutes. The solution was then dripped from a syringe into a 2% $CaCl_2$ solution, mixing periodically. The beads formed were left for 1 hour for curing and then washed in sterile saline solution.

Because immobilization normally reduces the number of bacteria in the beads, a second incubation step is performed by incubating the co-immobilized culture overnight in diluted nutrient broth (1:10), under the same conditions as *Azospirillum* culture. After the secondary incubation, the beads were washed three times with sterile saline solution.

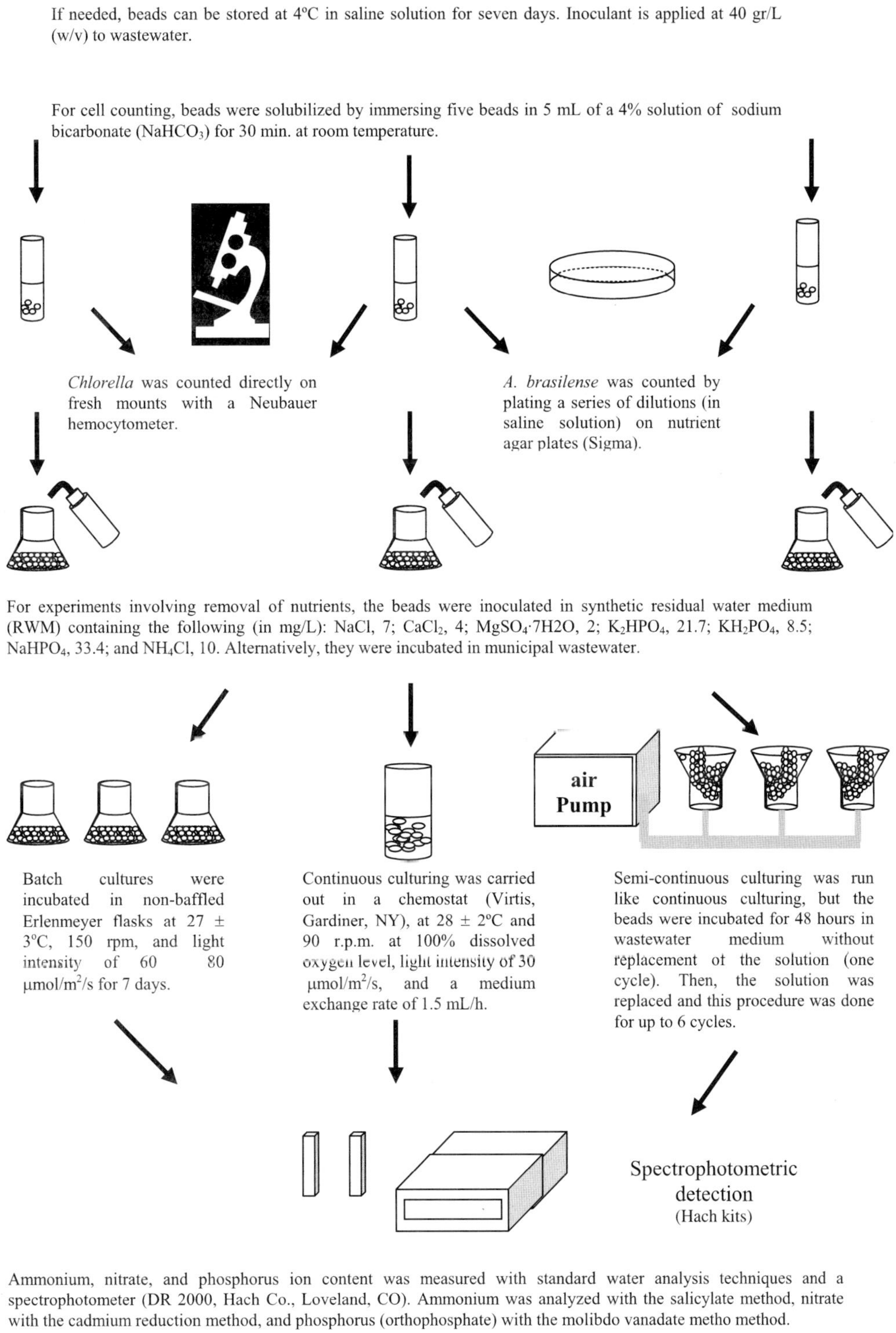

Figure 1. Flow chart describing methods and techniques used to immobilize, co-immobilize, count, and cultivate microalgae and MGPB for wastewater treatment.

wastewater treatment (Chevalier and De la Noüe, 1985; Lau et al., 1997) because it facilitates the major difficulty of collecting enormous populations of cells developed during the treatment, hampering regular microalgae treatments (Tam and Wong, 2000).

This study describes the methods used to co-immobilize the two microorganisms in small alginate beads and to show that this artificial biological association, enforced by close proximities of the microorganisms inside small polymer beads, improve the capacity of the microalgae in its main practical task – removing nitrogen and phosphorus from municipal wastewater.

Materials and methods

Microorganisms

Two species of unicellular microalgae *Chlorella vulgaris* Beijerinck (UTEX 2714) and *C. sorokiniana* Shih. et Krauss (UTEX 1602) were used. The microalgae growth-promoting bacterium *Azospirillum brasilense* Cd (DMS 1843) was used for co-immobilization experiments with each of the microalgae species.

Immobilization procedures

Various modifications of immobilizing these microorganisms in alginate beads and counting and growing them in cultures were published as the research evolved (Bashan, 1986; Bashan et al., 2002; Gonzalez and Bashan, 2000). To straighten contradictory details, the current procedures are summarized in details in Figure 1. *A. brasilense* was cultivated, prior immobilization, by standard techniques for this species (Bashan et al., 1993).

Municipal wastewater source

Wastewater was collected periodically, for every separate run of the bioreactors, at the municipal wastewater treatment plant of the city of La Paz, Baja California Sur, Mexico. Samples were collected from a stream of wastewater after the initial aerobic activated sludge treatment and immediately transferred to the laboratory. If necessary, debris in the wastewater was filtered through a gauze-cotton filter in a funnel. All wastewater were used as they arrived from the treatment plant. We stored wastewater at 4 °C for several days only as a precaution following the run of the bioreactors. Analyses of the wastewater content done by the Analytical Service Unit of CIB and by the municipal wastewater treatment plant of La Paz showed that the average content of the wastewater is: (mg/L) suspended solids, 0.978–80; BOD, 53.5–113; dissolved solids, 0.001; total nitrogen, up to 55; nitrates, 4–5.18; ammonium, 0.1–4.26; total phosphates, up to 5; orthophosphate, 4.1; NaCl 1.1; conductivity 1633 μS/cm; arsenic, 0.0013; cadmium, <0.005; cupper, 0.018; chrome, 0.004–0.018; mercury, 0.0013; nickel, 0.031; lead, 0.064: zinc, 0.118 and pH 6.3–7.9. The most notable variations observed among samplings were the presence of different nitrogen ions (ammonia or nitrate) and their concentration. Therefore, the values of the initial ion concentration are given in each figure.

Water analyses of treated wastewater

Standard water analyses techniques (APHA, AWWA, WPCF, 1992) were performed with a Hach DR/2000 spectrophotometer and Hach kits (Hach Co., Loveland, CO, USA) for nitrogen and phosphorus.

Experimental design and statistical analysis

The experiments were performed in inverted, 1000–mL conical, glass bioreactors containing 600 mL wastewater, equipped with bottom aeration controlled by a peristaltic pump (1.8 L air per min) at 26 ± 2 °C, with constant illumination of 60 μmole/m^2/s. Each experiment was performed in triplicate, where one bioreactor served as a replicate. The setup was of semi-continuous cultures, where wastewater was replaced every 48 h, as described earlier (de-Bashan et al., 2002b). Controls (beads without microorganisms, wastewater alone, and microalgae and bacteria alone) were routinely used. Three 50-mL samples were taken for each water analysis at each sampling time. Each experiment was repeated 3 times using 3 slightly-different natural municipal wastewater samples, since we could not control the output effluent of the municipal wastewater treat-

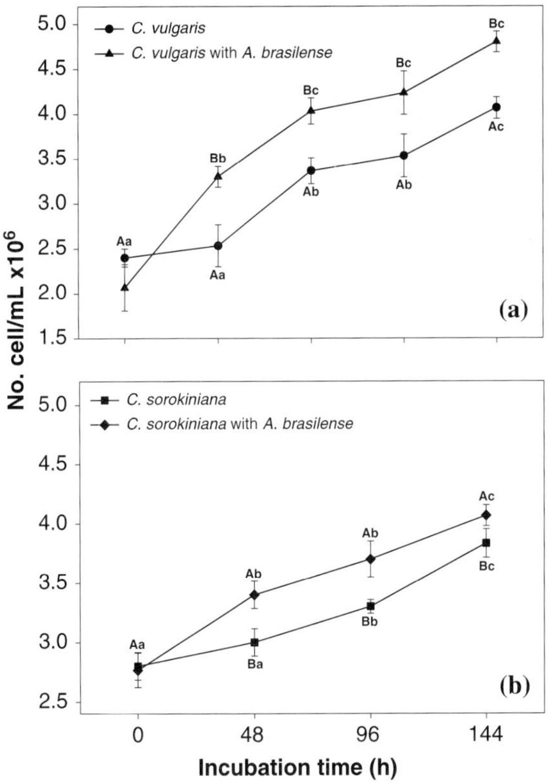

Figure 2. Growth promotion of *C. vulgaris* and *C. sorokiniana* by *A. brasilense* Cd growing in municipal wastewater. Points on curves denoted by a different lower case letter differ significance by ANOVA at $P \leq 0.05$. Points at each cycle denoted by a different capital letter differ significantly with Student's *t*-test at $P \leq 0.05$. Bars represent standard error.

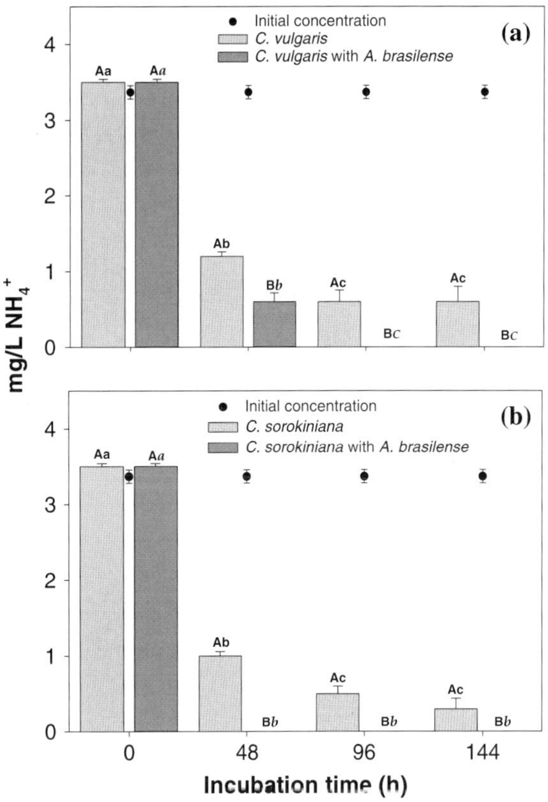

Figure 3. Removal of ammonium ions from municipal wastewater by *Chlorella* spp. co-immobilized with *A. brasilense* Cd. (a) *C. vulgaris*; (b) *C. sorokiniana*. Columns (of each gray scale) denoted by a different lower case letter or italics lower case letter, separately, differ significantly by ANOVA at $P \leq 0.05$. Pairs of columns for each cycle denoted by a different capital letter differ significantly with Student's *t*-test at $P \leq 0.05$. Bars represent standard error.

ment facility. Results were analyzed by ANOVA and Student's *t*-test, with significance at $P \leq 0.05$: using Statistica software (Statsoft, Inc. Tulsa, OK). As results were similar, only one representative experiment is presented.

Results and discussion

To define a practical, useful association between two microorganisms, it is essential to demonstrate that one (or both) microorganism affects the main practical function of the other. Demonstration of improved growth parameters, as was shown earlier for a bacteria–microalgae association (de-Bashan et al., 2002a; Gonzalez and Bashan, 2000; Mouget et al., 1995), is insufficient to that end since these growth promotions occurred under defined *in vitro* mixed cultivation. Therefore, the main purpose of this study was to show that when a microalgae is co-immobilized and co-cultured with a MGPB, with both submerged in "natural" municipal wastewater, the nutrient absorption capacity of the microalgae increases from the association, and the treated effluent wastewater is poorer in nitrogen and phosphorus.

Semi-continuous treatments performed for 4–5 cycles each, where the municipal wastewater but not co-immobilized microorganisms were replaced every 48 h, using both microalgae species. Four parameters were evaluated, growth promotion of the two microalgal species affected by the MGPB, and removal of ammonium, nitrate, and phosphorus from the wastewater.

A. brasilense Cd continuously and significantly enhanced the growth of both *Chlorella*

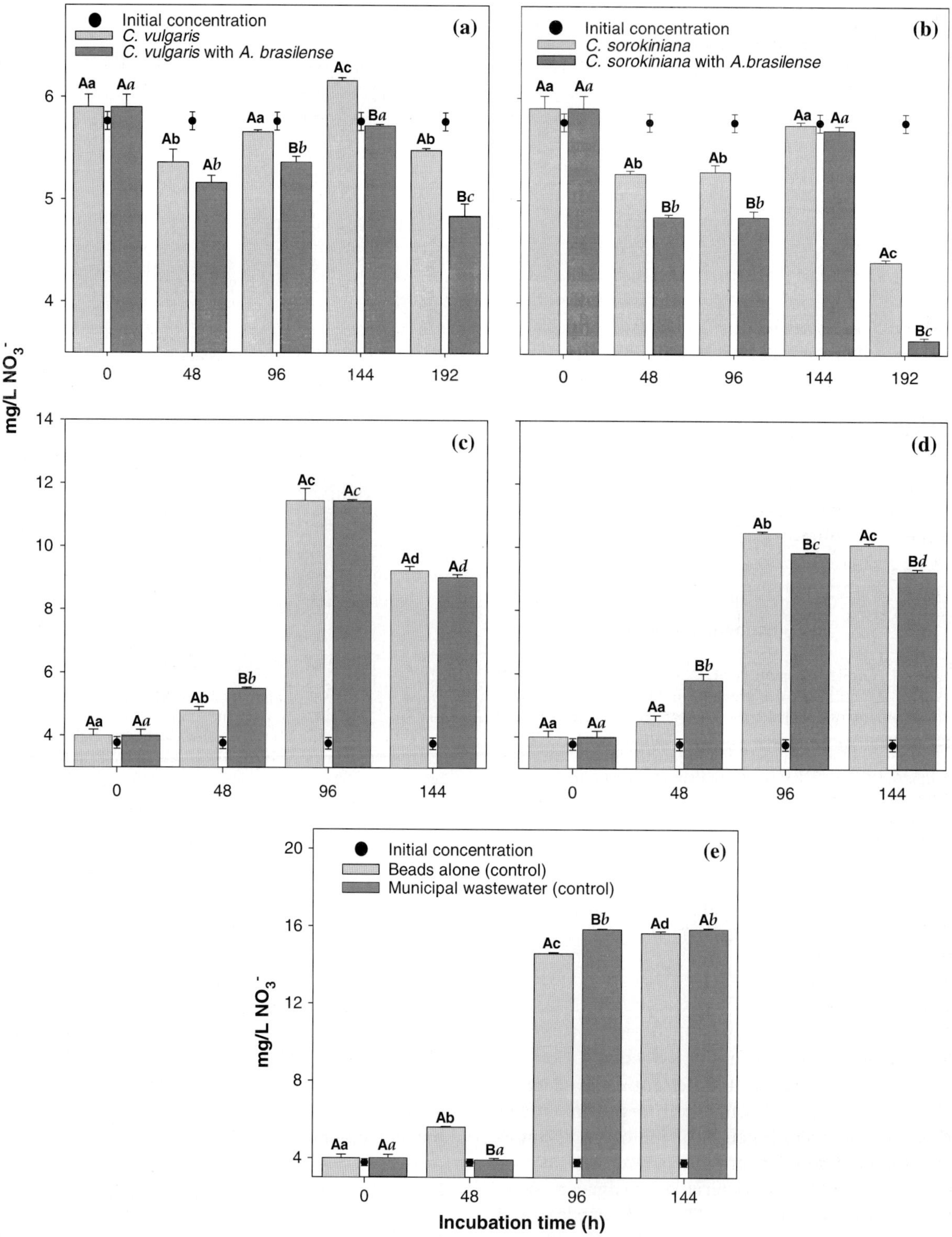
Initial concentration
C. vulgaris
C. vulgaris with A. brasilense
(a)
Initial concentration
C. sorokiniana
C. sorokiniana with A.brasilense
(b)
mg/L NO3-
(c)
(d)
Initial concentration
Beads alone (control)
Municipal wastewater (control)
(e)
mg/L NO3-
Incubation time (h)

Figure 4. Removal of nitrate ions from municipal wastewater by *Chlorella* spp. co-immobilized with *A. brasilense* Cd. Subfigures a and c represent *C. vulgaris*; b and d represent *C. sorokiniana*. Subfigures a and b represent original high levels of nitrate; Subfigures c and d represent elevation of nitrate during incubation. Subfigures e is the control without immobilized microorganisms. Columns in each subfigure (of each gray scale) denoted by a different lower case letter or italics lower case letter, separately, differ significantly by ANOVA at $P \leq 0.05$. Pairs of columns for each cycle denoted by a different capital letter differ significantly with Student's *t*-test at $P \leq 0.05$. Bars represent standard error.

species when the co-immobilized microorganisms were grown in the wastewater at 2.1×10^6–4.8×10^6 cells/mL (*C. vulgaris*) and at 2.8×10^6–4.0×10^6 cells/mL (*C. sorokiniana*) after 5 cycles (Figure 2a, b). Addition of beads without microorganisms (control) did not affect ammonium removal (data not shown), while incubation of the non-sterile wastewater in bioreactors removed some ammonium (from 0.08 to 0.07 mg/L after 3 to 5 cycles). *A. brasilense* Cd alone did not remove measurable quantities of ammonium or phosphorus (data not shown). However, co-immobilization of *C. vulgaris* with *A. brasilense* Cd significantly enhanced ammonium removal (Figure 3a), while this did not occur with *C. sorokiniana* (Figure 3b), although both *Chlorella* species were capable of eliminating most of the ammonium when immobilized separately in beads. Removal of nitrates were tested under two settings, when the wastewater arrived from the treatment plant were loaded with nitrates and during the incubation cycles in the bioreactors when the natural resident microflora of the wastewater converted ammonium to nitrate, increasing nitrate concentration (Figure 4e). In general, removal of high initial nitrate concentrations by co-immobilization with both microalgae species was superior to the removal of the microalgae alone (Figure 4a, b). However, in the case where nitrate was increased during the process, only co-immobilization with *C. sorokiniana* was more efficient than the removal of the microalgae alone (Figure 4c, d). The control treatments (beads without microorganisms and untreated wastewater incubated similarly) did not remove any nitrate during the 5 cycles tested (Figure 4e). Removal of phosphate from the wastewater was always better when microalgae were co-immobilized with *A. brasilense* Cd (Figure 5a, b), and the controls were incapable of removing any phosphate.

The indigenous microflora of the wastewater was not analyzed in detail in this study. There were a large number of bacteria and definitely a population of nitrifiers in some wastewater samples. However, the semi-continuous treatment scheme performed in this study did not yield significant removal of nutrients by the native microflora.

Removal of ammonium and phosphate from culture medium (synthetic wastewater lacking a carbon source) showed similar tendencies to this study, regarding the interaction of *C. vulgaris* and *A. brasilense* Cd (de-Bashan et al., 2002b). No report on removal of nutrients by co-immobilized *C. sorokiniana* and *A. brasilense* or with any other bacterial species is available. However,

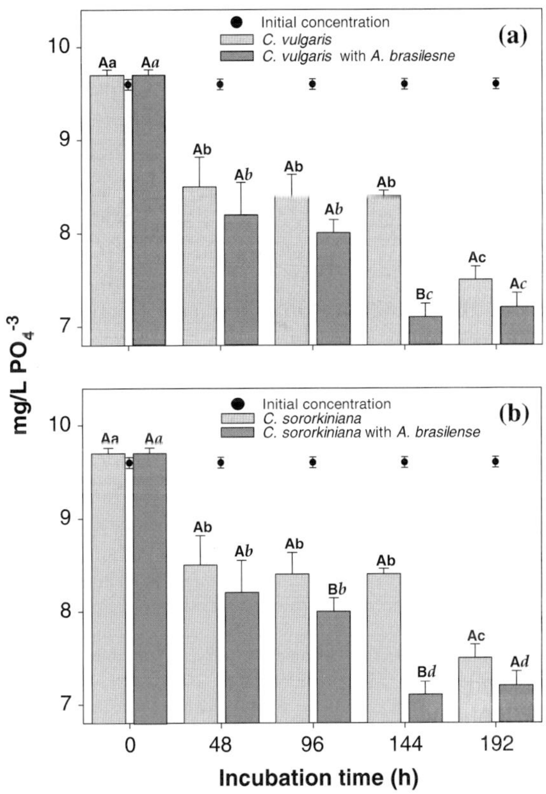

Figure 5. Removal of phosphate ions from municipal wastewater by *Chlorella* spp. co-immobilized with *A. brasilense* Cd. (a) *C. vulgaris*; (b) *C. sorokiniana*. Columns (of each gray scale) denoted by a different lower case letter or italics lower case letter, separately, differ significantly by ANOVA at $P \leq 0.05$. Pairs of columns for each cycle denoted by a different capital letter differ significantly with Student's *t*-test at $P \leq 0.05$. Bars represent standard error.

when *C. sorokiniana* was mixed and incubated as free suspended culture with *Rhodobacter sphaeroides*, neither microorganism could simultaneously remove acetate, propionate, ammonia, nitrate, and phosphate from synthetic wastewater, while a mixed culture could accomplish this (Ogbonna et al., 2000).

In summary, this study is the first report demonstrating that the new co-immobilization technology is capable of reducing nutrients (N and P) from regular municipal wastewater and might have a potential in devising new approaches to biological removing of nitrogen and phosphorus from wastewater.

Acknowledgements

This study was written in the memory of the late Avner Bashan of Israel. We thank Ivan Murrillo and Baudilio Acosta of CIB for analyses of wastewater, Abigail Solano and Juana Solano from the municipal wastewater treatment plant of La Paz for free access to their facility and analysis of the wastewater, and Ira Fogel for editing the English text. This study was partially supported by a travel grant to the First International Meeting on Phosphate Solubilizing Bacteria by the University of Salamanca, Spain and by the Bashan Foundation.

References

APHA, AWWA, WPCF (American Public Health Association, American Waterworks Association. Water Pollution Control Federation) 1992 Standard methods for the examination of water and wastewater 17th ed. Madrid, Spain. 1105 p.

Bashan Y 1986 Alginate beads as synthetic inoculant carriers for the slow release of bacteria that affect plant growth. Appl. Environ. Microbiol. 51, 1089–1098.

Bashan Y 1998 Inoculants of plant growth-promoting bacteria for use in agriculture. Biotechnol. Adv. 16, 729–770.

Bashan Y, Hernandez J -P, Leyva L A and Bacilio M 2002 Alginate microbeads as inoculant carrier for plant growth-promoting bacteria. Biol. Fertil. Soils 35, 359–368.

Bashan Y and Holguin G 1997 *Azospirillum*-plant relationships: environmental and physiological advances (1990–1996). Can. J. Microbiol. 43, 103–121.

Bashan Y, Holguin G and Lifshitz R 1993 Isolation and characterization of plant growth-promoting rhizobacteria. *In* Methods in Plant Molecular Biology and Biotechnology. Eds. B R Glick and J E Thompson. pp. 331–345. CRC Press, Boca Raton, FL.

Chevalier P and De la Noüe J 1985 Wastewater nutrient removal with microalgae immobilized in carrageenan. Enzyme Microbiol. Technol. 7, 621–624.

de-Bashan L E, Bashan Y, Moreno M, Lebsky V K and Bustillos J J 2002a Increased pigment and lipid content, lipid variety, and cell and population size of the microalgae *Chlorella* spp. when co-immobilized in alginate beads with the microalgae-growth-promoting bacterium *Azospirillum brasilense*. Can. J. Microbiol. 48, 514–521.

de-Bashan L E, Moreno M, Hernandez J -P and Bashan Y 2002b Ammonium and phosphorus removal from continuous and semi-continuous cultures by the microalgae *Chlorella vulgaris* coimmobilized in alginate beads with *Azospirillum brasilense*. Water Res. 36, 2941–2948.

De la Noüe J and De Pauw N 1988 The potential of microalgal biotechnology: a review of production and uses of microalgae. Biotechnol. Adv. 6, 725–770.

Gonzalez L E and Bashan Y 2000 Growth promotion of the microalgae *Chlorella vulgaris* when coimmobilized and cocultured in alginate beads with the plant growth-promoting bacteria *Azospirillum brasilense*. Appl. Environ. Microbiol. 66, 1537–1541.

Gonzalez L E, Cañizares R O and Baena S 1997 Efficiency of ammonia and phosphorus removal from a Colombian agroindustrial wastewater by the microalgae *Chlorella vulgaris* and *Scenedesmus dimorphus*. Biores. Technol. 60, 259–262.

Gonzalez-Bashan L E, Lebsky V, Hernandez J P, Bustillos J J and Bashan Y 2000 Changes in the metabolism of the microalgae *Chlorella vulgaris* when coimmobilized in alginate with the nitrogen-fixing *Phyllobacterium myrsinacearum*. Can. J. Microbiol. 46, 653–659.

Lau P, Tam N F Y and Wong Y S 1997 Wastewater nutrients (N and P) removal by carrageenan and alginate immobilized *Chlorella vulgaris*. Environ. Technol. 18, 893–899.

Lebsky V K, Gonzalez-Bashan L E and Bashan Y 2001 Ultrastructure of coimmobilization of the microalga *Chlorella vulgaris* with the plant growth-promoting bacterium *Azospirillum brasilense* and with its natural associative bacterium *Phyllobacterium myrsinacearum* in alginate beads. Can. J. Microbiol. 47, 1–8.

Mouget J L, Dakhama A and De Lavoie la Noüe M C J 1995 Algal growth enhancement by bacteria: is consumption of photosynthetic oxygen involved? FEMS Microbiol. Ecol. 18, 35–44.

Ogbonna J C, Yoshizawa H and Tanaka H 2000 Treatment of high strength organic wastewater by a mixed culture of photosynthetic microorganisms. J. Appl. Phycol. 12, 277–284.

Tam N F Y and Wong Y S 2000 Effect of immobilized microalgal bead concentration on wastewater nutrient removal. Environ. Pollut. 107, 145–151.

Valderrama L T, Del Campo C M, Rodriguez C M, de-Bashan L E and Bashan Y 2002 Treatment of recalcitrant wastewater from ethanol and citric acid production using the microalga *Chlorella vulgaris* and the macrophyte *Lemna minuscula*. Water Res. 36, 4185–4192.

Solubilization of hardly soluble iron and aluminum phosphates by the fungus *Aspergillus niger* in the soil

C.B. Barroso[1] & E. Nahas[2,3]

[1]*Pós-Graduanda em Microbiologia, UNESP, Universidade Estadual Paulista-UNESP, 14884-900, Jaboticabal, SP, Brazil.* [2]*Departamento de Produção Vegetal, FCAV, Universidade Estadual Paulista-UNESP, Rodovia Prof. Paulo Donato Castellane km 5, 14884-900, Jaboticabal, SP, Brazil.* [3]*Corresponding author**

Received 12 December 2002. Accepted in revised form 2 January 2003

Key words: hardly soluble phosphate, molasses, P-solubilization, respiratory activity, soil acidity

Abstract

Some Brazilian soils present high contents of hardly soluble iron and aluminum phosphates and a high capacity for fixation of soluble phosphates. This study evaluated the ability of the fungus *Aspergillus niger* F_{111} isolated from soil to solubilize Fe and Al phosphates. Iron, aluminum or calcium phosphate were added to soil samples and inoculated with the *A. niger* F_{111}. Sugar-cane molasses (2% v/w) was added as a carbon source on the 1st and 10th day of incubation. Soil samples without molasses, phosphates or fungus were used as control. Soil was incubated at 30 °C for twenty days and samples were removed every 5 days for determination of soil respiration (CO_2 production), pH, titratable acidity, soluble phosphate and total carbohydrate contents. Soil respiration increased early on the first day after molasses addition and decreased thereafter to the minimum level. The largest contents of soluble phosphorus were observed on the 5th and 15th day of incubation, with the following sequence of phosphate solubilization: aluminum phosphate > iron phosphate > calcium phosphate > control. Another experiment was performed under the same conditions as described above using only aluminum phosphate as a source of phosphorus, with evaluations performed daily for 15 days. Aluminum phosphate solubilization was related to CO_2 evolution, which increased on the 2nd and 12th day of incubation. Soluble phosphate increased on the 2nd and 11th day and titratable acidity increased on the 3rd and 11th day. Carbohydrates decreased after molasses application. The effect of solubilization of insoluble phosphates by the fungus depended on the addition of a carbon source (molasses) but decreased as soon as the carbon source was mineralized in the soil.

Introduction

For higher plants, the bioavailability of phosphorus in the soil is determined by factors like: (1) the quantities of $H_2PO_4^-$ and HPO_4^{2-} in the soil solution, (2) the solubility of Fe-, Al- and Ca-phosphate in either acid or in calcareous soils, (3) the amount and stage of decomposition of organic residues, (4) the activity of microorganisms in solubilizing insoluble P sources (Stevenson and Cole, 1999), and (5) plant roots (Ae et al., 1990). The content of soluble phosphorus is very low in most of the soils throughout the world (Jones, 2000) and do not satisfy the requirements for plants nutrition. Inorganic or organic insoluble phosphates represent from 95 to 99% of total soil phosphate (Hayman, 1975).

Soils from Brazil were formed from acid igneous rocks and are generally poor in phosphorus. The phosphorus deficiency is a consequence of the high weathering rate, decreased contents of macro and micronutrients and high solubilization of Al

* FAX No: +55-16-322-4275.
E-mail: enahas@fcav.unesp.br

E. Velázquez and C. Rodríguez-Barrueco (eds.),
First International Meeting on Microbial Phosphate Solubilization, 193–198

and Fe. One of the outstanding changes in soil due to these processes was the accumulation of hardly soluble iron and aluminum phosphates like strengite and variscite (Fageria and Baligar 2001). Soil pH from a number of these soils ranged from 3.8 to 6.4, presenting an average of 4.9 (Nahas et al., 1994). Additionally, Brazilian soils have a high capacity for fixation of soluble phosphates. Most of the phosphorus was found as aluminum phosphate (1.4 to 35.1 ppm) and iron phosphate (26.6 to 158.3 ppm) rather than calcium phosphate (0 to 19.2 ppm) (Nahas, 2002).

The population of phosphate solubilizing microorganisms in soil is generally very high (Illmer et al., 1995; Pal, 1998). In thirteen Brazilian soils the number of P-solubilizing microorganisms ranged from 0 to 79×10^5 bacteria and 2 to 58×10^5 fungi g^{-1} of soil representing from zero to 58% of the total soil microorganisms (Nahas et al., 1994).

The ability of these microorganisms to solubilize insoluble P sources has been studied in several soils. Several bacteria and fungi found in the soils were reported to solubilize different apatites and calcium phosphate (Cerezine et al., 1988; Nahas, 1996; Sundara et al., 2002). Strains from the genera *Penicillium* and *Pseudomonas* were among the most powerful isolates from forest soil having the ability to solubilize hydroxylapatite and calcium hydrogen phosphate dihydrate (Illmer et al., 1995). *Penicillium rugulosum* was reported to solubilize hardly soluble inorganic phosphates such as hydroxylapatite, strengite, variscite, and some rock phosphates ores (Reyes et al., 1999). However, a few reports have indicated the P-solubilizing ability of soil microorganisms to solubilize hardly soluble Fe- or Al-phosphates. The purpose of this study was to measure the ability of *A. niger* F_{111} to solubilize two hardly soluble phosphates ($FePO_4$, $AlPO_4$ and hydroxyapatite) during the incubation of soil with molasses as a carbon source.

Materials and methods

A. niger F_{111}, selected among 481 isolates for its great ability to solubilize insoluble phosphates ($FePO_4$, $AlPO_4$ and hydroxyapatite) in culture media, with a solubilization capacity above 1000 μg PO_4^{3-} ml^{-1} (Barroso and Nahas, 2002) and used throughout this study, was maintained on Sabouraud agar slants. For inoculum preparation, the organism was grown on Sabouraud agar slants for 7 days at 30 °C. Spores were suspended in sterilized water and filtered with a double layer of gauze to remove contaminating mycelia. The number of spores was determined with a Neubauer counting chamber and the inoculum was adjusted to 4.62×10^7 spores ml^{-1}.

A Kandiustalf soil, with pH 5.5 (1:1 in $CaCl_2$), collected from the surface layer (0–20 cm) of the Experimental Station of UNESP Jaboticabal (BR) was air dried and passed through sieves of 2-mm mesh. The soil contained 15 ppm P and 3.5 % organic matter.

The treatments used in the first experiment were described in Table 1. The P sources (200 kg P_2O_5 ha^{-1} or 36.4 mg P kg^{-1} soil, w/w) were added to the soil and this was mixed 1:3 (w/w) with sand. 200g of the soil mixture were placed in 2.5-liters flasks. Hardly insoluble P sources were Araxá rock phosphate (fluorapatite, CaP), iron phosphate (FeP), and aluminum phosphate (AlP). Molasses, a by-product of sugar cane processing consisting of 64% total sugars and 59% reducing sugars (Prado Filho et al., 1998) was used as the carbon source (2%, v/w). The molasses was added to the rehydration water and sterilized. A mixture was prepared under no sterile conditions by the addition to the molasses solution of the N source (200 mg NH_4Cl kg^{-1} soil, w/w) and 3.5 ml of the spore suspension per flask of the *A. niger* F_{111}. Water content of the soil was adjusted to 50% of the total water-holding capacity by the addition of the solution mixture. The molasses was added on the 1st and 10th day of incubation. Molasses added on the 10th day was also previously sterilized. Incubation was performed at 30 °C for up 20 days in 1st experiment and for up 15 days in 2nd experiment. Samples of soil (11 g) were collected from each flask every 5 days in the 1st experiment and collected daily for the 2nd experiment. Evolved CO_2 was collected in 40 ml traps of 1.0 *M* NaOH solution and titrated with 1.0 *M* HCl after 24 h of soil incubation.

Soluble phosphate was determined in water soil extracts by a method described by Ames (1966). Total carbohydrates were measured by the anthrone method (Angers and Mehuys, 1989)

Table 1. Solubilization of different insoluble phosphates by the fungus *Aspergillus niger* F_{111} in soil added with molasses

Additions	Incubation time (days)				
	0	5	10	15	20
	μg P solubilized g^{-1} dry soil†				
No addition	0.431^{NS}	0.395c	0.152de	0.501e	0.666c
F_{111}	0.412^{NS}	0.768c	1.064d	0.653e	0.752bc
Molasses	0.393^{NS}	0.952c	0.818def	2.123d	0.697bc
Molasses-F_{111}	0.350^{NS}	2.539b	1.451c	2.264cd	0.604c
Molasses-CaP	0.432^{NS}	0.766c	0.707ef	2.376cd	0.609c
Molasses-CaP-F_{111}	0.423^{NS}	2.461b	1.379c	2.816bc	0.797bc
Molasses-FeP	0.410^{NS}	0.948c	0.612f	3.325ab	0.929ab
Molasses-FeP-F_{111}	0.436^{NS}	4.783a	1.974b	3.082ab	1.054a
Molasses-AlP	0.432^{NS}	0.846c	0.654ef	1.952d	0.920ab
Molasses-AlP-F_{111}	0.425^{NS}	5.645a	2.714a	3.590a	1.048a
F test	2.02^{NS}	99.14**	110.05**	63.11**	11.47**

**Significant ($P < 0.01$).
NSNot significant.
†Values followed by different letters indicate significant differences ($P < 0.05$).
F_{111}, *Aspergillus niger* F_{111}; CaP, fluorapatite; FeP, iron phosphate; AlP, aluminum phosphate.

after extraction of the soil sample with H_2O (Metzger et al., 1987). Soil pH was measured with a glass electrode and titratable acidity was determined by titrating the samples according Cerezine et al. (1988).

The study was carried out in a completely randomized design and the data analyzed using the SAS statistic package for ANOVA (SAS Institute, 1990). All the data were the mean values of duplicate analyses of five samples. Least significant differences (LSD) were calculated from the ANOVA analysis at the 0.05 level of probability by using the Tukey's test. Simple correlation analysis (r) was performed to examine the relationships between individual soils properties.

Results and discussion

The soluble phosphate contents increased in the 5th and 15th day as a response to the molasses application and decreased thereafter (Table 1). Phosphate solubilization was 24% higher than of the control as a result of the fungus inoculation. When molasses was added to the soil, phosphate solubilization was 69% higher than of the control. Similarly, inclusion of P sources to the soil had a significant effect on the solubilization of phosphates when we compared it to the control. The addition of *A. niger* F_{111} alone or in combination with molasses or insoluble phosphates enhanced the amount of soluble phosphate from 44 to 179% in relation to the non-inoculated treatments. The ability of solubilization of the insoluble phosphates by the *A. niger* F_{111} decreased in the following order: P-Al > P-Fe > P-Ca > no phosphate. It was clear in this work that the solubilization of insoluble phosphates in the soil depended on five factors: incubation time, the addition of carbon source, the *A. niger* F_{111} inoculation, and the presence and type of P source added.

Because *A. niger* F_{111} was active in dissolving aluminum phosphate more than other phosphates, the factors that affected solubilization of P-Al were investigated (Fig. 1). Shortly after the molasses addition, total carbohydrates decreased (Fig. 1a) and the production of CO_2 was enhanced (Fig. 1b) due to the mineralization of C source from the molasses by the soil microorganisms (in the control) or effectively by the *A. niger* F_{111} in the inoculated treatments, then enhancing P solubilization (Fig. 1e). This result was corroborated by significant correlation among the respiratory activity and P solubilized ($r = 0.48^{**}$, Table 2). Molasses presents a high content of sugars, thus it appears clear that the production of acids (Figure 1c) and the consequent decrease in pH (Figure 1d) are related to oxidative metabolism of the C source and would account for the

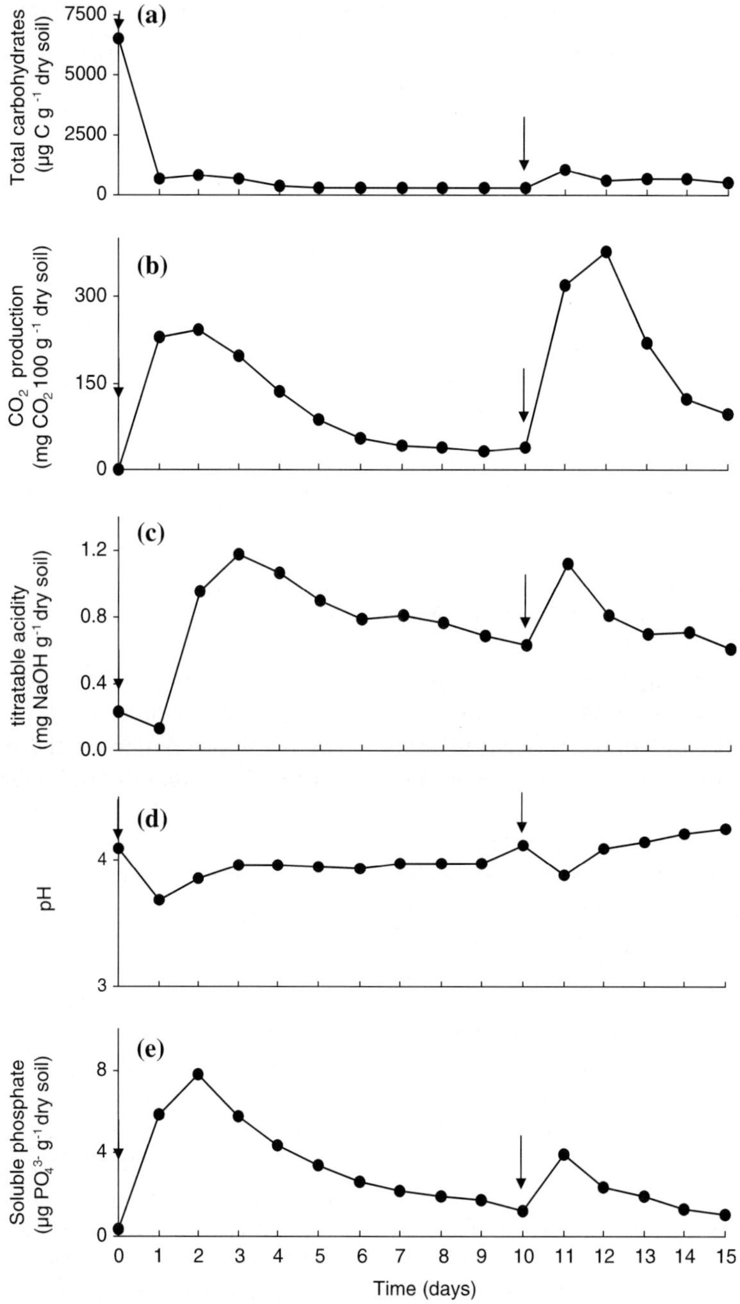

Figure 1. Solubilization of aluminum phosphates by the fungus *Aspergillus niger* F_{111} in the soil added with molasses at the first and tenth day of incubation (arrow).

solubilization of insoluble phosphates (Richardson, 2002). This relationship may be confirmed by the significant correlation between the P solubilized and the amount of total carbohydrates ($r = -0.28^{*}$), titratable acidity ($r = 0.36^{*}$), pH ($r = -0.73^{**}$) and the amount of total carbohydrates and titratable acidity ($r = -0.46^{**}$) (Table 2).

These findings suggest that the metabolism of sugars by the fungus would enhance CO_2 and acids production, and would decrease the pH and soil carbohydrates contents. Carbon sources

Table 2. Correlations (r) among CO_2 production, total carbohydrates, titratable acidity, pH, and soluble phosphate

Relationship	Soluble Phosphate	Total Carbohydrates	Titratable Acidity	pH
CO_2 production	0.48**	-0.22^{NS}	0.28^{NS}	-0.22^{NS}
Soluble phosphate		−0.28*	0.36*	−0.73**
Total carbohydrates			−0.46**	0.12^{NS}
Titratable acidity				-0.07^{NS}

*Significant ($P < 0.05$).
**Significant ($P < 0.01$).
^{NS}Not significant.
Degrees of freedom = 48.

easily assimilable were degraded producing CO_2 in the first days of incubation (Hattori, 1988; Bernal, 1998). After a new addition of molasses on the 10th day (Figure 1), the respiratory activity enhanced but the P solubilization decreased. Possibly, the microbiota and the *A. niger* F_{111}, already adapted in the soil had uptaken part of the solubilized P (Falih and Wainwright, 1996), increasing its growth and the CO_2 production and decreasing the carbohydrates contents more quickly. The correlations presented previously (Table 2) confirm these statements. In agreement with this work, Marstorp and Witter (1999) reported that the increase in CO_2 production after glucose addition was accompanied by exponential microbial growth. The decrease in the microbial activities can reflect the decrease of the microbial population, with consumption of the more easily assimilable nutrient (Ajwa et al., 1998), as we found in the present study.

The bioavailability of solubilized phosphate can be attributed to changes of the mineralization and immobilization processes in the soil (Singh and Amberger, 1991). The results in Table 2 show a significant and positive correlation between the soluble phosphate contents and the respiratory activity, thus demonstrating that the process of solubilization was related to the size and activity of the microbial population (Tardieux-Roche, 1966).

In conclusion, it seems clear that the *A. niger* F_{111} had the ability to solubilize aluminum and iron phosphates through the excretion of organic acids producing soluble phosphate. The proportion of soluble phosphate depend on the availability of an assimilable carbon source and the type of hardly soluble phosphate. Due to the distinguishable character of the Brazilian farmers to use vinasse (Nahas et al., 1990) as potassium fertilizer, the results of this study suggest the possibility of solubilization of soil phosphates by this practice, producing available phosphate and increasing plant productivity.

Acknowledgements

This research was supported by grant from the FUNDUNESP. We wish to thank CNPq for the fellowship granted to both authors. The authors thank Dr. Lawrence Mayhew who kindly revised the English version of this manuscript.

References

Ae N, Arihara J, Okada K, Yoshihara T and Johansen C 1990 Phosphorus uptake by pigeon pea and its role in cropping systems of the Indian subcontinent. Science 248, 477–480.

Ajwa H A, Rice C W and Sotomayor D 1998 Carbon and nitrogen mineralization in tallgrass prairie and agricultural soil profiles. Soil Sci. Soc. Am. J. 62, 942–951.

Ames B N 1966 Assay of inorganic phosphate and phosphatases. Meth. Enzymol. 8, 115–116.

Angers D A and Mehuys G R 1989 Effects of cropping on carbohydrate content and water-stable aggregation of a clay soil. Can. J. Soil Sci. 69, 373–380.

Barroso C B and Nahas E. 2002 Unpublished results.

Bernal M P, Sánchez-Monedero M A, Paredes C and Roig A 1998 Carbon mineralization from organic wastes at different composting stages during their incubation with soil. Agric. Ecosystems and Environ. 69, 175–189.

Cerezine P C, Nahas E and Banzatto D A 1988 Soluble phosphate accumulation by *Aspergillus niger* from fluorapatite. Appl. Microbiol. Biotechnol. 29, 501–505.

Fageria N K and Baligar V C 2001 Improving nutrient use efficiency of annual crops in Brazilian acid soils for sustainable crop production. Commun. Soil Sci. Plant Anal. 32, 1303–1319.

Falih A M K and Wainwright M 1996 Microbial and enzyme activity in soils amended with a natural source of easily available carbon. Biol. Fertility Soils 21, 177–183.

Hattori A 1988 Microbial activities in soil amended with sewage sludges. Soil Sci. Plant Nutr. 43, 221–232.

Hayman D S 1975 Phosphorus cycling by soil micro-organisms and plant roots. *In* Soil Microbiology. Ed. N Walker. pp. 67–91. Butterworths, London.

Illmer P, Barbato A and Schinner F 1995 Solubilization of hardly soluble $AlPO_4$ with P-solubilizing microorganisms. Soil Biol. Biochem. 27, 265–270.

Jones R D 2000 Phosphorus cycle. *In* Encyclopedia of Microbiology, Vol. 3, 2nd edn. Ed. J Lederberg. pp. 614–617. Academic, San Diego.

Marstorp H and Witter E 1999 Extractable dsDNA and product formation as measures of microbial growth in soil upon substrate addition. Soil Biol. Biochem. 31, 1443–1453.

Metzger L, Levanon D and Mingelgrin U 1987 The effect of sewage sludge on soil structural stability: microbiological aspects. Soil Sci. Soc. Am. J. 51, 346–351.

Nahas E 1996 Factors determining rock phosphate solubilization by microorganisms isolated from soil. World J. Microbiol. Biotechnol. 12, 567–572.

Nahas E, Banzatto D A and Assis L C 1990 Fluorapatite solubilization by *Aspergillus niger* in vinasse medium. Soil. Biol. Biochem. 22, 1097–1101.

Nahas E, Centurion J F and Assis L C 1994 Phosphate-solubilizing and phosphatase-producing microorganisms from various soils. Rev. Bras. Ci. Solo 18, 43–48.

Nahas E 2002 Factors affecting the solubilization of insoluble phosphates. *In* First International Meeting on Microbial Phosphate Solubilization. Ed. University of Salamanca IRNA-CSIC. pp. 20–22. Salamanca, Spain, 16–19 July 2002.

Pal S S 1998 Interactions of an acid tolerant strain of phosphate solubilizing bacteria with a few acid tolerant crops. Plant Soil. 198, 169–177.

Prado-Filho L G, Domingos R N and Silva S M G 1998 Acúmulo de cádmio por *Saccharomyces cerevisiae* fermentando mosto de melaço. Sci. Agric. 55, 128–132.

Reyes I, Bernier L, Simard R R and Antoun H 1999 Effect of nitrogen source on the solubilization of different inorganic phosphates by na isolate of *Penicillium rugulosum* and two UV-induced mutants. FEMS Microb. Ecol. 28, 281–290.

Richardson A E 2002 Making microorganisms mobilize soil phosphorus. *In* First International Meeting on Microbial Phosphate Solubilization. Ed. Univ. Salamanca IRNA-CSIC pp. 3–8. Salamanca, Spain. 16–19 July 2002.

SAS Institute 1990 Statistical Analysis System, SAS/STAT use's guide (Version 6), 3rd edn. SAS Institute, Cary N.C. 705 pp.

Singh C P and Amberger A 1991 Solubilization and availability of phosphorus during decomposition of rock phosphate enriched straw and urine. Biol. Agric. Hort. 7, 261–269.

Stevenson F J and Cole M A 1999 Cycles of Soil: Carbon, Nitrogen, Phosphorus, Sulfur, Micronutrients, 2nd edn. Wiley, New York. 448 pp.

Sundara B, Natarajan V and Hari K 2002 Influence of phosphorus solubilizing bacteria on the changes in soil available phosphorus and sugarcane and sugar yields. Field Crop Res. 77, 43–49.

Tardieux-Roche A 1966 Contribution a l'étude des interactions entre phosphates naturels et microflore du sol. Ann. Agron. 17, 403–471.

Fertilizers, food and environment

J. M. Igual[1] & C. Rodríguez-Barrueco
Instituto de Recursos Naturales y Agrobiología, CSIC, Apartado 257, 37071, Salamanca, Spain.
[1]*Corresponding author**

Received 20 December 2002. Accepted in revised form 2 January 2003

Key words: biofertilizers, environmental contamination, fertilization, food production, phosphate-solubilizing bacteria, phosphorus

Abstract

Although phosphoric fertilizers will continue to play a major role in intensive agriculture, depletion of natural resources and long-term unsustainability necessitate alternative strategies be investigated and implemented to buffer against food insecurity and environmental degradation. Phosphorus is not a renewable resource and its future use in agriculture will be impacted by declining availability and increased cost. Moreover, the striking increase in the use of fertilizers by intensive agriculture practices has led to degradation of air and water quality. This paper offers an overview on the sources and production process of phosphate fertilizers, the sources of environmental contamination due to their production and use, and finally focuses on the use of phosphate-solubilizing bacteria as an alternative to avoid the excessive use of such fertilizers.

Beginnings of chemical fertilizers

Man began to cultivate land in an organized way for food grains around 8000 BC. When man wanted to produce his food intensively instead of merely collecting it from places where it occurred naturally, it was soon realized that the same soil cannot sustain continuously plant growth and good productivity. Therefore, new areas were colonized and techniques for cultivation of plants were gradually introduced as manifested by the writings of ancient civilizations of Mesopotamia (about 3000 BC), the Nile (about 2000 BC), the Indus Valley (about 1500–2000 BC), and of the Roman (around 700 BC) and the Chinese (around 2000 BC) empires.

The first empirically known fertilizer was the excreta of human being and their animals. However, until some 300 years ago, only the Chinese used such residues in a systematic way. The rationale of plant nutrition came out only after the works of the French scientists Antonie Lavoisier in 1774 and J. B. Boussingault in 1834, and the German chemist Justus von Liebig in 1840, who undertook chemical analyses of plants and soils and arrived at the conclusion that chemical elements in plants came from soil and air. In the Von Liebig's book "Chemistry in Its Application to Agriculture and Physiology" was for the first time set down the value of mineral elements in plant nutrition and the importance of fertilization in maintaining soil fertility. Therefore, von Liebig's thinking on the value of mineral element in plant nutrition can be considered as the theoretical base for the development of the fertilizer industry.

Chemical salts of nitrogen and potassium were available commercially at the beginning of the 19th century. Later, in 1842, J. B. Lawes and his associate J. H. Gilbert produced superphosphate by chemical treatment of crushed bones and mined phosphate with sulphuric acid. Thus,

* E-mail: igual@usal.es

E. Velázquez and C. Rodríguez-Barrueco (eds.),
First International Meeting on Microbial Phosphate Solubilization, 199–202

nitrogen, phosphorus and potash fertilizers were available at this time. The next milestone in the area of fertilizer production came from the German chemist Fritz Haber, who synthesized ammonia from N_2 and hydrogen during the First World War. Therefore, the modern fertilizer industry is little more than 150 years old.

Fertilizers and agriculture production

Before 1950, the expansion of the area of cultivated land and mechanization accounted for the increase in agricultural production. Since that date, advances are associated with the heightened use of fertilizers and agro-chemicals, more irrigation and the introduction and adoption of hybrid varieties of maize and high-yielding wheat and rice (Isherwood, 1996). Thanks to the "green revolution", food production has kept up with the increase in population, although there are hundreds of millions of people in the world who are still under-nourished.

Global fertilizer consumption has increased substantially since 1950, while the world's population has grown from 2.5 to 6 billion. In 1960, developed countries accounted for 88% of the world fertilizer consumption. However, by 1999 their share had fallen to 39% whereas developing countries accounted for 61% (developing Asia alone accounts for 48%). In 1998/99, the world fertilizer consumption amounted to 136.93 million tons of nutrients, from which 82.18 tons were of nitrogen, 32.88 of P_2O_5 and 21.87 of K_2O (IFA, 2000). It is forecast a population growth up to 8–9 billion people by 2040 and agriculture, as currently practiced, will require and additional 40 and 20 million of metric tons of N and P fertilizer, respectively, to meet the food production needs (Vance, 2001).

Phosphate fertilizers

Phosphorus is eleventh in order of abundance in the earth's crust but its concentration in many rocks is very small. However, there are deposits which are sufficiently rich in phosphorus and extraction is commercially viable. Phosphate rock (27–38% P_2O_5) is the raw material source from which all types of phosphate fertilizers are produced, with the minor exception of basic slag (13–15% P_2O_5), which is a by-product of steel production.

Phosphate ores are of two major geological origins: Igneous or Sedimentaries. The phosphate minerals in both types of ore are of the apatite group. The most commonly encountered variants are Fluorapatite and Francolite, which predominate in igneous and in sedimentary phosphate rocks, respectively. Although phosphate rock deposits are found in over 30 countries throughout the world, the three major producing countries, i.e. the USA, China and Morocco currently produce approximately two-third of the global phosphorus requirement (EFMA, 2002).

To increase the availability of the phosphorus nutrient, and to obtain a more concentrated product, phosphate rock is processed using sulphuric acid, phosphoric acid or nitric acid. The acidulation by means of sulphuric acid produces either phosphoric acid as an intermediate for TSP (triple superphosphate) and compound fertilizer production, or single superphosphate as a fertilizer. The basic chemistry of this process is very simple. The tricalcium phosphate in the phosphate rock is converted by reaction with concentrated sulphuric acid into phosphoric acid and the insoluble salt calcium sulfate:

$$Ca_3(PO_4)_2 + 3H_2SO_4 \rightarrow 2H_3PO_4 + 3CaSO_4$$

The efficiency of the process ranges between 2.6 and 3.5 ton of rock phosphate used to obtain 1 ton of P_2O_5, and the energetic requirements are estimated at 120–180 kW h per ton of P_2O_5 produced (EFMA; http://www.efma.org).

Drawbacks of phosphate fertilizers

Phosphate rock contains various metals as minor constituents in the ores. Varying amounts of these elements are transferred to P fertilizers in production processes, and later are applied to soils with these fertilizers. Cadmium is the heavy metal potentially most dangerous because it is readily absorbed by plants and may seriously affect human health. Some other heavy metals contaminants in P fertilizers are arsenic, chromium, lead, mercury, nickel, and vanadium (Mortvedt, 1996).

Moreover, radioactive materials, like uranium and radium, are normal constituents of the earth's crust. As with all elements, the distribution of radioactive elements in the crust is not even. Geological processes have enhanced the radioactivity of sedimentary phosphate rock. As example, South Carolina's ore reaches the highest values of activity concentrations (in Bq kg^{-1}): 4800 of ^{238}U, 4800 of ^{226}Ra and 78 of ^{232}Th (Unscear, 1982). By processing phosphate rock to fertilizer the radioactivity of the ore is transferred to the product and to the waste products. Exposure of workers and the public to radiation from phosphate rock and fertilizer is therefore not unlikely.

The production of phosphoric acid from phosphate rock emits gaseous flourides, containing about 10 mg N m^{-3} as fluorine. A secondary emission is dust originating from the unloading, handling and grinding of phosphate rock, which contains about 3–4% water-insoluble fluoride (EFMA; http://www.efma.org). Moreover, most fertilizer plants are situated near estuaries. The by-product gypsum is released via surface water into the sea or disposed in ponds where it settles. Around of 5 tons of phosphogypsum are generated per ton of P_2O_5 produced. Gypsum slurry contains some of the heavy metals from the phosphate rock, including cadmium (Rutherford et al., 1994) and some radioactive elements, mainly ^{226}Ra (Martínez-Aguirre and García-León, 1994; Scholten and Timmermans, 1996). Disposal of improperly treated pond wastes may also produce a rapid change of pH in surroundings due to the acidity of the process water, which can affect most species of fish, aquatic life and vegetation.

From an environmental point of view, the main hazard associated with the use of phosphate fertilizers is water eutrophication. Because phosphorus is the most limiting nutrient in the fresh water, the increase of this nutrient disturbs the ecological structure by producing a rapid increase in the growth of phytoplankton. When a bloom of some species of algae is developing rapidly, there is a negative net oxygen balance and the deoxygenation of the water may lead to the suffocation of fish and higher animals. The loss of water-soluble phosphates to water has increased from less than 0.1 to over 0.3 kg P ha^{-1} $year^{-1}$. In Europe, phosphorus from agriculture could account for 40% of total phosphorus contained in water (EEA, 1999).

The manufacture of chemical fertilizers is dependent on the use of non-renewable forms of energy. In view of the escalating energy costs, energy will be a key limiting factor for increasing agricultural production in future. In the case of phosphate fertilizers, between 120 and 180 kW h are required to produce 1 ton of P_2O_5 (EFMA; http://www.efma.org). In the case of phosphate fertilizers the raw material, rock phosphate, is also a non-renewable resource. Estimating reserves is not easy for a number of reasons: confidentiality of such information for governments and companies due to its commercially sensitive character; uncertainty about future rates of consumption; or possible future changes in technology and cost of production, which would allow to exploit resources not economically viable at present (EFMA, 2002). In any case, phosphorus reserves are finite and some authors estimate that inexpensive rock phosphate reserves could be depleted in as little as 60–80 years (Runge-Metzger, 1995). As noted by Abelson (1999), a potential phosphate crisis looms for agriculture in the 21st century.

Phosphate-solubilizing bacteria as biofertilizers: an alternative

Bacterial involvement in the solubilization of inorganic phosphate is known since the first decade of the past century. Most of the studies on phosphate solubilization were done first by isolating the microorganisms from the soil and then studying the solubilization *in vitro*. The investigations on solubilization of phosphates under field conditions and on the uptake by plants were however started later. Ectorhizospheric strains from pseudomonads and bacilli, and endosymbiotic bacteria from rhizobia have been described as effective phosphate-solubilizing bacteria (PSB). Beneficial effects of the inoculation with PSB to many crop plants have been described by numerous authors. Rhizobia are, perhaps, the most promising group of PSB on account of their ability to fix nitrogen symbiotically with legumes and the capacity of some strains for solubilizing insoluble inorganic phosphate compounds. Several publications have demonstrated that phosphate-solubilizing strains of *Rhizobium* and *Bradyrhizobium* increase growth and P content of nonleguminous as well leguminous plants. An alternative approach for the use of PSB as microbial inoculants is the use of mixed cultures or co-inoculation with other microorganisms. In this regard, some results suggest a synergistic interaction between vesicular arbuscular mycorrhizae (VAM) and PSB, which allows for better utilization of poorly soluble P sources. Similarly, plant growth can be increased by dual inoculation with PSB and *Azospirillum* or *Azotobacter* (for references, see Igual et al., 2001; Rodriguez and Fraga, 1999).

Phosphate-solubilizing bacteria have been already used as biofertilizer for agricultural improvement. For example, in the former Soviet Union a commercial biofertilizer under the name "phosphobacterin" was first prepared by incorporating *Bacillus megaterium* var. *phosphaticum* and widely used in the Soviet Union, East European countries and India. In this last country, a carrier based preparation under the name "Microphos" was developed by the Indian Agricultural Research Institute, using efficient phosphate dissolving strains of *Pseudomonas striata*, *Bacillus poymyxa* and *Aspergillus awamori* packet in a wood charcoal and soil mixture. These cultures were tested in multilocal field trials and were found to be, in general, effective (Subba Rao, 1993). More recent findings are described in this same book (see contributions by J V Cross and M E Legget, respectively).

Despite of these promising results, PSB-based biofertilizers has not got wide spread application in agriculture mainly because of the response of plant species or genotypes to inoculation often varies according to the bacterial strain used. Differential rhizosphere effect of crops in harboring a target PSB strain or even the modulation of the bacterial phosphate solubilizing capacity by specific root exudates may account for the observed differences. On the other hand, good competitive ability and high saprophytic competence are the major factors determining the success of a bacterial strain as an inoculant. Therefore, studies to know the competitiveness and persistence of specific microbial populations in complex environments, such as the rhizosphere, should be addressed in order to obtain efficient inoculants. In this regards, research efforts in order to obtain appropriate formulations of microbial inoculants, which protect the microorganisms against environmental stresses and at the same time enhance and prolong its activity, may help in raising the use of such beneficial bacteria for agricultural improvement.

References

Abelson P H 1999 A potential phosphate crisis. Science 283, 2015.

EEA, European Environment Agency, Denmark 1999 Nutrients in European Ecosystems. Environmental Assessment Report No 4.

EFMA (European Fertilizer Manufacturer Association) 2002 Phosphorus Essential Element for Food Production. Bruxelles.

IFA, International Fertilizer Industry Association, 2000 Fertilizer Indicators. Bull. April. Paris.

Igual J M, Valverde A, Cervantes E and Velázquez E 2001 Phosphate-solubilizing bacteria as inoculants for agriculture: use of updated molecular techniques in their study. Agronomie 21, 561–568.

Isherwood K F 1996 The present situation of fertilizer production and use in the world. *In* Fertilizers and Environment. Ed. C Rodríguez-Barrueco. pp. 13–18. Kluwer Academic Publishers, Dordrecht, The Netherlands.

Martínez-Aguirre A and García-León M 1994 The distribution of U, Th and ^{226}Ra derived from the phosphate fertilizer industries on a estuarian system in southwest Spain. J. Environ Radioactivity 22, 155–177.

Mortvedt J 1996 Heavy metal contaminants in inorganic and organic fertilizers. Fertil. Res. 43, 55–61.

Rodríguez H and Fraga R 1999 Phosphate solubilizing bacteria and their role in plant growth promotion. Biotechnol. Adv. 17, 319–339.

Runge-Metzger A 1995 Closing the cycle: obstacles to efficient P management for improved global security. *In* Phosphorus in the Global Environment. Ed. H Tiessen. pp. 27–42. John Wiley and Sons Ltd, Chichester, UK.

Rutherford P M, Dudas M J and Samek R A 1994 Environmental impacts of phosphogypsum. Sci. Total Env. 149, 1–38.

Scholten L C and Timmermans C W M 1996 Natural radioactivity in phosphate fertilizers. *In* Fertilizers and Environment. Ed. C Rodríguez-Barrueco. pp. 171–175. Kluwer Academic Publishers, Dordrecht, The Netherlands.

Subba Rao N S 1993 Biofertilizers in Agriculture and Forestry. Oxford and IBH Publishing Co. Pvt. Ltd, New Delhi; p 242.

Unscear 1982 Ionizing Radiation: Sources and Biological Effects. United Nations, New York.

Vance C P 2001 Symbiotic Nitrogen Fixation and Phosphorus Acquisition. Plant Nutrition in a World of Declining Renewable Resources. Plant Physiol. 127, 390–397.

Phosphate solubilizing microorganisms vs. phosphate mobilizing microorganisms: What separates a phenotype from a trait?

A.H. Goldstein[1,3] & P.U. Krishnaraj[2]
[1]*Center for Biomaterials Research, Alfred University, 2 Pine Street, Alfred, New York 14802, USA.* [2]*Department of Agricultural Microbiology, College of Agriculture U.A.S., Dharwad, 580 005, Karnataka India.* [3]*Corresponding author**

Received 14 October 2002. Accepted in revised form 2 January 2003

Key words: gluconate, mineral phosphate solubilization, phosphate mobilization

Abstract

Soils are often high in insoluble mineral phosphates but deficient in the soluble orthophosphate (Pi) essential for the growth of most plants and microorganisms. In agricultural crop production, phosphorous is second only to nitrogen in importance as a fertilizer amendment so that phosphorus fertilizers are the world's second largest bulk agricultural chemical and, therefore, the second most widely applied chemical on Earth. There is a broad spectrum of mineral phosphate chemistries; but in arid to semiarid soils the predominant forms are the calcium phosphates. Calcium phosphates are soluble to varying degrees in the presence of the wide array of organic acids produced by microorganisms. Other biosolubilization mechanisms exist as well, so that conversion of mineral phosphates to Pi is generically attributed to microorganisms in most representations of global P cycling. With respect to plant growth, some workers have postulated that associations between plant roots and mineral phosphate solubilizing (MPS) microorganisms could play an important role in phosphorus nutrition in many natural and agroecosystems. As a result, an enormous amount of research has been conducted over the last 100 years involving isolation and characterization of MPS microorganisms from many soils with the goal of developing P biofertilizers that would accomplish much the same function as biological nitrogen fixation. To date, the results of these efforts have been problematic. In this review, we will attempt to identify the variables of state with respect to the MPS phenomenon in bacteria and briefly summarize the challenges that confront this field of research. Finally we will discuss our observation that, in Gram-negative rhizobacteria, extracellular oxidation of glucose to gluconic acid and 2–ketogluconic acid via the direct oxidation pathway provides the biochemical basis for highly efficacious calcium phosphate solubilization and may, in fact, be the basis for the evolution of mutualistic plant–bacteria relationships in some phosphate-limited soil ecosystems.

Introduction

Phosphorus is an essential macronutrient required for every aspect of cell biology from energy metabolism to the structure of the genetic material. In agriculture, Phosphorus is second only to Nitrogen as an applied fertilizer for crop production and, as such, represents the world's second largest bulk agricultural chemical product (Goldstein, 1995, 2000; Larsen, 1967). Soluble orthophosphate (Pi) availability is also growth-limiting in many natural ecosystems (Hausenbuiller, 1972; Jackson, 1973; Stevenson, 1986). The ecophysiological paradox is that, while most organisms can only assimilate Pi or low

* FAX No: +1-607-871-2354.
E-mail: fgoldste@alfred.edu

E. Velázquez and C. Rodríguez-Barrueco (eds.),
First International Meeting on Microbial Phosphate Solubilization, 203–213

molecular weight soluble organic phosphates (e.g. glycerol-P), most of the P 'pool' in the soil is not available as a nutrient for either microbial or plant biomass accumulation. Most inorganic soil P is made up of poorly soluble mineral phosphate precipitates. Likewise, organic P (phytate and other common organic P molecules) cannot be transported across biological membranes. Therefore, P must be cycled from these unavailable forms back into available forms. In our work, we use the term insoluble mineral phosphate in recognition of the fact that Pi is also a mineral form of phosphate. It should further be noted that, for minerals, the use of the term 'insoluble' is a relative since if these compounds were truly insoluble there would be nothing left to say on the topic. Historically, the use of this terminology is based on the equilibrium solubility of the compound in a pure aqueous solution. Many of the calcium phosphates, including rock phosphate ores (fluoroapatite, francolite), are insoluble in the soil with respect to the release of Pi at rates necessary to support agronomic levels of plant growth (see below and refs. Goldstein, 1995 and 2000).

The ability of some Gram-negative bacteria to dissolve poorly soluble forms of calcium phosphate to Pi may be the oldest phenotype known to soil microbiology (Goldstein, 1986). While most reviews of MPS bacteria start with the landmark paper by Gerretson in the first volume of Plant and Soil (Gerretson, 1948), the first published MPS paper encountered by this author (Goldstein, 1986) is by Sackett and co-workers published in 1908. This work, entitled "The solvent action of soil bacteria upon the insoluble phosphates of raw bone meal and natural raw rock phosphate" appeared in the obscure journal, Central bl. Bakteriol. 20, 688–703. Given that this is probably not the first paper on this topic, the MPS literature is certainly a century old. But the observation of dissolution of materials such as limestone and bonemeal by the soil and by soil extracts is more correctly credited to Justus von Liebig, a distinguished legacy that places the birth of the MPS phenomenon somewhere in the mid-nineteenth century. Since that time, the role of microbes in the MPS process has passed like a stealth aircraft through every representation of the Phosphorus cycle. We have all seen these flowchart cartoons of global P cycling in virtually every microbiology and ecology textbook. These figures invoke, without mechanistic definition (i.e. off the radar screen), a significant role for microorganisms in the 'flow' of mineral P to Pi.

It is essential to recognize the crucial difference between N and P with respect to the definition of the term *fixation*. Nitrogen *fixation*, is the process whereby microorganisms make an unavailable inorganic mineral N (N_2 gas) bioavailable. In the Phosphorus literature, '*fixation*' has just the opposite meaning; available Pi in *fixed* into unavailable mineral precipitates (Goldstein, 1986). To further confuse matters, the term 'mineralization' is commonly used in the P cycling literature to describe the liberation of Pi from organic P in the soil. This mineralized Pi can, of course, be re-precipitated as insoluble mineral P. The stealth configuration of the MPS field covers not only the phenomenon itself but current paradigms with respect to the role of microorganisms in plant mineral nutrition. Unlike the aircraft, the result has not been enhanced performance (at least in the opinion of these authors).

Microbial nitrogen fixation has achieved well-deserved prominence on the radar screen of microbial physiology and ecology in studies involving both agro- and natural ecosystems. Conversely, research to elucidate specific mechanisms and pathways of bacterial mineral phosphate solubilization has not really emerged as a coherent and recognized field. This is not to say that global P metabolism in microorganisms has been overlooked. ASM has published two books on the subject (Torriani-Gorini et al., 1987, 1993), and the bacterial pho regulon and P stimulon continue as experimental systems of paramount importance with respect to bacterial physiology, genetics (and now, undoubtedly, genomics). However, the *bona fides* of mineral phosphate solubilization (MPS) as a true bacterial phenotype remain in question. Given our current understanding of the need to develop renewable and sustainable approaches to agricultural crop production, and given recent progress in our mechanistic understanding of at least one biochemical mechanism for the MPS phenomenon, it is appropriate after a century of research, to briefly review the evidence for a MPS phenotype. While this field lacks the coherence of nitrogen fixation, an enormous amount of work has been done, so that we can only touch on selected aspects here. A true review of this field would require a book, perhaps entitled "One Hundred Years of Insolubility."

We will limit this review to the availability of P in one component of the biosphere, namely soils. As discussed above, most of the Phosphorus in soils is not bioavailable. Therefore, poorly soluble mineral P and most forms of organic P must be converted to either Pi ($H_2PO_4^-$ or HPO_4^{2-}) or an extremely limited number of low molecular weight organic phosphates (e.g. glycerol-P in *E. coli*). The molecular genetics and enzymology of bacterial biotransformations of organic phosphates, occurring through the activity of alkaline or acid phosphatases, phosphodiesterases, and other enzymes has received a great deal of attention. As a result, our understanding of the biochemistry and genetics (if not the ecology) of these systems is advanced (cf. Torriani-Gorini et al., 1987). This minireview will describe some of the variables of state with respect to the effective dissolution of an important family of poorly soluble mineral phosphates, the calcium phosphates. This group of related compounds makes up the major fraction of unavailable mineral P soils with a pH of 7 or greater which includes most arid and semi-arid ecosystems. By comparison with organic phosphates, the molecular genetic and biochemical bases of bacterial transformations of poorly soluble calcium phosphates has been somewhat difficult to resolve into a systematic field of study. Paradoxically, the reason may be the simplicity of the solubilization mechanism. Calcium phosphates are dissolved by acidification (proton substitution). Therefore, any bacterium that acidifies its external medium will show some level of mineral phosphate solubilization. In most soils, proton substitution reactions are driven by production of organic acids; represented generically by the equation (1)

$$(Ca)_n(PO4)_m + HA = H(PO4)^x + (Ca)A \tag{1}$$

While the stoichiometry is intentionally left vague to account for the complexity of calcium phosphate chemistry, it is understood that ionic calcium and phosphate are in their normal di- and trivalent forms so that the net result is the formation of Pi and the Calcium salt of a conjugate organic acid.

It may be seen that any number of organic acids (HAs) may be substituted into equation (1) with varying efficacies depending on the solubility of the mineral phosphate and the strength (pK_a) of the organic acid. Therefore, one is immediately confronted with the question of how to investigate the MPS phenomenon in a way that can determine if some bacteria produce acids as a strategy to make calcium phosphates bioavailable. In that case the organic acid production could be considered a *phenotype*, defined as 'a characteristic of the organism'. The alternative is the physicochemical reality that the proximity of organic acid-producing bacteria and calcium phosphates within the soil will result in some solubilization phenomena which, for the purposes of this essay, we will call a MPS *effect*, as in 'to cause Pi to come into being'. As previously mentioned, an enormous amount of applied microbiology research has been conducted in this field, mainly in developing countries, where the practice of application of excess chemical fertilizers is not common and the interest in biofertilization is much greater than in developed countries where high levels of soluble chemical fertilizers are routinely applied (12, 51).

A thorough review of this agricultural microbiology literature as well as 20 years worth of directed studies leads these authors to propose that some type of relationship exists between highly efficacious MPS bacteria and the expression of the direct oxidation pathway (Goldstein, 1994, 1995, 2000 and Goldstein et al., 1999). Invariably, during the screening of any population of Gram-negative soil bacteria on glucose minimal + insoluble P media (Goldstein, 1986), the superior strains are shown to express the direct oxidation pathway at a high level. Having made this metabolic connection, we now confront the matter of causality: is high-efficacy calcium phosphate solubilization a byproduct of direct oxidation-based bioenergetics in these bacteria, or do these bacteria express the direct oxidation pathway in order to access the pool of insoluble calcium phosphates with greater efficiency. Upon posing this question, one is immediately led to the related question of mutualism and even symbiosis between efficacious MPS bacteria and higher plants within the rhizosphere (Goldstein et al., 1999). Because the microbial ecology and agricultural microbiology implications are equally important, we will focus on the rhizosphere component of the soil ecosystem. The rhizosphere is most generally defined as the plant root and its immediate environment (Tinker, 1980).

Mineral phosphate cycling in the rhizosphere

It is well established that the pool of soluble P is quickly depleted in the region around a growing plant root so that Pi is the growth-limiting mineral nutrient in both natural and agroecosystems (Babu-Khan et al., 1995; Goldstein, 1986; Stevenson, 1986, and references therein). Conversely, most soils with a pH of 7.0 or greater contain substantial levels of poorly soluble calcium phosphates (Goldstein et al., 1999; Hausenbuiller, 1972). While bulk soil itself is a dynamic system, the rhizosphere is a unique ecological niche where complex interactions between the soil, its microbial population, and the plant roots create a dynamic system of great complexity. Within this system, Pi levels depend not only on the total amount of phosphorus in the environment, but on numerous phosphate-based chemical and biochemical reactions. With respect to availability as a biological mineral nutrient the soil contains three P 'pools'; (1) Available soluble P (Pi and a limited group of low molecular weight organics), (2) Unavailable organic P, and (3) Unavailable insoluble mineral (a.k.a. inorganic) P. In soils, the pool of available P is depleted by biological uptake and by two nonbiological reactions; fixation (defined above) and immobilization. Immobilization involves short-term reversible adsorption phenomena, whereas phosphate fixation into insoluble precipitates plays a role of major importance to the economics of world food production. In many soils, more than 70% of the applied phosphorus fertilizers get fixed in the soil rendering them unavailable for plant uptake (Bagyaraj et al., 2000; Holford, 1997; Kadrekar, 1977; Larsen, 1967; Stevenson, 1986).

Mineral phosphates have a wide range of solubilities, which in general, follow an inverse relationship with the Ca/P ratios. For example, monocalcium phosphate [$Ca(H_2PO_4)_2$, Ca/P = 0.50] has an equilibrium water solubility of 150 000 ppm at pH 7 whereas the fluoroapatite in commercial rock phosphate ore [$Ca_{10}(PO_4)_6F_2$, Ca/P = 1.66] is functionally insoluble with an equilibrium water solubility of 0.003 ppm (Goldstein, 2000). Poorly soluble mineral phosphates such as fluoroapatite or hydroxyapatite can only be effectively dissolved in aqueous solution under acidic conditions, they are useless as biological nutrients in the short term, i.e. for annual crop plants or many plant species in natural ecosystems.

It is fair to say that, at some point, almost all soil P gets fixed, i.e. passes through a mineral precipitate phase. Even Pi that has successfully cycled through biomass is released by autolysis, directly into the soil solution as Pi or organic P. The former may precipitate directly, while the latter ultimately gets fixed as well after encounters with a requisite number of enzymes usually ending with some type of phosphatase. Soluble P (both Pi and low molecular weight organics) temporarily exit the soil solution by adsorption to soil particles (immobilization). For example, the highly polar compound phytate (inositol hexaphosphate), usually a major component of the soluble P pool, is frequently immobilized and therefore unavailable for phosphatase cleavage and plant growth (Richardson, 1994). The form in which Pi exists also changes according to the soil pH. The net result of all these processes is that the average orthophosphate concentration in most soil solutions is around 10^{-6} M (Jackson, 1973). This is near the limit at which plants can transport phosphate, even in hydroponic solutions. Given the barriers to diffusion and fluid flow in the soil solution, it is no surprise numerous studies have shown that Pi is quickly depleted around the growing root *in vivo*.

In the agroecosystems of the developed world, chemical phosphorus fertilizers are regularly applied to get maximum yields. Continuous application of these fertilizers will result in increased concentration of mineral phosphorus in the soil over time due to the reactions described above, resulting in large reserves of fixed P. This is an important sink that needs to be tapped for phosphorus nutrition. Often, less than 10% of commercially applied fertilizer P is absorbed by the target crop plant (Goldstein, 1986, and references therein) while the rest is fixed *in situ* or washed out of the soil during the brief post-application period when Pi levels in the soil solution are high. It is now generally accepted that current fertilizer application methods and the associated N and P runoff play a major role in the well characterized eutrophication phenomena occurring in ecosystems such as the Gulf of Mexico. This combination of fixation and runoff means that, in spite of decades of fertilization, P deficiency remains a widespread problem in developed agroecosystems. Annual application of P fertilizers is almost universally required for efficient crop production in these soils. However, increasing ecological awareness has brought with it the call for a sustainable, eco-rational approach to the production of food and fiber. Many workers have attempted to use microbial recycling of mineral P as part of a sustainable system to support plant growth and soil conservation (cf. Bagyaraj, 2000; Blake, 1993; Gaur, 1990; Krishnaraj, 1987, 1996; Kucey et al., 1989; Sundara Rao, 1963). While we know microorganisms play a vital role in this cycling, we know very little about the applied aspects of this phenomenon. What knowledge we do have has largely accumulated from attempts to apply microorganisms as 'bioinoculants'. The strategy, as successfully exemplified by N_2 fixation, is to use microorganisms as a 'green alternative' to chemical fertilizers. Due to the complexity of the soil and rhizosphere ecosystems, the results of such efforts with MPS microorganisms have been problematic. Rhizobacteria can cause the efficacious release of nutrients into the soil solution (Blake, 1993) and exert beneficial effects on plant development (Glick, 1995). However, for every report of successful P biofertilization there has surely been a failure. In the final analysis, lack of commercial products in this area testify to the state-of-the-art. However, the potential remains. Development of reproducible microbiology-based P biofertilizers, would provide the basis for a crop production system that is largely organic, sustainable, and reduces reliance on expensive phosphatic fertilizers whose production and application have significant environmental consequences.

The MPS effect: acidification comes in many flavors

After 100 years of insolubility, there can be little doubt that, leaving aside unique ecosystems with significant populations of mineral acid-forming bacteria, most microbial bioconversion of mineral phosphates is accomplished by organic acid-mediated dissolution. There have been innumerable studies on this topic. The reader is referred to recent reviews by Goldstein (Goldstein, 1995; Goldstein and Liu, 1987; Krishnaraj and Gowda, 1990; Sundra Rao and Sinha, 1963).

The standard experimental approach is for soil and rhizosphere samples to be screened for MPS bacteria, which, in turn, are found to produce the usual array of organic acids common to microbial metabolism. In general, bacteria are isolated from soils and plated on an insoluble mineral phosphate medium. The MPS effect is visualized by a zone of clearing around the bacterial colony or fungal hyphae and may be seen in many of the experimental papers contained within this volume as well as in most of these authors' publications. Depending on the source of the sample, up to 20% or more of isolates can show some degree of clearing with glucose as the carbon source (cf. Sperber, 1957). The amount of acids liberated by these solubilizing bacteria is usually more than 5% of the carbohydrate consumed (Banik and Dey, 1982).

As expected, there is generally direct correlation between pH and calcium phosphate solubilization in the culture media (Agnihotri, 1970; Liu et al., 1992; Sperber, 1957). To help clarify the mechanism of P solubilization, studies were conducted using MPS^- mutants for the first time in 1996 by Krishnaraj (1996). The derived MPS^- mutants were compared with their wild type parents with respect to the Pi release in the tricalcium phosphate broth with respect to pH and organic acid production. The final pH remained the same but the MPS effect was not observed, indicating that there was some specificity with respect to the organic acids involved in solubilization. The Goldstein laboratory has tested a direct oxidation minus mutant of *Burkholderia cepacia* generously supplied by Lessie (Lessie et al., 1979). As with the work cited above, the final pH of the batch culture was similar (~3.3 to 3.5) but the MPS effect was not observed (unpublished). These results are not surprising given the buffering capacity of released orthophosphate. The interesting (but experimentally complicating) facts of life in 'MPS-world' are that dissolution of calcium phosphates releases buffering hydrogen phosphates into the medium (HPO_4^-, $H_2PO_4^{2-}$); not to mention a whole host of byproduct calcium phosphate hydroxides and hydrous oxides. As these buffers are released, the pH is simultaneously being driven up by H^+ depletion because protons must replace Ca^{2+} to generate the dissolution of the primary mineral P substrate. Therefore, a MPS bacterium producing both strong and weak organic acids might equilibrate to approximately the same pH as its MPS knockout derivative precisely because the MPS trait was the result of production of the strong organic acid. Strong acid-mediated dissolution of mineral P, in turn, raised the pH and generated buffering capacity in the closed batch culture system (more on this later). A MPS knockout, by definition, does not dissolve significant mineral P so that weak acidification controls the pH directly in unbuffered medium.

It is generally accepted that the mechanism for MPS activity with respect to calcium phosphates is the acidification of the medium via biosynthesis and release of organic acids (Goldstein, 1986; Krishnaraj, 1996). As discussed in the preceding paragraph, the degree of solubilization was not always correlated to a decrease in pH (Asea et al., 1988; Kadrekar and Talashilkar, 1977; Mehta and Bhide, 1970; Wani et al., 1979). This is not surprising given the complex acid–base equilibria generated by dissolution of a compound such as fluoroapatite, where numerous forms of amorphous calcium phosphates may exist simultaneously. Changes in pH generated by an initial burst of organic acid production, especially in experiments using batch cultures, may vanish in a veritable maze of solubilization and re-precipitation reactions well known to fertilizer process engineers (cf. Goldstein, 2000). While outside the scope of this article, organic acid production is not the only mechanism that may result in release of Pi from poorly soluble mineral phosphates *in vitro*. For example, both H^+ excretion originating from NH_4^+ assimilation (Illmer and Schinner, 1992, 1995; Parks et al., 1990), and respiratory H_2CO_3 production (Juriank et al., 1986) have been proposed as alternative metabolic bases for acid-mediated calcium phosphate solubilization.

This wide array of experimental conditions will foil even the most dedicated reductionist. One must deal with data generated using media containing mineral phosphates with solubilities ranging over nine orders of magnitude (fluoroapatite vs. monocalcium phosphate), minimal medium vs. a multiplicity of defined/buffered media, batch vs. continuous culture, etc. This situation has historically hampered any consensus on a working definition (or even agreement on the existence) of a bacterial MPS phenotype. As with N_2 fixation, there is obviously no single approach or universal organism that will provide world agriculture with a microbiology-based P fertilizer system. Both bacteria and fungi are known to dissolve mineral phosphates (Halder et al., 1991; Kucey et al., 1989; Maheshkumar, 1997; Maheshkumar et al., 1998; Martin and Cunninghum, 1973; Mehta and Bhide, 1970; Parks et al., 1990; Pereira, 1990 and Wani et al., 1979) and increase plant growth via enhanced P nutrition (Pikovskaya, 1948; Richardson, 1994 and Tinker, 1980). In fact, there are literally hundreds of published agronomic studies showing enhanced productivity in the presence of various microbial inoculants selected in vitro for the MPS effect. Unfortunately, there are an equal number of studies that yielded no fertilizer effect. One can imagine the range of experimental design problems inherent in these types of studies. For example, the effect of inoculation of MPS microbes on the growth and yield of crop plants was been reviewed by Sundara Rao and Sinha (1963) who noted that only ten out of the 18 field experiments produced a significant increase in the yield due to bacterial inoculation under diverse agroclimatic conditions in the country. Contrasts abound with inoculation using a MPS bacterium resulting in significant increases in the yield of *Phaseolus mungo* (Tomar et al., 1993), whereas inoculation with another MPS microbe, *Penicillium digitatum*, failed to show significant increase of dry matter over control in several crop plants (Wani et al., 1979). To further complicate matters, the positive plant growth effects observed with MPS bacteria may not always be due to the *in vitro* trait for which they were originally selected. For example, Chabot et al. (1993) attributed some growth enhancement to the production of siderophores.

Variations in microbial survival, soil-buffering capacity, types of mineral phosphates, and microenvironment effects are only the tip of the iceberg. It is also important recall that many plant root systems have highly complex, often symbiotic endo- or ectomycorrhizal associations that modify everything from root architecture to the effective surface area for uptake of nutrients and water (Tinker, 1980). Yet, in spite of these great experimental challenges, the benefits are greater still. As with the bioconversion of nitrogen, microorganisms have the potential to provide humanity with the phosphorus biofertilizer system necessary for sustainable agricultural production of food and fiber which, in turn, necessary for the long-term ecological stability of the planet.

A specific MPS phenotype mediated by the direct oxidation pathway?

The wide range of MPS mechanisms discussed above, has made it impossible to develop a unified approach to the microbiology of P solubilization such as is available to workers in the field of N_2 fixation where all pathways must converge at some variation of the nitrogenase system. However, at least for the Gram-negative rhizobacteria, workers have observed a recurring pattern when screening isolates for the ability to dissolve calcium phosphates on minimal medium with glucose as the carbon source (Goldstein, 1994, 1995, 2000; Goldstein and Liu, 1987; Goldstein et al., 1993, 1999 and references therein). Literally thousands of bacterial isolates have been evaluated and, when using these screening criteria, Gram-negative bacteria expressing the direct oxidation pathway far surpass other isolates in their ability to dissolve calcium phosphates. These bacteria have been designated as having a MPS^+ phenotype (Babu-Khan et al., 1995; Goldstein, 1995). These MPS^+ bacteria have the capacity to dissolve highly insoluble phosphates such as fluoroapatite because of the extremely low pK_as of the glucose oxidation products; gluconic acid, and 2-ketogluconic acid (−pK_as of 3.4 and −2.6, respectively). In addition, since these acids are produced in the periplasmic space, acidic protons are efficiently released into the extracellular medium, or rhizosphere space *in vivo*. An interesting corollary to the observed efficacy of the direct oxidation pathway in solubilizing calcium phosphates may be found in a report that expression of this pathway may be induced by Pi starvation conditions in the medium (Bagyaraj et al., 2000). Levels of other nutrients can modify expression of this pathway as well (cf. Hardy et al., 1993).

Direct oxidation (a.k.a. nonphosphorylating oxidation) is one of the four major metabolic pathways for glucose (aldose) utilization by bacteria. For many bacterial species, the direct oxidation pathway is the primary mechanism for aldose sugar utilization (Duine, 1991; van Schie et al., 1985, and references therein). The quinoprotein glucose dehydrogenase is so named because it is a member of the group of bacterial enzymes that utilize the redox cofactor PQQ (2,7,9-tricarboxyl-1H-pyrrolo[2,3-f]quinoline-4,5-dione. PQQGDH is a membrane-bound aldose dehydrogenase whose catalytic domain is located on the outer face of the cytoplasmic membrane. This enzyme transfers electrons from aldose sugars directly to ubiquinone in the cytoplasmic membrane via two electrons, two proton oxidations mediated by the cofactor PQQ (Duine, 1991). Direct oxidation of glucose to gluconic acid generates a transmembrane proton motive force (PMF) that may be used for bioenergetic and/or membrane transport functions (van Schie et al., 1985). In many Gram-negative bacteria gluconic acid may undergo one or two additional periplasmic oxidations. The second oxidation, catalyzed by gluconate dehydrogenase results in the production of 2-ketogluconic acid, one of the strongest naturally occurring organic acids known (pK_a ~ 2.6) and one which has historically been associated with bacteria selected for extremely high levels of calcium phosphate solubilization (Goldstein, 1986; Pereira, 1990).

Different bacterial species utilize different direct oxidation products. If uptake occurs after direct oxidation(s), all catabolic pathways converge to gluconate-6-phosphate. It has recently been shown that both fermentative bacteria such as *E. coli*, aerobic bacteria such as pseudomonads, and acetic acid bacteria can have direct oxidative and phosphorylative routes of metabolism. Fermentative bacteria, having the glycolytic pathway, usually take up glucose by way of a phosphotransferase system, but *E. coli* is now known to constitutively synthesize the apoglucose dehydrogenase protein and use the direct oxidative pathway in the presence of PQQ (Lessie et al., 1979). Furthermore, activation of the direct oxidation pathway and/or the presence of gluconic acid induces the Entner–Douderoff pathway in *E. coli* (Egan et al., 1992). Protons generated from these oxidations contribute directly to the transmembrane PMF. Evidence exists to suggest that PQQGDH plays a bioenergetic role in this crucial aspect of energy metabolism. In several bacterial species, it has further been shown that the efficiency of uptake of solutes such as alanine, lactose and proline is modified by PQQGDH-mediated electron transfer (van Schie et al., 1985, and references therein). Little is known, however, about the molecular mechanisms whereby these responses are regulated or about the biochemical or genetic regulatory mechanisms by which the cell switches between the phosphorylative and periplasmic oxidative mode.

While the direct oxidation pathway has sometimes been called a dissimilatory bypass (van Schie et al., 1985) because of the 'wasted' extracellular oxidation of glucose, gluconate can provide a highly efficient energy source while also providing a potential competitive advantage by allowing the bacterium expressing this pathway to acidify its microenvironment. The bioenergetics of such a strategy was described for *Acetobacter diazotrophicus* by Luna et al. (2000), based on the assumption that gluconate produced is further metabolized through the hexose monophosphate pathway. Extracellular oxidation of aldose sugars via the direct oxidation pathway can lead to an increase in energy generation due to both direct generation of a proton motive force and the reducing power provided by donation of electrons from PQQ to the respiratory electron transport system (56).

Future directions

The direct oxidation pathway is the basis for the highly efficacious MPS^+ phenotype: a testable hypothesis

Our hypothesis that the direct oxidation pathway is the metabolic basis for the superior MPS^+ phenotype in gram negative bacteria provides workers with both a unifying metabolic strategy, and a set of biochemical and genetic probes with which to systematically identify and evaluate the role of a specific subpopulation of rhizosphere bacteria in P cycling. In addition, the demonstrated efficacy of the direct oxidation pathway for the dissolution of fluoroapatites has provided a potential strategy for large-scale bioprocessing of rock phosphate ores. This industrial application, while certainly applied microbiology, is outside the scope of this minireview. Interested readers are referred to reviews by Goldstein et al. (1993) and Goldstein (2000).

In terms of future research, successful molecular cloning of direct oxidation pathway genes provides a powerful set of tools with which to study causal relationships between the population dynamics of MPS^+ bacteria in the rhizosphere and Pi in the soil solution. Available probes include several apoglucose dehydrogenase genes, PQQ biosynthesis genes, and at least one gluconate dehydrogenase gene. One of the authors (AHG) has used such genetic probes to help demonstrate the presence of unique populations of MPS^+ rhizobacteria in two alkaline desert soil environments where the levels of poorly soluble calcium phosphates are extremely high but Pi is undetectable in bulk soil extracts (Goldstein, 1994; Goldstein et al., 1999). Both unique MPS^+ populations were capable of high levels of direct oxidation of glucose, and the presence of the quinoprotein glucose dehydrogenase was confirmed by both enzymatic and molecular biology assays. In one case (Goldstein et al., 1999), a unique rhizobacterial population of *Enterobacter cloacae* expressed the direct oxidation pathway only in the presence of compounds washed from the root of the host plant, *Helianthus* sp. These data provided preliminary evidence for mutualism in this highly alkaline soil environment (pH 10.0).

The purpose of this review is to bring the possibility and potential of a MPS^+ phenotype to the attention of the applied and environmental microbiology community and to point out that, at least for soils where calcium phosphates predominate, a metabolic basis for this phenotype in Gram-negative rhizobacteria has now been hypothesized. Furthermore, the tools are now available with which to rigorously test this hypothesis. But most importantly, the need to develop sustainable agriculture systems to feed a hungry world demands that the MPS phenotype and/or the MPS effect be harnessed to put an end to 100 years of insolubility.

Proposed unified nomenclature for the study of microbial transformations of soil P

As discussed at the beginning of this review, research to elucidate specific mechanisms and pathways of bacterial mineral phosphate solubilization has not really emerged as a coherent and recognized field. While many microbiologists working on P solubilization indicate a direct role in P nutrition for plant growth, most soil scientists and ecologists remain skeptical of a direct role and are more likely to consider that phosphate-solubilizing microorganisms (PSMs) are indirectly contributing to plant nutrition via increasing Pi availability in the soil solution. In the opinion of these authors, a prerequisite for bringing coherence to this field requires unified nomenclature and definitions that will provide the bases for both testable hypotheses and precise discussions by workers in the field. Therefore, we propose the following terminology and definitions.

Name	Acronym	Operational definition
Phosphate solubilizing microorganism	PSM	Any microorganism capable of transforming insoluble organic or mineral (inorganic) phosphate into soluble orthophosphate (Pi) in a manner that may be shown to make a significant contribution to the phosphate nutritional status of a specific plant or plant population within the microorganism's native soil ecosystem. Compare with PMM.
Mineral phosphate solubilizing microorganism	MPS	Any microorganism capable of transforming insoluble mineral (inorganic) phosphate into soluble orthophosphate (Pi) in a manner that may be shown to make a significant contribution to the Pi nutritional status of a specific plant or plant population within the microorganism's native soil ecosystem. Compare with MPM.
Organic phosphate solubilizing microorganism	OPS	Any microorganism capable of transforming insoluble organic phosphate into soluble orthophosphate (Pi) in a manner that may be shown to make a significant contribution to the Pi nutritional status of a specific plant or plant population within the microorganism's native soil ecosystem. Compare with OPM.
Phosphate mobilizing microorganism	PMM	Any microorganism capable of transforming insoluble organic or mineral (inorganic) phosphates into soluble orthophosphate (Pi) in a manner that may be shown to make a significant contribution to pool of available orthophosphate (Pi) in the native soil ecosystem. Compare with PSM.
Mineral phosphate mobilizing microorganism	MPM	Any microorganism capable of transforming insoluble mineral (inorganic) phosphate into soluble orthophosphate (Pi) in a manner that may be shown to make a significant contribution to the pool of available soluble orthophosphate (Pi) in the native soil ecosystem. Compare with MPS.
Organic phosphate mobilizing microorganism	OPM	Any microorganism capable of transforming insoluble organic phosphate into soluble orthophosphate (Pi) in a manner that may be shown to make a significant contribution to the pool of available soluble orthophosphate (Pi) in the native soil ecosystem. Compare with OPS.

While some might argue that this terminology is cumbersome, it will provide both heuristic clarity and more stringent experimental design parameters in a field of tremendous importance both to sustainable agriculture and the ecology of nutrient cycling. The author suggests that workers in the field adopt this unified terminology and that it be ratified at the next International Conference on PSMs.

Acknowledgements

Preparation of this review was supported, in part, by the Fierer Endowed Chair to AHG. PKU is thankful to the Department of Biotechnology, Ministry of Science and Technology, Government of India for granting a BOYCAST fellowship to pursue MPS research.

References

Adamowicz M, Conway T and Nickerson K W 1991 Nutritional complementation of oxidative glucose metabolism in *Escherichia coli* via pyrroloquinoline quinone-dependent glucose dehydrogenase and the Entner–Doudoroff pathway. Appl. Environ. Microbiol. 57, 2012–2015.

Agnihotri V P 1970 Solubilization of insoluble phosphates by some soil fungi isolated from nursery seed beds. Can. J. Microbiol. 16, 877–880.

Asea P, Kucy R M N and Stewart J.W.B. J W B 1988 Inorganic phosphate solubilization by two *Penicillium* species in solution culture and soil. Soil. Biol. Biochem. 20, 459–464.

Babu-Khan S, Yeo T-C, Martin W L, Duron M, Rogers R D and Goldstein A H 1995 Cloning of a mineral phosphate solubilizing gene from *Pseudomonas cepacia*. Appl. Environ. Microbiol. 61, 972–978.

Bagyaraj D J, Krishnaraj P U and Khanuja S P S 2000 Mineral phosphate solubilization: agronomic implications, mechanism and molecular genetics. Proc. Indian Natl. Sci. Acad. (PINSA) 66, 69–82.

Banik S and Dey B K 1982 Available phosphate content of an alluvial soil as influenced by inoculation of some isolated phosphate-solubilizing microorganisms. Plant Soil 69, 353–364.

Blake F 1993 Organic Food Production in World Agriculture. Cartwright Sterling Publication Ltd, Hong-Kong p. 284 .

Chabot R, Antoun H and Cescas M P 1993 Stimulation de la croissance du mais et de la laitue romaine par des microorganisms dissolvant de phosphate inorganique. Can. J. Microbiol. 39, 941–947.

Cosgrove D J 1977 Microbial transformations in phosphorus cycle. Adv. Microb. Ecol. 1, 95–128.

Duine J A 1991 Quinoproteins: enzymes containing the quinoid cofactor pyrroloquinoline quinone, topaquinone or tryptophan-tryptophan quinone. Eur. J. Biochem. 200, 271–284.

Egan S E, Fliege R, Tong S, Shibata A, Wolfe R E Jr and Conway T 1992 Molecular characterization of the Entner–Doudoroff pathway in *Escherichia coli*: Sequence analysis and localization of promoters for edd-eda operon. J. Bacteriol. 174, 4638–4646.

Gaur A C 1990 Phosphate Solubilizing Microorganisms as Biofertilizer. Omega Scientific Publications, New Delhi p. 176.

Gerretsen F C 1948 The influence of microorganisms on the phosphate intake by the plant. Plant Soil 1, 51–81.

Glick B R 1995 The enhancement of plant growth by free living bacteria. Can. J. Microbiol. 32, 145–148.

Goldstein A H 1986 Bacterial mineral phosphate solubilization: Historical perspective and future prospects. Am J. Alt. Agric. 1, 57–65.

Goldstein A H and Liu S-T 1987 Molecular cloning and regulation of a mineral phosphate solubilizing gene from *Erwinia herbicola*. Bio/Technology 5, 72–74.

Goldstein A H, Rogers R D and Mead G 1993 Separating phosphate from ores via bioprocessing. Bio/Technology 11, 1250–1254.

Goldstein A H 1994 Involvement of the quinoprotein glucose dehydrogenase in the solubilization of exogenous phosphates by gram-negative bacteria. *In* Phosphate in Microorganisms: Cellular and Molecular Biology. Eds. A. Torriani-Gorini, E. Yagil and S. Silver. pp. 197–203. ASM Press, Washington, D.C. p. 223.

Goldstein A H 1995 Recent progress in understanding the molecular genetics and biochemistry of calcium phosphate solubilization by gram negative bacteria. Biol. Agric. Hortic. 12, 185–193.

Goldstein A H, Braverman K E and Osorio N 1999 Evidence for mutualism between a plant growing in a phosphate-limited desert environment and a mineral phosphate solubilizing (MPS) rhizobacteria. FEMS Microbiol. Ecol. 30, 295–300.

Goldstein A H 2000 Bioprocessing of rock phosphate ore: essential technical considerations for the development of a successful commercial technology. *In* Proceedings of the 4th International Fertilizer Association Technical Conference. IFA, Paris, p. 220.

Halder A K, Mishra A K and Chakraborthy P K 1991 Solubilizing of inorganic phosphates by *Bradyrhizobium*. Indian J. Expt. Biol. 29, 28–31.

Hardy G P, Texeria M A, de Mattos M J and Neijssel O M 1993 Energy conservation by pyrroloquinoline quinone-linked xylose oxidation in *Pseudomonas putida* NCTC10936 during carbon limited growth in chemostat culture. FEMS Microbiol. Lett . 107, 107–110.

Hausenbuiller R L 1972 Soil Science, Principles and Practices. Wm. C. Brown Company, Dubuque, Iowa p. 501.

Holford I C R 1997 Soil phosphorus its measurement and its uptake by plants. Aust. J. Soil Res. 35, 227–239.

Holt J G, Krieg N R, Sneath P H A, Stanley J T and William S T 1994 Bergey's Mannual of Determinative Bacteriology, 9th Edition, Williams and Willikins, Baltimore.

Illmer P. and Schinner F 1992 Solubilization of inorganic phosphates by microorganisms isolated from forest soils. Soil Biol. Biochem. 24, 389–395.

Illmer P and Schinner F 1995 Solubilization of inorganic calcium phosphates- solubilization mechanisms. Soil Biol. Biochem. 27, 257–263.

Jackson ML 1973 Soil Chemical Analysis. Prentice Hall of India (P) Ltd, New Delhi, India. p. 318.

Juriank J J, Dudley S C, Allen M F and Knight W G 1986 The role of calcium oxalate in the availability of phosphorus in soils of semiarid regions: Thermodynamic study. Soil Sci. 142, 255–261.

Kadrekar S B and Talashilkar S C 1977 Efficiency of applied phosphorus in relation to its saturation in lateritic soils of Konkan. J. Indian Soc. Soil Sci. 25, 269–273.

Krishnaraj P U 1987 Studies on beneficial microorganisms in crop plants. M. Sc. (Agri) Thesis, U.A.S., Bangalore.

Krishnaraj PU and Gowda T K S 1990 Occurrence of phosphate solubilizing bacteria in the endorhizosphere of crop plants. Curr. Sci. 59, 933–934.

Krishnaraj P U 1996 Genetic characterization of mineral phosphate solubilization in *Pseudomonas* sp. Ph.D. Thesis, I.A.R.I., New Delhi.

Kucey R W N, Tanzen H H and Leggett M E 1989 Microbially mediated increases in plant available phosphorus. Adv. Agron. 42, 199–228.

Larsen S 1967 Soil phosphorus. Adv. Agron. 19, 151–210.

Lessie T G, Berka T and Zamanigian S 1979 *Pseudomonas cepacia* mutants blocked in the direct oxidative pathway of glucose degradation. J. Bacteriol. 139, 323–325.

Liu S-T, Lee L-Y, Tai C-Y, Horng C-H, Chang Y-S, Wolfram J H, Rogers R D and Goldstein A H 1992 Cloning of an *Erwinia herbicola* gene necessary for gluconic acid production and enhanced mineral phosphate solubilization in *Escherichia coli* HB101. J. Bacteriol. 174, 5814–5819.

Luna M F, Mignone C F and Boiardi J L 2000 The carbon source influences the energetic efficiency of the respiratory chain of N_2-fixing *Acetobacter diazotrophicus*. Appl. Microbiol. Biotechnol. 54, 564–569.

Maheshkumar K S 1997 Studies on microbial diversity and their activity in soil under bamboo plantation. M.Sc. Thesis, Univ. Agricultural Sciences, Dharwad, India.

Maheshkumar K S, Alagawadi A R, Krishnaraj P U, Patil V C 1998 Microbial diversity in the rhizosphere and inoculation effect of selected rhizobacteria on growth of bamboo seedlings. 39th Annual Conference, Association of Microbiologists of India. Dec. 5–7:1998. Mangalore, India.

Martin J K and Cunninghum R B 1973 Factors controlling the release of phosphorus from decomposing wheat roots. Aust. J. Biol. Sci. 26, 715–727.

Mehta Y R and Bhide V P 1970 Solubilization of tricalcium phosphate by some soil fungi. Indian J. Exp. Biol. 8, 228–229.

Parks E J, Olson G J, Brickman F E and Baldi F 1990 Characterization of high performance liquid chromatography (HPLC) of the solubilization of phosphorus in iron ore by a fungus. J. Indian Microbiol. 5, 183–190.

Pereira AT 1990 Endorhizosphere bacteria of wetland rice, their H_2-dependent chemolithotrophy, N_2 fixation, P-solubilization and interaction with rice genotypes. Ph.D. Thesis, University of Agricultural Sciences, Bangalore, p. 192.

Pikovskaya R I 1948 Mobilization of phosphorus in soil in connection with vital activity of some microbial species. Mikrobiologiya 17, 362–370.

Richardson A E 1994 Soil microoganisms and phosphorus availability. Soil Biota 50, 35–39.

Sperber J I 1957 Solution of mineral phosphates by soil bacteria. Nature 180, 994–995.

Stevenson F J 1986 Cycles of Soil Carbon, Nitrogen Phosphorus Sulphur and Micronutrients. Wiley, New York. p. 428.

Sundara Rao W B and Sinha M K 1963 Phosphate dissolving microorganisms in the soil and rhizosphere. Indian J. Agric. Sci. 33, 272–278.

Tinker P B 1980 The role of rhizosphere microorganisms in phosphorus uptake by plants. *In* The role of phosphorus in agriculture. Ed. F E Khasawneh, American Society for Agronomy Press, Madison. p. 909.

Tomar S S, Pathan M A, Gupta K P and Khandar U R 1993 Effect of phosphate solubilizing bacteria at different levels of phosphate on blackgram (*Phaseolus mungo*). Indian J. Agron. 38, 131–133.

Torriani-Gorini A, Rothman F G, Silver S, Wright A and Yagil E 1987 Phosphate Metabolism and Cellular Regulation in Microorganisms. ASM Press, Washington, D.C p. 307.

Torriani-Gorini A, Yagil E and Silver S 1993 Phosphate in Microorganisms: Cellular and Molecular Biology. ASM Press, Washington, D.C. p. 427.

van Schie K J, Hellingwerf J P, van Dijken M G L, Elferink J M, van Dijl L S, Kuenen J G and Konings W N 1985 Energy transduction by electron transfer via pyrroloquinoline quinone-dependent glucose dehydrogenase in *Escherichia coli, Pseudomonas aeruginosa,* and *Acinetobacter calcoaceticus* (var lwoffii). J. Bacteriol. 163, 493–499.

Wani P V, More B B and Patil P L 1979 Physiological studies on the activity of phosphorus solublizing microorganisms. Indian J. Microbiol. 19, 23–31.

Challenges in commercializing a phosphate-solubilizing microorganism: *Penicillium bilaiae*, a case history

M. Leggett[1,2], J. Cross[1], G. Hnatowich[1] & G. Holloway[1]
[1]*Philom Bios, Inc., # 318 – 111 Research Drive, Saskatoon, SK, Canada S7N 3R2.* [2]*Corresponding author**

Received 17 December 2002. Accepted in revised form 2 January 2003

Key words: Phosphate-solubilizing microorganisms, fungi, commercializing, market, inoculant commercialization, JumpStart, *Penicillium bilaiae*, phosphate solubilization

Abstract

The commercialization of a phosphate inoculant is a challenging process. The active ingredient of the phosphate inoculant *JumpStart*® (*P. bilaiae*) was isolated in 1982. Although the concept of P solubilization was proven, much additional research was required. Full-scale, cost-effective manufacturing, packaging and QA systems; easy-to-use, shelf-stable formulations needed to be developed. Extensive field research to confirm efficacy and comprehensive data on compatibility with seed-applied pesticides were required. In addition, we needed to develop and refine the product positioning and branding to ensure we were delivering value to the farmer. Development continues to be an on-going process with the use of the product on new crops, improved production methods and formulations, new applications, and continuing market research to monitor changing farmer needs.

Introduction

Commercializing any product is a difficult process. Commercializing a biologically based product presents additional challenges, and commercializing a biological product aimed at enhancing phosphorus nutrition is even more difficult. This paper describes the scientific, technical, and marketing challenges involved in commercializing and marketing *JumpStart*® (a phosphate inoculant based on the phosphate-solubilizing fungus *Penicillium bilaiae*).

Kucey (1983) isolated *Penicillium bilaiae* from Canadian prairie soils in 1983. He demonstrated that the organism could solubilize phosphate (P) on agar plates and in liquid culture (Kucey, 1983). He also demonstrated that inoculating soil with the fungus could increase the growth and P uptake in wheat and beans in greenhouse and field trials. Philom Bios acquired the rights to commercialize this product in 1986. Agriculture and Agri-Food Canada (AAFC) obtained a patent (Canadian patents 1,308,270 and 1,308,566) and a royalty agreement was signed between Philom Bios and AAFC. This agreement was crafted such that the royalty funds would go back into research rather than into general government funds. Philom Bios has paid AAFC $1.7 million from 1990 to 2002.

The concept of *P. bilaiae* as a phosphate inoculant had been demonstrated (Asea et al. 1988; Kucey, 1983, 1987, 1988). However, a tremendous amount of work was required to make this into a commercial product. Extensive field trials were conducted to prove that the organism was effective over a wide range of soil, environmental conditions, and crop types to support the registration of the product. We had to develop a production method, a commercially acceptable formulation, a quality assurance plan, and information on application strategies. Initial

* FAX No: +1-306-975-1215.
E-mail: mleggett@philombios.ca

E. Velázquez and C. Rodríguez-Barrueco (eds.),
First International Meeting on Microbial Phosphate Solubilization, 215–222

knowledge of the behavior and mode of action was limited.

We had to develop a marketing plan. The research needed to bring the product to market was expensive and so we had to find financing. It could not all be completed at Philom Bios and so we found cooperators to assist with the work. All of these operations were interconnected and the process required coordination and teamwork.

The phosphate inoculant was registered for use on wheat in 1990 and is now registered on most major crops in Western Canada. From a small number of hectares inoculated in the first year, some one million hectares were inoculated in 2002. These advances have resulted from an increasing commitment to research and a constant willingness to develop cooperative projects with researchers across Canada.

This paper will discuss the challenges involved in developing and marketing a commercially viable phosphate inoculant.

Field research

Inoculant organisms are affected by many soil and environmental factors and must be thoroughly tested under conditions that will exist where they will be used.

Pre-commercialization

Field trials had been done on two crops (wheat and beans) at one site (Kucey, 1988). The inoculated bran media had been applied in furrow (by hand) in these studies.

Commercialization

Philom Bios did not have the resources to carry out all the field tests needed for registration. Philom Bios therefore joined with Dow Elanco Canada (DEC) to carry out field tests across the Canadian Prairies. The Saskatchewan Wheat Pool (SWP) also conducted independent trials to ensure an unbiased source on information for the registration package. As an in-furrow application using bran material was impractical for large scale farming operations, a seed treatment was used. Thirty-eight trials were established across the three Canadian Prairie Provinces (Manitoba, Saskatchewan and Alberta) in 1987 and 1988. All trials were arranged in a split plot experimental design. The control (untreated) and *P. bilaiae* treatments were compared over four rates of P_2O_5 (0, 10, 20, 30 kg P_2O_5 ha^{-1}). *P. bilaiae* increased yield and P uptake in wheat (Table 1) at the lower rates of P application.

Subsequent field-work proved the fungus could also increase P uptake and yield in pea, lentil, bean, canola, and alfalfa.

On-going

The field program continues to increase as the market expands and sales move into the Northern U.S. states. In addition, as each new crop or new formulation is added field trials are conducted to ensure inoculant effectiveness.

Production and formulation

A good quality inoculant must be able to survive storage, desiccation after inoculation onto the seed, and natural competition in the rhizosphere (Maurise et al., 2001). Formulation development is a complex process and is still more of an art than a science (Daigle and Connick, 1990).

Pre-commercialization

Kucey (1988) used a straw substrate to produce spores for greenhouse and field trials. This was effective for the small trials but impractical on a commercial scale. The production process was cumbersome and only a limited amount of material could be produced. Twenty-three pyrex dishes were needed to produce the inoculum for

Table 1. Effect of *P. bilaiae* on the yield of wheat. Multiple year field summary of 38 locations

Phosphate applied	Mean yield (kg ha^{-1})			
	Untreated	*P. bilaiae*	Difference	Statistical significance
0	2771	2827	56	0.01
10	2909	2958	49	0.04
20	2930	2962	32	0.15
30	2962	2948	(14)	0.54

From Hnatowich et al. 1990.

five research sites (approximately 1.0 ha). The material was applied by hand as it could not be applied through a commercial seeder.

Commercialization

A liquid fermentation method was developed which produced sufficient spores to inoculate 25,000 ha of wheat per batch. The spores were collected and processed into a dry powder, which had to be kept frozen to maintain viability for an effective inoculant. This frozen powder, *PB50*®, was introduced to the market place in 1990. The economics were prohibitive however, as up to 80% of the viability was lost during the drying process. The development of a frozen liquid formulation, *Provide*®, solved the problem and raised the number of hectares treated to 126,000 per batch. The batch-to-batch variability of the production process was large. Attempts to adjust the titre by changing the conditions in response to a low titre in one batch resulted in an overall decrease in average yield in the fermentor and an increase in variability. Implementation of statistical process control procedures that monitored yields but did not change conditions until detailed laboratory research showed it was warranted, resulted in an increase in yield (increase in treated ha to 235,000 per batch) and a

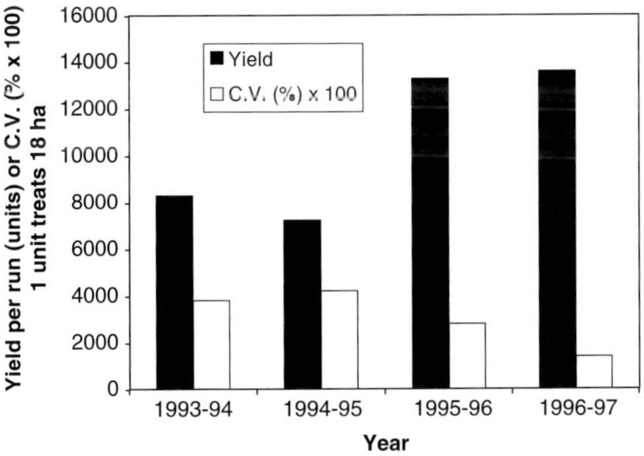

Year	Yield	C.V. (%) x 100
1993-94	8290	3800
1994-95	7222	4200
1995-96	13268	2800
1996-97	13583	1400

Figure 1. Improvements in production after adoption of Statistical Process control practices, in 1994–95.

decrease in variability (Figure 1, Philom Bios unpublished data, 1999). The stability of a formulation on seed is also an important criterion when assessing the acceptability of a formulation. An improvement in the formulation allowed for the introduction of a room temperature stable powder, *JumpStart*®. The move from the *Provide*® frozen liquid formulation to the dry room temperature stable *JumpStart*®, increased the half-life of the fungus on seed from 10 to 35 days.

On-going research

The search for improved formulations is ongoing. The current formulation has a half-life of 4 weeks at 28 °C. In the Canadian prairies the product will rarely be subjected to long periods above 25 °C so this is an acceptable shelf-life. As we move into warmer climates, increased stability at higher temperatures is a commercial imperative.

Quality assurance

Once an inoculant is developed for any crop, or area, there must be strict adherence to quality standards (Hedge et al., 1999). Substandard materials restrict the popularity and acceptability of the product (Hedge et al., 1999).

Pre-commercialization

Before commercialization, the *P. bilaiae* spores had been produced and applied with the rate based on the amount of dry material added per meter of furrow (Kucey, 1983). This did not allow for the development of a quality control procedure as the quality and the amount of fungus in a gram of substrate varied from batch to batch.

Commercialization

The first step in the development of a quality assurance program was to set parameters for the product. In Canada, rhizobium products are regulated by the Fertilizers Act and the number of bacteria per seed is pre-determined based on seed size (Olsen et al., 1994). As the phosphate inoculant was a new product, Philom Bios assisted

Ottawa in establishing a standard (the minimum number of spores per seed required for efficacy) by submitting field data. Once this level was determined, a quality assurance system could be developed. As the quality (cfu per g) varies from batch to batch, each batch must be evaluated separately. Enough samples must be taken from each batch to ensure that a statistically valid number could be obtained. We chose a dilution plating method and cfu count as the basis for our quality assurance system because a large number of samples could be assayed without expensive analytical tools. This assay is, however, a tedious process that becomes more cumbersome as production volume increases. The number of agar plates for the quality assurance program for the phosphate inoculant at Philom Bios has increased from 3000 in 1992 to 46,000 in 2002.

On-going

The increase in production is increasing our need to develop a fast reliable method to assess the number of viable spores in our product to replace the cfu plate assay.

Application

The inoculant use must conform to standard application practices used on-farm. An inoculant can be applied either as a seed treatment or as in-furrow application. Seed inoculation is the most commonly used method of inoculation. However, seed-applied pesticides are also commonly used and many seed treatment chemicals contain fungicides that reduce the survival of *P. bilaiae* on seed. We therefore must be concerned about the ability of the inoculant organism to survive on fungicide treated seed long enough to be effective in the field.

Pre-commercialization

Fungicide use on wheat was not universally practiced in the 1980's. All of the early trials were conducted with untreated seed.

Commercialization

The use of seed treatment fungicides increased in the 1990s. Farmers needed these materials to protect their crop from increased disease pressure and could not omit the fungicide in order to use the phosphate inoculant. We therefore developed a system to test the compatibility of *P. bilaiae* with commonly used seed applied chemicals. The chemical and the inoculant are applied to seed and the population of the *P. bilaiae* on the seed is determined. The seed is stored at room temperature and the population of *P. bilaiae* is monitored for up to 4 weeks. A regression line is determined using SigmaPlot software. The intersection of this regression line with the minimum number of organisms required on seed at planting determines the planting window (the minimum time allowed between inoculating and seeding for that seed treatment–inoculant combination). The application methods used to apply the two materials mimic the application methods a farmer would use. The fungicide and the inoculant may be mixed together in a slurry (tank-mix), applied to seed at the same time through separate hoses (simultaneous), or the chemical may be applied to the seed first and allowed to dry before the inoculant is added (sequential). Although the tank-mix is usually the most damaging to the fungus, farmers prefer this method, as it is quick and easy (Table 2).

Generally, the planting window is the longest when the two materials are applied sequentially (Table 2). Each chemical formulation must be analyzed separately as it is often the formulation ingredients, rather than the active ingredient that affects the fungus. A change in the formulation of either the chemical or the inoculant will alter the planting window.

Table 2. Planting windows for use of *JumpStart*® with commonly used fungicides on wheat

Seed treatment*	Planting window (days)		
	TankMix	Simultaneous	Sequential
Bare Seed	15	15	15
Baytan® 30	10	10	10
DB Green	Do not use	2	7
Proseed®	Do not use	10	10
Vitavax® Single	Do not use	10	10
Vitaflo® 280	Do not use	1	4

Baytan 30, Vitaflo 280 and Vitavax Single Solution, are registered trademarks of Uniroyal Chemical Ltd.
DB Green is a registered trademark of Agsco Ltd.
Proseed is a registered trademark of Zeneca Agro.

P. bilaiae must also remain alive throughout any application process. Some air-seeders are equipped with tanks that allow seed to be inoculated as it is sown. It can take 6–8 h to empty one tank. We know that the fungus population in slurry does not drop ($P < 0.05$) in 8 h so we are confident that the number of viable spores is still adequate for effective inoculation during the entire planting operation. We tested each inoculant formulation in a range of commercial air-seeders to ensure that the material would not clog the hoses or screens or lose viability during application.

On-going

We must continue to test our materials with new chemicals or formulations. Our current compatibility tests look at loss in viability on seed due to chemicals but do not look directly at efficacy. We plan to add a greenhouse or field-screening component to these tests. Seed coating companies are constantly looking at polymers to improve seed flow in seeders, protect rhizobial inoculants from environmental stresses, and manipulate seed germination. We constantly evaluate these materials to determine if they will reduce or increase the survival of the *P. bilaiae* on seed.

Registration

Inoculants in Canada are registered as fertilizer supplements under the Fertilizers Act administered by the Canadian Food Inspection Agency (CFIA). This process requires proof that the organism is safe and that it will perform according to the claim on the label. All new inoculants must go through this process and any claim on the label (including pesticide compatibility) must be reviewed by the CFIA.

Pre-commercialization

The product could not be marketed and sold until it was registered. The product was first registered for use on wheat in 1990 under the name *PB50*®.

Commercialization

Every new crop must be registered and so field and compatibility data was submitted and reviewed before canola, pea lentil and alfalfa were added to the label in 1992, 1993 and 1996, respectively. The amount of data Philom Bios submits to the CFIA is increasing rapidly as new crops, and chemical compatibility information are added to the label and new formulations are developed. This is beginning to present a problem as the large volume of material that has to be reviewed creates a backlog in the system and delays the introduction of new applications. The market introduction of a granular formulation was delayed by one year due to delays in the registration system.

On-going

Philom Bios continues to work with the CFIA to try to streamline the registration process. We ensure that they approve of the format of our reports and the statistical analysis we use before we send in large submissions.

Mode of action and behavior

The more we understand about the mode of action and behavior of the *P. bilaiae* the more we are able to manipulate it so we can maximize the effectiveness of the inoculant.

Asea et al. (1988) used the ^{32}P-dilution method and found that greenhouse-grown wheat inoculated with *P. bilaiae* obtained 18% of its P from sources unavailable to non-inoculated plants. Often this work requires equipment and areas of expertise that are not available at Philom Bios. We therefore collaborated with university and government scientists to provide this information.

Pre-commercialization

Philom Bios, DEC, and SWP conducted field studies to gather data to support registration and monitored P uptake as well as yield (Gleddie et al., 1991) to show that the fungus increased the phosphate nutrition of the plants. This information was backed up by greenhouse experiments with ^{32}P using wheat, and flax conducted by Chambers and Yeomans (1990) of the University of Manitoba. They found that plants inoculated with *P. bilaiae* increased tissue P concentration, primarily through increased soil P contributions (as opposed to fertilizer P).

Commercialization

Researchers continue to discover that the effects of our phosphate solubilizing inoculant are more complex than a simple solubilization of P. Recent work at the University of Manitoba has shown that root growth (Vessey and Heisinger, 2001) and root hair development are increased by *P. bilaiae* (Gulden and Vessey 2000). *P.bilaiae* may increase the absorptive capacity of roots, which may lead to increased P, other nutrients, or even water uptake (Vessey and Heisinger, 2001).

We need to study the ecology of the fungus *in situ* if we are to fully understand the behavior and the limitations of the inoculant. A polymerase chain reaction (PCR) assay developed by O'Gorman et al. (1998) demonstrated that *P. bilaiae* was able to effectively colonize the roots of six plant species in non-sterile soil.

On-going

We will continue exploration of factors that affect the ability of *P. bilaiae* to colonize roots, solubilize P, and increase yield. We will continue to access expertise at university and government labs to help us to develop and use techniques that will help with these investigations.

Marketing

It is extremely important to have a coordinated marketing plan with clear descriptions of the benefits of the inoculant.

Pre-commercialization

This was especially important for our phosphate-inoculant, as it was a new concept in the marketplace. This concept had to be clearly linked to the research plan. Research results were continually used to clarify and refine the benefit statement and, just as importantly, to define the limitations of the use of the inoculant. Growers had to be educated on the value of phosphate to crop growth and on the value of inoculants. We had limited internal resources and marketing expertise, so a marketing partnership with DowElanco Canada was particularly beneficial.

Commercialization

The benefit statement has been constantly refined as we discover more about *P. bilaiae*. As we learned more about the inoculant and the need to fully support the product with timely information, we discovered we needed to be closer to our customers. Each farmer is different, and his or her needs must be addressed separately. In 1996 we assumed direct responsibility for marketing our products, and developed a one-to-one marketing strategy. This strategy allows us to provide individual customers with specific information about the product, its benefits, and how to use it to optimum value. At the same time this approach ensures a rapid feedback system. Questions or concerns of farmers that require more research can therefore be incorporated into the research plan. A thorough response to customer concerns and questions is crucial to the successful commercialization of an inoculant.

It is also important to continually demonstrate the value of the inoculant. Every year Philom Bios puts in Inoculant Performance Trials (IPT). These are large-scale (5–50 ha) on-farm trials that directly compare inoculated and non-inoculated portions of fields. Over 300 trials have been done in the last 10 years demonstrating an average yield increase of 7% in crop yield due to inoculation. The success of this approach can be seen in market surveys conducted every year. In 1990, most farmers had not heard of a phosphate inoculant. By 2001, *JumpStart®* was the most recognized inoculant product across the Canadian Prairies with 72% of farmers aware of its use.

Financing

The commercialization of a new inoculant is not a trivial matter and as such it requires significant funding. This need for research funding does not end with the commercialization of the product. It must continue, as there is always a requirement for product support and improvement. The discovery and initial testing of *P. bilaiae* took about 6 person years. The pre-commercialization stage required 32 person years, and since then, Philom Bios has invested over 120 person years to continue to improve the inoculant by adding new crops, new formulations, and continual support to the use of *JumpStart*®. These figures do not include the time and money used by external researchers. We are fortunate to have received assistance in funding some of the work from AAFC Matching Investment Initiatives (MII) and NRC Industrial Research Assistance Program (IRAP) grants.

Pre-commercialization

During this phase Philom Bios did not have any products or revenue and financing had to be sought based on a clear description of the potential value of the inoculant to the Western Canadian market. The value of the product to growers had to be clearly understood. This was made more difficult as the value of phosphate itself was poorly understood by Prairie farmers. Inoculants were not recognized as important crop inputs, and new biological technologies were perceived as "snake-oil". Financing was obtained, however, and development advanced.

Commercialization

There is a constant need for ongoing research investment to drive market expansion to new crops and new regions, to develop new formulations, and to assess new equipment and chemical seed treatments as they become available to farmers.

On-going

Over the past 10 years we have built a successful business by maintaining a tight strategic focus and practicing a responsive business model. We focus solely on developing, manufacturing, and marketing high-end inoculants, and we believe we do this better than anyone else. The business model is a seamless link from Research to Manufacturing to Marketing, overlaid with Corporate systems. Superior R&D generates product improvements and competitive advantages, and process improvements which enable Manufacturing to achieve cost leadership with rewarding gross margins. Disciplined Marketing has grown the business consistently over the past 5 years, steadily increased market share, and built the highest brand strength of all inoculants in our marketplace. It is not enough to have superior R&D; it must be combined with high market share and economies of scale to deliver better products to establish sustained market leadership. Market share only matters if it generates high returns through cost leadership in manufacturing.

This focused business approach is creating value for our customers and our shareholders. We have been financially self-sustaining since 1991. The Company is essentially debt-free, is demonstrating appealing returns on capital, is profitable, and boasts positive retained earnings. We intend to continue to increase delivery of high-end inoculants to more farmers in an expanding market in the years ahead.

Conclusion

The development of a commercial phosphate inoculant has been a challenging and rewarding process. The procedures had to be developed (and mistakes made) as we went as there was no "users manual" to lead us through the process. The challenges do not end with the commercialization process but continue to arise as we improve our product and processes, keep pace with new developments in agriculture and expand our market.

References

Asea P E A, Kucey R M N and Stewart J W B 1988 Inorganic phosphate solubilization by two *Penicillium* species in solution culture. Soil Biol. Biochem. 20, 450–464.

Chambers J W and Yeomans J C 1990 The influence of PB-50 (*Penicillium bilaji* inoculant) on yield and phosphorus uptake by wheat. Proc. Ann. Manitoba Soc. Soil Sci. Mtg. 33, 283–293.

Daigle D J and Connick W J Jr 1990 Formulation and application technology for microbial weed control. *In* Microbes and Microbial Products as Herbicides. Ed. E Hoagland. pp. 288–304. American Chemical Society, Washington.

Gleddie S C, Hnatowich G L and Polonenko D R 1991 A summary of spring wheat response to *Penicillium bilaji*, a phosphate inoculant. Proceedings for the Western Phosphate/Sulfur Workshop Conference, Colorado State University, Fort Collins, Colorado, March 21–22, 1991.

Gulden R H and Vessey J K 2000 *Penicillium bilaii* inoculation increases root hair production in field pea. Can. J. Plant Sci. 80, 801–804.

Hedge D M, Dwived B S and Sudkahara Babu S N 1999 Biofertilizers for cereal production in India –A review. Can. J. Agric. Sci. 69, 73–83.

Hnatowich G L, Gleddie S C and Polonenko D R 1990 Wheat responses to PB-50 (*Penicillium bilaji*), a phosphate-inoculant. Proceedings for the 1990 Great Plains Soil Fertility Conference, Denver, Colorado, March 6–7.

Kucey R M N 1983 Phosphate-solubilizing bacteria and fungi in various cultivated and virgin Alberta soils. Can. J. Soil Sci. 63, 671–678.

Kucey R M N 1987 Increased phosphorus uptake by wheat and field bean inoculated with phosphorus-solubilizing *Penicillium bilaji* strain and with vesicular arbuscular mycorrhizal fungi. Appl. Environ. Microbiol. 53, 2699–2703.

Kucey R M N 1988 Effect of *P. bilaji* on the solubility and uptake of P and micronutrients form soil by wheat. Can. J. Soil Sci. 68, 261–270.

Maurice S, Beauclair P, Giraud J-J, Sommer G, Hartmann A and Catroux G 2001 Survival and change in physiological state of *Bradyrhizobium japonicum* in soybean (*Glycine max* L. Merril) liquid inoculants after long-term storage. World J. Microbiol. Biotechnol. 17, 635–643.

O'Gorman D, Leggett M and Lévesque C A 1998 Molecular approach to monitor a beneficial *Penicillium* species. Can. J. Plant Pathol. 20, 212.

Olsen P E, Rice W A, Bordeleau L M and Biederbeck V O 1994 Analysis and regulation of legume inoculants in Canada; the need for an increase in standards. Plant Soil 161, 127–134.

Vessey J K and Heisinger K G 2001 Effect of *Penicillium bilaii* inoculation and phosphorous fertilization on root and shoot parameters of field grown pea. Can. J. Plant Sci. 81, 361–366.

The use of ^{32}P isotopic dilution techniques to evaluate the interactive effects of phosphate-solubilizing bacteria and mycorrhizal fungi at increasing plant P availability

J.M. Barea[1], M. Toro & R. Azcón

Departamento de Microbiología y Sistemas Simbióticos, Estación Experimental del Zaidín. CSIC, Profesor Albareda 1, 18008, Granada, Spain. [1]*Corresponding author**

Received 12 December 2002. Accepted in revised form 2 January 2003

Key words: ^{32}P, arbuscular mycorrhiza, isotopic dilution, phosphate-solubilizing rhizobacteria, rock phosphate

Abstract

Isotopic (^{32}P) dilution approaches have been used to evaluate the extent at which inoculated phosphate-solubilizing bacteria (PSB) and arbuscular mycorrhizal (AM) fungi improve plant use of soil P sources of low bioavailability, either endogenous or added as rock phosphate (RP). This paper firstly examines the conceptual background, the main achievements and the state of the art on the related topics. Then, a model own experiment is described and discussed to offer a comprehensive view on the effects and mechanisms involved, and to propose the appropriate methodological approaches. Measurements of the specific activity ($^{32}P/^{31}P$) of P in plants grown in ^{32}P-labelled soil, and the subsequent calculations of the amount of plant P derived from either the bio-available (isotopically labelled) soil sources or from the added RP, allow to conclude that the dually (AM + PSB)-inoculated plants were able to use otherwise unavailable P sources, resulting in an improvement of plant P acquisition. The proposed mechanism is that the inoculated PSB actually release phosphate ions (^{31}P) from sparingly soluble phosphates, ions which are taken up by the external AM mycelium to be transferred to the plant.

Introduction

Soil microbial populations are immersed in a framework of interactions able to affect plant developments, being accepted that certain beneficial microbial activities can be exploited, as a low-input biotechnology, with regard to sustainability issues (Kennedy and Smith, 1995). Particularly relevant to these concerns are the microbiologically mediated processes involved in nutrient cycling, as those responsible for increasing the phosphate availability in soil (Kucey et al., 1989; Richardson, 2001). Both saprophytes and mutualistic symbionts are involved in microbial reactions to improve P acquisition by plant (Barea et al., 2002a). The saprophytes include both bacteria and fungi able to solubilize sparingly soluble phosphates in soil (Kucey et al., 1989; Whitelaw, 2000). The mutualistic symbionts, refer to arbuscular mycorrhizal (AM) fungi (Barea et al., 1997). These fungi, upon root colonization, develop an external mycelium which is a bridge connecting the root with the surrounding soil microhabitats to explore phosphate ions from soil solution, to transfer them to the plant (Smith and Read, 1997). In addition, AM fungi have been proposed to mobilize sparingly soluble inorganic phosphate in soil (Yao et al., 2001). All in all, and whatever the mechanisms involved, it is accepted that, by linking the biotic and geochemical portions of the ecosystem, the AM fungi can contribute to P capture and supply,

* E-mail: josemiguel.barea@eez.csic.es

E. Velázquez and C. Rodríguez-Barrueco (eds.),
First International Meeting on Microbial Phosphate Solubilization, 223–227

thereby affecting P cycling rates and patterns in both agro-ecosystem and natural environments (Jeffries and Barea, 2001).

Phosphate solubilizing bacteria (PSB) inoculants have been assayed but their effectiveness in the soil-plant system is still unclear (Catroux, 2002; Igual and Rodriguez-Barrueco, 2002). First of all, the inoculated PSB must establish in the root-associate soils habitats. This is why it has been recommended to select the inoculant PSB from the subset of plant-growth-promoting rhizobacteria (PGPR) populations (Glick, 1995). In addition, the role of the inoculated PSB, as supplying P to the plant, seems limited because the transient nature of the compounds released by PSB responsible for phosphate solubilization, and because the possible re-fixation of phosphate ions on their way to the root surface, if any solubilization does take place (Kucey et al., 1989). However, it was proposed that if the phosphate ions, as released by the inoculated PSB, these ions would be taken up by a tailored AM mycelium, resulting in a synergistic microbial (mycorrhizosphere) interaction to improve P acquisition by the plant (Barea et al., 1983). The feasibility of this hypothesis has been investigated in several studies which included the application of less expensive, but poorly reactive in non-acidic soils, rock phosphate (RP) as a P source (Barea et al., 1997).

Since radioactive P (^{32}P) can be used for evaluating the exchange rates governing phosphate equilibrium between the soil solution and the solid phases of the soil (Fardeau, 1993), ^{32}P-based techniques were recommended to measure P availability in RP materials (Zapata and Axmann, 1995). Labelling of the so-called "isotopically exchangeable soil P" is carried out with phosphate ions labelled with ^{32}P, being assumed that all 'labile' P, and only this fraction, attain isotopic exchange within a short-term experimental period (Fardeau, 1993). The isotopic composition, or "specific activity", i.e. the $^{32}P/^{31}P$ ratio, is then determined in the plant tissues, as shown in Figure 1.

The specific activity (SA) in plants growing in ^{32}P-labelled soils is the basis for calculations to ascertain the P sources that a plant is actually using (Zapata and Axmann, 1995), therefore, it was suggested that it could be used to ascertain whether or not AM and non-AM plants are using the same P sources (Raj et al., 1981). Bolan (1991) reviewed the topic and concluded that, in

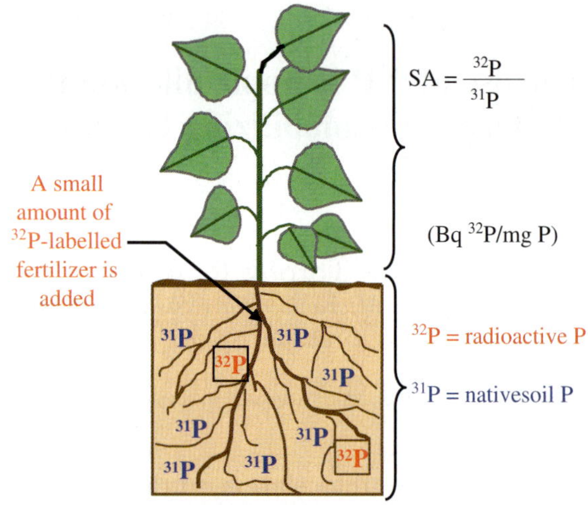

Figure 1. Specific Activity (SA) of P in plant.

most cases, no differences in the SA were found between AM and non-AM plants. This conclusion supports that both AM and non-AM plants were using similarly labelled P from soils, but the possibility that AM-plants can use P forms which are not available to non-AM controls cannot be ruled out (Joner and Jakobsen, 1994). The use of ^{32}P in AM relationships was considered, therefore, an open research topic. Experiments carried out in this Laboratory (Toro et al., 1997, 1998; Barea et al., 2002b) further investigate whether PSB + AM inoculation affect the SA of plants in RP added, ^{32}P-labelled, soils. These studies found that dual (AM + PSB)-inoculation induced a lowering in the SA of the host plant indicating that these plants used extra ^{31}P solubilized from otherwise unavailable P sources, either endogenous or added as RP.

Once the conceptual background, the main achievements and the state of the art of the related topics have been analysed, this paper aims at discussing some selected results from model own experiments to deliberate general conclusions on both effects and mechanisms, and to propose the appropriate methodological approaches to investigate PSB + AM interactions on P capture, cycling and supply.

Material and methods

An experiment from those described by Barea et al. (2002b) was taken as a model to have a basis for discussion. This experiment involved a factorial combination of four microbial treatments [mycorrhiza inoculation (AM); phosphate-solubilizing rhizobacteria inoculation (PSB); the AM + PSB dual inoculation; and a un-inoculated control, but having the naturally existing AM fungi and PSB (Control)], and two chemical treatments [a un-amended control without P application, and rock phosphate (RP) application]. These 8 treatments were replicated five times giving a total of 40 pots (in fact microcosms based on unstelized soil) that were arranged in the greenhouse in a randomised block design. In this assay, alfalfa (*Medicago sativa* L., cultivar Aragón) was the test plant. Seedlings were inoculated with a specific rhizobial strain and transplanted into 1 L pots containing an agricultural soil in which alfalfa had never been grown. The test non-acidic soil, which does not contain active $CaCO_3$, was collected in the province of Granada, Spain. The AM fungus was *Glomus mosseae* (BEG 12) and the phosphate solubilizing rhizobacterium an *Enterobacter* sp. The source of RP was from Riecito (Venezuela) and contained 11.4% total P with 6.64% of neutral ammonium citrate-soluble P (Casanova, 1995). The RP was applied as finely ground material (less than 100 mesh) at a rate of 100 mg of total P per kg soil. Plants were harvested after 55 d of growth (see Barea et al., 2002 for more details).

The isotope dilution technique (Zapata and Axmann, 1995) was used for ^{32}P studies. An aliquot containing 1850 K Bq ^{32}P pot^{-1} was added to obtain sufficient activity in the plant material. To prepare the ^{32}P-labelled carrier solution the total activity required for the experiments was added as ^{32}P carrier-free to a known volume of KH_2PO_4 carrier solution with 10 ppm P. Labelling was done by mixing the soil thoroughly with 10 mL of the solution containing ^{32}P phosphate ions. Seedlings were transplanted one day after soil labelling. The ^{32}P activity in the plant material was measured by liquid scintillation (Packard Tri-Carb 300) counting of the ^{32}P, by the Cerenkov effect. Counts were corrected for isotope decay and counting efficiency (50%). The specific activity of P was then calculated by considering the radioactivity per amount of total P content in the plant and expressed in Bq mg P^{-1} (Zapata and Axmann, 1995), as indicated in Figure 1. The percentage of P in plant derived from the bioavailable (labelled) P fraction (PdfL) was then obtained by using isotope dilution concepts (Zapata and Axmann, 1995), as follows:

$$\%\text{PdfL} = \frac{\text{SA plant in presence of RP}}{\text{SA plant in absence of RP}} \times 100$$

The %Pdf RP is obviously (100−%PdfL).

Data were processed by ANOVA and Duncańs test ($P = 0.05$).

Results and discussion

As Figure 2 shows, both RP addition and microbial inoculation improve biomass production and P accumulation in alfalfa plants, with dual PSB + AM inoculation as the most effective treatment. Whether or not RP was added, AM-inoculated plants showed a lower specific activity ($^{32}P/^{31}P$) than did their comparable non-AM inoculated controls (Figure 2). This contrasts with previous findings (Bolan, 1991) where similar SA values were found for both AM- and non-AM inoculated plants. If the $^{32}P/^{31}P$ ratio in soil solution is uniform both spatially and temporally, this will produce a similar SA in the plant whether AM-inoculated or not. Conversely, in the reported experiments, an because PSB seem to be effective at releasing ^{31}P from sparingly soluble sources, the SA values were lower in AM-inoculated plants than in the corresponding non-AM inoculated controls, whether or not inoculated with PSB (Figure 2). This means that AM plants were taking soil P which is labelled differentially from that taken up by non-AM inoculated controls, suggesting that AM-plants used otherwise unavailable P sources. Since the lowest values of SA were found in dually inoculated AM + PSB plants, the effectiveness of the inoculated PSB in co-operating with the inoculated AM fungi appears evident.

The SA-based calculations (Zapata and Axmann, 1995) allow to evaluate the extent of plant use of RP material, as affected by microbial inoculation. The SA values in plants grown either in presence or absence or RP (Figure 2) were the basis to calculate the proportion of plant P according to its origin (%PdfL *vs* %PdfRP). From these data the total amount of P in plant derived from either the available (labelled) soil fraction (PdfL) or from the added

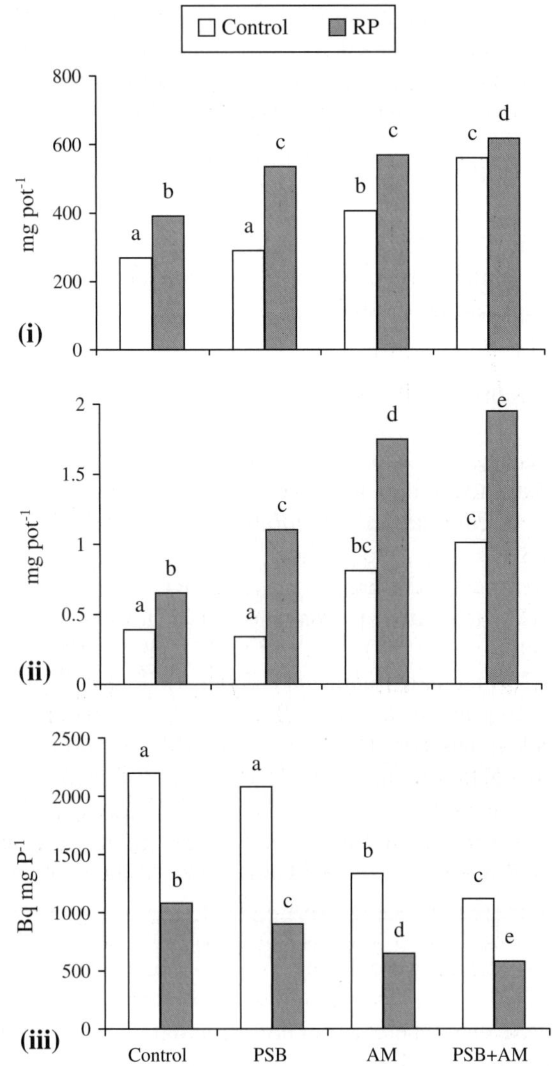

Figure 2. Shoot dry weight (i); shoot P content (ii); and specific activity (iii) of alfalfa plants receiving several microbial inoculation treatments with and without rock phosphate (RP) application. For each response variable, means ($n = 5$) not sharing a letter in common differ significantly ($P = 0.05$) from each other (Duncańs multirange test).

RP (PdfRP) were, in turn, calculated. As Figure 3 shown, plants took P from both of these sources, but clearly, the total P uptake was far higher in AM-plants, particularly in those also inoculated with selected PSB (Figure 3).

To conclude on the effects and mechanisms involved to account for the increased use of sparingly soluble phosphates, like RP, by dually inoculated plants it can be stated that PSB actually release phosphate ions from these low-available P sources. The release of P ions will constitute a part of the total ^{31}P pool from which the AM mycelium taps phosphate to contribute to plant nutrition. Such microbial activities could results in the lowest SA in dually (PSB + AM)-inoculated plants. Deficiency in active Ca may benefit the solubilization of P ions by PSB from the RP particles in the non-acidic test soil (Khasawneh and Doll, 1978; Rajan et al., 1996).

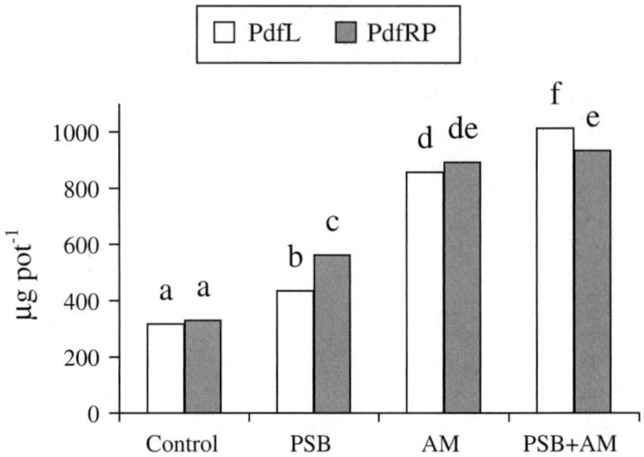

Figure 3. Total amount ($\mu g\ pot^{-1}$) of plant P derived from the “bioavailable” (labelled) P (PdfL) or from rock phosphate (PdfRP) in alfalfa plants receiving several microbial inoculation treatments and rock phosphate (RP). For each response variable, means ($n = 5$) not sharing a letter in common differ significantly ($P = 0.05$) from each other (Duncańs multirange test).

Acknowledgements

The authors thank the Joint FAO/IAEA Division. United Nations, Vienna (Phosphate CRP), particularly Dr. F. Zapata for his comments and advice, and Mr N. Algaba (EEZ) for technical help in isotope application. This study was supported by the ECO-SAFE (QLRT-1999-31759 project) UE.

References

Barea J M, Azcón R and Azcón-Aguilar C 1983 Interactions between phosphate solubilizing bacteria and VA mycorrhiza to improve plant utilization of rock phosphate in non acidic soils. *In* 3rd International Congress on Phosphorus Compounds. pp. 127–144. Brussels, October 4–6.

Barea J M, Azcón R and Azcón-Aguilar C 2002a Mycorrhizosphere interactions to improve plant fitness and soil quality. Antonie van Leeuwenkoek 81, 343–351.

Barea J M, Azcón-Aguilar C and Azcón R 1997 Interactions between mycorrhizal fungi and rhizosphere microorganisms within the context of sustainable soil-plant systems. *In* Multitrophic interactions in terrestrial systems. Ed. A C Gange and V K Brown. pp. 65–77. Blackwell Science, Cambridge.

Barea J M, Toro M, Orozco M O, Campos E and Azcón R 2002b The application of isotopic ^{32}P and ^{15}N-dilution techniques to evaluate the interactive effect of phosphate-solubilizing rhizobacteria, mycorrhizal fungi and *Rhizobium* to improve the agronomic efficiency of rock phosphate for legume crops. Nutr. Cycling Agroecosyst. 63, 35–42.

Bolan N S A 1991 critical review on the role of mycorrhizal fungi in the uptake of phosphorus by plants. Plant Soil 134, 189–207.

Casanova E F 1995 Agronomic evaluation of fertilizers with special reference to natural and modified phosphate rock. Fertil. Res. 41, 211–218.

Catroux G 2002 Problems associated with the selection and practical use of PSB. *In* First International Meeting on Microbial phosphate Solubilization. pp. 34. Universidad de Salamanca, IRNA, CSIC, Salamanca July 16–19.

Fardeau J C 1993 Le phosphore assimilable des sols: sa représentation par un modèle fonctionnel á plusieurs compartiments. Agronomie 13, 317–331.

Glick B R 1995 The enhancement of plant growth by free-living bacteria. Can. J. Microbiol. 41, 109–117.

Igual J M and Rodríguez-Barrueco C 2002 Phosphate solubilizing bacteria as inoculants for agriculture. *In* First International Meeting on Microbial phosphate Solubilization. pp. 32. Universidad de Salamanca, IRNA, CSIC, Salamanca July 16–19.

Jeffries P and Barea J M 2001 Arbuscular Mycorrhiza – a key component of sustainable plant-soil ecosystems. *In* The Mycota. Vol. IX. Fungal Associations. Ed. B Hock. pp. 95–113. Springer-Verlag, Berlin, Heidelberg.

Joner E J and Jakobsen I 1994 Contribution by two arbuscular mycorrhizal fungi to P uptake by cucumber (*Cucumis sativus* L.) from ^{32}P-labelled organic matter during mineralization in soil. Plant Soil 163, 203–209.

Kennedy A C and Smith K L 1995 Soil microbial diversity and the sustainability of agriculture soils. Plant Soil 170, 75–86.

Khasawneh FE and Doll EC 1978 The use of phosphate rock for direct application to soils. *In* Advances in Agronomy, vol 30. Ed. N C Brady. pp. 159–206. Academic Press, New York.

Kucey RMN, Janzen HH and Leggett ME 1989 Microbiologically mediated increases in plant-available phosphorus. *In* Advances in Agronomy, vol 42. Ed. N C Brady. pp. 199–228. Academic Press, New York.

Raj J, Bagyaraj D J and Manjunath A 1981 Influence of soil inoculation with vesicular-arbuscular mycorrhiza and a phosphate-dissolving bacterium on plant growth and ^{32}P-uptake. Soil Biol. Biochem. 13, 105–108.

Rajan SSS, Watkinson JH and Sinclair AG 1996 Phosphate rocks for direct application to soils. *In* Advances in Agronomy, vol 57. Ed. D L Spaks. pp. 78–159. Academic Press, New York.

Richardson A E 2001 Prospects for using soil microorganisms to improve the acquisition of phosphorus by plants. Aust. J. Plant Physiol. 28, 897–906.

Smith S E and Read D J 1997 Mycorrhizal Symbiosis. Academic Press, S. Diego 605 pp.

Toro M, Azcón R and Barea J M 1997 Improvement of arbuscular mycorrhizal development by inoculation with phosphate-solubilizing rhizobacteria to improve rock phosphate bioavailability (^{32}P) and nutrient cycling. Appl. Environm. Microbiol. 63, 4408–4412.

Toro M, Azcón R and Barea J M 1998 The use of isotopic dilution techniques to evaluate the interactive effects of *Rhizobium* genotype mycorrhizal fungi phosphate-solubizing rhizobacteria and rock phosphate on nitrogen and phosphorus acquisition by *Medicago sativa*. New Phytol 138, 265–273.

Whitelaw MA 2000 Growth promotion of plants inoculated with phosphate-solubilizing fungi. Adv. Agron. 69, 99–151.

Yao Q, Li X L, Feng G and Christie P 2001 Mobilization of sparingly soluble inorganic phosphates by the external mycelium of an arbuscular mycorrhizal fungus. Plant Soil 230, 279–285.

Zapata F and Axmann H 1995 ^{32}P isotopic techniques for evaluating the agronomic effectiveness of rock phosphate materials. Fertil. Res. 41, 189–195.

Distribution pattern and role of phosphate solubilizing bacteria in the enhancement of fertilizer value of rock phosphate in aquaculture ponds: state-of-the-art

B. B. Jana*

Aquaculture and Applied Limnology Research Unit, Department of Zoology and International Centre of Ecological Engineering, University of Kalyani, Kalyani, 741235, West Bengal, India

Received 23 December 2002. Accepted in revised form 2 January 2003

Key words: aquaculture ponds, fertilizers, phosphate solubilizing bacteria, rock phosphate, water

Abstract

Phosphorus, though required in small quantities, has often been implicated as the most limiting element controlling biological productivity in natural waters. As a result, aquaculture ponds demand for frequent application of phosphate fertilizer for enhanced fish production. It is estimated that about 10% of the fertilizer applied caused increase in soluble phosphate in the water phase, which is absorbed by the phytoplankton within few minutes of fertilizer application, whereas the rest is rapidly precipitated and settled at the bottom and converted into insoluble compounds. Thus, the pond bottom acts as a sink of phosphorus in fertilized ponds, whereas a source of P in unfertilized ponds. Increasingly high cost of chemical phosphate fertilizers has been the main stimulus for searching alternative cheap, effective and dependable source of phosphorus from natural sources for pond fertilization. Rock phosphate is trade name of mineral phosphates, which denotes the product obtained from mining and subsequent metallurgical processing of phosphorus containing ores. India has a vast reserve of 126.90 million tones of rock phosphate. Though the available form of phosphorus obtainable from rock phosphate is very little, it contains essential nutrients like calcium, magnesium, zinc, molybdenum, silica, organic carbon and potash, which are useful in biological production. It has proved to be an important phosphate fertilizer for agriculture soils under acidic conditions. A major problem encountered in the direct application of rock phosphate in fish ponds is that it is sparingly soluble in water. The association of tricalcium phosphate and calcium fluoride forming a mineral fluorapatite had made it more resistant to weathering. In the biogeochemical cycle of phosphorus, a mixed population of microbes is essential to promote enzymatic degradation of naturally occurring organic phosphorus compounds. Extracellular products of the microbial community such as enzymes and chelating agents (organic acids) have substantial effect, respectively on phosphorus mobilization from organic P esters and inorganic salts. Phosphatases are stated to promote the degradation of complex phosphorus compounds into orthophosphate, which can be readily utilized by phytoplankton. Alkaline phosphatase can catalize the liberation of orthophosphate from organic P compounds and inorganic pyrophosphate and tripolyphosphate. Synthesis of external alkaline phosphatases is often repressed by high concentrations of phosphate and depressed at low phosphate concentration. This enables to use phosphatase activity as a good indicator of the degree of nutrient regeneration in surface sediments. Solubilization of insoluble inorganic phosphate by bacteria is of considerable importance in the anthropogenically-managed system. A large number of phosphate solubilizing microorganisms such as bacteria, fungi, and cyanobacteria occurring in water and sediments of fish ponds are capable of assimilating insoluble inorganic phosphate

* FAX No: +091-33-2582-8282.
E-mails: bbj@cal2.vsnl.net.in; bbj_icee@yahoo.com

E. Velázquez and C. Rodríguez-Barrueco (eds.),
First International Meeting on Microbial Phosphate Solubilization, 229–238

like hydroxyapatite, tricalcium phosphate, and rock phosphate and make a large portion soluble by the production of organic and inorganic acids. It is suggested that solubilization of rock phosphate during the process of its composting with organic substances is accelerated by the liberation of organic acids in the first step and proliferation of phosphate solubilizing microorganisms at a later stage. It is reported that amongst the microbes, *Bacillus* is the most effective one in respect of phosphate solubilization and phosphates production. Invertebrates can contribute to the mineralization of dissolved and particulate compounds in the sediment and their burrowing activity can affect several exchange processes by increasing the mixing of the sediment surface. A number of bottom grazing fishes are also able to increase the fertilizer value of rock phosphate through their bioturbation effects in bottom sediments. This has been clearly demonstrated in experimental studies. Exogenous introduction of phosphate solubilizing bacteria is of considerable interest in solubilization of rock phosphate in fish culture ponds because of extremely low natural microbial solubilization of rock phosphate in fish ponds. Exogenous introduction of phosphate solubilizing bacteria with compost resulted in the highest concentrations of different species of phosphate in water or sediments among all treatments. This was attributable to the combined effects of the phosphate solubilizing bacteria population of both exogenous and compost origin with short generation time. The present paper reviews the state-of-the-art of various approaches for using rock phosphate as direct source of phosphate fertilizer and the dynamics of phosphate solubilizing bacteria and bacteria induced solubilization of rock phosphate in the fish culture ponds of India.

Introduction

Phosphorus was discovered by Henning Brandt in Germany about 333 years ago and the name has been derived from the Greek word (P*hos* = light and *Phorus* = bringing) meaning light production. Being important constituent of all nucleic acids and a backbone of the Kreb's cycle, phosphorus is the most important single element regulating the biological productivity in aquatic environments though nitrogen and carbon have been appreciated as important fertilizer (Boyd, 1990). Further, phosphorus is not exchangeable with other elements in biological systems (Pierrou, 1976; Arson et al., 1984).

Biogeochemical cycle of phosphorus

In typical biogeochemical cycle, phosphorus is transferred to consumers and decomposers as organic phosphate and subsequently converted into inorganic phosphate by phosphatising bacteria functioning through enzymatic processes. Particulate organic P occurs in suspension in living and dead protoplasm and is insoluble, whereas the dissolved organic P is derived from the particulate matter by excretion and decomposition. Bacteria have a profound role in the transformation. Microbial biomass contributed in the sediment as humified P (Figure 1), which may again contributory to the overlying water. An important phenomenon in the cycle of phosphorus is that deposit of rock phosphate and guano deposits cyclic pool either through weathering/erosion or though the action of chelating agents i.e. organic acid secretion by microbial community and solubilization thereafter. Excretion of grazing animals also contributed significantly to the inorganic P, which is most important in the cycle. In general, the P cycle is influenced by bacteria, actinomycetes, algae and zooplankton in aquatic environment (Jansson et al., 1981).

Phosphorus occurs in natural waters in different forms: orthophosphates, condensed phosphates (pyro-, meta-, and polyphosphates) and organically bound phosphates. These may occur in the soluble form, in particles of detritus, or in the bodies of aquatic organisms. Condensed phosphates are mainly man made, discharged with domestic and industrial wastes and also generated by all living organisms are unstable in water, and are slowly hydrolysed to the orthophosphate form.

Mobilization of phosphorus

The amount of phosphorus compounds, which are not easily soluble in water are absorbed by the pond soil and converted into insoluble compounds or it may be lost from the circulation through leaching into deep sediment (Boyd, 1995). Of various transport mechanism by which

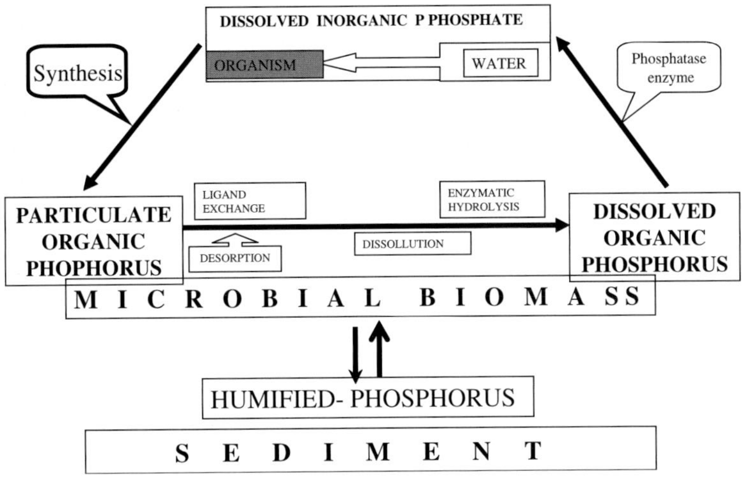

Figure 1. A basic phosphorus cycle depicting the influence of humified phosphorus on microbial biomass.

mobilization of P takes place in water particularly from the sediment to the overlying water (Wetzel, 2001), molecular diffusion is a main process of transport into the overlying anaerobic water. The role played by bioturbation of the biota including fish role in the transport of P from sediment into water has been emphasized.

Physical and chemical mobilization includes desorption, dissolution particularly associated by microbial mediated acidity and ligand exchange mechanism between phosphate and hydroxide ions or organic chelating agents. Microbial mobilization processes include hydrolysis of phosphate–ester bonds. A portion of the phosphorus released in decomposition is absorbed by plants, but the remainder reacts with Fe_3^{+}, Al_3^{+}, Ca_2^{+}, and soil colloids and is fixed in the soil (Boyd, 1995).

Distribution of phosphate solubilizing bacteria

A number of phosphate solubilizing bacteria (*Psedumonas*, *Micrococcus*, *Bacillus*, *Mycobacterium*, *Flavobacterium*, *Penicillium*, *Sclerotium*, *Fusarium* and *Aspergillus*) has been reported to occur in nature (Alexander, 1978). Phosphate solubilizing bacteria are most common in aquatic (Jana and Patel, 1984) and terrestrial environments and influence the P cycling in a number of direct and indirect ways. In aquatic environment, a mixed population of microbes is essential to promote enzymatic degradation of naturally occurring organic phosphorus compounds. Enumeration of Phosphate solubilizing bacteria in water and sediment samples of six fish growing ponds that were managed under different farming systems (data not shown) was highly variable. Culture of a single species of fish (monoculture) resulted in the occurrence of maximum counts in the monoculture system followed by polyculture and traditional systems reflecting the distribution of PSB as direct function of a function of nutrient enrichment of the ponds. Among different causative factors examined, ambient phosphate contributed markedly in the population growth of PSB in water and bottom sediment of the pond (Jana and Patel, 1984).

The distribution pattern of PSB popualtion was also affected by the experimental conditions of the quality of input fertilizers (organic and inorganic) and soil conditions (laterite and alluvial); the counts were maximum in the tank treated with mixed combination of organic and inorganic fertilizers followed by poultry manure, cattle manure and inorganic fertilizer.

It is further evident from the experimental studies of Jana et al. (2001b) that distribution of PSB was highly regulated by the CN and NP ratios of input fertilizers (Figure 2a); the CN ratio of 11.8 (88.6:7.5) and NP ratio of 7.5(7.5:1) of input fertilizers favoured the growth of PSB populations than the remaining ratios employed. The CNP ratio of mixed fertilizer combination used in the CNP ratio of 124:10:1 was comparable with that of 75.8:6:1 and 101:8:1 of the subsequent experiment (Figure 2b). The system efficiency in terms of fertilizer mineralization index in the

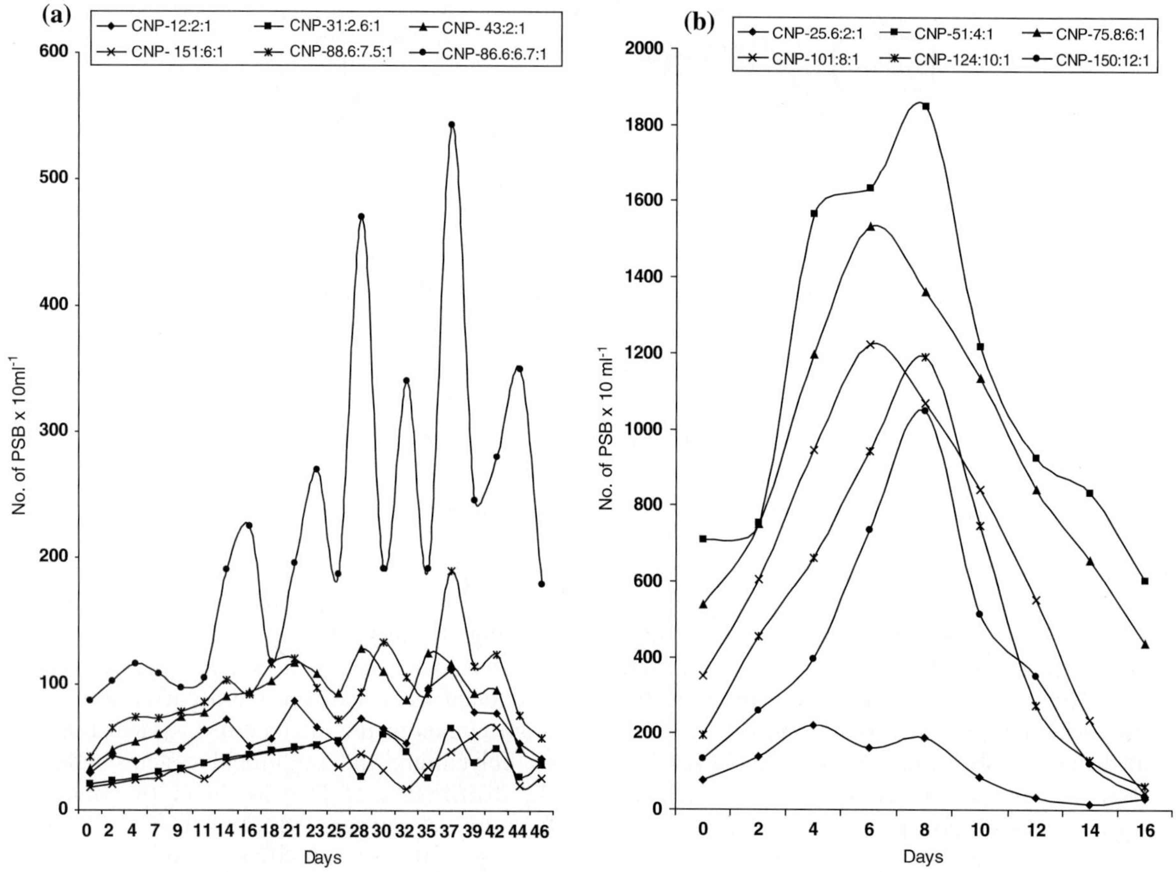

Figure 2. Regulation of the distribution of PSB by the CN and NP ratios of input fertilizers.

former was as high as 46, 52 and 58% for carbon, nitrogen and phosphorus, respectively. The results further revealed that the CN (11.9) and NP (3.34) ratios of ambient water (Figure 3) induced the growth of PSB populations (Jana et al., 2001b).

So far as the temporal response was concerned, the PSB populations responded more to the initial phase of manure application than later, implying that pond fertilization was microbiologically more dynamic in the earlier phase of pond fertilization. The same conclusion can be drawn from the results of the studies carried out in natural ponds treated with qualitatively different fertilizers which exhibited immediate response to first instalment of tank fertilization followed by gradual decline in growth rate.

Among large number of heterotrophic and autotrophic microorganisms, *Bacillus* has been stated to be most effective in the solubilization of insoluble phosphate, such as hydroxyapatite, tricalcium phosphate and rock phosphate. The potential of *Bacillus* in increasing the fertilizer value of rock phosphate has been considerably studied, and was found to be direct function of the abundance of PSB population.

Experimental studies of Sahu and Jana (2000) have shown that exogenous introduction of *Bacillus* as PSB in the compost with rock phosphate consistently resulted in higher mean concentration of phosphate (1.98 mg l^{-1}) followed by rock phosphate in presence of bacteria free compost (1.65 mg l^{-1}), rock phosphate with compost (1.18 mg l^{-1}). The possible mechanism by which rock phosphate has been mineralised is perhaps through organic acid secretion and enzyme activities of the bacteria (Figure 4), as evident from changes in depletion of oxygen, and organic carbon contents of water, lowering of water pH. Extracellular products of the microbial community such as enzymes and chelating agents (organic acids) have substantial effect respectively on phosphorus mobilization from organic P

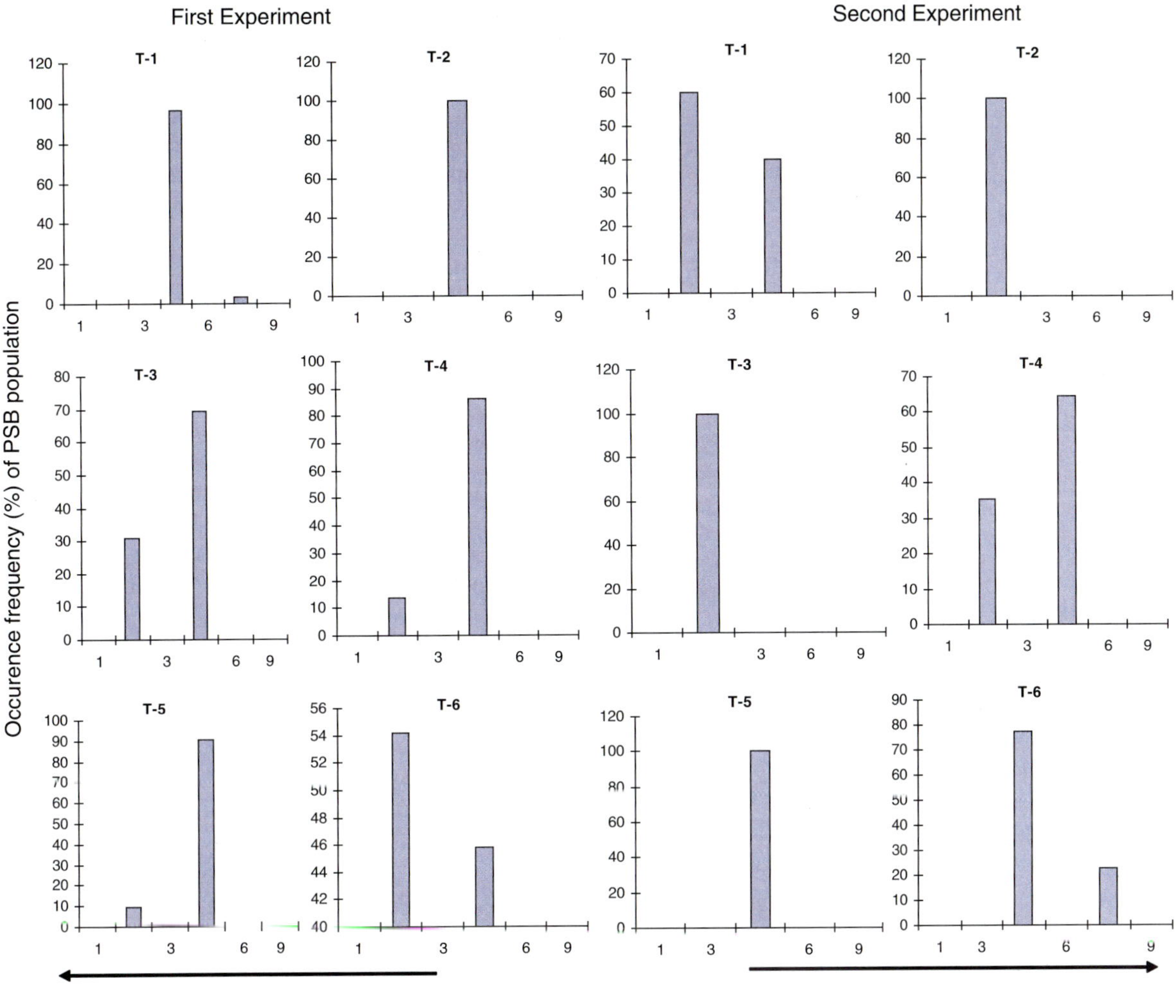

Figure 3 Occurrence frequency of PSB population against NP ratio of water treated with different CNP ratios of input fertilizers.

esters and inorganic salts (Jansson et al., 1988). A major implication of the study was that phosphatase producing bacteria can relate as phosphatic biofertilizer for aquaculture systems, as an economical and eco-friendly approach for induction of phosphorus release from the autochthonous and allochthonous organic sources.

Mechanism of phosphatase-substrate reactions

In the mechanism of enzymatic processes, phosphatases are enzymes, which promote the degradation of naturally occurring complex organic phosphorus compounds into orthophosphate and an organic moiety. The reaction mechanism for the phosphatase-catalysed hydrolysis of organic phosphorus phosphomonoester may occur under four major steps:

1. non-covalent binding of the substrate to the enzyme,
2. release of alcoholic group from the complex and orthophosphate becomes covalently bound to the enzyme forming a phosphoryl enzyme compound,
3. uptake of water by the phosphoryl enzyme compound to form a non-covalent complex and
4. release of orthophosphate and regeneration of free enzyme.

Though not well documented, phosphatases such as phosphomonoesterase apart from organic phosphorus, can also act on inorganic phosphorus

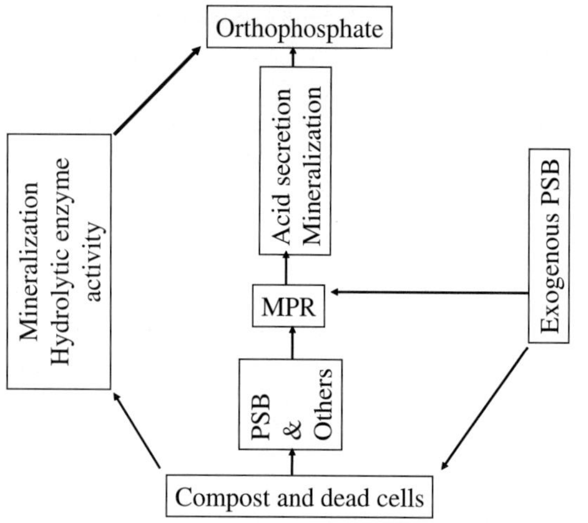

Figure 4. A suggested model for rock phosphate solubilization in the aquatic environment by exogenous introduction of PSB and other bacteria of compost origin.

bound with metal ions and convert into soluble inorganic phosphate.

Phosphatase activity is a good indicator of the degree of nutrient regeneration in surface sediments. Phosphatases have been typically classified into alkaline and acid phosphatases according to their maximum hydrolysing capacity at different pH values. Phosphatases fall into the category of extracellular enzymes, which are secreted and actively pass through the cytoplasmic membrane, and are associated with the producers. Phosphatases released extracellularly by aquatic microorganisms often form complexes with humic compounds that are released from the decomposing tissues of the plants and imported in dissolved or colloial forms to aquatic bodies (Boavida and Wetzel, 1998). Hence it has an essential role to play in the phosphorus dynamics of the aquatic environment.

The term phosphatase is commonly used for the enzymes, which catalyse the hydrolysis of a variety of phosphomonoesters. Alkaline phosphatase permits the biota to hydrolyse dissolved phosphomonoester substrates present in the water column thereby providing an additional source of orthophosphate for biotic assimilation (Berman, 1969; Heath, 1986). It is demonstrated that the properties of phosphatases varied according to zonation and depth (Hadas and Pinkas, 1997) of water body, as the enzymes of photic zone have inductive, whereas the enzymes of profundal zone have constitutive properties in a lake (Chrost et al., 1984).

APase activity is always repressed when subjected to orthophosphate enrichment, whereas the total APase activity (free activity + cell associated or particulate) is not always repressed. External lake water phosphatase usually have pH optima in the alkaline region. Acid phosphatases generally active in the internal cell metabolism. Synthesis of external alkaline phosphatases is often repressed by high concentrations of phosphate and depressed at low phosphate concentration. The concentration of free enzymes in lake water is apparently correlated with the amounts of phytoplankton and bacteria (Richard et al., 1967).

The total counts of viable aerobic heterotrophic phosphatase producing bacteria were in the range of $0.03–3.0 \times 10^{-3}$ cfus ml^{-1} in case of water, and 3.0 to 80.0×10^{3} cfus g^{-1} dry weight of sediments of some tropical aquaculture ponds (Barik et al., 2001). Identification of bacterial isolates shwed 66% Gram positive bacilli, 32% Gram negative bacilli and 2% Gram positive cocci. Gram positive bacilli were found to be dominant in both water and sediment, and were represented by *Bacillus*, and *Corynebacterium*, whereas Gram negative bacilli comprised the genera *Pseudomonas*, *Alcaligens*, *Vibrio*, and *Enterobacter*. The Gram positive cocci were *Micrococcus*, and *Staphylococcus*. *Bacillus* and *Pseudomonas* were the most common genera comprising 64 and 23%, respectively, of the total bacterial isolates (Barik et al., 2001). The predominance of *Bacillus* spp. over the other species reveals the fact that this genus is more important in freshwater ecosystem as well as in the marine environment with respect to phosphatase enzyme (Thompson and MacLeod, 1974).

Fertilizer value of phosphorus

Since phosphorus has often been implicated as the most limiting element in natural ponds, phosphatic fertilizers are extensively used to augment fish production in most of the managed ponds of India. Application of fertilizers stimulates pond productivity largely through autotrophic pathways and also through heterotrophic pathways (Green et al., 1989; Debeljak et al., 1990). Thus the influence of fertilizers on fish production is indirect. Several studies have evaluated the growth performance and production of fish in the ponds fertilized with rock phosphate. Among different forms of phosphate fertilizers, rock phosphate is significant because it is less expensive, but are potentially high in P_2O_5 content (21.2%), carbonate (13%), total-P (8.1%), SiO_2 (6.6%), MgO (5.6%), Fe (4.4%), sulphur (4%), calcium, magnesium, zinc, molybdenum, silica, organic carbon (1.14%) and potash (0.25%) (PPCL, 1987).

Increasingly high cost of chemical phosphate fertilizer led to search for alternative inexpensive, effective and dependable source of phosphorus from natural sources. In India, there is around 165 million metric tons of rock phosphate (Jain and Swaminathan, 1985). Rockphosphate has become an important fertilizer in agriculture specially in acid soils since 1970s (Debnath and Basak, 1984; Mishra and Banger, 1986; Motsara and Datta, 1976), but has not been evaluated adequately in aquaculture ponds.

A major problem encountered in the direct application of rock phosphate to fish ponds is that it is sparingly soluble in water due to the presence of crystalline apatite, and therefore is largely precipitated to the pond bottom. The association of tricalcium phosphate and calcium fluoride, forming a mineral fluroapatite, has made it more resistant to weathering (PPCL, 1987). The solubility of the rock and the receptibility are, however, dependent upon the inherent reactivity of the rock and the receptibility factors of the soil.

Direct application of phosphate rock in ordinary fish ponds with alkaline to neutral pH often results in accumulation in the pond bottom without exhibiting its fertilizer value to the overlying water for stimulation plankton production.

The fertilizer value of rockphosphate has been evaluated by Jana and Das (1992b) by conducting experiment in the outdoor tanks 6 treatment combinations: rockphosphate in low and high doses, single super phosphate (SSP), SSP mixed with rockphosphate, composted rockphosphate and compost of water hyacinth and cow manure. Forty five fry of *Labeo rohita*, *Catla catla*, and *Cirrhinus mrigala* (1:1:1) were introduced into each tank (4.2 m^2) and reared for 12 months. Fertilizers were applied at monthly intervals and no artificial feeding was given. Examination of carp growth revealed maximum weight for all three species in RP treatments (composted rockphosphate, rockphosphate high dose and rockphosphate mixed with SSP) during the major part of the study. The average body weight for all three species of carp was lowest in the compost treatment. Increase in fish yield was related to the orthophosphate level of the water rather than the input of P_2O_5. The net fish yield as well as net primary productivity tended to rise with an increase of concentration of orthophosphate in the water up to 0.33–0.34 mg l^{-1}, but declined with a further rise of orthophosphate to 0.52 mg l^{-1} (Jana and Das, 1992b).

While examining the fertilizer value of rockphosphate in simulated fish ponds as well as in natural fish ponds under three treatment combinations of low dose, high dose and rockphosphate mixed with SSP, Das et al. (1999) observed that net yield of carps after 1 year did not differ much significantly between the high dose and mixed rockphosphate treatments (Figure 5). Two-fold increase in rockphosphate dose resulted in only 38% increased production compared with low dose in fish ponds.

The data on frequency of rock phosphate application is hardly available. Jana and Sahu (1994) have, however, demonstrated that net primary production of water as well as growth of mrigal (*Cirrhinus mrigala*) were distinctly higher (23–76%) in the weekly application of rock phosphate than those for fortnightly and monthly application. Different species of phosphate in water (orthophosphate, total hydrolysable phosphate, total phosphate) and sediment (available phosphate and citrate soluble phosphate), net primary productivity as well as fish growth were maximum for the weekly system, followed by the fortnightly and were lowest in the monthly system (Jana and Sahu, 1994).

Rock phosphate application in fish pond often resulted in its accumulation in pond sediments which gradually increases over time resulting in decline in sediment capacity to absorb more nutrients from the water phase. On the other hand good amount of rock phosphate has been accumulated in the sediment, which would may have important fertilizer value in the aquaculture ponds. Das and Jana (1996) examined the residual effect of rock phosphate by suspending the rock phosphate applications for 1 year and

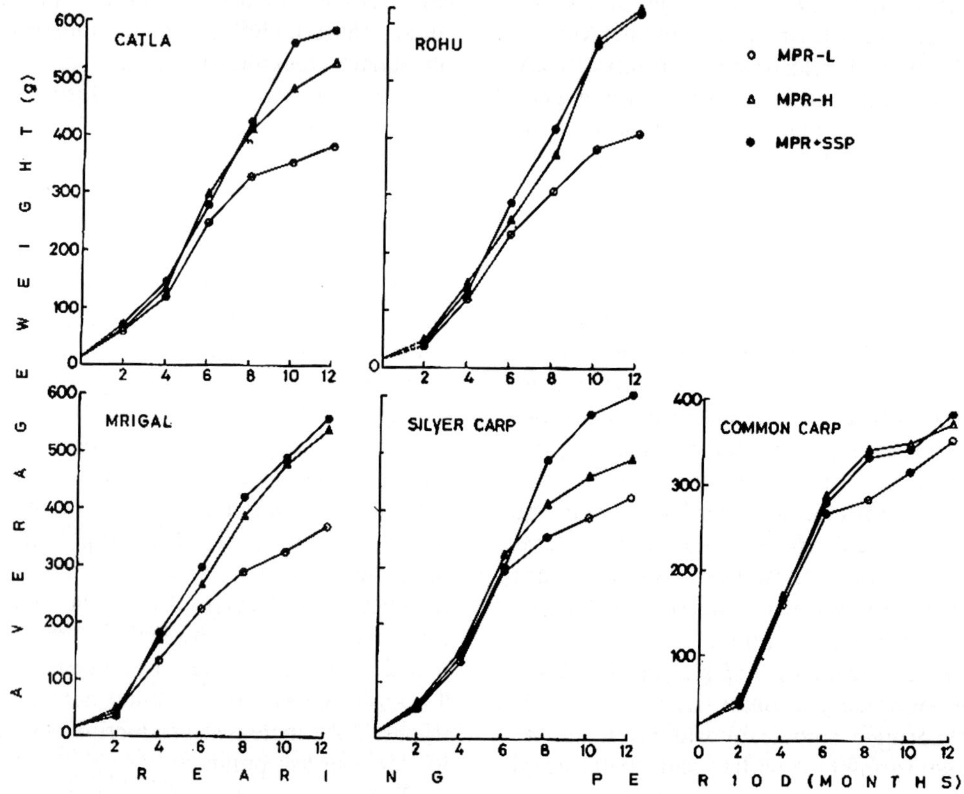

Figure 5. Average weight of different species of test carps in different treatments of fish ponds.

then compared with that of regular rock phosphate applications for 2 consecutive years under 6 different treatments such as direct application of rock phosphate at low (100 kg P_2O_5/ha/month) and high doses (200 kg P_2O_5/ha/month), composted rock phosphate (100 kgP_2O_5/ha/month), single super phosphate (SSP) mixed with rock phosphate (1:1) (50 kg P_2O_5/ha/month), SSP (50 kg P_2O_5/ha/month) and compost (100 kg P_2O_5/ha/month).

The average weight of three species of carp (catla, rohu, mrigal) as well as total production of fish in the residual treatments were only 2.5–34% and 6–23% less compared to their counter parts with continued fertilization in the second year. Among the residual treatments composted RP was the best in maintaining sustained growth and production, whereas, SSP exhibited maximum growth reduction.

The rate of return on investment (Figure 6) tended to increase as the doses of P_2O_5 increase from 50 kg/ha to 100 kg/ha but there was a decline with further rise in doses of P_2O_5 application, suggesting application of 100 kg/ha was profitable in fiscal term.

The bioturbation-induced fertilizer value of rockphosphate by common carp (*Cyprinus carpio*),

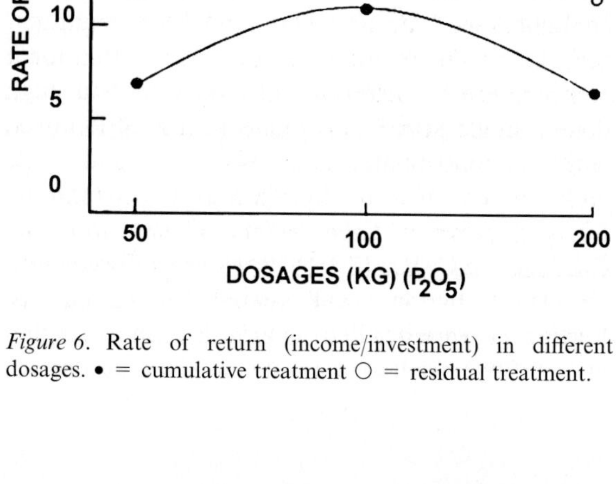

Figure 6. Rate of return (income/investment) in different dosages. • = cumulative treatment ○ = residual treatment.

mrigal (*Cirrhinus mrigala*) and singhi (*Heteropneustes fossilis*) was examined by Jana and Sahu (1993). The bioturbation activity of the common carp was highest, even though its body mass was three times less than mrigal, and about the same as that of singhi. Bioturbation resulted in the amount of citrate-soluble phosphate and available phosphate of bottom soil to be decreased by 15–33%, respectively, whereas the level of orthophosphate of water was increased by 27%. This suggests that citrate-soluble phosphate and/or available phosphate of soil served as a source of phosphate in overlying water. There was a strong correlation between the orthophosphate of water and citrate soluble phosphate or available phosphate of soil.

Further studies revealed the effect of bioturbation of common carp fry in increasing the fertilizer value of rock phosphate. Jana and Das (1992a) conducted in a series of experiments using glass jars containing 3 cm layer of dry soil, phosphate rock at the rate of 3.33 g l^{-1} and common carp fry introduced per jar in the range of 1–12. The level of phosphate coupled with alkaline phosphatase activity tended to rise in a logistic manner with an increase in the number of common carp fry introduced into the system. In any given treatment, alkaline phosphatase activity of the water was directly correlated with phosphate level to a certain extent, beyond which an inverse relationship between them was indicated.

A pertinent issue regarding the distribution of different species of phosphate in three layers (0–2.5 cm, 2.6–5.0 cm and 5.1–7.5 cm) of bottom sediment has been addressed (Sahu and Jana, 1994). The results of study conducted in carp culture tank fertilized with varying levels (43.66 kg and 87.32 kg P) and frequencies 7,15 and 30 days of rock phosphate showed that there were marked layer differences in phosphate and phosphatase activity in any of the treatments employed. Whereas, significant treatment differences were restricted to first and second layer, but not in third layer implying that upper most layer was the most active site of treatment action. The variations of alkaline phosphatases in the first layer of sediment in each treatment was strongly influenced by Al–P (99.78%), available-P (0.16%) and Fe–P (0.05%), whereas the variations in second layer was mainly affected by Al–P (99.02%).

It is concluded that rock phosphate as P fertilizer can be used as a direct phosphate fertilizer in aquaculture ponds.

References

Alexander M 1978 Introduction to Soil Microbiology. Wiley Eastern Limited, New Delhi, India.

Arson O M, Solomatina V D and Romanenko V D 1984 Role of phosphorus in aquous medium in the regulation of bioenergatic processes in fish. Gidrobiol. Zh. 20(1), 53–57.

Barik S K, Purushothaman C S and Mohanty A N 2001 Phosphatase activity with reference to bacteria and phosphorus in tropical freshwater aquaculture pond systems. Aquac. Res. 32, 819–832.

Berman T 1969 Phosphatase release of inorganic phosphorus in lake Kinneret. Nature 224, 1231–1232.

Boavida M J and Wetzel R G 1998 Inhibition of phosphatase activity by dissolved humic substances and hydrolytic reactivation by natural ultraviolet light. Freshwater Biol. 40, 285–293.

Boyd C E 1990 Water Quality in Ponds for Aquaculture. Albama Agricultural Experiment Station, Auburn University, Alabama.

Boyd C 1995 Soil and water quality management in aquaculture ponds. Infosh Fish International. 5/95.

Chrost R J, Siuda W and Halemejko G 1984 Longterm studies on alkaline phosphatase activity (APA) in a lake with fish aquaculture in relation to lake eutrophication and phosphorus cycle. Arch. Hydrobiol. 70, 1–32.

Das S K and Jana B B 1996 Does rockphosphate fertilization adequately sustain fish production in the next year following its application during the first?. J. Inland Fish. Soc. India 28, 67–75.

Das S K, Chakrabarty D and Jana B B 1999 Growth responses of carps to phosphate rock fertilizer in simulated fish ponds and *in situ*. Proc. Zool. Soc. Calcutta 52, 16–28.

Debeljak L, Turk M, Fasaic K and Popovic J 1990 Mineral fertilizers and fish production in carp ponds. Proc. FAO-EIFAC symposium on production enhancement in still water pond culture. Prague, pp 187.

Debnath N C and Basak R K 1984 Effeciencies of some rockphosphates and basic slag as phosphatic fertilizers for rice, wheat and greengram in acid laterite soil of West Bengal. Indian J. Agric. Sci. 54, 722–726.

Green B W, Phelps R P and Alvarenga H R 1989 The effects of manures and chemical fertilizers on the production of *Oreochromis niloticus* in earthen ponds. Aquaculture 76, 37–42.

Hadas O and Pinkas R 1997 Arylsulfatase and alkaline phosphatase (Apase) activity in sediments of lake Kinneret, Israel. Water Air Soil Pollut. 99, 671–679.

Heath R T 1986 Dissolved organic phosphorus compounds: do they satisfy planktonic phosphate demand in summer?. Can. J. Fish. Aquat. Sci. 43, 343–350.

Jain B K and Swaminathan B 1985 Handbook on Fertilizer Technology. In Alexander T M (ed.), The Fertilizer Association of India. New Delhi, India, pp 133.

Jana B B and Patel G N 1984 Spatial and seasonal variations of phosphate solubilizing bacteria in fish ponds of varying fish farming managements. Arch. Hydrobiol. 101, 555–568.

Jana B B and Das S K 1992a Bioturbation induced changes of fertilizer value of phosphate rock in relation to alkaline phosphatase activity. Aquaculture 103, 321–330.

Jana B B and Das S K 1992b The fertilizer value of phosphate rock in carp culture. Israeli J. Aquacult.–Bamidgeh. 44, 13–23.

Jana B B and Sahu S N 1993 Relative performance of three bottom grazing fishes (*Cyprinus carpio*, *Cirrhinus mrigala*, *Heteropneustes fossilis*) in increasing the fertilizer value of phosphate rock. Aquaculture 115, 19–29.

Jana B B and Sahu S N 1994 Effects of frequency of rock phosphate application in carp culture. Aquaculture 122, 313–321.

Jana B B, Chatterjee J, Ganguly S and Jana T 2001a Responses of phosphate solubilizing bacteria to qualitatively different fertilization in simulated and natural ponds. Aquacult. Int. 9, 17–34.

Jana B B, Chakraborty P, Biswas J K and Ganguly S 2001b Biogeochemical cycling bacteria as indices of pond fertilization: importance of CNP ratios of input fertilizers. J.Appl. Microbiol. 90, 733–740.

Jansson M, Olsson H and Broberg O 1981 Characterisation of acid phosphatase in acidified lake Gardsjon. Arch. Hydrobiol. 92, 377–395.

Jansson M, Olsson H and Pettersson K 1988 Phosphatases: origin, characteristics and function in lakes. Hydrobiologia 170, 157–175.

Mishra M M and Banger K C 1986 Proceedings of the National Seminar on Organic Waste Utilization and Vermicomposting. School of life sciences, Sambalpur University, Jyotivihar-768019, India.

Motsara M R and Datta N P 1976 Bull. Indian Soc. Soil Sci. 11, 271.

Pierrou U 1976 The Phosphorus Cycle: Quantitative Aspects and the Role of Man. Institute of Limnology, University of Uppsala.

PPCL 1987 Mussoorie-phos. A Natural Phosphatic Fertilizer for Direct Application, Pyrites, Phosphate and Chemicals Ltd., New Delhi.

Richard W, Overbeck J and Steubing L 1967 Free dissolved enzymes in lake waters. Nature 216, 1345–1347.

Sahu S N and Jana B B 1994 Phosphate and phosphatase distribution in sediment depths of rock phosphate treated carp culture system. Fertilizer Res. 39, 123–131.

Sahu S N and Jana B B 2000 Enhancement of the fertilizer value of rock phosphate engineered through phosphate solubilising bacteria. Ecol. Eng. 15, 27–39.

Thompson L M N and MacLeod R A 1974 Biochemical localisation of alkaline phosphatase in the cell wall of marine pseudomonad. J. Bacteriol. 117, 819–825.

Wetzel R G 2001 Limnology: Lake and River Ecosystems. Academic Press, London, U. K. 1006 p.

Vector for chromosomal integration of the *phoC* gene in plant growth-promoting bacteria

R. Fraga-Vidal[1], H. Rodríguez Mesa & T. González-Díaz de Villegas
Cuban Research Institute on Sugarcane By-Products, P.O. Box 4026, CP 11 000, Havana, Cuba. [1]*Corresponding author**

Received 9 December 2002. Accepted in revised form 2 January 2003

Key words: acid phosphatases, GMO, mini Tn5, PGPB, phosphate solubilization, plant growth promotion

Abstract

This work describes the subcloning of the gene encoding the PhoC acid phosphatase from *Morganella morganii* (*phoC* gene) in a vector that permits stable chromosomal integration of this gene in plant growth-promoting bacteria (PGPB). A plasmid was constructed using the suicide delivery vector pJMT6 (a pUT/mini Tn*5* derivative vector) and the plasmid pLR1, the latter harboring the *phoC* gene. The recombinant construction pLF17, which contains a non-antibiotic resistance selection marker, was transformed and expressed in *Escherichia coli* CC118λ*pir*. A transformant clone, *E. coli* CC118λ*pir* F17 was selected and further characterized, showing *phoC* gene expression through an histochemical assay and zymograms developed to detect phosphatase activity. With this technique, it was possible to detect, in the whole cell extract, the 25-kDa polypeptidic component responsible for acid phosphatase activity. Acid phosphatase activity was quantified in the whole cell and in the supernatant of the culture as being higher in the transformant *E. coli* CC118λ*pir* F17 than in *E. coli* CC118λ*pir* without plasmids along the cultivation time.

Introduction

Phosphorus is an essential element for plant growth. However, a considerable portion of organic and inorganic phosphate is in a poorly soluble state in soil (Goldstein, 1996). The capacity of some microorganisms to solubilize mineral and organic phosphorus in soil, making this compound available for plant growth, has been a focus of research for many years (Rodríguez and Fraga, 1999). Particularly, organic phosphates can be found in the humus in soil, and the solubilization of part of this phosphorus can be carried out by means of phosphatase enzymes produced by certain rhizobacterial strains.

Plant growth-promoting bacteria (PGPB) are bacteria that can exert a beneficial impact on plant growth and development in a direct or indirect way. The direct promotion generally entails providing the plant with a compound that is produced by the bacterium or facilitating the uptake of nutrients from the environment (Glick, 1995). This is the case of phosphate-solubilizing bacteria. Genetic manipulation by means of recombinant DNA technology seems to offer feasible approaches for obtaining improved phosphate-solubilizing strains (Fraga et al., 2001; Rodríguez and Fraga, 1999). The objective of this work was the subcloning of the *phoC* gene from *Morganella morganii* that codes for PhoC acid phosphatase in a mini-Tn5 derivative transposon-vector in order to achieve stable chromosomal integration of this gene in the recipient strain. Advantages of this delivery system are: stability without selection,

* FAX No: +53-7-988243.
E-mail: reynaldofraga@yahoo.com

E. Velázquez and C. Rodríguez-Barrueco (eds.),
First International Meeting on Microbial Phosphate Solubilization, 239–244

non-antibiotic selection markers, prevention of risk of metabolic load, small size, and minimal horizontal transfer of cloned genes to indigenous microorganisms.

Materials and methods

Plasmids, bacterial strains and cultivation conditions

Escherichia coli strains were grown in LB broth (in g L^{-1}: Bacto-tryptone, 10; yeast extract, 5; NaCl, 10), at pH 7, and in LB agar (LB supplemented with 1.5% agar). Ampicillin (Ap) was used at 100 μg ml^{-1} and potassium tellurite (K_2TeO_3) was used at 60 μg ml^{-1}. Growth was carried out at 37 °C in an orbital shaker at 175 rpm. Growth was followed by the measurement of absorbance at 600 nm. Statistical and regression analysis were carried out with Statgraphics Statistical Graphics System, version 5.0, with a 95% level of significance. Plasmids and bacterial strains are listed in Table 1.

Recombinant DNA techniques

Plasmids were purified using the alkaline lysis method, essentially as described by Sambrook and Russell (2001). Restriction enzymes and T_4 DNA ligase (New England Biolab, Ltd., Ontario, Canada) were used according to the manufacturer's instructions. Preparation of competent cells and transformation were carried out according methods described by Sambrook and Russell (2001).

Cloning strategy for the construction of the integrating plasmid

To construct the delivery vector for the integration of the *phoC* gene, plasmid pPM12R was digested with *Eco*RI and the released fragment cloned at the *Eco*RI site of vector pUC18*Not* to flank the *phoC* gene with the restriction site *Not*I. The ligation product was transformed in *E. coli* MC1061 and plated on LB medium supplemented with Ap. This construction (pLR1) was then digested with *Not*I to release the *phoC* gene, and this fragment was ligated to the vector pJMT6, previously digested with *Not*I, to generate a minitransposon harboring the *phoC* gene and with potassium tellurite resistance (Tel^r) as the only selection marker (carried by pJMT6) (Figure 1). The ligation product (pLF17) was transformed in *E. coli* CC118λpir and plated on LB medium supplemented with Ap and K_2TeO_3.

Table 1. Strains and plasmids used in this study

Strain or plasmid	Relevant characteristics	Reference
Strains		
E. coli MC1061	F' *araD139 (ara-leu)7696 galE15 galK16 (lac)X74 rpsL* (Str^r) *hsdR2* (r_k-m_k) *mcrA mcrB1*	Sambrook and Russell (2001)
E. coli CC118λpir	Δ*(ara-leu) araD* Δ*lacX74 galE galK phoA thi-1 rpsE rpoB argE*(Am) *recA1* lysogenized with λ*pir* phage	Herrero et al. (1990)
Plasmids		
pPM12R	Ap^r, 4.158 kb; derivative of pBluescript SK+/− (Stratagene), harboring a 1.2-kb fragment from a library of *M. morganii* that codes for the PhoC acid phosphatase	Thaller et al. (1994)
pUC18*Not*	Ap^r, as pUC18 but multiple cloning site flanked by *Not*I sites	Sánchez-Romero et al. (1998)
pLR1	Ap^r, 3.8 kb; derivative of pUC18*Not* with the *phoC* gene ligated to the *Eco*RI site	This study
pJMT6	Ap^r, Tel^r, 8.2-kb pUT/mini-Tn*5* Tel (*Not*I site free)	Sánchez-Romero et al. (1998)
pLF17	9.4 kb, identical to pJMT6 but with the *phoC* gene ligated to the *Not*I site	This study

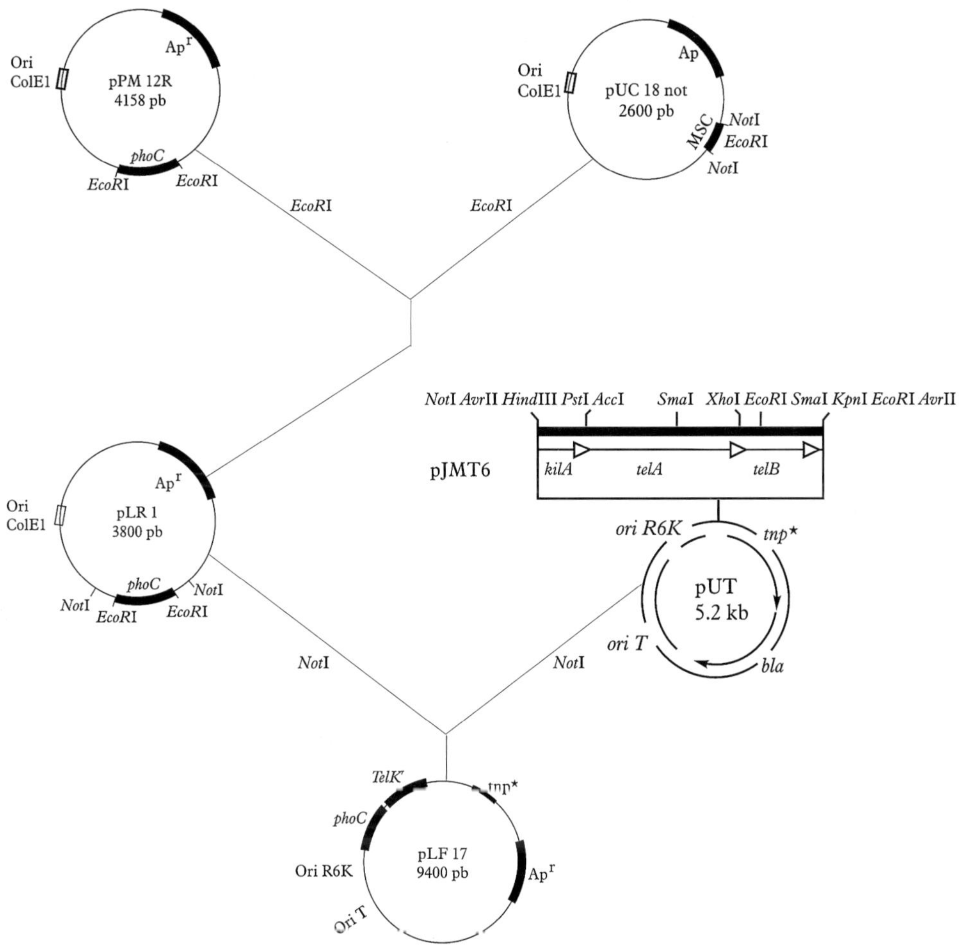

Figure 1. Cloning strategy for the construction of the integrating vector (pLF17).

Detection of acid phosphatase activity and SDS-PAGE (Zymogram)

To detect the expression of the *phoC* gene on plates, a modification of the phosphatase indicator medium, based on an histochemical detection system developed by Thaller et al. (1994), was used. This medium was LB agar supplemented with phenolphthalein diphosphate 0.2% (PDP, disodium salt, Sigma) as substrate for the enzyme and methyl green 0.005% (MG, Sigma) as stain. For the electrophoretic separation, the intact cells were washed with normal saline solution and resuspended in this solution to an optical density (OD) (600 nm) of 40. An aliquot of 40 μl of this suspension was mixed with 10 μl of the loading buffer and 20 μl from that were submitted to SDS-polyacrylamide gel electrophoresis (15%) (SDS-PAGE), according to the method of Laemmli (1970). To visualize the bands corresponding to the total proteins, the gels were stained with Coomassie Brilliant Blue R-250. For the detection of bands with phosphatase activity (Zymogram), the technique described by Thaller et al. (1994), was used.

Phosphatase activity from liquid cultures was evaluated in intact cells and supernatant fractions as described by Fraga et al. (2001).

Results and discussion

*Expression of the phoC gene in E. coli CC118*λpir

Characterization of transformant clones

Some putative transformant clones were selected for further characterization. Plasmids were extracted and digested with different restriction enzymes (Figure 2). One of the transformants, designated F17, harbored a plasmid (pLF17), which showed the expected size for the construction resulting from the union of vector pJMT6 and the *Not*I *phoC* fragment (9.4 kb). After digestion with *Not*I, plasmid pLF17 was split in two corresponding elements: a fragment of approximately 8.2 kb (pJMT6) and a 1.2 kb (*phoC* gene) fragment. After digestion with *Sma*I, pLF17 yielded a fragment of approximately 7.8 kb and a 1.6-kb fragment. This confirmed the presence of the expected recombinant plasmid in the selected transformant.

Phosphatase activity

After the growth on plates with the phosphatase indicator medium, a dark green color was observed in the F17 recombinant strain, showing the *pho*$^+$ phenotype, in contrast to the white-yellowish color of *E. coli* CC118λpir (data not shown). This qualitative method indicated that the *M. morganii phoC* gene was expressed in the host *E. coli* CC118λpir.

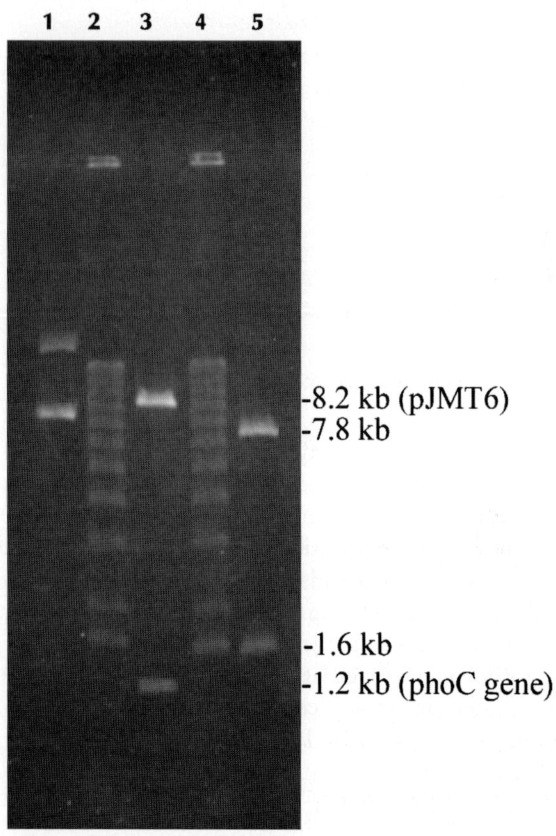

Figure 2. Restriction pattern of the plasmid pLF17 digested with different restriction enzymes. Lane 1: pLF17 not digested. Lane 3: pLF17 digested with the restriction enzyme *Not*I. Lane 5: pLF17 digested with the restriction enzyme *Sma*I. Lanes 2 and 4: DNA molecular weight markers (Marker X).

Phosphatase activity in whole cells and supernatant fractions of the recombinant clone F17 is shown in Figures 3 and 4, in comparison with the *E. coli* CC118λpir strain without any plasmid. During the exponential phase of growth, a much higher level of acid phosphatase activity was detected in intact cells compared with the supernatant fraction (Figure 3). This is consistent with the periplasmic localization of the PhoC enzyme reported by Thaller et al. (1994).

Escherichia coli CC118λpir showed a significantly smaller level of cell-bound acid phosphatase activity, in comparison with the F17 recombinant strain. This basal level of phosphatase activity in the host *E. coli* CC118λpir could be related to a low expression rate of the gene *aphA*, encoding the class B acid phosphatase/phosphotransferase reported for *E. coli* MG 1655 by Thaller et al. (1997). However, the high level of acid phosphatase activity detected in the F17 strain shows the expression of the *phoC* gene present in pLF17, and that the gene is being expressed under its own promoter.

The F17 clone showed increased levels of activity in the culture supernatant after the stationary phase of growth (Figure 4), probably a result of

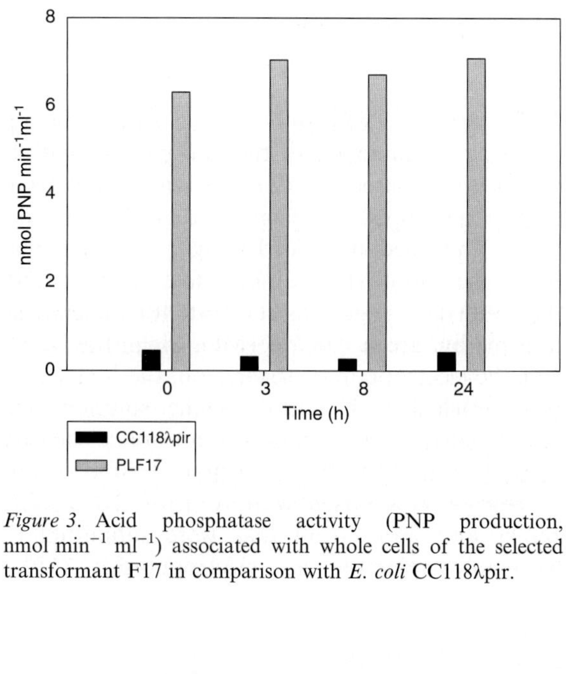

Figure 3. Acid phosphatase activity (PNP production, nmol min^{-1} ml^{-1}) associated with whole cells of the selected transformant F17 in comparison with *E. coli* CC118λpir.

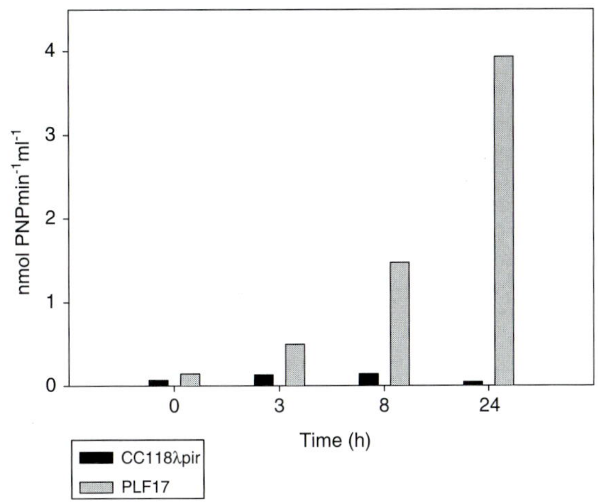

Figure 4. Acid phosphatase activity (PNP production, nmol min^{-1} ml^{-1}) associated with the supernatant of the selected transformant F17 in comparison with *E. coli* CC118λpir.

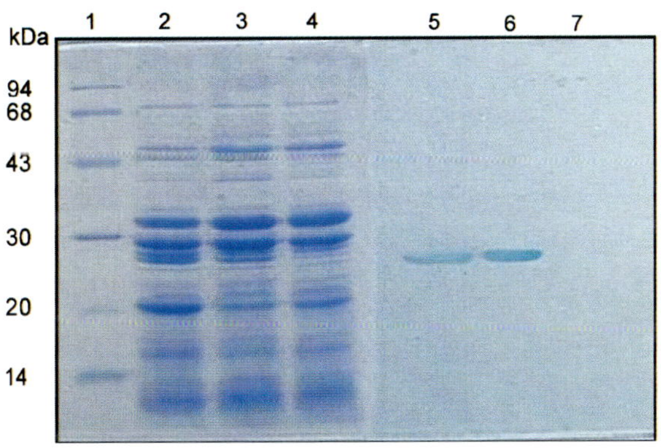

Figure 5. SDS-PAGE analysis of proteins and zymogram for phosphatase activity from *E. coli* transformants. Lane 1: Protein size markers in kDa. Lanes 2–4: Coomassie Blue-stained whole cell protein preparation of *E. coli* MC1061 PM12R (pPM12R), *E. coli* CC118λpir F17 (pLF17), and *E. coli* CC118λpir without plasmids. Lanes 5–7: Zymogram developed for phosphatase activity against PDP at pH 6.0 of *E. coli* MC1061 PM12R (pPM12R), *E. coli* CC118λpir F17 (pLF17), and *E. coli* CC118λpir without plasmids.

the cellular lysis typical of this stage of growth. The same behavior was reported by Thaller et al. (1994) for *E. coli* DH5α PM12R harboring and expressing the *phoC* gene originally cloned. Without plasmids, strain *E. coli* CC118λpir showed very low detectable activity in the supernatant.

Figure 5 shows the result of the SDS-PAGE of total proteins, as well as the zymogram pattern for the detection of the phosphatase activity. The high intensity of the color bands from *E. coli* MC1061 PM12R (pPM12R) and *E. coli* CC118λpir F17 (pLF17) (Figure 5, lanes 5 and 6) suggests that these are the product of the expression of the *phoC* gene in both cases.

Strain F17 was able to produce a band of approximately 25 kDa, which corresponds to the PhoC band of *E. coli* MC1061 (pPM12R). This result corroborates that the *M. morganii* DNA sequences located upstream of the *phoC* gene promote transcription of the *phoC* gene in *E. coli* CC118λpir F17. As expected, no band of phosphatase activity was detected in *E. coli* CC118λpir without plasmid.

Conclusions

- An integrating suicide vector (pLF17), harboring a gene encoding the PhoC acid phosphatase of *M. morganii* was constructed.
- A transformant clone (*E. coli* CC118λpir F17) harbouring the integrating vector pLF17 and expressing the gene *phoC*, was obtained.

Acknowledgements

Plasmid pPM12R was kindly supplied by Gian M. Rossolini, from Siena University, Italy. We are grateful to Victor de Lorenzo at the National Center of Biotechnology, Madrid, Spain, for the kind gift of plasmids pUC18Not, pJMT6, and the strain *E. coli* CC118λ*pir*, as well as useful technical advice. We thank Ira Fogel, from CIB-NOR Mexico, for correcting the English text.

References

Fraga R, Rodríguez H and González T 2001 Transfer of the gene encoding the NapA acid phosphatase of *Morganell morganii* to a *Burkholderia cepacia* strain. Acta Biotechnol 21, 359–369.

Glick B R 1995 The enhancement of plant growth by free-living bacteria. Can. J. Microbiol. 41, 109–117.

Goldstein A H 1996 Involvement of the quinoprotein glucose dehydrogenase in the solubilization of exogenopus phosphates by gram-negative bacteria. *In* Phosphate in Microorganisms: Cellular and Molecular Biology. Eds. A Torriani-Gorini, E Yagil and S Silver. pp. 197–203. ASM Press, Washington D.C.

Herrero M, de Lorenzo V and Timmis K N 1990 Transposon vectors containing non-antibiotic resistance selection markers for cloning and stable chromosomal insertion from foreign genes in gram-negative bacteria. J. Bacteriol. 172, 6557–6567.

Laemmli U K 1970 Cleavage of structural proteins during the assembly of the head of bacteriophage T4. Nature (London) 227, 680–685.

Rodríguez H and Fraga R 1999 Phosphate solubilizing bacteria and their role in plant growth promotion. Biotechnol. Adv. 17, 319–339.

Sambrook J and Russell D W 2001 Molecular Cloning: A Laboratory Manual. 3rd edition, Cold Spring Harbor Laboratory Press, Cold Spring Harbor, New York.

Sánchez-Romero J M, Díaz-Orejas R and de Lorenzo V 1998 Resistance to tellurite as a selection marker for genetic manipulation of *Pseudomonas* strains. Appl. Environ. Microbiol. 64, 4040–4046.

Thaller M C, Berlutti F, Schippa S, Lombardi G and Rossolini G M 1994 Characterization and sequence of PhoC, the principal phosphate-irrepressible acid phosphatase of *Morganella morganii.* Microbiology 140, 1341–1350.

Thaller M C, Schippa S, Bonci A, Cresti S and Rossolini G M 1997 Identification of the gene (*aph*A) encoding the class B acid phosphatase/phosphotransferase of *Escherichia coli* MG 1655 and characterization of its product. FEMS Microbiol. Lett. 146, 191–198.

Microorganisms with capacity for phosphate solubilization in Dão red wine (Portugal)

L. R. Silva[1], R. Rivas[2], A. M. Pinto[1], P. F. Mateos[2,3], E. Martínez-Molina[2] & E. Velázquez[2]
[1]*Agrarian College of Viseu, Viseu, Portugal.* [2]*Departamento de Microbiología y Genética, Universidad de Salamanca, Salamanca, Spain.*

Received 13 November 2002. Accepted in revised form 2 January 2003

Key words: acetic bacteria, *Gluconobacter*, phosphate solubilization microorganisms, wine

Abstract

The red wine production is a complex microbiological process involving several microorganisms. Yeasts are the responsible of alcoholic fermentation and bacteria develop the malo-lactic fermentation. These two processes are essential for the red wine production. However, other bacteria develop process that cause spoilage in the wine. The acetic bacteria oxidize the ethanol to acetic acid. Within these bacteria genera *Acetobacter* and *Gluconobacter* are the most important producers of wine spoilage. We have identified the strain 39PCAac1 as *Gluconobacter oxydans* subsp. *oxydans* in Dão red wine (Portugal). This strain shows high ability to solubilize phosphate.

Introduction

The microbiology of winemaking has been extensively studied and these studies revealed the complexity of the wine ecology (Fleet, 1993). As it is well known, the microorganisms are significant in winemaking because: (i) they developed the alcoholic fermentation; (ii) they can spoil wines during conservation in the cellar and after packaging, and (iii) they affect wine quality (Bidan et al., 1995).

During the winemaking process the microorganisms may produce alterations that diminish the quality and acceptability of the final product. During the wine production, any uncontrolled microbial growth can change the chemical composition of the wine and alterations in the sensory properties by the action of these microorganisms (moulds, yeasts, acetic acid bacteria and lactic acid bacteria) may be produced. A wine spoilage is a serious problem for the wine industry because it renders a product unacceptable for human consumption (Loureiro, 2000).

The yeasts are responsible for the alcoholic fermentation of grape juice into wine and certain species of bacteria could grow in wine causing its spoilage. One of the main causes of wine spoilage is the production of acetic by several species of bacteria. The acetic acid bacteria belong to the family Acetobacteraceae and to genera *Gluconobacter* and *Acetobacter* (Holt et al., 1994).

The genus *Acetobacter* oxidize lactic and acetic into CO_2 and can be non-motile or motile by peritrichous flagella. The genus *Gluconobacter* does not oxidize lactic or acetic acid, and is non-motile or motile by polar flagella. *Gluconobacter oxydans*, *Acetobacter aceti*, *A. pasteurianus*, *A. liquefaciens* and *A. hansenii* are normally associated with grapes and wine (Du Toit and Lambrechts, 2002). The main species observed on unspoiled grapes and grape juice is *G. oxydans* (Joyeux et al., 1984).

Family Acetobacteraceae belongs to alpha subclass of Proteobacteria which includes several phosphate solubilizing bacteria (PSB) such as

* E-mail: pfmg@usal.es

E. Velázquez and C. Rodríguez-Barrueco (eds.),
First International Meeting on Microbial Phosphate Solubilization, 245–248

Table 1. Mineral elements in wines (adapted by Cabanis et al., 1995)

Majors elements	Minimum (g/L)	Maximum (g/L)	Media (g/L)
Potassium	0.40	1.84	0.97
Calcium	30×10^{-3}	0.20	70×10^{-3}
Magnesium	40×10^{-3}	0.16	90×10^{-3}
Sodium	$3 \times 10^{-3} - 30 \times 10^{-3}$	$50 \times 10^{-3} - 0.35$	$25 \times 10^{-3} - 0.10$
Silicium	20×10^{-3}	90×10^{-3}	35×10^{-3}
Phosphate (PO_4^{3-})	0.10	0.80	0.40
Sulphate (SO_4^{2-})	40×10^{-3}	0.60	0.10
Chlorides (Cl^-)	$10 \times 10^{-3} - 60 \times 10^{-3}$	0.2–0.8	0.058

rhizobia. However, the ability of bacteria to solubilize phosphate has been studied in habitats different to soil. Although the importance of PSB in wine is already unknown, the anion phosphate is the second most abundant in the wine after the sulphate. For this reason we have analysed red wine samples from Dão DO Region to detect PSB.

Materials and methods

Wine sampling

In this work, we have analysed ten samples of bottled red wine of Dão DO Region, after a minimum stage of 18 months in oak wood barrels. Table 1 shows the mineral composition of wine according to Cabanis et al. (1995).

The samples were inoculated (100 μL) in PCA plates (plate count agar) and TJA plates (tomato juice agar), after two days of incubation, we have isolated several colonies of bacteria.

We have inoculated these bacteria in YED-P (yeast extract glucose phosphate) plates to observe the phosphate solubilization (Peix et al., 2001). The colonies surrounded by a clear halo higher than 15 mm (de Freitas et al., 1997) were counted, isolated and identified.

Amplification and determination of nucleotide sequences of the 16S rRNA gene and analysis of the sequence data

DNA extraction was carried out as previously described (Rivas et al., 2001). The amplification of 16S rDNA and its sequencing was performed according to the method already described (Rivas et al., 2002). The sequence obtained was compared with those from the GenBank using the FASTA program (Pearson and Lipman, 1988). Sequences were aligned using the Clustal W software (Thompson et al., 1997). The distances were calculated according to Kimura's two-parameter method (Kimura, 1980). Phylogenetic trees were inferred using the neighbour-joining method (Saitou and Nei, 1987). Bootstrap analysis was based on 1000 resamplings. The MEGA2 package (Kumar et al., 2001) was used for all analyses. The trees were rooted using *Rhizobium leguminosarum* as outgroup.

Results and discussion

Count of PSB

From the samples of red wine, 68 bacterial strains were isolated and 12 of them showed solubilization of phosphate in YED-P plates (Figure 1).

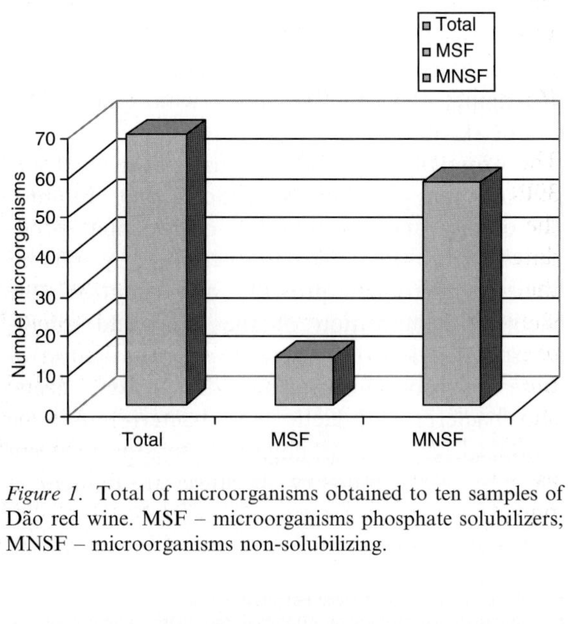

Figure 1. Total of microorganisms obtained to ten samples of Dão red wine. MSF – microorganisms phosphate solubilizers; MNSF – microorganisms non-solubilizing.

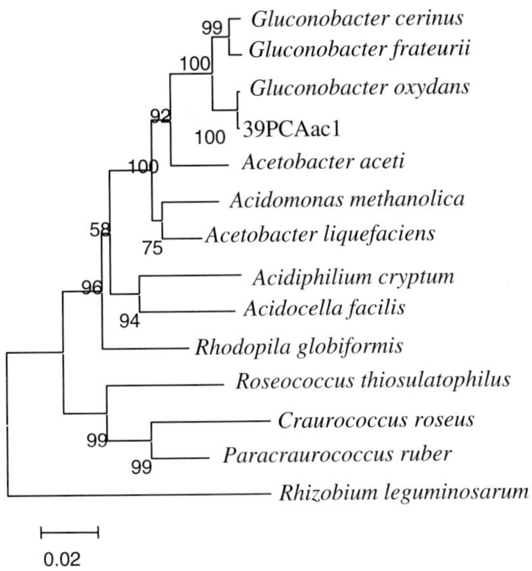

Figure 2. Comparative sequence analysis of 16S rDNA from *Xylanimonas cellulosilytica* XIL07^{T} and representative strains from the GenBank. The significance of each branch is indicated by a bootstrap value calculated for 1000 subsets. Bar, 2 nt substitutions per 100 nt. The GenBank accession numbers for the sequences used to generate the phylogenetic tree are the following: strain 39PCAac1, AY206688, *Gluconobacter cerinus* IFO 3267, AB063286, *G. frateurii* IFO3264^{T}, X82290, *G. oxydans* DSM 3503^{T}, X73820, *Acetobacter aceti* DSM 3508^{T}, X74066, *Acidocella facile* ATCC 35904^{T}, D30774, *Acidiphilium cryptum* DSM 2389^{T}, Y18445, *Acidomonas methanolica* IMET 10945^{T}, D30770, *Craurococcus roseus*, NS130^{T}, D85828, *Acetobacter liquefaciens* IFO12388^{T}, X75617, *Paracraurococcus ruber* NS89^{T}, D85827, *Rhodopila globiformis* DSM161^{T}, D86513, *Roseococcus thiosulatophilus* RB-3^{T}, X72908 and *Rhizobium leguminosarum* ATCC10004^{T}, U29386.

One of them, designated as 39PCAac1, showed an unusually high solubilization of phosphate.

16S rDNA sequence analysis

The complete 16S rDNA sequence for isolate 39PCAac1 was obtained. A comparison against the 16S rDNA sequences held in the GenBank database indicated that the organism belongs to the species *Gluconobacter oxydans* and that shows a 99.66% homology with the subspecies *oxydans*. In Figure 2 the phylogenetic location of the isolate within family Acetobacteraceae is shown.

This is the first description of this bacterium as PSB and therefore the significance of the phosphate solubilization phosphate in species *Gluconobacter oxydans* should be studied in the future. At the moment this species is known to be responsible for acetic acid formation in wine and its isolation in bottled wine indicates that these wine are subject of spoilage.

In future works we will analyse other red wines from different geographical origins to establish if this bacterium isolated from several sources and if all of the isolates have the ability to solubilize phosphate.

Acknowledgement

This work was supported by the Junta de Castilla y León (Spain).

References

Bidan P, Divies C and Cachon R 1995 Micro-organismes d'alteration des vins. *In* Œnologie: Fondements Scientifiques et Technologiques. Ed. C Flanzy. pp. 536–553. Lavoisier Tec & Doc, France.

Cabanis J C, Cabanis M T, Cheynier V and Teissedre P L 1995 Tables de composition. *In* Œnologie: Fondements Scientifiques et Technologiques. Ed. C Flanzy. pp. 315–336. Lavoisier Tec & Doc, France.

de Freitas J R, Banerjee M R and Germida J J 1997 Phosphate-solubilizing rhizobacteria enhance the growth and yield but not phosphorus uptake in canola (*Brassica napus* L.). Biol. Fertil. Soil 24, 358–364.

Du Toit W J and Lambrechts M G 2002 The enumeration and identification of acetic acid bacteria from South African red wine fermentations. Int. J. Food Microbiol. 74, 57–64.

Fleet G H 1993 Wine Microbiology and Biotechnology Harwood Academic Publishers, Sydney, Australia.

Holt J G, Krieg N R, Sneath P H A, Staley J T and Williams S T 1994 Genus *Acetobacter* and *Gluconobacter*. *In* Bergey's Manual of Determinative Bacteriology. Eds. J G Holt, N R Krieg, P H A Sneath, J T Staley and S T Williams. pp. 71–84. 9th edition, Williams and Wilkins, MD, USA.

Joyeux A, Lufon-Lufourende S and Ribéreau-Gayon P 1984 Evolution of acetic acid bacteria during fermentation and storage of wine. Appl. Environ. Microbiol. 48, 153–156.

Kimura M 1980 A simple method for estimating evolutionary rates of base substitutions through comparative studies of nucleotide sequences. J. Mol. Evol. 16, 111–120.

Kumar S, Tamura K, Jakobsen I B and Nei M 2001 Molecular Evolutionary Genetics Analysis Software. Arizona State University, Tempe, AZ, USA.

Loureiro V 2000 Spoilage yeasts in food and beverages: Characterisation and ecology for improved diagnosis and control. Food Res. Int. 33, 247–256.

Pearson W R and Lipman D J 1988 Improved tools for biological sequence comparison. Proc. Natl. Acad. Sci. USA 85, 2444–2448.

Peix A, Rivas-Boyero A A, Mateos P F, Rodríguez-Barrueco C, Martínez-Molina E and Velázquez E 2001 Growth promotion of chickpea and barley by a phosphate solubilizng strain of *Mesorhizobium mediterraneum* under growth chamber conditions. Soil Biol. Biochem. 33, 103–110.

Rivas R, Velázquez E, Valverde A, Mateos P F and Martínez-Molina E 2001 A two primers random amplified polymorphic DNA procedure to obtain polymerase chain reaction fingerprints of bacterial species. Electrophoresis 22, 1086–1089.

Rivas R, Velázquez E, Willems A, Vizcaíno N, Subba-Rao N S, Mateos P F, Gillis M, Dazzo F B and Martínez-Molina E 2002 A new species of *Devosia* that forms a nitrogen-fixing root-nodule symbiosis with the aquatic legume *Neptunia natans* (L. f.) Druce. Appl. Environ. Microbiol. 68, 5217–5222.

Saitou N and Nei M 1987 A neighbour-joining method: A new method for reconstructing phylogenetics trees. Mol. Biol. Evol. 44, 406–425.

Thompson J D, Gibson T J, Plewniak F, Jeanmougin F and Higgins D G 1997 The clustalX windows interface: Flexible strategies for multiple sequence alignment aided by quality analysis tools. Nucleic Acid Res. 24, 4876–4882.

Phosphate solubilizing microorganisms in the rhizosphere of native plants from tropical savannas: An adaptive strategy to acid soils?

M. Toro*

Facultad de Ciencias, Laboratorio de Estudios Ambientales, Instituto de Zoología Tropical, Universidad Central de Venezuela, Apartado 47.058, Caracas, 1041-A, Venezuela.

Received 25 November 2002. Accepted in revised form 2 January 2003

Key words: acid soil, P fractionation, phosphate solubilizing microorganisms, savanna, *Trachypogon plumosus*

Abstract

Savannas are natural ecosystems that predominate in the tropics. These systems usually have acid soils with low fertility in which nutrients, specially phosphorus, are scarce. Phosphorus is generally fixed in insoluble forms that cannot be rapidly incorporated by plants. In acid soils phosphorus is fixed as aluminum and iron phosphates; in calcareous soils phosphorus is fixed as calcium phosphate. In both cases these phosphorus forms need to be solubilized in order to make phosphate ions available to the plant in soil solution. Besides the natural soil acidity, organic acids produced through microbial mechanisms or plant roots have been proved to solubilize these phosphates. I investigated if native plant rhizospheres of acid or calcareous soils are enriched with phosphate solubilizing microorganisms, in order to find mechanisms to improve plant nutrition and agrosystem sustainability. The rhizosphere of a typical and dominant grass from savannas, *Trachypogon plumosus* Ness, was studied in order to corroborate the former hypothesis. Furthermore different phosphorus forms in rhizospheric soil were determined applying Hedley et al., (1982) P-fractionation method. *T. plumosus* growing in acid and neutral-calcareous soil rhizospheres were compared in terms of microbial populations and phosphate fractions. My results show that in many of the rhizospheres considered P–Al and P–Fe solubilizing organisms predominate when P–Al and P–Fe are important P fractions present in soil. This was not the case in calcareous soils where P–Ca solubilizing organisms P–Ca fractions predominate in soil. Through this approach I elucidate mechanisms operating in plant rhizospheres to make hardly soluble phosphates available to plants. The implications of such mechanisms on biotechnological and agricultural approaches are discussed.

Introduction

The availability of P for plants uptake is very low in soil solution, specially in acid and/or neutral calcareous soils, in which it is firmly fixed in insoluble forms of iron, aluminum and/or calcium (López-Hernández, 1977). It has been proven that fungi and bacteria have the capacity to solubilize these compounds (Illmer, 1995; Kucey et al., 1989). Phosphorus is also unavailable when organic. Around 50% of total P in soil is in organic form and needs to be mineralized through microorganisms and/or plants enzymatic activities to release inorganic ions (Mc Laughlin et al., 1990).

We are interested in studying the microbial populations existing in the rhizosphere of native plants able to grow in acid and/or slightly acid soils with low P contents. We will test microbial populations of an autocthonous plant of savannas such as *Trachypogon plumosus* NESS, and its

* FAX No: + 58-212-605 1204.
E-mail: mtoro@strix.ciens.ucv.ve

E. Velázquez and C. Rodríguez-Barrueco (eds.),
First International Meeting on Microbial Phosphate Solubilization, 249–252

capacities to solubilize predominant phosphates in these soils: calcic, iron and aluminum phosphates. We will relate the proportions of different phosphate solubilizing organisms with those P fractions (Hedley et al., 1982) dominating each rhizosphere.

Materials and methods

Test plant and locations studied

Trachypogon plumosus NESS native grass from two acid (Calabozo and Parupa) and two slightly acid (Jarillo and Loma) savanna soils from Venezuela. Three of them: Calabozo, Jarillo and Loma are located in the central northern part of the country whereas Parupa is located in the eastern zone of the country. Main soil orders in these savanna systems are Entisols and Ultisols.

Microbiological analyses

Five rhizospheric soil samples from *T. plumosus* were collected at each location. In order to quantify total bacteria and fungi, serial dilutions were made of 1 g of rhizospheric soil in saline solution (NaCl, 0,82%). Dilutions were placed on Petri dishes to count total fungi and bacteria according to Varma (1998). Growing media contained 0.2% of calcic, iron or aluminum phosphates in order to evaluate the microorganisms solubilization capacity over those insoluble sources. The plates were incubated at 28 °C for 7–14 days. The number of bacteria and fungi was registered, as well as the number of solubilizing bacteria and fungi of each one of the insoluble phosphates, which were recognized by a clear zone or halo surrounding the colony. Data are presented as colony forming units per gram of rhizospheric soil (ufc/g of rhizospheric soil). Percentage of solubilizers referred to total Fungi and Bacteria were calculated to obtain proportions of specialized groups that solubilize each phosphate source.

Soil analysis

Samples were sieved (2 mm) for the chemical parameters analysis of the soils: pH (H_20) and phosphorus fractions. In order to determine P pools a fractionation following Hedley et al. (1982) procedure with modifications of Tiessen and Moir (1993) was applied. Organic P (Po) content was calculated as the difference between Pt and Pi.

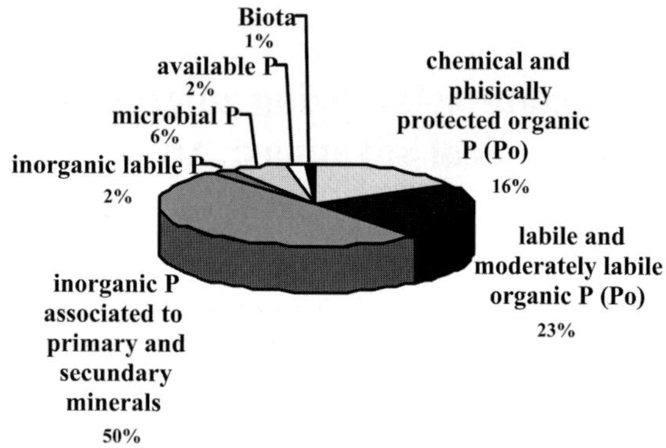

Figure 1. P fractions of an acid savanna soil.

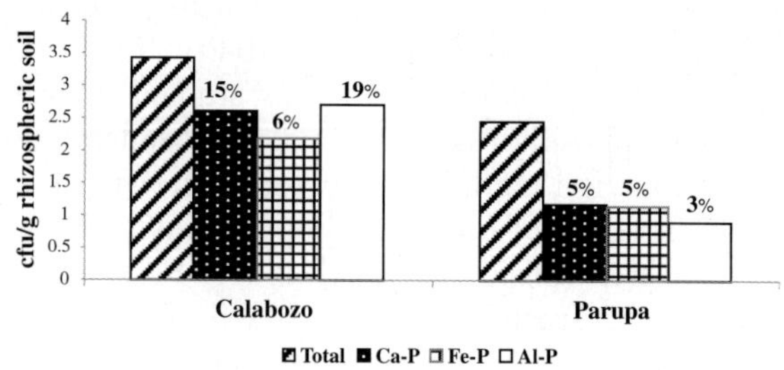

Figure 2. Total and phosphate solubilizing fungi in the rhizosphere of *T. plumosus* in acid savanna soils.

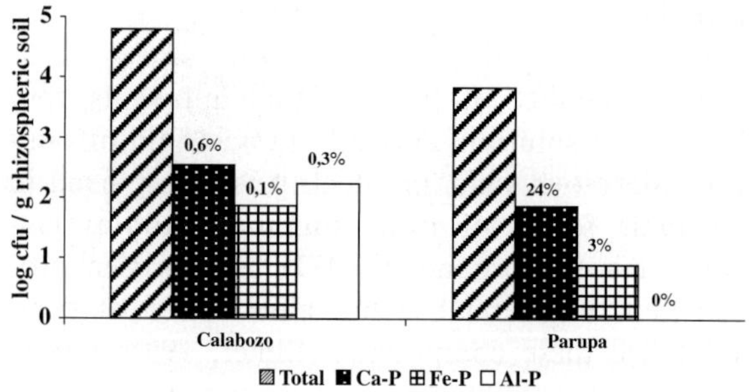

Figure 3. Total and phosphate solubilizing bacteria in the rhizosphere of *T. plumosus* in acid savanna soil.

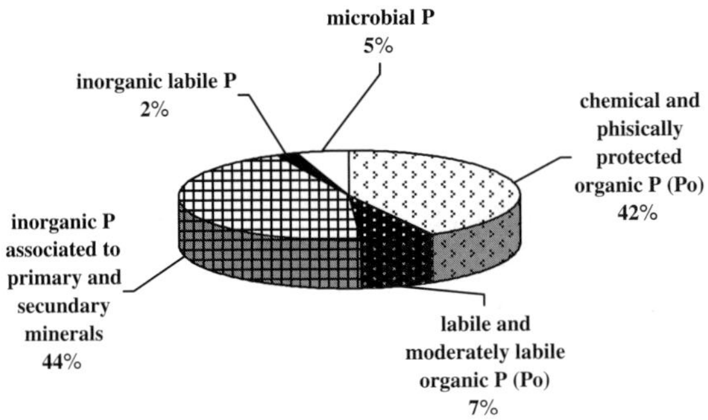

Figure 4. P fractions of a slightly acid savanna soil.

Statistical analysis

Tests were carried out in order to verify the normal data distribution. A comparison of the population averages was also carried out in order to verify the differences among them using the U Test from Wilcoxon–Mann Whitney.

Results and discussion

Acid soil rhizospheres show unavailable inorganic phosphate forms dominating over unavailable organic forms, typical of savanna acid soils (Hernandez-Valencia and Lopez-Hernandez, 1999), being the inorganic forms (as Fe-P and Al-P) the most important in these soils (Figure 1). Available P (readily taken by the plants) ranges between 2% and 4%, considered a low soil solution concentration. Phosphate solubilizing fungi (Figure 2) were more abundant in Calabozo soil (pH = 5.0) than in Parupa soil (pH = 4.8). That is not the case for bacteria (Figure 3), which are present in very low amounts (< 1%) in Calabozo's rhizospheres, but are higher (mainly Ca–P and Fe–P) in Parupa's. Evidently, both Ca–P solubilizing fungi and bacteria are more abundant in the studied rhizospheres of acid soils, but the proportion of P–Fe and P–Al solubilizing fungi seem to be important according to phosphates determined in soil.

For slightly acid rhizospheres, unavailable inorganic phosphate forms are similar to unavailable organic forms (Figure 4). In this case, inorganic forms as Ca–P, dominate over Fe–P and Al–P forms. Available P (readily taken by the plants) is 2% lower than the amount detected for acid soil solution rhizosphere. Phosphate solubilization fungi were more abundant in Jarillo soil (pH H_2O = 5.8) than in Loma soil (pH H_2O = 6.0). For both soils, solubilizing fungi (16–9%) were present in similar or higher amounts than bacteria (12–0%) (Figures 5 and 6). It has to be shown that P–Al groups are more abundant in slightly acid than in acid soils, even being P–Al less abundant in slightly acid than in acid soils.

The abundance of microorganisms with such capacities is important in this kind of soils, where interacting with mycorrhiza will constitute a synergistic mechanism for plant nutrition (Barea et al., 1975; Linderman, 1988; Toro et al., 1997). More has to be investigated in this area to

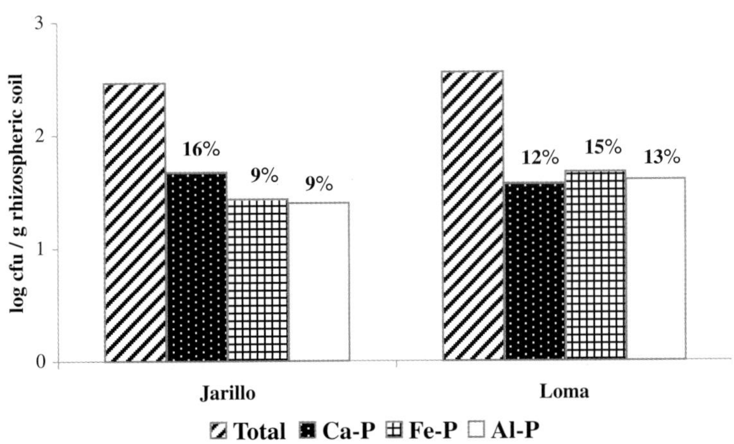

Figure 5. Total and solubilizing fungi in the rhizosphere of *T. plumosus* in slightly acid soil.

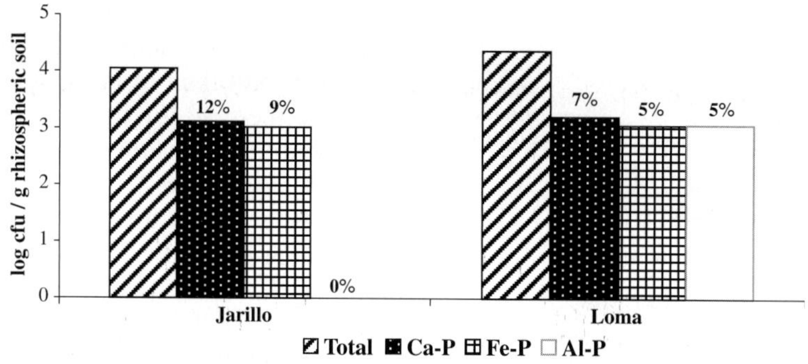

Figure 6. Total and solubilizing bacteria in the rhizosphere of *T. plumosus* in slightly acid savanna soil.

improve the actual knowledge on the role of these microorganisms in agricultural and natural ecosystems, as Vázquez et al. (2000) have pointed out.

Acknowledgements

This work was sponsored by the *Consejo de Desarrollo Científico y Humanístico* (CDCH) of the Universidad Central de Venezuela. Project No. 03-31-4109-2000.

References

Barea J M, Azcón R and Hayman D 1975 Possible synergistic interactions between *Endogone* and phosphate-solubilizing bacteria in low-phosphate soils. *In* Endomycorrhizas. Eds. F E Sanders, B Mosse and P B Tinker. pp. 409–417. Academic Press, London.

Hedley M J, Stewart J W B and Chauhan B S 1982 Changes in inorganic and organic soil phosphorus fractions induced by cultivation practices and by laboratory incubations. Soil Sci. Soc. Am. J. 46, 970–976.

Hernández-Valencia I and López-Hernandez D 1999 Allocation of phosphorous in a tropical savanna. Chemosphere 39, 199–207.

Illmer P 1995 Solubilization of hardly soluble $AlP0_4$ with P-solubilizing microorganisms. Soil Biol. Biochem. 27, 265–270.

Kucey R M N, Janzen H H and Leggett M E 1989 Microbially mediated increases in plant-available phosphorus. Adv. Agron. 42, 199–228.

Linderman R 1988 Mycorrhizal interaction with the rhizosphere microflora: The mycorrhizosphere effect. Phytopathology 78, 488–505.

López-Hernández D 1977 La química del fósforo en suelos ácidos. Ediciones de la Biblioteca, Caracas.

Mc Laughlin M J, Baker T G, James T R and Rundle J A 1990 Distribution and forms of phosphorus and aluminum in acidic topsoils under pastures in south-eastern Australia. Aust. J. Soil Res. 26, 371–385.

Tiessen H and Moir J O 1993 Characterization of available P by sequential extraction. *In* Soil Sampling and Methods of Analysis. Ed. M R Carter. pp. 75–86. CRC Press, Boca Ratón.

Toro M, Azcón R and Barea J M 1997 Improvement of arbuscular mycorrhizal development by inoculation with phosphate-solubilizing rhizobacteria to improve rock phosphate bioavailability (^{32}P) and nutrient cycling. Appl. Environ. Microbiol. 63, 4408–4412.

Varma A 1998 Mycorrhiza Manual. Springer-Verlag, Berlin.

Vázquez P, Holguin G, Puente M E, Lopez-Cortes A and Bashan Y 2000 Phosphate solubilizing microorganisms associated with the rhizosphere of mangroves in a semiarid coastal lagoon. Biol. Fert. Soils 30, 460–468.

Effects of solarization on phosphorus and on other chemical constituents of soil

A.F.M.A. Pinto[1], L.R. da Silva[1,4,5], E. Velázquez[2] & A. César[3]
[1]*Agrarian College of Viseu, Viseu, Portugal.* [2]*University of Salamanca, Salamanca, Spain.* [3]*Agrarian College of Bragança, Bragança, Portugal.* [4]*Escola Superior Agrária de Viseu, Estrada de Nelas – Quinta da Alagoa – Ranhados, 3500-606, Viseu, Portugal.* [5]*Corresponding author**

Received 19 December 2002. Accepted in revised form 2 January 2003

Key words: chemical soil constituents, electrical conductivity, exchangeable bases, organic matter, phosphorus, soil solarization

Abstract

The results showing the effects of soil solarization on phosphorus and other chemical constituents of the soil were obtained, in Portugal, during one year. In the case of the available phosphorus we observed a little decrease in solarized soils, but this difference was not significant. Similar results were obtained in the organic matter and the calcium and sodium exchangeable bases. We obtained a significant increase in the levels of nitric and amoniacal nitrogen, electrical conductivity, available and exchangeable potassium and magnesium in the solarized soils. On the other hand, we observed a decrease of the pH levels in the solarized soils.

Introduction

Since its conception, in Israel in 1976, the soil solarization has been revealed as an efficacious control method of different weed strains and some plant diseases caused by various soil pathogenic agents such as fungi, nematodes, bacteria and soil arthropods (Katan and DeVay, 1991; McGorven et al., 2000). During the hot season, soil solarization consists of a disinfestation method of soils by mulching with transparent polyethylene sheets, that results in temperature increase and killing plant pathogens. Plant growth in solarized soil was enhanced even in the absence of known plant pathogens, similar to situations with other disinfestation methods. Several mechanisms have been proposed, such as release of plant growth substances, destruction of phytotoxic materials, killing of unknown parasites, stimulation of beneficial micro-organisms and increased macro- and micro-element nutrients in the soil solution.

The results presented in this investigation were obtained in Portugal during one year and they represent the effects of soil solarization on phosphorus and other chemical constituents of soil. In the case of the available phosphorus we observed a little decrease in solarized soils, but this difference was not significant. Similar results were obtained in the organic matter and the calcium and sodium exchangeable bases. We obtained a significant increase in the levels of nitric and amoniacal nitrogen, electrical conductivity, available and exchangeable potassium and magnesium in the solarized soils. On the other hand, we observed a decrease of the pH levels in the solarized soils.

* E-mail: luis@esav.ipv.pt

E. Velázquez and C. Rodríguez-Barrueco (eds.),
First International Meeting on Microbial Phosphate Solubilization, 253–256

Material and methods

In Portugal, a field experiment on a sandy loam soil showed the effect of solarization on chemical soil constituents. In 1977, during 75 days in mid-summer, soil solarization was carried out using transparent polyethylene (TP) of 0.050 mm thickness. Sixteen plots of about 12 m^2 each constituted the trial experiment. 8 plots were covered by (TP) and the others were uncovered (control), Figure 1.

Soil samples were collected along 25 cm depth immediately after solarization and a mixed sample was formed, one for each plot, from 10 isolated samples. The different soil constituents were evaluated by adequate methods at the Soil Laboratory of Agrarian College of Coimbra.

Results and discussion

Results of available phosphorus are seen in Figure 2. Soil solarization produced a little decrease on phosphorus level, but this effect was not significant. Similar results were obtained by Chauhan et al. (1988), Meron et al. (1989), Pinto (1992, 1998) and Pinto and César (1999). Recently Arora (1998), obtained a significant decrease of available phosphorus in solarized soils. Relatively to the available potassium, we observed a significant increase, about 23%, in solarized plots. Similar results were obtained by Grunzweig et al. (1998), Pinto (1998) and Pinto and César (1999) on potassium. However, Pinto (1992) obtained no significant differences on potassium levels by solarization.

Analysing the results of nitric nitrogen and amoniacal nitrogen in Figure 3, show a significant increase in both nitric and amoniacal nitrogen. The bigger increase was obtained in nitric

Figure 1. Soil solarization using a transparent polyethylene.

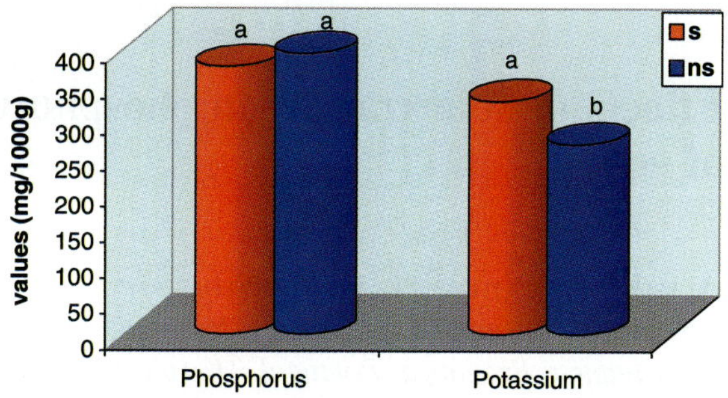

Figure 2. Available phosphorus and potassium levels in solarized and no solarized soils. For each pair columns with different letter are significantly different ($P < 0.05$).

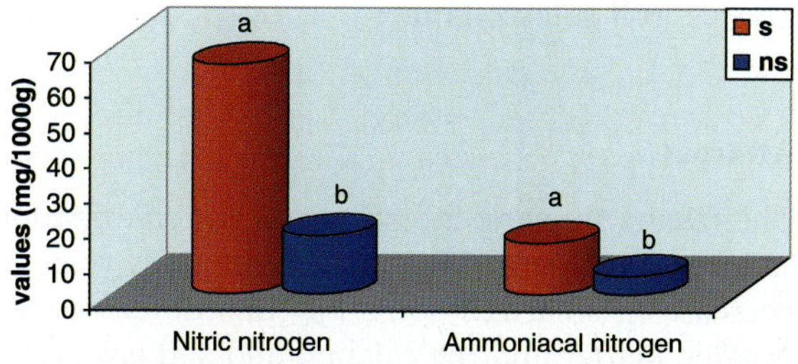

Figure 3. Nitric nitrogen and amoniacal nitrogen levels in solarized and no solarized soils. For each pair columns with different letter are significantly different ($P < 0.05$).

nitrogen, about 300%. Similar results were obtained by Stapleton and DeVay (1982), Chauhan et al. (1988), Pinto (1998) and Pinto and César (1999). However, Daelemans (1989) and Meron et al. (1989) observed no significant difference in this soil constituent. Yet recently, Arora (1998) obtained a significant increase on nitric nitrogen in solarized soils. We can observe a significant increase, about 180% in amoniacal nitrogen level in solarized soils. Similar results were obtained by Arora (1998), Pinto (1998) and Pinto and César (1999). However in earlier studies we can show no significant differences for this constituent (Chauhan et al., 1988; Daelemans, 1989; Meron et al., 1989).

Results of pH are seen in Figure 4. Soil solarization decreased significantly the pH values. Similar results were obtained by Gamliel and Katan (1991), Pinto (1998) and Pinto and César (1999). However, we cannot show any significant difference for this parameter, in the study made by Stapleton and Devay (1982), Pinto (1992) and Arora (1998).

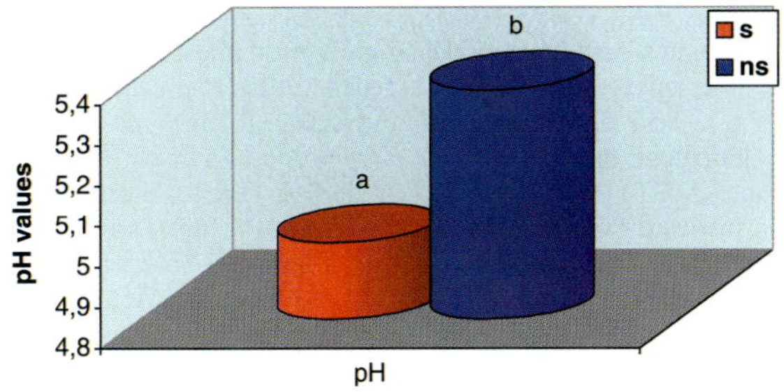

Figure 4. pH values in solarized and no solarized soils. For each pair columns with different letter are significantly different ($P < 0.05$).

Results of total organic matter were placed in Figure 5. No significant differences were observed for this constituent. Similar results were obtained by Daelemans (1989), Pinto (1992, 1998), Arora (1998) and Pinto and César (1999).

Results of electrical conductivity were placed in Figure 6. We can observe a significant increase in electrical conductivity values in solarized soils. Similar results were obtained by Gamliel and Katan (1991), Pinto (1992, 1998) and Pinto and César (1999). Recently Arora (1998) obtained a marginal decrease in solarized soils.

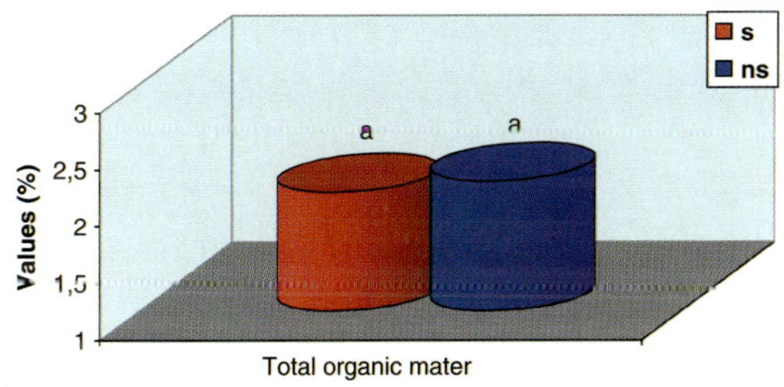

Figure 5. Total organic matter levels in solarized and no solarized soils. For each pair columns with different letter are significantly different ($P < 0.05$).

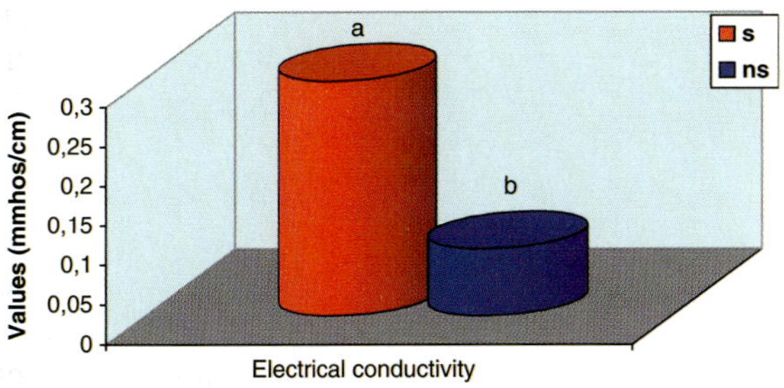

Figure 6. Electrical conductivity levels in solarized and no solarized soils. For each pair columns with different letter are significantly different ($P < 0.05$).

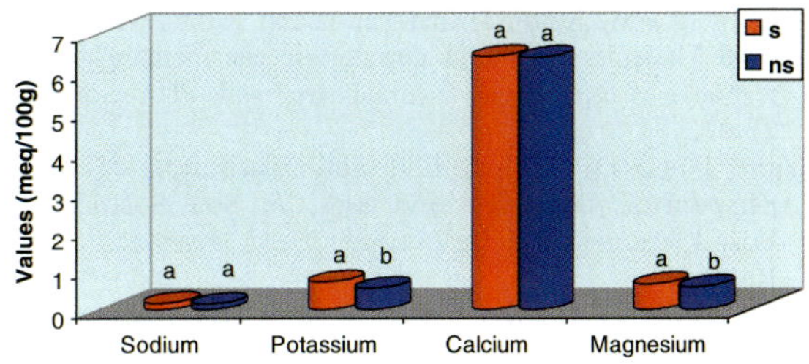

Figure 7. Exchangeable bases levels in solarized and no solarized soils. For each pair columns with different letter are significantly different ($P < 0.05$).

The results of exchangeable bases in Figure 7, show a significant increase in both magnesium and potassium, about 12 and 30%, respectively, in solarized soils. Similar results were obtained by Gamliel and Katan (1991), Grunzweig et al. (1998), Pinto (1998) and Pinto and César (1999). In the other exchangeable bases no significant differences were observed.

Conclusions

The maximum temperature value recorded in solarized soil was 51 °C at 5 cm depth; no significant effects were obtained on available phosphorous, total organic matter and exchangeable bases (calcium and sodium) by solarization; a very significant increase on nitric nitrogen, amoniacal nitrogen and electrical conductivity in solarized soils was found; a significant increase on available potassium, exchangeable potassium and magnesium in solarized soils was recorded; a significant decrease in pH values in solarized soils, was detected.

References

Arora A 1998 Soil solarization: effects on soil moisture, soil temperature and chemical properties of the soil. Crop Res. Hisar. 16, 41–49.

Chauhan Y S, Nene Y L, Johansen C, Haware M P, Saxena N P, Sardar S, Sharma S B, Sahrawat K L, Burford J R, Rupela O P, Kumar-Rao J K and Sithanatham S 1988 Effects of soil solarization on pigeonpea and chickpea. Patancheru, Res. Bull. 11, ICRISAT, pp. 17.

Daelemans A 1989 Soil solarization on West-Cameroon: effect on weed control, some chemical properties and pathogens of the soil. Acta Horticult. 255, 169–175.

Gamliel A and Katan J 1991 Involvement of fluorescent pseudomonads and other microorganisms in increase growth response of plants in soils. Phytopathology 81, 494–502.

Grunzweig J M, Katan J, Ben-Tal Y and Rabinowitch H D 1998 The role of mineral nutrients in the increase growth response of tomato plants in solarized soil. Plant Soil 206, 21–27.

Katan J and DeVay J E 1991 Soil solarization: historical perspectives, principles, and uses. *In* Soil Solarization. Eds. J Katan and J DeVay. pp. 23–37. Press Inc., Boca Raton.

McGovern R J, McSorley R and Urs R R 2000 Reduction of Phytophtora blight of Madagascar periwinkle in Florida by soil Solarization in autumn. Plant Dis. 84, 185–191.

Meron M, Filler Y, Cohen Y, Galilee M and Grinstein A 1989 Solarization and fumigation for reclamation of organic soils. Acta Horticult. 255, 117–224.

Pinto A F M A 1992 Solarização do solo em estufa: efeitos em algumas características biológicas e químicas do solo, nas infestantes e na cultura de feijão verde (*Phaseolus vulgaris* L.). Tese de Mestrado, Universidade Técnica de L.isboa, Portugal, pp. 79.

Pinto A F M A 1998 Solarização do Solo: Efeitos Secundários. Instituto Politécnico de Viseu, ESAV, Portugal, pp. 80.

Pinto A F M A and César A 1999 Efeitos da Solarização Em Alguns Constituintes Químicos do Solo. *In* Actas do V Encontro Nacional de Protecção Integrada, Bragança. 23 p.

Stapleton J J and DeVay J E 1982 Effects of soil solarization on populations of selected soilborne microorganisms and growth of deciduous fruit tree seedlings. Phytopathology 72, 323–326.

Tricalcium-phosphate solubilizing efficiency of rhizosphere bacteria depending on the P-nutritional status of the host plant

A. Deubel[1], A. Gransee & W. Merbach
Institute of Soil Science and Plant Nutrition, Martin Luther University Halle-Wittenberg, Adam-Kuckhoff-Str. 17b, D-0618, Halle/Saale, Germany. [1]*Corresponding author**

Received 8 November 2002. Accepted in revised form 2 January 2003

Key words: microbial transformation of root deposits, phosphate mobilization, rhizodeposition, rhizosphere bacteria

Abstract

This study assesses the influence of saccharides in the rhizodeposition on the phosphate solubilizing ability of rhizosphere bacteria. Water-soluble rhizodeposits were analysed of ^{14}C labelled pea plants (*Pisum sativum*) which were grown at two different levels of P-nutrition. The sugars produced were fed *in vitro* either as single compounds or as synthetic mixtures to three bacterial strains and the ability of the bacteria to mobilize $Ca_3(PO_4)_2$ was measured.The ability of *Pseudomonas fluorescens* (PsIA12) to dissolve tertiary calcium phosphate was lower with sugars of the P-deficient plants than with glucose. On the other hand, sugars of P-deficient plants increased the P-mobilizing ability of *Pantoea agglomerans* (D 5/23) and *Azospirillum* sp. (CC 322) considerably. The production of different carboxylic acids was responsible for these effects.

Introduction

Root deposits of higher plants contain ecological relevant C amounts (11–20% of net CO_2 assimilation e.g., 13–32% of ^{14}C incorporated into the plant). About 75% of the organic root deposits are water-soluble (Merbach et al., 1999). Sugars, amino acids and carboxylic acids are main components, with sugars usually representing the largest part. P-mobilizing effects of carboxylic acid exudation are well documented, but less is known about the influence of sugar exudates on phosphate availability.

This study should answer following questions:

- Which sugars can be utilized by different bacterial strains as a substrate and source for P-mobilizing substances?
- Do changes in the sugar exudation of plants influence the phosphate mobilizing processes in the rhizosphere?
- What are the major mechanisms in the bacterial mobilization of calcium phosphate?

Materials and methods

Plant growth, extraction and analysis of organic root deposits

Plants (*Pisum sativum* cv. 'Grapis') were grown in sand culture at different P-nutritional levels. Before harvest they were labelled with $^{14}CO_2$ (3–6 days). Roots with the adhering sand were dipped for 2 min into cold water to collect water-soluble exudates. After filtration and freeze-drying of the suspension, water-soluble compounds were separated by ion exchange

* E-mail: annette.deubel@landw.uni-halle.de

E. Velázquez and C. Rodríguez-Barrueco (eds.),
First International Meeting on Microbial Phosphate Solubilization, 257–260

chromatography into sugars, amino acids/amides and carboxylic acids. Spectrum of individual compounds was identified by HPLC and GC (Gransee and Wittenmayer, 2000).

Cultivation of bacteria

Three P-solubilizing strains were used: D 5/23 (*Pantoea agglomerans*), PsIA12 (*Pseudomonas fluorescens*), CC 322 (*Azospirillum* sp.). Bacteria were cultivated in standing liquid culture (120 mL glass bottles, 30 mL MUROMCEV medium containing 1% glucose and 0.1% asparagine as C and N sources, 30 mg $Ca_3(PO_4)_2$, 28 °C). In order to determine the influence of root exudates on P-solubilizing ability of the strains, glucose was replaced either by single other sugars that were found in root deposits or by synthetic sugar mixtures which corresponded to the saccharides released by optimal nourished or P-deficient plants (equal quantitative C supply). After an incubation time of 1 weak, pH, protein concentration and the concentration of solved P were measured in the solution. Carboxylic acids produced by bacteria were determined with the same technique as in root exudates (Deubel et al., 2000).

Results and discussion

P-deficient pea plants have a lower sugar exudation (59% of water-soluble organic rhizodeposition) in comparison to well nourished plants (70%) combined with a higher release of carboxylic acids (0 mg P: 18%, 80 mg P: 8%). But there are also remarkable qualitative changes in the composition of the sugar fraction (see Figure 1). The relative glucose proportion of pea exudates decreased under P-deficiency while the content of galactose, ribose, xylose and fucose increased.

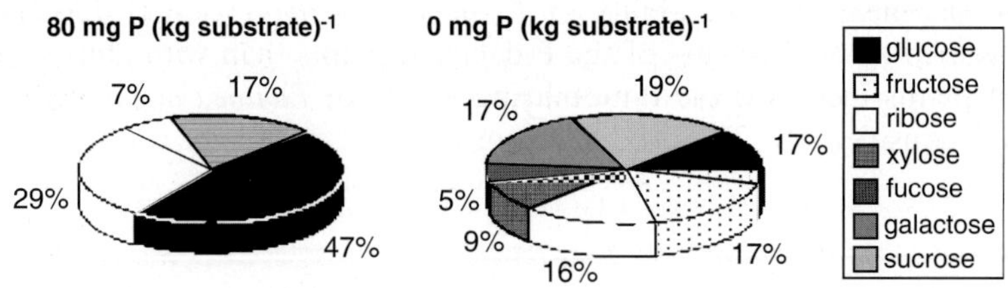

Figure 1. Composition of the sugar fraction of organic rhizodeposition of pea plants (cv. 'Grapis', 3 weeks old) depending on phosphate nutrition.

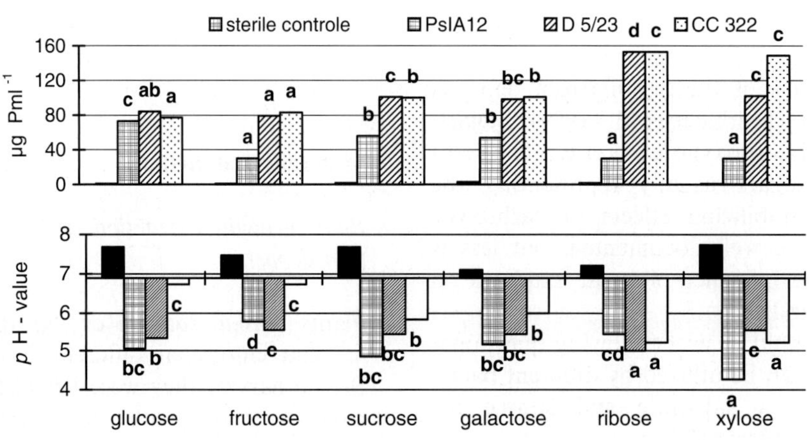

Figure 2. Mobilization of $Ca_3(PO_4)_2$ by three bacterial strains compared to sterile solution (control) in dependence of the kind of sugar as C source (incubation time: 7 days). The lower part shows the changes in pH values of the nutrient media during the incubation (initial pH 6.9). Statistical comparison between the sugars within a bacterial strain, different letters indicate significant differences (Tukey, $P \leq 0.05$).

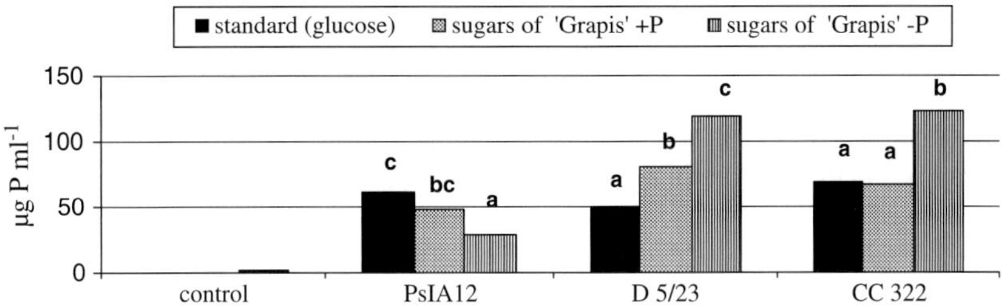

Figure 3. Influence of synthetic sugar mixtures (in analogy to root deposits of pea plants with and without P supply) on the $Ca_3(PO_4)_2$ – solubilizing ability of *Pseudomonas fluorescens* (PsIA12), *Pantoea agglomerans* (D5/23) and *Azospirillum* sp. (CC 322). Incubation time: 7 days. Statistical comparison between the sugar treatments within a bacterial strain, different letters indicate significant differences (Tukey, $P \leq 0.05$).

P-solubilizing ability of rhizosphere bacteria depends on the kind of sugar as C source. While it was comparable in all 3 strains with glucose (standard medium), it was lower in PsIA12 with all other sugars but higher in D 5/23 and CC 322, particularly with ribose and xylose, (Figure 2). The lower part of this figure shows the change of the pH value during the incubation.

In vitro feeding of single sugars and sugar mixtures showed, that the ability of *Pseudomonas fluorescens* (PsIA12) to dissolve tertiary calcium phosphate was lower with the mixed sugars corresponding to pea root exudates, especially of P-deficient plants, than with glucose. On the other hand, that shift in the sugar pattern observed under P-deficiency increased the P-mobilizing ability of *Pantoea agglomerans* (D 5/23) and *Azospirillum* sp. (CC 322) considerably (see Figure 3).

There was no close relation between decrease of pH and P mobilization (Figure 2). Bacterial solubilization of $Ca_3(PO_4)_2$ can only partly be explained by the acidification of the nutrient

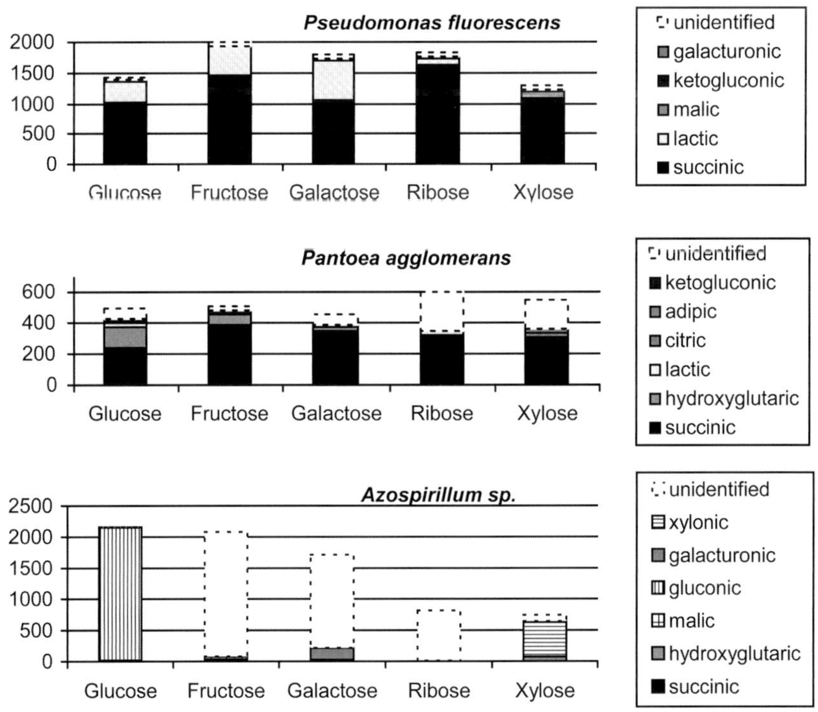

Figure 4. Production of carboxylic acids by rhizosphere bacteria ($\mu g\ mL^{-1}$ nutrient solution) depending on the kind of sugar as C source (incubation time 7 days).

medium. Bacteria also produced different carboxylic anions depending on sugar supply (Figure 4). In addition to low-molecular mono-, di-, and tricarboxylic acids which are known as P-solubilizing substances, sugar acids also played an important role, particularly in strain CC 322.

Conclusion

The P-nutritional status of the host plant has an influence on the P-mobilizing effectiveness of rhizosphere bacteria.

References

Deubel A, Gransee A and Merbach W 2000 Transformation of organic rhizodepositions by rhizosphere bacteria and its influence on the availability of tertiary calcium phosphate. J. Plant Nutr. Soil Sci. 163, 387–392.

Gransee A and Wittenmayer L 2000 Qualitative and quantitative analysis of water-soluble root exudates in relation to plant species and development. J. Plant Nutr. Soil Sci. 163, 381–385.

Merbach W, Mirus E, Knof G, Remus R, Ruppel S, Russow R, Gransee A and Schulze J 1999 Release of carbon and nitrogen compounds by plant roots and their possible ecological importance. J. Plant Nutr. Soil Sci. 162, 373–383.

Phosphate-solubilizing microorganisms in the rhizosphere of *Pinus pinaster* and in the mycosphere of associated *Lactarius deliciosus*

B. Ramos[1,3], J. Barriuso-Maicas[1], J.A. Lucas García[1], T. Pereyra de la Iglesia[1], A. Daza[2] & F.J. Gutierrez Mañero[1]

[1]*Facultad de Farmacia, Universidad San Pablo CEU, P.O. Box 67, Boadilla del Monte, 28668, Madrid, Spain.* [2]*CIFA Las Torres-Tomejil, C/Sevilla-Alcalá del Río Km.12,200. Apto. Oficial, C.P. 41200, Alcalá del Río, Sevilla, Spain.* [3]*Corresponding author**

Received 9 December 2002. Accepted in revised form 2 January 2003

Key words: genetic diversity, mycosphere, PGPR, phosphate solubilisation, rhizosphere

Abstract

Plant hormone production, nutrient mobilization and plant protection are the generally accepted mechanisms for plant growth promotion. The rhizosphere of plants seems to be the most suitable place to search for these beneficial microorganisms since it's a widely accepted fact that plants select those bacteria that may be more beneficial for their health. A screening for PGPRs was carried out in the rhizosphere of *Pinus pinaster* and in the mycosphere of associated *Lactarius deliciosus*. A total of 204 strains, approximately 100 strains from each plant species were isolated, purified, grouped by morphological criteria and tested for ACC degradation, auxin and siderofore production and phosphate solubilization. About 32% of strains showed at least one of these activities *in vitro*, and 47.7% of these were phosphate solubilizers. Most of them were gram-positive bacilli, esporulated if isolated from the rhizosphere, and non-esporulated if isolated from the mycosphere. All isolates that showed at least one of these activities were analyzed by PCR-RAPDs with six random primers. From this analysis and considering data from phosphate-solubilizing gram-postive bacilli only, five groups appeared at 85% similiarity: three for espore forming bacilli, and two for non-espore forming bacilli. Results from PCR-RAPDs of phosphate-solubilizing strains indicate the low diversity existing in the selective environments under the influence of plants or fungi. It's striking that some strains isolated from different areas and plant species showed 100% similarity indexes.

Introduction

The mechanisms by which plant growth-promoting rhizobacteria (PGPR) promote plant growth are not fully understood, but are thought to include (i) the ability to produce or change plant hormones concentration, (ii) asymbiotic N_2 fixation, (iii) antagonism against phytopathogenic microorganisms and (iv) solubilisation of mineral phosphates and other nutrients (Bowen and Rovira, 1999). The rhizosphere of plants seems to be the most suitable place to search for these beneficial microorganisms since it is a widely accepted fact that plants select as a result of coadaptation and coevolution processes those bacteria that may be more benefitial for their health (Marilley and Aragno, 1999).

Lactarius deliciosus naturally mycorrhizates *Pinus pinaster* improving pine health and helping its establishment in degraded lands. Therefore, improvement of this relationship appears as an encouraging alternative for successful reforestation. Under the rationale that those beneficial microorganisms that are present in the rhizosphere of *P. pinaster* and in the mycosphere of

* FAX No: +34-913-510-496.
E-mail: bramsol@ceu.es

E. Velázquez and C. Rodríguez-Barrueco (eds.),
First International Meeting on Microbial Phosphate Solubilization, 261–264

L. deliciosus will be beneficial for both species and probably enhance the mycorrhization process (Mycorrhizal Helper Bacteria, MHB) (Dupponois and Garbaye, 1991), a screening for PGPRs was carried out in the rhizosphere of *P. pinaster* and in the mycosphere of associated *L. deliciosus*. The screening for putative PGPRs was based on their ability to affect the plant's hormonal balance (Hirsch et al., 1997) or improve plant nutrition. A selection of genetically different bacterial strains was carried out to minimize genetic redundancy maintaining maximum genetic variability.

Materials and methods

The rhizosphere and mycosphere of *Pinus pinaster* and *Lactarius deliciosus* were sampled in two different areas in Sierra de Aracena (Southwest Spain), in November 2000. In each area, six individuals of each species were sampled and the soil under the direct influence of each plant was pooled. Ten-fold dilutions were prepared and plated on standard culture media agar and incubated at 28 °C. Isolates were selected after 36 h and 4 days of incubation.

A total of 204 strains, approximately 100 strains from each plant species were isolated, purified, grouped by morphological criteria and tested for ACC degradation (Glick et al., 1995), auxin (Benizri et al., 1998) and siderophore production (Alexander and Zuberer, 1991) and phosphate solubilisation (de Freitas et al., 1997).

All isolates that showed at least one of these activities were analyzed by PCR-RAPDs with six different primers in a Perkin Elmer Cetus DNA Thermal Cycler. DNA isolation and amplification conditions have been described elsewere (Lucas et al., 2001).

Results and discussion

Gram-positive isolates predominated in both environments (Table 1). Consistent with other authors (Gutierrez Mañero et al., 2003), esporulated bacilli predominated over non-esporulated in the rhizosphere opposite to what was found in the mycosphere. Gram-negative isolates were present only in the mycosphere.

As regards to the biochemical activities putative of PGPR traits, 32% of the isolates showed at least one of these activities *in vitro* (data not shown), being phosphate solubilisation the most abundant PGPR trait (Table 1). A higher and more diverse biochemical potential was found in the micosphere, where ACC degraders were present as well as isolates that showed more than one PGPR trait *in vitro*. However, no individual

Table 1. Frequency of isolates and biochemical activities putative of PGPR traits in each morphological group from the rhizosphere of *Pinus pinaster* and from the mycosphere of associated *Lactarius deliciosus*

	Gram-positive esporulated bacilli	Gram-positive non-esporulated bacilli	Gram-positive cocci	Gram-negative bacilli	Total%
Pinus pinaster					
IAA	8.3	5.55	0	0	13.88
CAS	8.3	22.22	0	5.55	36.11
PDYA	44.44	2.70	0	0	47.22
PDYA + ACC	0	2.7	0	0	2.77
Total %	61.11	33.33	0	5.55	
Lactarius deliciosus					
IAA	0	0	0	0	0.00
CAS	19.35	12.90	0	6.45	38.70
PDYA	6.45	22.58	3.22	6.45	38.70
ACC	3.22	9.67	0	0	12.90
IAA + CAS	3.22	0	0	0	3.22
IAA + PDYA + CAS	0	0	0	3.22	3.22
CAS + PDYA + ACC	0	0	3.22	0	3.22
Total %	32.26	45.16	6.45	16.13	

IAA: auxin producers, CAS: siderophore producers, PDYA: phosphate solubilisers, ACC: aminocyclopropanecarboxylic degraders.

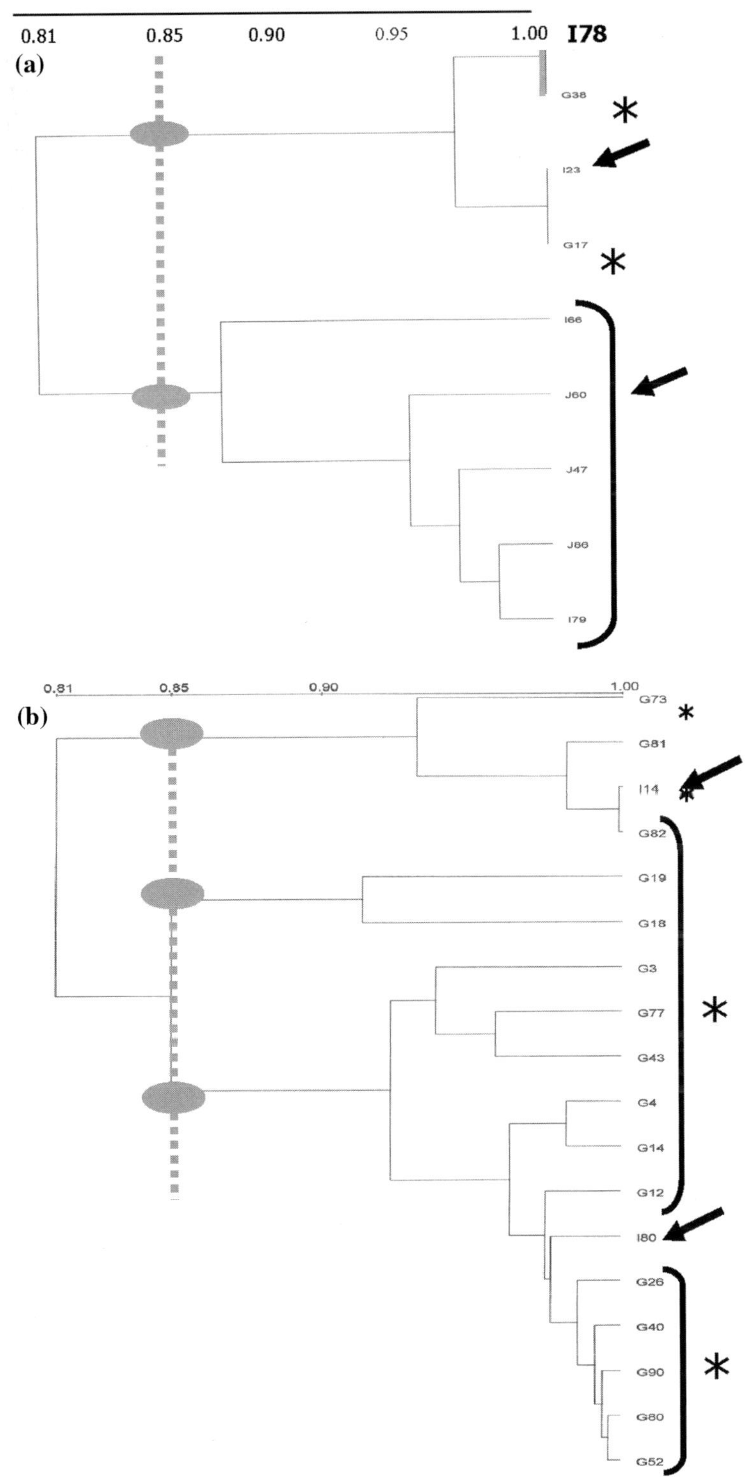

Figure 1. Dendrogram showing the groups obained in this study.

auxin producers were detected while this trait was well represented in the rhizosphere. Presence of ACC degraders in the mycosphere may indicate elongation of the micelium to explore a larger surface, probably searching for pine roots, while auxin producers in the rhizosphere may speak of increasing root surface to improve chances of mycorrhization. Interestingly, the only two isolates with three potential activities (I26 & I68) belong to the morphological groups with the lowest representation, cocci and gram-negative bacilli (data not shown).

All phosphate-solubilising gram-positive isolates were analysed by PCR-RAPDs with six random primers. Figure 1a, b show data from gram-positive non-espore-forming and espore-forming bacilli. At 85% similiarity, only two and three groups appear indicating the low diversity existing in the selective environments under the influence of plants (Lucas et al., 2001; Marilley and Aragno, 1999) or fungi. Interestingly, some isolates from rhizosphere (asterisks) and micosphere (arrows) showing 100% similarity appear among the espore forming and among the non-espore forming bacilli. It is also striking that among the non-espore-forming gram-positive bacilli, all isolates from the rhizosphere and the micosphere with 100% similarity (Figure 1a) gather in one group.

These results reveal the selection that has taken place between the plant and the fungi, from adaptation and coevolution processes (Marilley and Aragno, 1999), resulting in these isolates that being probably mycorrhizae helper bacteria, able to improve the establishment of mycorrhization.

Acknowledgements

This study has been supported by project FEDER 1FD97-1441.

References

Alexander D B and Zuberer D A 1991 Use of chrome azurol S reagents to evaluate siderophore production by rhizosphere bacteria. Biol. Fertil. Soils 12, 39–45.

Benizri E, Courtade A, Picard C and Guckert A 1998 Role of maize root exudates in the production of auxins by *Pseudomonas fluorescens* M.3.1. Soil. Biol. Biochem. 30, 1481–1484.

Bowen G D and Rovira A D 1999 The rhizosphere and its management to improve plant growth. Adv. Agron. 66, 1–103.

Clark A G and Lanigan C M S 1993 Prospects for estimating nucleotide divergence with RAPDs. Mol. Biol. Evol. 10, 1096–1111.

de Freitas J R, Banerjee M R and Germida J J 1997 Phosphate-solubilizing rhizobacteria enhance the growth and yield but not phosphorus uptake of canola (*Brassica napus* L.). Biol. Fertil. Soils 24, 358–364.

Duponnois R and Garbaye J 1991 Mycorrhization helper bacteria associated with the Douglas fir-*Laccaria laccata* symbiosis: effects in vitro and in glasshouse conditions. Ann. Sci. For. 48, 239–251.

Glick B R, Karaturovic D M and Newell P C 1995 A novel procedure for rapid isolation of plant growth promoting pseudomonads. Can. J. Microbiol. 41, 533–536.

Gutierrez Mañero F J, Ramos B, Lucas García J A, Probanza A and Barrientos Casero M L 2003 Systemic induction of biosynthesis of terpenic compounds in *D. lanata*. J. Plant Physiol. 160, 105–113.

Hirsch A M, Fang Y, Asad S and Kapulnik Y 1997 The role of phytohormones in plant–microbe symbioses. Plant Soil 194, 171–184.

Lucas García J A, Probanza A, Ramos B and Gutierrez Mañero F J 2001 Genetic variability of rhizobacteria from wild populations of four *Lupinus* species based on PCR-RAPDs. J. Plant Nutr. Soil Sci. 164, 1–7.

Marilley L and Aragno M 1999 Phylogenetic diversity of bacterial communities differing in degree of proximity of *Lolium perenne* and *Trifolium repens* roots. Appl. Soil. Ecol. 13, 127–136.

Characterization of a strain of *Pseudomonas fluorescens* that solubilizes phosphates *in vitro* and produces high antibiotic activity against several microorganisms

M.E. Trujillo[1], E. Velázquez[1], S. Miguélez[1], M.S. Jiménez[2], P.F. Mateos[1,3] & E. Martínez-Molina[1]

[1]*Departamento de Microbiología y Genética, Edificio Departamental de Biología, Pl. Doctores de la Reina, Universidad de Salamanca, Salamanca, 37007, Spain.* [2]*Laboratorio de referencia de Micobacterias, Servicio de Bacteriología, Centro Nacional de Microbiología, Madrid, Spain.* [3]*Corresponding author**

Received 28 October 2002. Accepted in revised form 2 January 2003

Key words: phosphate solubilizing bacteria, antibiotic, *Pseudomonas fluorescens*

Abstract

The genus *Pseudomonas* is known as a P-solubilizer and antibiotic producer. In this work we have characterized a strain, PFBV1, which actively solubilizes phosphate *in vitro* and produces several antibiotics with high activity against several microorganisms. Using LMW RNA profile analysis we identified this strain as *Pseudomonas fluorescens*. Antimicrobial activity was detected against a wide spectrum of microorganisms, which included Gram negative bacteria such as *Proteus vulgaris*; Gram positive bacteria such as *Bacillus subtilis* and yeasts such as *Candida albicans*.

Introduction

The former genus *Pseudomonas* included several phosphate-solubilizing species (Peix et al., 2001a; Rodríguez and Fraga, 1999). Recently, the genus has been separated into several new genera at the same time that new species have been described in the current genus *Pseudomonas sensu stricto* (Kersters et al., 1996). *Pseudomonas fluorescens*, which is commonly isolated from rhizospheric soils, is known as a phosphate solubilizing bacterium, which promotes plant growth (Deubel et al., 2000; Di-Simine et al., 1998). Besides solubilizing phosphate, this species is known as an antibiotic producer. Many strains of *P. fluorescens* have been reported to produce antibiotics such as pyoluteorin that suppress plant pathogens (Nakata et al., 2000; Whistler et al., 2000).

The identification of *Pseudomonas* species is difficult because of the high number of very closely related species included in this genus. Stable Low Molecular Weight (LMW RNA) molecules separated by staircase electrophoresis (Cruz-Sánchez et al., 1997) are molecular signatures of prokaryotic and eukaryotic microorganisms (Velázquez et al., 2001) and can be used to identify new isolates.

In this work, a strain isolated from a plant extract was studied for its ability to solubilize phosphate and for the production of antimicrobial metabolites.

Material and methods

Evaluation of bicalcium phosphate solubilization of rhizobial strains

The ability to solubilize bicalcium phosphate was tested on YED-P plates (yeast extract 0.5%;

* FAX No: +34-923-224876.
E-mail: pfmg@usal.es

E. Velázquez and C. Rodríguez-Barrueco (eds.),
First International Meeting on Microbial Phosphate Solubilization, 265–268

glucose 1% and agar 2%; of bicalcium phosphate 0.2%). The inoculated plates were incubated for seven days or until a solubilization zone surrounding the colonies was observed (Peix et al., 2001b).

Phenotypic tests

Isolate PFBV1 was grown in YED plates during 48 h. The ability to oxidize and ferment glucose was checked using the Hugh–Leifson medium. Catalase production was assayed by using 0.3% hydrogen peroxide with one colony taken from YED agar plates. Oxidase activity was detected by using *N,N,N′,N′*-tetramethyl-1,4-phenylenediamine dihydrochloride. Other physiological and biochemical tests were performed using the API 20NE strip (BioMérieux, France) following the manufacturer's instructions and the data obtained were analysed with the software API-LAB 3.3. version.

LMW RNA extraction and SCE LMW RNA profiling

RNA extraction was accomplished following the phenol/chloroform method described by Höfle (1988), with cells grown on tryptone-yeast agar, TY (Beringer, 1974). LMW RNA profiles were obtained using staircase electrophoresis (Cruz-Sánchez et al., 1997), which was performed in 400× 360 × 0.4 mm gels in a vertical slab unit (Poker Face SE 1500 Sequencer, Hoeffer Scientific Instruments, San Francisco, CA, USA). The separating gel contained 14% acrylamide/Bis (acrylamide: *N,N*-methylene bisacrylamide 29:1 (w/w), 7 M urea in TBE buffer: 100 m*M* Tris, 83 m*M* boric acid, 1 m*M* EDTA, pH: 8.5) in TBE buffer, pH: 8.5. Samples containing 3 μg were added to 5 μL of a loading solution (300 mg/mL of sucrose, 460 mg/mL of urea, 10 μL/mL 20% SDS, 1 mg/mL xylene cyanol) and, after 10 min of heating at 70 °C, applied to each well. The commercial molecules from Boehringer Manheim (Manheim, Germany) and Sigma (St. Louis, MO, USA) were used as reference: 5S rRNA from *Escherichia coli* MRE 600 (120 and 115 nucleotides) (Bidle and Fletcher, 1995), tRNA specific for tyrosine from *E. coli* (85 nucleotides) and tRNA specific for valine from *E. coli* (77 nucleotides) (Sprinzl et al., 1985). Before running the pre-electrophoresis (30 min at 100 V), the system was stabilized at 50 °C. The running buffer (TBE, ×1.2) was recycled at a flow rate of 300 mL/min with a peristaltic pump (MasterFlex, Cole Parmer Instruments, Chicago, Illinois, USA). After electrophoresis, gels were silver-stained according to Haas et al. (1994).

Detection of antibiotic activity in vitro

The production of antimicrobial substances by strain PFVB1 was determined using various target fungal and bacterial strains (Table 1). Isolate PFVB1 was grown in nutrient broth amended with Cu^{2+} (25 ppm) for 48 h at 28 °C and 180 rpm. The culture broth obtained was centrifuged at 7000 × *g* and filtered through a 42 μm diameter filter (Millipore, France) and the resultant supernatant used to impregnate Whatman paper discs with 200 μL aliquots. The test strains were inoculated on antibiotic medium No. 1 (Difco) plates using a sterile swab; the discs were placed on the surface agar of the inoculated plates and the activity was detected after 24 h of incubation. The activity against *B. subtilis* was determinated semiquantitatively on plates by comparison with different penicillin concentrations.

Results and discussion

Evaluation of bicalcium phosphate solubilization of rhizobial strains

The diameter of the clearing halo zones after 5 days of incubation on YED-P plates was

Table 1. Sensibility of different microbial species to the antibiotics produced by the strain PFBV1

	Sensibility
Gram positive bacteria	
Bacillus subtilis	+
Mycobacterium phlei	+
Gram negative bacteria	
Escherichia coli	–
Proteus mirabilis	w
Yeasts	
Candida albicans	+

(+), sensible; (–), resistant; (w), weakly sensible.

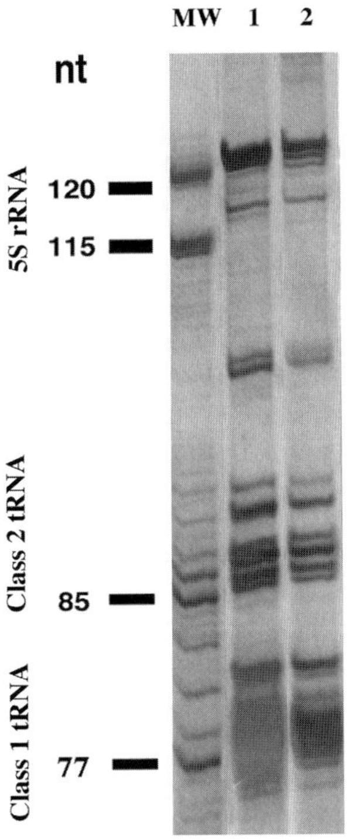

Figure 1. LMW RNA profiles of type strain of *P. fluorescens* CECT378^T (lane 1) and strain PFBV1 (lane 2).

approximately 10 mm. Therefore, according to the criterion of de Freitas et al. (1997) this strain showed a moderate activity as a P-solubilizer.

Phenotypic tests

Strain PFVB1 is a Gram negative rod unable to ferment glucose in Hugh–Leifson medium. Catalase and oxidase were positive. The results obtained using the API20NE strip gave a 99.9% identification index with *P. fluorescens* when the results were compared to those in the API database (Biomerieux, France). The strain was positive for arginine dehydrolase and it was able to grow in arabinose, gluconate, caprate, malate and citrate as sole carbon sources. Therefore based on these results isolate PFVB1 was identified as *P. fluorescens.*

LMW RNA profiles

LMW RNA profiling was used to confirm the identification of strain PFBV1, as *P. fluorescens* as suggested by the phenotypic data. Figure 1 shows the LMW RNA profiles of strain (lane 1) and that of *P. fluorescens* CECT378^T (lane 2), respectively. As can be seen, the LMW RNA profiles were identical.

Detection of antibiotic activity in vitro

The results obtained for the antimicrobial activity presented by PFVB1 are shown in Table 1. This microorganism produced a wide range of secondary metabolites with antimicrobial activity against an array of microorganisms including *Bacillus subtilis, Mycobacterium phlei, Proteus mirabilis* and *Candida albicans.* These results suggest that isolate PFVB1 may be used as a potential phosphate solubilizer as well as a biological control agent.

Acknowledgements

This work was supported by the Junta de Castilla y León to PFM.

References

Beringer J E 1974 R factors transfer in *Rhizobium leguminosarum.* J. Gen. Microbiol. 84, 188–198.

Bidle K D and Fletcher M 1995 Comparison of free-living and particle-associated bacterial communities in the Chesapeak Bay by stable low-molecular-weight RNA analysis. Appl. Environ. Microbiol. 61, 944–952.

Cruz-Sánchez J M, Velázquez E, Mateos P F, Velázquez E and Martínez-Molina E 1997 Enhancement of resolution of low molecular weight RNA profiles by staircase electrophoresis. Electrophoresis 18, 1909–1911.

de Freitas J R, Banerjee M R and Germida J J 1997 Phosphate-solubilizing rhizobacteria enhance the growth and yield but not phosphorus uptake in canola (*Brassica napus* L.). Biol. Fertil. Soils. 24, 358–364.

Deubel A, Gransee A and Merbach W 2000 Transformation of organic rhizodepositions by rhizosphere bacteria and its influence on the availability of tertiary calcium phosphate. J. Plant Nutr. Soil Sci. 163, 387–392.

Di-Simine C D, Sayer J A and Gadd G M 1998 Solubilization of zinc phosphate by a strain of *Pseudomonas fluorescens* isolated from a forest soil. Biol. Fertil. Soils 28, 87–94.

Haas H, Budowle B and Weiler G 1994 Horizontal polyacrylamide gel electrophoresis for the separation of DNA fragments. Electrophoresis 15, 153–158.

Höfle M G 1988 Identification of bacteria by low molecular weight RNA profiles: A new chemotaxonomic approach. J. Microbiol. Methods 8, 235–248.

Kersters K, Ludwig W, Vancannegt M, de Vos P, Gillis M and Schleifer K H 1996 Recent changes in the classification of the pseudomonads: An overview. System. Appl. Microbiol. 19, 465–477.

Nakata K, Harada N, Sumitomo K and Yoneda K 2000 Enhancement of plant stem growth by flocculation of the antibiotic-producing bacterium, *Pseudomonas fluorescens* S272, on the roots. Biosci. Biotechnol. Biochem. 64, 459–465.

Peix A, Mateos P F, Rodríguez-Barrueco C, Martínez-Molina E and Velázquez E 2001a Growth promotion of common bean (*Phaseolus vulgaris* L.) by a strain of *Burkholderia cepacia* under growth chamber conditions. Soil Biol. Biochem. 33, 1927–1935.

Peix A, Rivas-Boyero A A, Mateos P F, Rodríguez-Barrueco C, Martínez-Molina E and Velázquez E 2001b Growth promotion of chickpea and barley by a phosphate solubilizing strain of *Mesorhizobium mediterraneum* under growth chamber conditions. Soil Biol. Biochem. 33, 103–110.

Rodríguez H and Fraga R 1999 Phosphate solubilizing bacteria and their role in plant growth promotion. Biotechnol. Adv. 17, 319–339.

Sprinzl M, Moll J, Meissner F and Hatmann T 1985 Compilation of tRNA sequences. Nucleic Acid Res. 13, 1–49.

Velázquez E, Trujillo M E, Peix A, Palomo J L, García-Benavides P, Mateos P F, Ventosa A and Martínez-Molina E 2001 Stable low molecular weight RNA analyzed by staircase electrophoresis, a molecular signature for both prokaryotic and eukaryotic microorganisms. Syst. Appl. Microbiol. 24, 490–499.

Whistler C A, Stockwell V O and Loper J E 2000 Low protease influences antibiotic production and UV tolerance of *Pseudomonas fluorescens* Pf-5. Appl. Environ. Microbiol. 66, 2718–2725.

Phosphate solubilizing bacteria isolated from the inside of *Glomus mosseae* spores from Cuba

Loreli Mirabal-Alonso[1] & Eduardo Ortega-Delgado[2,3]
[1]*INCA (National Institute of Agriculture Science), PO Box 1, San José de las Lajas, Havana, Cuba.* [2]*Lab. Plant Physiology, Dpt Plant Biol, Biology Fac, University of Havana, Havana, Cuba.* [3]*Corresponding author**

Received 24 December 2002. Accepted in revised form 2 January 2003

Key words: phosphate solubilizing bacteria, mycorrizhae, poli P granules

Abstract

From 13 bacterial isolates obtained from *Gomus mosseae* spores, 9 have capacity to solubilize calcium phosphate. The isolates EndoGm1, 5, 10 and 11, presented higher phosphate solubilizing capacity from all isolates. Tetrad formation was observed by the isolates EndoGm1 and EndoGm11 and also capacity of poliP granules accumulation. All of these characteristics would be important to the interaction mycorrhizas-plant-bacteria, which may provide phosphorus to the plant metabolism.

Introduction

Arbuscular mycorrhizal fungi (AMF) are obligated endosymbiont. They are in numerous natural and agricultural ecosystems and they have a great responsibility in physiological integrity of plants. Recent studies have described endosymbiotic bacteria isolated from arbuscular mycorrhizal fungi (Bianciotto et al., 2000). Many bacteria synthesize different reserve polymers with poliphosphate (Poly P) being of special interest.

In this report, we determined the comparative phosphate solubility capacity from different endosporic bacteria isolated from *Glomus clarum* and we characterized these bacteria.

Materials and methods

Isolation of phosphate solubilizing bacteria from G. mosseae *spores*

Glomus mosseae spores (Figure 1) were isolated from the root system of sorghum plant growing in the Cuban soil type mineral substrate.

Spores were disinfected externally step by step with solutions of Cloramine T 5%, Tween 40 and Cephalexine (2.5 g L^{-1}); the last one during 6 h, as previously reported by Mirabal et al. (2002).

Spores were treated as following:

(a) mixed with sterile distilled water in a mortar and macerated.
(b) mixed with sterile distilled water in a mortar but non macerated.

Thirty microliters of broth from (a) or (b) were inoculated in the mediums LGIP, Nutrient Agar, and SYP

Evaluation of the ability to solubilize phosphate

Halo formation capacity was observed inoculating 10 μL of bacterial culture in exponencial phase (SYP media) in plate with tricalcium phosphate as phosphorus supplementation source, according to Illmer and Schinner (1992).

The concentration of the soluble phosphate was determinated in a EndoGm1 culture in NBRIP medium as proposed by Nautiyal (1999).

* E-mail: eortega@fq.uh.cu

E. Velázquez and C. Rodríguez-Barrueco (eds.),
First International Meeting on Microbial Phosphate Solubilization, 269–272

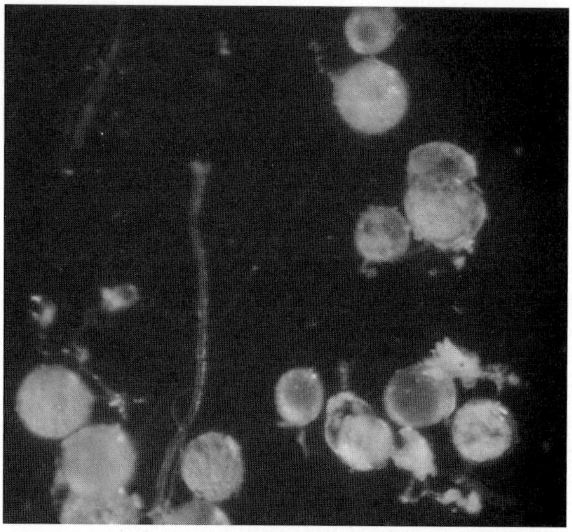

Figure 1. Spores of *Glomus mosseae* from which PSB bacteria were isolated.

Morphological and biochemical characterization

Biochemical and morphological characteristics of the isolates were determinated according to Harrigan and McCance (1968).

Results and discussion

Thirteen isolates were obtained in the present study from *G. mosseae* spores surface totally desinfected and macerated (a). Any bacterial growth was detected from treatment (b).

Nine of the thirteen isolates were able to solubilize phosphate. The capacity of the isolates to solubilize phosphate is relevant (Table 1 and Figure 2); it is better understood comparing with the results of Nautiyal (1999) who found halo diameters between 2 and 8 mm for bacteria

Table 1. Biochemical characteristics of the 13 isolates obtained from the inside of *Glomus mosseae* spores

Isolated	Indol production	Starch hydrolysis	Growth in NaCl 3%	Nitrate reductase⁻	Acid production from carbohydrate				
					Arabinose	Maltose	Glucose	Mannose	Sucrose
EndoGm1	+	–	+ (orange)	–	–	+	+	+	+
EndoGm2	+	–	–	–	+	–	+	–	+
EndoGm3	+	–	+ (white)	– (a lot of burbles)	–	–	+	–	–
EndoGm4	–	–	+ (white)	–	+	–	+	–	–
EndoGm5	+	–	+ (white)	– (a lot of burbles)	+	–	–	+	–
EndoGm6	+	+ (2mm)	–	No growth	+	–	–	+	–
EndoGm7	–	–	+ (white)	+	–	–	+	+	+
EndoGm8	+	–	+ (white)	–	+	–	+	–	+
EndoGm9	+	–	+ (white)	–	+	–	+	–	–
EndoGm10	–	+ (5mm)	–	–	+	–	+	+	–
EndoGm11	+	–	+ (white)	–	+	+	+	+	+
EndoGm12	–	+	–	–	+	–	–	–	–
EndoGm13	+	–	–	–	–	–	+	–	–

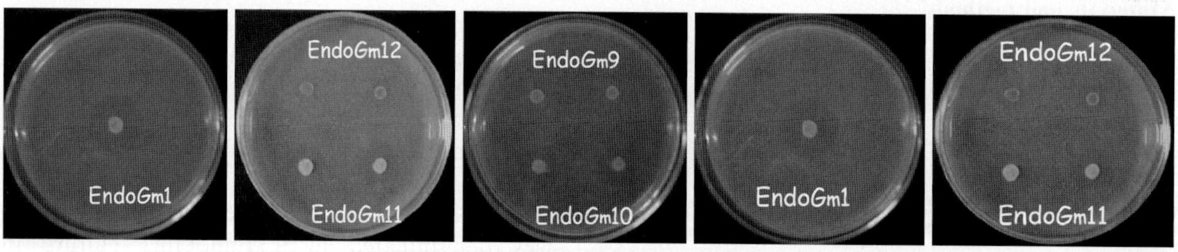

Figure 2. Characteristics of the halo formation by some isolates with different phosphate solubilizing activity in Illmer and Schinner growing medium. Petri dishes are 9 cm in diameter.

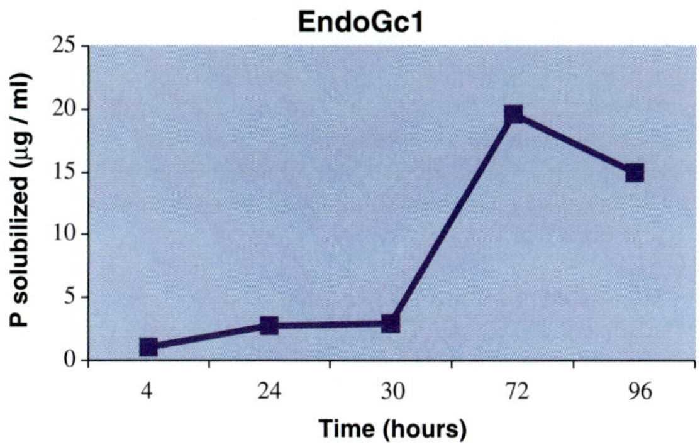

Figure 3. Kinetic of phosphate solubilization by isolate EndoGm1.

belonging to genus *Pseudomonas* and *Bacillus* growing during 14 days. The isolates obtained in this work produce halos between 4 and 52 mm after 8 days growing. The kinetic of phosphate solubilization by isolate EndoGm1 is showed in Figure 3.

The morphological and biochemical characteristics of the isolates are shown in Tables 2 and 3.

Bacterium-like – organisms (BLOs) were identified inside the arbuscular mycorrhizal (AM) fungi *Glomus caledonius* (MacDonald and Chandler, 1981). Later on the group of Paola Bonfante has reported (Bianciotto et al., 2000) the identification of *Burkholderia* sp. as endo-symbiont of the AM fungi *Gigaspora margarita*, with the identification of a putative P-transporter operon (Ruiz-Lozano and Bonfante, 1999) in the genome of the bacteria. This added new information to the knowledge of the participation AM fungi in the plant phosphorus nutrition.

Table 3. Phosphate solubility (halo radio per colony in cm)

Isolated	24 h	72 h	192 h (8 days)
EndoGm1	No halo	1.4	2.4
		1.6	2.6
EndoGm2	No halo	0.6	0.6
		0.6	0.6
EndoGm3	No halo	No halo	No halo
EndoGm4	0.6	0.8	1.0
EndoGm5	0.6	2.4	3.6
EndoGm6	No halo	0.2	0.2
EndoGm7	No halo	No halo	No halo
EndoGm8	0.4	1.2	1.6
	0.4	1.2	1.4
EndoGm9	No halo	No halo	No halo
EndoGm10	1.8	3.0	5.0
	1.4	3.0	4.2
EndoGm11	No halo	1.0	2.4
		1.2	2.6
EndoGm12	No halo	0.2	0.6
EndoGm13	No halo	No halo	No halo

Table 2. Other biochemical and microbiological characteristics of the isolates

Isolated	Morphologic characteristic and cell mean length	Gram staining reaction	Catalase	Oxygen requirement
EndoGm1	Similar to a big cocci, in some cultures is similar to G bacteria (Mazenan, 1998)	Difficult to determine	+	Aerobic
EndoGm2	Bacillus, long and thin, 1.2 μm	−	+	Aerobic
EndoGm3	Chain of bacillus, 1.1 μm	−	−	Aerobic
EndoGm4	Bacillus, 1.15 μm	−	−	Aerobic
EndoGm5	Cocci-bacillus, 0.6 μm	−	−	Aerobic
EndoGm6	Bacillus with ending spore, 1.65 μm	+	+	Aerobic
EndoGm7	Small bacillus, 0.8 μm	−	−	Aerobic
EndoGm8	Bacillus, long and thin, 1.15 μm	−	+	Aerobic
EndoGm9	Bacillus, 0.85 μm	−	+	Aerobic
EndoGm10	Bacillus, 0.85 μm	−	+	Aerobic
EndoGm11	Similar to a big cocci, in some cultures is similar to G bacteria (Mazenan, 1998)	Difficult to determine	+	Aerobic
EndoGm12	Bacillus, 1.25 μm	−	+	Aerobic
EndoGm13	Bacillus, 0.9 μm	−	+	Aerobic

To our knowledge, there are no previous reports of phosphate solubilizing bacteria in *Glomus mosseae* or other AM fungi. This information could be a contribution to understand the relationship between endosymbiotic bacteria and AM fungi and its role in plant nutrition.

References

Bianciotto V, Lumini E, Lanfranco L, Minerdi D, Bonfante P and Perotto S 2000 Detection and identification of bacterial endosymbionts in arbuscular mycorrhizal fungi belonging to the family *Gigasporaceae*. Appl. Environ. Microbiol. 66, 4503–4509.

Mirabal L, Ortega E, Rodés R and Fernández F 2002 Método efectivo para la desinfección total de esporas de hongos micorrizogenos arbusculares (HMA). Aislamiento y caracterización de bacterias endospóricas de *Glomus clarum*. Cultivos Tropicales 23, 21–24.

Illmer P and Schinner F 1992 Solubilization of inorganic phosphates by microorganisms isolated from forest soil. Soil Biol. Biochem. 24, 389–395.

Nautiyal C S 1999 An efficient microbiological growth medium for screening phosphate solubilizing microorganisms. FEMS Microbiol. Lett. 170, 265–270.

Harrigan W F and McCance M E 1968 Métodos Básicos en Microbiología. Editorial Academia, 7 p.

MacDonald R M and Chandler M R 1981 Bacterium-like organelles in the vesicular-arbuscular mycorrhizal fungus *Glomus caledonius*. New Phytol. 89, 241–246.

Mazenan A M 1998 The G Bacteria puzzle. La Trobe University, Bendigo. http//www.scitrav.com/wwater/asp1/gbact.htm.

Ruiz-Lozano J M and Bonfante P 1999 Identification of a putative P-transporter operon in the genome of a *Burkholderia* strain living inside the arbuscular mycorrhizal fungus *Gigaspora margarita*. J. Bacteriol. 181, 4106–4109.

Polyphasic characterization of phosphate-solubilizing bacteria isolated from rhizospheric soil of the north-eastern region of Portugal

A. Valverde, J.M. Igual & E. Cervantes[1]
Instituto de Recursos Naturales y Agrobiología de Salamanca, IRNA-CSIC, Cordel de Merinas, 40-52, 37008 Salamanca, Spain. [1]*Corresponding author**

Received 23 December 2002. Accepted in revised form 2 January 2003

Key words: phosphate solubilizing bacteria, *Pseudomonas*, rhizosphere, soil

Abstract

Phosphate-solubilizing microorganisms are often used as plant growth promoters. In the frame of a project focused on the isolation of endosymbionts and ectorhizospheric bacteria associated with *Phaseolus vulgaris* plants growing in a mountain soil at the north-east of Portugal, we obtained six bacterial strains with high phosphate-solubilizing capability, as demonstrated by the formation of 'halos' when plated in YED medium supplemented with tricalcium-phosphate. These bacterial strains were characterized by a polyphasic approach using both phenotypic (API 20 NE) and molecular assays (RAPD, with M13 primer, TP-RAPD, and 16S rDNA sequencing). TP-RAPD yielded an identical band patterns in the six strains indicating that they belong to the same bacterial species. The 16S rRNA sequence analysis of a group-representative strain (P4-22) revealed a sequence similarity value of 99.27% with *Pseudomonas jessenii*. Moreover, the RAPD fingerprints of these six strains did not show differences, indicating that they form a highly homogenous group. This high homogeneity could be a consequence of the recurring agronomical practices used in this region from the antiquity that include organic fertilization and monoculture of *Phaseolus vulgaris*.

Introduction

The phosphorous is an essential plant nutrient which is added to soil as soluble inorganic phosphates that, in a large portion, becomes insoluble and, therefore, unavailable to plants (Singh and Kapoor, 1994). Many species of bacteria are able to solubilize phosphates *in vitro* and the most of them live in the plant rhizosphere. At the present, bacilli, rhizobia and pseudomonas are the most studied P-solubilizers (Rodríguez and Fraga, 1999). Nevertheless, only a few number of species belonging to the current genus *Pseudomonas* are known as P solubilizers. Within them, *Pseudomonas putida* (Kumar and Singh, 2001; Manna et al., 2001; Villegas and Fortin, 2002; Viveganandan and Jaurhi, 2000), *P. aeruginosa* (Musarrat et al., 2000), *P. corrugata* (Pandey and Palni, 1998), *P. stutzeri* (Vázquez et al., 2000) *Pseudomonas fluorescens* are the most studied (Deubel et al., 2000; Di-Simine et al., 1998). Nevertheless, many rhizospheric phosphate solubilizing bacterial species remain unknown and more studies are needed to reveal the high biodiversity of these bacteria. Although the study of rhizospheric bacteria is difficult because the high number of bacteria present in the soil, the characterization and identification of these bacteria are needed for wide ecological studies of the plant rhizosphere.

* FAX No: +34-923219609.
E-mail: ecervant@usal.es

E. Velázquez and C. Rodríguez-Barrueco (eds.),
First International Meeting on Microbial Phosphate Solubilization, 273–276

Materials and methods

Bacterial isolation and evaluation of their tricalcium-phosphate solubilizing capability

Phosphate-solubilizing bacteria were isolated from rhizospheric samples by plating serial dilutions of rhizospheric soil extracts in YED (yeast extract 0.5%, glucose 1% and agar 2%) supplemented with 0.2% of tricalcium-phosphate (YED-P). The plates were incubated at 28 °C for 7 days. After this time the colonies surrounded by a clear zone, that indicates the phosphate solubilization capability, were selected to obtain pure cultures.

To measure the ability to solubilize tricalcium-phosphate and to test the persistence of this capability, each strain was plated in YED-P medium and incubated for 7 days. The criterium for strain selection was the size of clearing zone surrounding the colonies of each strain and the stability of its phosphate-solubilizing capability after five subcultures (Igual et al., 2001).

Phenotypic test

The selected strains were first subjected to the commercial phenotypic assay API 20 NE. The API 20 NE test was carried out under the conditions described by the manufacturer and identification was done by the Apilab Plus v.3.3.3 software.

TP-RAPD and RAPD

The TP-RAPD profiles were obtained by using the primers and conditions described by Rivas et al. (2001). RAPD profiles were obtained by using primer M13 (Schonian et al., 1993) under the conditions previously described.

16S sequencing and analysis

DNA extraction was carried out as previously described (Rivas et al., 2001). The amplification of 16S rDNA and its sequencing was performed according to the method already described (Rivas et al., 2002). The sequence obtained was compared with those from the GenBank using the FASTA program (Pearson and Lipman, 1988).

Mobilization of phosphorous in plants

Experiments to study the P mobilization in common bean plants were carried out in plots containing vermiculate supplemented with insoluble tricalcium-phosphate (0.2% w:w) under greenhouse conditions. Six plots, containing six plants per plot, were inoculated as described below and other six plots were maintained as uninoculated controls.

For inoculation, strain P4-22 was grown in Petri dishes with YED for 2 days. After that, sterile water was added to the plates to obtain a suspension with approximately 10^{11} cells ml^{-1} and 1 ml of this suspension was added to each seedling.

At harvest, 30 days after inoculation, shoot dry weights of common bean plants were determined. Plant phosphate content was measured according to the A.O.A.C. methods (Johnson, 1990). The data obtained were analysed by one-way analysis of variance, with the mean values compared using the Fisher's Protected LSD (Least Significant Differences) ($P = 0.05$).

Results and discussion

Bacterial isolation and evaluation of their tricalcium-phosphate solubilizing capability

Several bacterial strains were isolated from rhizospheric soil taken at Arcos de Valdevez (Portugal), where *P. vulgaris* is traditionally cultivated. Only six strains (P4-19 to P4-24) were able to solubilize actively phosphate *in vitro* and showed persistence of this capability after five or more subcultures. These phosphate-solubilizing strains were selected for further characterization.

Phenotypic test

The results of this test showed that all these bacterial strains belong to the species *Pseudomonas fluorescens*. All strains isolated in this study utilize arabinose, mannose, mannitol, N-acethyl-D-glucosamine, gluconate, caprate, malate or citrate as sole carbon sources. By contrast, they do not grow in phenil-acetato nor adipate. This species actively produce urease, β-glucosidase, arginine dihydrolase and β-galactosidase. They do not reduce nitrate to nitrogen and not produce triptophan deaminase nor arginine dehidrolase.

TP-RAPD and RAPD

TP-RAPD yielded an identical band patterns in the six strains indicating that they belong to the same bacterial species. Moreover, the RAPD fingerprints of these six strains did not show differences, indicating that they form a highly homogenous group. Therefore, only the strain P4-22 was selected for further experiments.

This high homogeneity could be a consequence of the recurring agronomical practices used in this region from the antiquity that include organic fertilization and monoculture of *Phaseolus vulgaris*.

16S rDNA sequence analysis

The strain P4-22 was identified at genus level using 16S rDNA complete sequence. The complete sequence of 16S rDNA was compared with those from databanks using the FASTA program (Pearson and Lipman, 1988). This comparison showed a 99.27% whit *P. jessenii* (AF068259) that is in disagreement with those results obtained by using the commercial system API 20 NE, due likely to the fact that this kind of identification systems have been designated for identification of clinical isolates but not for soil bacteria.

Mobilization of phosphorous in plants

The results of the inoculation assays are shown in Table 1. According to these results, no significant differences in dry weight and total P were observed between plants inoculated with strain P4-22 and the uninoculated control plants.

Table 1. Total P per plot

Strain	Dry weight (mg plot^{-1})	Total P (mg plot^{-1})
P4-22	6115^{a}	3.04^{a}
Control	6305^{a}	3.56^{a}

Values followed by the same letter are no significantly different from each other at $P = 0.05$ according to Fisher's Protected LSD (Least Significant Differences).

Moreover, it is observed a decrease in shoot dry weight and P content of the inoculated plants with respect to the uninoculated control plants that may indicate some deleterious effects of this *Pseudomonas* strain on plant growth.

In conclusion, the phosphate-solubilizing ability *in vitro* is not always correlated with phosphate mobilization to plant. In order to obtain efficient inoculants it is very important to identify correctly the target strain and to test its performance with the selected crops to avoid potential deleterious effect on plant growth.

References

Deubel A, Gransee A and Merbach W 2000 Transformation of organic rhizodepositions by rhizosphere bacteria and its influence on the availability of tertiary calcium phosphate. J. Plant Nutr. Soil Sci. 163, 387–392.

Di-Simine C D, Sayer J A and Gadd G M 1998 Solubilization of Zinc phosphate by a strain of *Pseudomonas fluorescens* isolated from a forest soil. Biol. Fertil. Soils 28, 87–94.

Igual J M, Valverde A, Cervantes E and Velázquez E 2001 Phosphate-solubilizing bacteria as inoculants for agriculture: use of updated molecular techniques in their study. Agronomie 21, 561–568.

Jonhson F J 1990 Detection method of nitrogen (total) in fertilizers. *In* Methods of Analysis of the Association of Official Analytical Chemists. Ed. K Elrich. pp. 17–19. Association of Official Analytical Chemists, USA.

Kumar V and Singh K P 2001 Enriching vermicompost by nitrogen fixing and phosphate solubilizing bacteria. Bioresour. Technol. 76, 173–175.

Manna M C, Ghosh P K, Ghosh B N and Singh K N 2001 Comparative effectiveness of phosphate-enriched compost and single superphosphate on yield, uptake of nutrients and soil quality under soybean–wheat rotation. J. Agric. Sci. 137, 45–54.

Musarrat J, Bano N and Rao R A K 2000 Isolation and characterization of 2,4-dichlorophenoxyacetic acid-catabolizing bacteria and their biodegradation efficiency in soil. World J. Microbiol. Biotechnol. 16, 495–497.

Pandey A and Palni L M S 1998 Isolation of *Pseudomonas corrugata* from Sikkim Himalaya. World J. Microbiol. Biotechnol. 14, 411–413.

Pearson W R and Lipman D J 1988 Search for DNA homologies was performed with the FASTA program. Proc. Natl. Acad. Sci. USA 85, 2444–2448.

Rivas R, Velázquez E, Valverde A, Mateos P F and Martínez-Molina E 2001 A two primers random amplified polymorphic DNA procedure to obtain polymerase chain reaction fingerprints of bacterial species. Electrophoresis 22, 1086–1089.

Rivas R, Velázquez E, Willems A, Vizcaíno N, Subba-Rao N S, Mateos P F, Gillis M, Dazzo F B and Martínez-Molina E 2002 A new species of *Devosia* that forms a nitrogen-fixing root-nodule symbiosis with the aquatic legume *Neptunia natans* (L. f.) Druce. Appl. Environ. Microbiol. 68, 5217–5222.

Rodríguez H and Fraga R 1999 Phosphate solubilizing bacteria and their role in plant growth promotion. Biotechnol. Adv. 17, 319–339.

Schonian G, Meusel O, Tietz H J, Meyer W, Graser Y, Tausch I, Presber W and Mitchell T G 1993 Identification of clinical strains of *Candida albicans* by DNA fingerprinting with polymerase chain reaction. Mycoses 36, 171–179.

Singh S and Kapoor K K 1994 Solubilization of insoluble phosphates by bacteria isolated from different sources. Environ. Ecol. 12, 51–55.

Vázquez P, Holguin G, Puente ME, López-Cortez A and Bashan Y 2000 Phosphate solubilizing microorganisms associated with the rhizosphere of mangroves in a semiarid coastal lagoon. Biol. Fertil. Soils 30, 460–468.

Villegas J and Fortín J A 2002 Phosphorous solubilization and pH changes as a result of the interactions between soil bacteria and arbuscular mycorrhizal fungi on a medium containing NO_3^- as nitrogen source. Can. J. Bot. 80, 571–576.

Viveganandan G and Jaurhi K S 2000 Growth and survival of phosphate-solubilizing bacteria in calcium alginate. Microbiol. Res. 155, 205–207.

Effects of plant community composition on total soil microbiota and on phosphate-solubilizing bacteria of ex-arable lands

I. Santa-Regina[1], A. Peix[1,3], T. Díaz[1], C. Rodríguez-Barrueco[1] & E. Velázquez[2]
[1]*Instituto de Recursos Naturales y Agrobiología (IRNA. CSIC), C/Cordel de Merinas, 40-52, 37008, Salamanca, Spain.* [2]*Departamento de Microbiología y Genética, Facultad de Farmacia, Universidad de Salamanca, Salamanca, 37007, Spain.* [3]*Corresponding author**

Received 12 December 2002. Accepted in revised form 2 January 2003

Key words: ex-arable lands, phosphate solubilizing bacteria, plant cover

Abstract

It is well known that trophic interactions between plant cover and soil organisms are key factors in ecosystems balance. Plant–microbe relationships occurring naturally in soil play a central role in composition of plant communities as well as microbial populations in soil. Many agricultural lands from Europe which were abandoned because of Agricultural Common Policy, are being converted into more natural ecosystems in order to enhanced and preserve biodiversity. The aim of this study was to investigate the effects of different experimental plant cover composition on soil microbiota of set-aside land, and the evolution of microbial populations within the different seasons, focusing on microorganisms important in nutrient cycling, and so phosphate solubilizing bacteria were used as indicator for testing these effects and processes. The results showed that generally high plant diversity treatment was the one with more total soil microbiota, but this was not so clear for phosphate-solubilizing bacteria. It was observed that phosphate-solubilizing bacteria was increasing from Autumn to Spring, and had a great fall in Summer, whereas total soil microbiota was more constant in the different seasons.

Introduction

The soil is a complex ecosystem in which microorganisms developed a fundamental role because they established relationships with other organisms and participate in biogeochemical cycles. Intensification of land use, specially agricultural practices in natural ecosystems may change the composition and reduce the diversity of biological communities (Schläpfer and Schmid, 1999a, b). The implication of bacteria in the nitrogen cycle is better known than in the phosphorous cycle. Although it is know that several bacteria are able to solubilize phosphate and that some of these bacteria mobilize phosphorous to plants, the role of these bacteria in the soil ecosystem has been few studied. Nevertheless, several studies have been performed to analyze the biodiversity of phosphate solubilizing bacteria in diverse soils (Ahmad and Jha, 1968, Gupta et al., 1986, Kucey, 1983, Yahya and Al-Azawi, 1989). Some of these studies were carried out in soils subjected to treatments with fungicides or insecticides (Wainwright and Pugh, 1975, Wainwright and Sowden, 1977). Normally, the studies have been performed in cultivated soils and in the most of the cases their aims were to analyze the ability to increase the growth of plants used in animal or human nutrition. Several authors have been also studied the populations present in soils cultivated with trees (Gupta et al., 1986, Subba-Rao, 1982). However, few studies are carried out in abandoned lands to compare the

* FAX No: +34-923-224876.
E-mail: alvarp@usal.es

E. Velázquez and C. Rodríguez-Barrueco (eds.),
First International Meeting on Microbial Phosphate Solubilization, 277–280

microbiota present in rhizospheric soil with diverse plant cover.

Therefore, the aim of this work was to analyze the populations of phosphate solubilizing bacteria present in a soil subjected to different treatments.

Material and methods

Soils and collection of soil samples

Soil samples were collected from plots in a soil in Northern Spain, subjected to different agronomic practices and with different plant cover. Soil analyses were performed according to the guidelines of the Soil Conservation Service (1972). The soils were classified according to their morphology and analytical data following the U.S. Soil Taxonomy (Soil Survey Staff, 1994).

In this soil we have established several experimental plots of 10 m × 10 m. Borderstrips between plots were of 10 × 10 m. Before installing plots, the field was harrowed and equalized. In bloc 1 we tried to promote the ecosystem development in order to check the evolution of bacterial microbiota. The effects of different plant cover were compared. For that, the following treatments have been established in twenty 10 × 10 m plots.

Five plots with natural colonization by plants growing spontaneously were used as control.

About 10 plots were sown with 1250 seeds/m^2 of Gramineae (*Festuca rubra, Phleum pratense, Bromus inermis, Poa trivialis* and *Poa pratensis*), 500 seeds/m^2 of *Leguminosae* (*Lotus corniculatus, Trifolium pratense, Medicago lupulina, Trifolium subterraneum and Trifolium fragiferum*) and 500 seeds/m^2 of other species (*Plantago lanceolata, Sangisorba minor, Achillea millefolium, Galium verum* and *Matricaria chamomilae*). In five plots only five of these species were used (low diversity) and in the other five plots all species were used (high diversity).

Finally, five plots were subjected to cultivation with barley applying the agricultural practices commonly used in this crop.

For bacterial counts, the soil samples were taken from the rhizospheric soil at a depth of 15–20 cm from three sites in each plot. Soil samples were placed in a cool box for transport, stored at 5 °C, and then used for microbial counts within 2 days of collection. The sampling was made after 2 years of treatment from Autumn to Spring.

Enumeration and isolation of phosphate solubilizing bacteria (PSB)

For counts of bacteria phosphate solubilizers we have used the method of Thomas and Shantaram (1986) modified as follows: for each site, the pooled soil was sieved (2 mm) and mixed thoroughly. A 10 g sample from each soil was

Table 1. Characteristics of the soil studied

Coarse sand (%)	Fine sand (%)	Slime (%)	Clay (%)	pH (water)	Carbonate (%)	Organic Matter (%)	N (%)	C/N	Assimilable P (ppm)	Assimilable K (ppm)	Assimilable Ca (ppm)
15.6	59.3	7	18.1	5.9	ND	1.5	0.08	11.0	26.0	158.0	1070.0

ND: Not detected.

Table 2. Counts of total bacteria and PSB per gram of soil

Plant cover	Autum				Winter				Spring				Summer			
	Total	SD	PSB	SD	Total	SD	PSB	SD	Total	SD	PSB	SD	Total	SD	PSB	SD
Natural colonization	212^{a}	61	7^{a}	3	550^{b}	50	24^{b}	1	623^{b}	25	17^{b}	2.5	750^{cd}	25	17^{b}	3
Low diversity	340^{b}	89	9^{a}	4	800^{d}	80	27^{b}	3	600^{b}	50	23^{b}	2.6	550^{a}	78	25^{c}	5
High diversity	354^{b}	110	14^{a}	5	673^{c}	80	30^{b}	2	486^{a}	75	15^{b}	5	640^{b}	78	28^{c}	3
monocrop	200^{a}	100	10^{a}	1	270^{a}	83	2^{a}	1	680^{b}	10	9^{a}	1.1	706^{bc}	11	9^{a}	1

Values followed by the same letter are no significantly different from each other at $P = 0.05$ according to Fisher's Protected LSD (Least Significant Differences)

emulsified in 90 mL of sterile water. Serial decimal dilutions were made from this suspension up to 1:10^7. Five aliquots of 0.1 mL of each dilution were used to inoculate petri dishes with YED (yeast extract 0.5%; glucose 1% and agar 2%) supplemented with a 0.2% of tricalcium-phosphate (YED-P). The plates were incubated at 28 °C for 7 days. After this time the number of

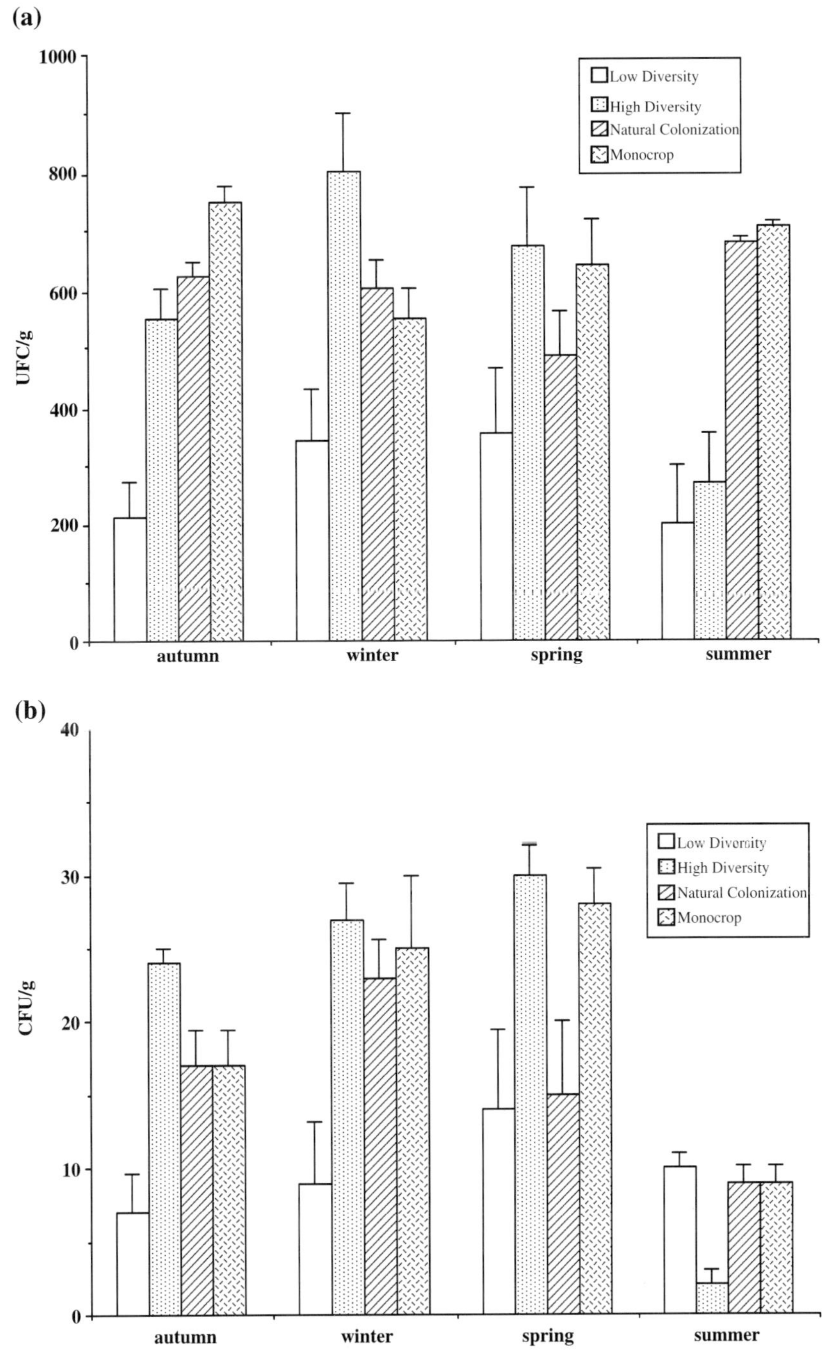

Figure 1. Evolution of the total microbiota and PSB in different soils along a year expressed in cfu/g ($\times 10^5$). (a) Total bacteria. (b) PSB .

colonies surrounded by a clear zone indicating the phosphate solubilization was counted. The number of total bacteria in each sample was also counted. The counts were quintupled. The data obtained were analyzed by one-way analysis of variance, with the mean values compared using Fisher's Protected LSD (Low Significative Differences) ($P = 0.05$).

Results and discussion

Soil analysis

The characteristics of soil studied are recorded in the Tables 1 and 2. According to the results obtained the soil has a low pH and is composed by sand and clay and contains low amounts of assimilable P.

Analysis of total bacteria and PSB in the different experimental parcels studied

The number of total and phosphate solubilizing bacteria (PSB) in each experimental block is recorded in Figure 1. According to the results obtained, the number of total bacteria in the blocks studied was higher than 1×10^5 cfu/g of soil in all cases. The number of PSB was 10 times lower than the number of total bacteria in each sample. This result is in agreement with those obtained by other authors (Kucey, 1983).

The results of the counts of total microbiota and PSB during a year in the soils subjected to the different treatments are recorded in Table 1. The number of total bacterial (Figure 1a) in natural colonization and in monocrop treatments in Autum and Winter was lower than in low diversity and high diversity treatments. These results may be related with the great abundance of gramineae in the cover plant from natural colonization and mainly in the monocrop treatment with barley. Barley is cultivated from winter to summer and as expected the number of rhizospheric bacteria is high in these two seasons. On the contrary, the number of PSB (Figure 1b) was significantly lower in monocrop treatment in Winter, Spring and Summer. The higher values were obtained in low and high diversity treatments although in the most of the cases the differences were no significant with respect to the natural colonization treatment.

The results found in the literature are very variable and some authors have found that cultivated soils have a higher number of PSB than in forest soils (Subba-Rao, 1982, Yahya and Al-Azawi, 1989), whereas other authors checked that forest and pasture soils contained higher PSB population than cultivated soils (Gupta et al., 1986). According our results the monocrop negatively affects to the rhizospheric PSB whereas the presence in the soil of plants during all year seasons, such as occurs in the low and higher diversity treatments, lead to a high develop of rhizospheric PSB.

References

Ahmad N and Jha K K 1968 Solubilization of rock phosphate by microorganisms isolated from Bihar. J. Gen. Appl. Microbiol. 14, 89–95.

Gupta R D, Bherdwaj K R, Morwan B C and Tripathi B R 1986 Occurrence of phosphate dissolving bacteria in soils of North-West Himalayas under biosequence and climosequence. J. Indian Soc. Sci. 34, 498–504.

Kucey R M N 1983 Phosphate solubilizing bacteria and fungi in various cultivated and virgin Alberta soils. Can. J. Soil Sci. 63, 671–678.

Schläpfer F and Schmid B 1999a Ecosystem effects of biodiversity: a classification of hypotheses and exploration of empirical results. Ecol. Appl. 9, 893–912.

Schläpfer F and Schmid B 1999b Expert estimates about effects of biodiversity on ecosystem processes and services. Oikos 84, 346–352.

Soil Conservation Service 1972 Soil Survey Laboratory Methods and Procedures for Collecting Soil Samples. U.S.D.A, Washington.

Soil Survey Staff 1994 Keys to Soil Taxonomy. 6th ed. Soil Conservation service, USA.

Subba-Rao N S, 1982 Phosphate solubilization by soil microorganisms. *In* Advances in Agricultural Microbiology. Ed. N S Subba-Rao. Butterworth Sci, pp. 295–305. London.

Wainwright M and Pugh G J F 1975 Effect of fungicides on numbers of microorganisms and frequency of cellulolytic fungi in soil. Plant Soil 43, 561–572.

Wainwright M and Sowden F J 1977 Influence of fungicide treatment on $CaCl_2$–extractable phosphorus and phosphate-solubilizing microoganisms in soil. Plant Soil 48, 335–345.

Yaha A I and Al-Azawi S K 1989 Ocurrence of phosphate-solubilizing bacteria in some Iraqui soils. Plant Soil 117, 135–141.

Population dynamics of P-solubilizers in the rhizosphere of major weed species from a tropical delta soil

S. Seshadri[1,3] & C. Lakshminarasimhan[2]
[1]*Shri AMM Murugappa Chettiar Research Center (MCRC), Taramani, Chennai 600 113, India.* [2]*P.R. Group of Institutions, Thanjavur 613 007, India.* [3]*Corresponding author**

Received 31 December 2002. Accepted in revised form 2 January 2003

Key words: phosphate solubilization, population dynamics, rhizosphere, tropical soil

Abstract

Microbes are found ubiquitous in all environments and rhizosphere and weed rhizosphere is not an exception. In the present approach, rhizospheres of weeds found in Paddy, Sugarcane, Garden, Riverbed and Wasteland soils were screened for enumerating the P-solubilizer population. P-solubilizer population was found high in Paddy and Sugarcane soils and poor in the Riverbed soils. The total available population ranged from 0 to 74×10^3 /gm dry rhizosphere soils. The study throws light on knowledge of the occurrence of P-solubilizers in the weed rhizosphere environment, evidencing the sustenance of native populations through indirect means in the soil ecosystem during off-season periods and stresses the need for studying the weed rhizosphere microflora in future for better understanding.

Introduction

Microorganisms are influenced in many ways by growing plants. Plant root surfaces and soil in contact with roots comprise a site of intense microbial activity and this environment under the influence of plant roots referred to as a rhizosphere and is under the continuous influence of plant roots. Microbial processes are more rapid in rhizosphere than in non-rhizosphere (Gaur, 1990).

Occurrence of phosphate solubilizing microbes in the rhizosphere of most of the crop plants was extensively studied and were reported by various workers (Sperber, 1958; Katznelson and Bose, 1959; Bardiya and Gaur, 1974; Ocampo et al., 1975; Banik and Dey, 1982; Craven and Hayasaka, 1982; Illmer and Schinner, 1992). However, information on the occurrence of such microbes in weed rhizosphere is rare or less and has not been reported elsewhere. In this study, we present the distribution of P-solubilizing rhizosphere microflora of certain selected weed species found in the Paddy, Sugarcane, Garden, River bed and Wasteland soils.

* E-mail: tsseshadri@rediffmail.com

Materials and methods

Site selection and soil sampling

Soil and rhizosphere samples were collected from respective fields in sterile plastic bags and were taken to the laboratory, puddled, sieved and dried in shade and were packed in a tight sterile polyethylene bag and kept in room temperature. The Physico-chemical characteristics of the soils were analyzed and are presented (Table 1).

Weed species

Fifteen weed species (Table 2) commonly found in all the selected environments were selected for the analysis and screened for further studies.

E. Velázquez and C. Rodríguez-Barrueco (eds.),
First International Meeting on Microbial Phosphate Solubilization, 281–284

Table 1. Soil analysis

Soil type	pH	N%	P%	K%	OM%	WHC %	EC	Texture
Paddy field	7.37 ± 0.15	0.063 ± 0.002	0.006 ± 0.001	2.23 ± 0.12	0.600 ± 0.100	40.33 ± 1.53	0.080 ± 0.010	Sandy clay loam
Sugarcane field	7.57 ± 0.25	0.083 ± 0.006	0.007 ± 0.001	2.53 ± 0.15	0.733 ± 0.153	45.33 ± 1.53	0.090 ± 0.010	Sandy clay loam
Garden soil	8.17 ± 0.15	0.057 ± 0.015	0.005 ± 0.002	1.43 ± 0.12	0.400 ± 0.100	31.67 ± 2.52	0.073 ± 0.012	Sandy clay loam
Waste land	8.27 ± 0.15	0.050 ± 0.010	0.007 ± 0.001	1.17 ± 0.21	0.300 ± 0.100	25.33 ± 0.58	0.067 ± 0.006	Sandy clay loam
Riverbed	8.00 ± 0.20	0.033 ± 0.006	0.005 ± 0.001	0.72 ± 0.03	0.093 ± 0.012	18.67 ± 2.08	0.063 ± 0.006	Sandy clay loam

(Values are mean ± standard deviation).

Enumeration of P-solubilizers

The P-solubilizing microbial population of respective soils and rhizospheric soils of each weed species was determined using Pikovskaya's medium (Pikovskaya, 1948). One-gram lots of soil samples were shaken with 9 ml sterile distilled water for 1 h. at 150 rpm and suitable dilutions were inoculated onto agar plates containing solid Pikovskaya's medium. Colonies showing clearing zones around them indicting the dissolution of Tri-calcium phosphate were identified as P-solubilizers (Seshadri, 1995). The size of the clear zones around the colonies showing P solubilization was noted and the results were expressed as solubilization efficiency (E = solubilization diameter (S)/ Growth diameter $(G) \times 100$) (Ngyuen et al., 1992).

Results and discussion

The results on the population of the P-solubilizers enumerated from different weed rhizospheres are presented in Table 2. All the soils studied differed in harboring P-solubilizers population. In

Table 2. P-solubilizers in the rhizosphere region of certain weed species from different soils and their native population

Sl. No	Weed species	Paddy field	Sugarcane field	Garden land	Waste land	River bed
I.	Native population	25.3 ± 3.5	50.7 ± 5.03	20.1 ± 2.67	15.4 ± 2.42	9.1 ± 2.17
II.	Weed Rhizosphere Population					
1	Acalypha indica	51.67 ± 4.33	63.33 ± 4.67	11.33 ± 1.67	11.67 ± 4.33	6.33 ± 7.33
2	*Achyranthus aspera*	4.33 ± 0.58	4.67 ± 1.53	1.67 ± 0.58	4.33 ± 1.53	7.33 ± 2.08
3	*Amaranthus viridis*	74.67 ± 44.67	73.67 ± 15.67	15.33 ± 9.33	11.33 ± 11.67	5.67 ± 2.60
4	Boerhaavia diffusa	44.00 ± 0.33	54.00 ± 2.00	2.67 ± 1.33	11.67 ± 5.67	1.33 ± 2.50
5	*B. hispida*	22.00 ± 2.67	12.50 ± 2.33	4.50 ± 1.33	3.50 ± 3.33	2.00 ± 0.00
6	*Cleome viscosa*	2.67 ± 1.53	2.33 ± 0.58	1.33 ± 0.58	3.33 ± 0.58	0.00 ± 0.00
7	*Commelina benghalensis*	44.00 ± 3.61	54.00 ± 2.00	2.67 ± 1.15	11.67 ± 3.06	1.33 ± 1.15
8	*Croton sparsiflorus*	0.33 ± 0.58	2.00 ± 1.00	1.33 ± 1.15	5.67 ± 1.53	2.50 ± 0.71
9	*Cyanotix axilaris*	27.33 ± 4.16	63.33 ± 0.58	3.67 ± 0.58	17.67 ± 3.51	2.67 ± 1.53
10	*Eclipata alba*	42.67 ± 1.53	3.67 ± 1.53	22.67 ± 1.53	21.33 ± 1.53	3.67 ± 2.08
11	*Euphorbia hirta*	7.33 ± 1.53	9.33 ± 3.21	3.67 ± 1.15	4.33 ± 1.53	2.67 ± 1.15
12	*Gomphrena decumbens*	52.00 ± 4.36	72.33 ± 3.21	12.33 ± 2.52	2.33 ± 0.58	6.00 ± 2.00
13	*Mimosa pudica*	13.80 ± 3.65	40.00 ± 1.73	6.67 ± 1.15	11.67 ± 2.31	3.33 ± 0.58
14	*Phyla nodiflora*	32.67 ± 0.58	14.00 ± 10.44	1.33 ± 0.58	31.67 ± 2.08	5.33 ± 2.08
15	*Phyllanthus niruri*	12.67 ± 3.06	45.67 ± 3.21	1.67 ± 1.15	12.00 ± 2.65	3.33 ± 1.53

(Values are mean ± standard deviation).

almost all the samples studies the population was found to be very less. Occurrence of low P solubilizer population could be attributed to the source of samples; that were drawn from tropical ecosystem, where soils are reported to contain very low P-solubilizer population (Sundara Rao and Sinha, 1963; Raghu and Mac Rae, 1966). Moreover collection time would have also played a role. All the collections were made during off-season periods when soils started supporting weed growth and after fields were left fallow for some time.

While, weed rhizosphere samples from Paddy and Sugarcane fields showed a considerable amount of population, in the other fields the population was very less. This could be related to various factors viz. possible weed root exudates, organic matter content of the soil (Gaur and Ramendra singh, 1982), soil moisture and texture and physico-chemical properties. Venkateswarlu et al. (1984) while studying the Aridsols soil population recorded a relatively low population even in a crop plant rhizosphere. Kundu and Gaur (1984) while working on PSM inoculants with rice under green house conditions enumerated the population as 1 to 10×10^6 per gram of soil after inoculation. Occurrence of less population (25.3×10^3 /gm dry weight) in the rice fields during this study could be attributed to the lack of viable host (crop) plants during sample collection. In the present study, the P-solubilizing rhizosphere microflora of weed species also echoed the same results. Among the fifteen weed species screened the *Amaranthus viride* rhizosphere soil from Paddy field recorded a maximum number of solubilizers (74.67×10^3 /gm dry soil). *Cleome viscosa* and *Croton sparsiflorus* were found to be the poor weeds to encouraging the P-solubilizer population.

From the results, it is concluded that the weed rhizosphere, irrespective of their host, supports very poor population of P-solubilizers. However, alternatively it can also be presumed that all the soils, irrespective of nature, support a considerable amount of P-solubilizers. It is also proved that the weed species have the great potential in nurturing soil microflora during off-season timings. This would be a boon, if the same weed were mulched for manuring purpose; the rhizosphere supported beneficial microflora could multiply further and aid the crop growth. It is stressed that in future, it is also necessary to study the weed rhizosphere for the beneficial microbes, which may have an indirect effect on the plant yield. Where mulching and intercropping is practiced in a major level, this kind of study would add data to the existing knowledge to draw suitable measures on nutrition. In addition, encouraging the weed species viz. *Amaranthus viride, Boerhavia diffusa, Acalypha indica, Commelina benghalensis, Cyanotis axilaris, Eclipta alba, Gomphrena decumbens and Phyla nodiflora* in any fields that are left fallow will aid in retaining the microflora of the native soils. This forms the first report on the P-solubilizer population enumeration from the weed species rhizosphere.

References

Banik S K and Dey B K 1982 Solubilization of inorganic phosphate and production of organic acids by microorganisms isolated in sucrose tricalcium phosphate agar plates. Zentralblat. Bakterol. Parasitenkl. Infeksionskr. Hyg. II. 136, 478–486.

Bardiya M C and Gaur A C 1974 Isolation and screening of microorganisms dissolving low grade rock phosphate. Folia Microbiol. 19, 386–389.

Craven P A and Hayasaka S S 1982 Inorganic phosphate solubilization by rhizosphere bacteria in *Zostera* marina community. Can. J. Microbiol. 28, 605–610.

Gaur A C 1990 *In* Phosphate solubilizing microorganisms, Omega Publications, 176 p.

Gaur A C and Singh R 1982 Integrated Nutrient Supply. Fertilizer News. 27, 87–98.

Illmer P and Schinner F 1992 Solubilization of inorganic phosphates by microorganisms isolated from forest soils. Soil Biol. Biochem. 24, 389–395.

Katznelson H and Bose B 1959 Metabolic activity and phosphate dissolving ability of bacteria isolates from wheat roots rhizosphere and non-rhizosphere soil. Can. J. Microbiol. 5, 79–85.

Kundu B S and Gaur A C 1984 Rice response to inoculation with N_2 fixing and phosphate solubilizing microorganisms. Plant Soil. 79, 227–234.

Nguyen C, Yan W, Le Tacon F and Lapayrie F 1992 Genetic variabillity of phosphate solubilizing activity by monocaryotic and dicaryotic mycelia of the ectomyccorhizal fungus *Laccaria bicolor* (Maire) PD Orton. Plant Soil 143, 193–199.

Ocampo J A, Barea J M and Montoya E 1975 Interactions between Azotobacter and phospho-bacteria and their establishment in the rhizosphere as affected by soil fertility. Can. J. Microbiol. 21, 1160–1165.

Pikovskaya R I 1948 Mobilization of phosphorus in soil in connection with vital activity of some microbial species. Mikrobiologiya 17, 362–396.

Raghu K and Mac Rae I C 1967 The effect of gamma isomer of BHC upon the microflora of submerged rice soils. II. Effect on nitrogen mineralization and fixation and selected bacteria. Can. J. Microbiol. 13, 621–627.

Seshadri S 1995 Phosphate solubilizing fungi from Thanjavur Delta, South India, Ph.D. Thesis, Bharathidasan University, Tiruchirappalli, India.

Sperber J I 1958 The incidence of apatite dissolving organisms in the rhizosphere and soil. Aust. J. Agric. Res. 9, 778–781.

Sundara Rao W V B, Bajpai P D, Sharma J P and Subbaiah B V 1963 Solubilization of phosphates by phosphorus solubilizing organisms using P^{32} as tracer and the influence of seed bacterization on the uptake by the crop. J. Indian Soc. Soil Sci. 11, 209–219.

Venkateswarlu B, Rao A V and Raina P 1984 Evaluation of phosphorus solubilization by microorganisms isolated from arid soils. J. Indian Soc. Soil Sci. 32, 273–277.

Malic acid mediated aluminum phosphate solubilization by *Penicillium oxalicum* CBPS-3F-Tsa isolated from Korean paddy rhizosphere soil

R.S. Gadagi, W.S. Shin & T.M. Sa[1]
Department of Agricultural Chemistry, Chungbuk National University, Cheongju, 361-763, South Korea.
[1]*Corresponding author**

Received 9 December 2002. Accepted in revised form 2 January 2003

Key words: $AlPO_4$, $FePO_4$, *Penicillium oxalicum*, Solubilization

Abstract

Penicillium oxalicum was isolated from acidic paddy rhizosphere soil, Milyang, Korea using modified media with $AlPO_4$ and bromo cresol green (BCG). Ability of fungus to solubilize mineral phosphates and its cause for solubilization were studied in vitro. The fungus was grown in liquid medium cultures containing $AlPO_4$/$FePO_4$/$Ca_3(PO_4)_2$ (1000 mg P L^{-1}). Concentration of organic acids, soluble phosphate and pH were determined periodically during 7 to 78 h incubation. *Penicillium oxalicum* demonstrated higher levels of $Ca_3(PO_4)_2$(129.10 mg L^{-1}) and $AlPO_4$ (119.80 mg L^{-1}) solubilization than that of $FePO_4$ (54.70 mg L^{-1}) solubilization. Soluble phosphate concentrations in the culture medium were directly proportional to the organic acids and inversely related to pH. Malic acid production may have contributed the higher Al–P solubilization in culture media. Evidence from abiotic solubilization using sodium malate and HCl to solubilize P also indicated that malic acid may have been a main organic acid involved in the solubilization of $AlPO_4$ and $Ca_3(PO_4)_2$.

Introduction

Phosphate utilization efficiency in soils is very low because applied phosphorus is mostly fixed to aluminum and iron in acidic soils and to calcium in alkaline soils (McLaughlin et al., 1988; Sample et al. 1980). Several bacteria and fungi were isolated from the soil and their mineral phosphate solubilizing activity was evaluated using various P sources such as $Ca_3(PO_4)_2$ (Gadagi and Alagawadi, 1999; Illmer and Schinner, 1995), $FePO_4$ (Whitelaw et al., 1999) and $AlPO_4$ (Illmer et al., 1995; Whitelaw et al., 1999).

Inorganic P solubilization by microorganisms has been attributed to processes involving acidification, chelation and exchange reactions in the growth environment (Molla and Chawdhury, 1984). The particular acids released differed between the fungi, as did the rate of P solubilization. Several processes can work in conjunction with each other, i.e. organic acids can act as chelating agents and have a direct acidifying effect on the surrounding environment (Gaur et al., 1973; Surange, 1985).

The mechanisms for inorganic solubilization by fungi have not been examined. In fact, few studies have rigorously addressed factors that could affect the rate at which microbial mediated inorganic phosphate solubilization occurs (Beever and Burns, 1981). The mineral phosphate solubilizing phenotype was designated as Mps^{+} by Goldstein (1995), and found to be the result of gluconic acid mediated dissolution of hydroxyapatite (HA) in Gram-negative bacteria. Gluconic acid has reported to be involved in the solubilization of Ca phosphate minerals by *Erwinia herbicola* (Liu

* E-mail: tomsa@chungbuk.ac.kr

E. Velázquez and C. Rodríguez-Barrueco (eds.),
First International Meeting on Microbial Phosphate Solubilization, 285–290

et al., 1992), *Penicillium* sp. (Illmer and Schinner, 1995), *P. chrysogenum* (Nielson et al., 1994), *P. notatum* (Pitt et al., 1983) and *Aspergillus niger* (Vassilev et al., 1995; Moksia, et al., 1996). Gluconic acid production in phosphate solubilizing microorganisms is being considered as main the mechanism of phosphate solubilization.

Higher Pi release was reported from Ca phosphate minerals solubilization compared to Pi release from the Al and Fe phosphate minerals solubilization by microorganisms (Stumm and Moragan, 1995). *Penicillium* sp. LAF2 was solubilizing $Ca_3(PO_4)_2$ very efficiently but $AlPO_4$ to a lesser degree (Banik and Dey, 1983). This was probably due to the fact that most of the strains were primarily isolated based on the halo zone formation in Pikovskaya's medium containing $Ca_3(PO_4)_2$ and subsequently, those strains were tested for $AlPO_4$ and $FePO_4$ solubilization. Halo zone was only formed on solid medium containing $Ca_3(PO_4)_2$ while halo zone formation on solid media containing $AlPO_4$ or $FePO_4$ has not yet been reported. *P. oxalicum* was isolated using modified media with $AlPO_4$ as P source and bromo cresol green and it demonstrated high solubilization of $AlPO_4$ and $FePO_4$ (Gadagi and Sa, 2002). In the present study, we tried to find cause for higher Al and Fe phosphates solubilization by *P. oxalicum* and mechanisms of $AlPO_4$ and $FePO_4$ solubilization were examined in comparison with Ca phosphate solubilization by *P. oxalicum*.

Materials and methods

Culture and media

We isolated *P. oxalicum* from rice rhizosphere soil using modified media with $AlPO_4$ as P source plus bromo cresol green (Gadagi and Sa, 2002). These fungus culture was multiplied in potato dextrose broth and maintained on potato dextrose agar.

Organic acid secretion, pH drop and phosphate solubilization by P. oxalicum

Penicillium oxalicum isolate was examined for phosphate solubilizing activity in Reyes's basal media (Reyes et al., 1999). Aliquot of 100 mL of Reyes's basal media with 1000 mg L^{-1} $AlPO_4/FePO_4/Ca_3(PO_4)_2$ were apportioned into 250 mL EM flask and autoclaved. The flasks were inoculated with three agar plugs with diameter of 7 mm drawn through cork borer along the margin of actively growing culture of fungi on potato dextrose agar. At the end of the 7, 14, 21, 28, 35 and 72 h of incubation at 30°C triplicate flasks of fungi *P. oxalicum* were withdrawn, the contents were filtered and the soluble P-content in the filtrate was analyzed colorimetrically following the method of Murphy and Riley (1962). The percentage P solubilized by *P. oxalicum* was estimated after subtracting the amount of P solubilized by autoclave.

The pH of the filtrate was also recorded. Organic acid was determined by HPLC (Waters™ 600S Model with 486 Tunable Absorbance Detector) with organic acid column (Pacina et al., 1984). The mobile phase was H_2SO_4 (0.1%) with flow rate of 0.5 mL min^{-1}. Organic acid standards included acetic, citric, formic, fumaric, galactouronic, gluconic, glucuronic, glycolic, keto gluconic, lactic, malic, maleic, oxalic, propionic, pyruvic, succinic, tartaric acids.

Abiotic solubilization

Phosphate solubilized by *P. oxalicum* was compared to the P dissolved in the uninoculated media by the addition of the following.

Addition of malate, oxalate and gluconate alone or mixed at a concentrations equivalent to that in the *P. oxalicum* culture media at the time of maximum P solubilization. Addition of HCl alone to reduce a pH as close as to pH observed in culture media at the time of maximum P solubilization. HCl (0.1 M) was added to the media until pH was close to 4.15 for $Ca_3(PO_4)_2$, 3.96 for $AlPO_4$ and 3.82 for $FePO_4$. To know the organic acid efficiency in abiotic $AlPO_4$ and $FePO_4$ solubilization, we used the standard concentration (200 mM L^{-1}) for all three organic acids in abiotic phosphates solubilization. All the treated flasks were incubated for 60 min on rotary shaker. The P-content was analyzed following the method as mentioned above.

Results and discussion

A fungus CBPS-3-Tsa was identified following the descriptions of Pitt (1980). The fungus CBPS-3F-Tsa was identified as *Penicillium* sp. subsequently as *P. oxalicum* CBPS-3F-Tsa by IMI, CABI, Bioscience, UK, (CABI accession number for our strain is IMI 387080). We are the first to report the phosphate solubilizing activity of *P. oxalicum*. This is new addition to Penicillium group of phosphate solubilizers (Asea et al., 1988; Cunningham and Kuiack, 1992; Illmer and Schinner, 1995; Kucey et al., 1989; Nielson et al., 1994; Pitt et al., 1983; Reyes et al., 1999).

Mechanisms of phosphate solubilization by P. oxalicum

The malic, oxalic and gluconic acids were detected by HPLC in the *P. oxalicum* media containing different source of phosphate. However, their concentrations varied with growth stages and source of phosphate. Malic acid production by *P. oxalicum* at 7 h of incubation was 10 mM L^{-1} and it reached maximum 145 mM L^{-1} at 72 h of incubation in $Ca_3(PO_4)_2$ culture media. Similarly, it was ranged from 27 to 148 mM L^{-1} in $FePO_4$ culture media and 14–240 mM L^{-1} in $AlPO_4$ culture media. Oxalic acid production was not detected in $FePO_4$ and $AlPO_4$ culture media at 7 and 72 h of incubation, it was ranged from 9 to 17 and 1 to 58 mM L^{-1} respectively during 14 to 48 h of incubation. However, oxalic acid concentration was 12 to 160 mM L^{-1} in $Ca_3(PO_4)_2$ culture media during 7–35 h of incubation. Gluconic acid production was not much differed with respect to source of phosphate and its concentration was also low compared to other two above mentioned acids. Accumulation of organic acids in liquid media resulted into acidification due to fungal inoculation (Iwase, 1992). In the present study, we also recorded the pH decrease from initial 6.5 pH due to fungal inoculation in $Ca_3(PO_4)_2$, $FePO_4$ and $AlPO_4$ culture media and they ranged from 0.22 to 2.35, 0.34 to 2.54 and 1.0 to 2.68 during 7 to 72 h of incubation respectively (Figure 1).

Phosphate solubilization ranged from 3.88 to 129.10, 0.09 to 54.70 and 0.81 to 111.80 mg L^{-1} in $Ca_3(PO_4)_2$, $FePO_4$ and $AlPO_4$ culture media, respectively, due to *P. oxalicum* growth during 7 to 72 h of incubation (Figure 1).

Penicillium oxalicum was able to solubilize high amounts of insoluble phosphate in $Ca_3(PO_4)_2/AlPO_4/FePO_4$ in liquid culture, which were correlated with organic acid production and pH drop in respective culture media. The similar positive correlations among the P-solubilization, pH drop and organic acid production have been reported (Reyes et al., 1999; Whitelaw et al., 1999).

The correlation coefficient for linear regression equations linking P solublization with pH drop for $Ca_3(PO_4)_2$, $AlPO_4$ and $FePO_4$ provided an evidence that lowering pH was a key mechanism for the solubilization of $Ca_3(PO_4)_2$, $AlPO_4$ and $FePO_4$. Similarly, insoluble mineral phosphates solubilization due to pH decrease was reported elsewhere (Stumm and Morgan, 1995). However, in the present, the tendency for

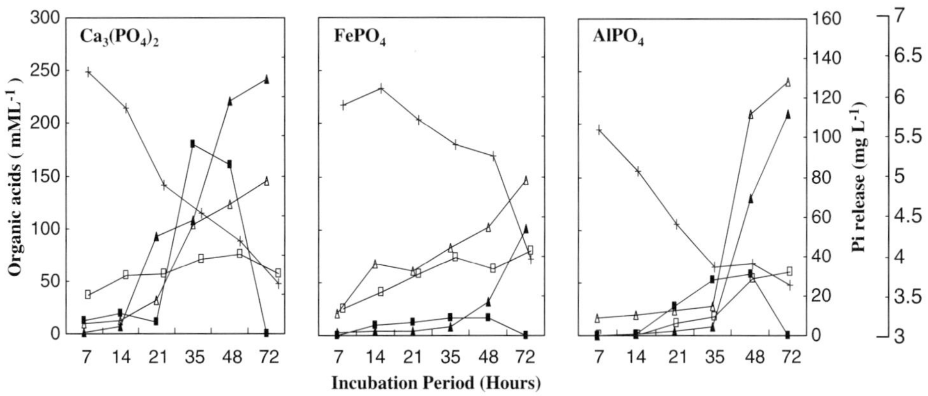

Figure 1. Changes in phosphate solubilization, organic acid production and pH over period of incubation time by *P. oxalicum*. (▲ Pi release; Δ Malic acid; ■ Oxalic acid; □ Gluconic acid; + pH).

correlation between P solubilization and malic acid was highly significant compared to correlation between P solubilization and pH (Table 1). This indicates that soluble P is probably due to some factors other than pH, i.e. may be metal chelation with organic acids. Although, there was also correlation between P solubilization and gluconic acid, however, its concentration is negligible compared to malic acid production by *P. oxalicum* with respect to P solubilization. There was also no significant correlation between P solubilization and oxalic acid production. Mechanisms of $Ca_3(PO_4)_2$ solubilization due to gluconic acid production by several fungi *P. chrysogenum*, *P. notatum*, *P. rugulosum*, *P. radicum*, *P. aurantiogriseum*, *P. bilajii* and *A. niger* were reported elsewhere (Asea et al., 1988; Illmer and Schinner, 1995; Nielson et al., 1994; Pitt et al., 1983; Whitelaw, et al., 1999). In the present study, solubilization of $Ca_3(PO_4)_2$ is attributed to higher concentration of malic and gluconic acids production by *P. oxalicum*. Gluconic acid production was less compared to malic acid production, indicating the existence of different mechanisms for solubilization of $Ca_3(PO_4)_2$ in *P. oxalicum*.

Solubilization of $AlPO_4$ and malic acid production significantly correlated in *P. oxalicum* culture (Table 1). Increased malic acid production was coupled with decrease in oxalic acid production in $AlPO_4$ culture media compared to $Ca_3(PO_4)_2$ culture media. Secretion of malic acid was stimulated in the presence of Al in *P. oxalicum* culture media. Malic acid is secreted from the roots of plants in response to Al (Basu et al. 1994; Delhaize et al. 1993; Ma, 2000). Basu et al. (1994) also reported that exposure to 100 μM Al increased the production of malic acid from the roots by 100–120%. All these results hypothesized that malic acid production is key factors in solubilization $AlPO_4$ solubilzation and followed different mechanisms from gluconic acid mediated mineral phosphates solubilization. Malic acid is known to be responsible for high complex ability with Al (Ma, 2000). Therefore, it seems to be likely that $AlPO_4$ was solubilized by acidolysis followed by complex formation. This hypothesis was also supported by our abiotic solubilization results, which will be presented later.

Illmer and Schinner (1995) also opined that Al–P and Ca–Ps solubilization follow two different solubilization mechanisms. They found that *A. niger* and *P. simplicissimum* are more effective in $AlPO_4$ solublization due to *A. niger* and *P. simplicissimum* are known to produce organic acids mainly citrate in considerable amounts (Burgstaller et al., 1992).

Solubilization of $FePO_4$ was also correlated with malic acid production compared to gluconic acid production. However, concentration of malic acid in $Ca_3(PO_4)_2$ culture media was almost similar compared to $FePO_4$ culture media. Oxalic acid concentration was very low and non-correlated to $FePO_4$ solubilization in *P. oxalicum* culture. Malic acid also complex with Fe, in turn involved in $FePO_4$ solubilization. This also supported by abiotic solubilization.

Table 1. Correlation coefficient (*r*) for the linear regression equations relating the amount of phosphate solubilized (Pi) by *P. oxalicum* from insoluble P with organic acid and pH

P source	Correlation coefficients (*r*) for entire incubation period			
	Pi vs. Malate	Pi vs. Oxalate	Pi vs. Gluconate	Pi vs. pH
$Ca_3(PO_4)_2$	0.94	0.31	0.61	0.84
$AlPO_4$	0.99	NC	0.82	0.80
$FePO_4$	0.84	NC	0.34	0.71

Abiotic solubilization

Malic, oxalic and gluconic acids concentration close to that of the *P. oxalicum* culture at the time of maximum P solubilization were added to media containing 1000 mg L^{-1} insoluble phosphates. Addition of malic acid, oxalic acid and gluconic acid solubilized 13, 11, and 4% of $Ca_3(PO_4)_2$, $FePO_4$ and $AlPO_4$ respectively. Organic acid addition reduced the pH more than that of decreased in *P. oxalicum* culture (Table 2). However, phosphate solubilization was almost in the same trend as that in *P. oxalicum* culture. Addition of malic acid alone resulted solubilization of 13.2, 11.0 and 2.2% of $Ca_3(PO_4)_2$, $AlPO_4$ and $FePO_4$, respectively. However, addition of gluconic acid alone solubilized 11.8, 1.7 and 2.4% of $Ca_3(PO_4)_2$, $AlPO_4$ and $FePO_4$ respectively. In contrast, addition of oxa-

Table 2. Abiotic phosphate solubilization (mg L^{-1}) and pH decrease by addition of malate, oxalate, gluconate and HCl

	$Ca_3(PO_4)_2$	$AlPO_4$	$FePO_4$
Malic + oxalic +	118.64	92.10	26.54
gluconic acids*	(1.05)**	(2.48)	(2.93)
Malic acid			
a	132.27	133.02	22.80
	(1.98)	(1.90)	(3.29)
b	134.46	112.00	31.71
	(2.18)	(1.89)	(3.02)
Oxalic acid			
a	122.00	49.34	45.21
	(1.26)	(1.19)	(2.89)
b	132.98	36.23	26.84
	(1.10)	(1.20)	(2.82)
Gluconic acid			
a	118.90	17.24	34.23
	(2.75)	(2.63)	(3.66)
b	122.64	5.43	22.99
	(2.81)	(2.30)	(3.54)
pH	3.39	0.89	0.12
	(2.59)***	(2.63)	(2.11)

*Concentration of organic acids mixture equivalent in *P. oxalicum* culture at maximum Pi release.
a Concentration of organic acid alone equivalent in *P. oxalicum* culture at maximum Pi release.
b – Standard conc. of organic acids (200 mM L^{-1}).
**Parenthesis values are pH of the media.
***pH adjusted with HCl.

lic acid solubilized 12.2, 4.9 and 4.5% $Ca_3(PO_4)_2$, $AlPO_4$ and $FePO_4$, respectively.

We tested efficiency of phosphate solubilization with these organic acids at standard concentration (200 mM L^{-1}). Addition of malic and oxalic acids solubilized same amount of $Ca_3(PO_4)_2$ (13%) while gluconic acid solubilized 12% of $Ca_3(PO_4)_2$. Addition of malic acid solubilized higher $AlPO_4$ (11.2%) when compared to oxalic acid (3.6%) and gluconic acid (0.5%). Oxalic acid addition solubilized higher $FePO_4$ (4.5%) compared gluconic acid addition (2.2%) and malic acid addition (2.6%). Addition of HCl alone at a pH close to 2.59 in $Ca_3(PO_4)_2$, 2.63 in $AlPO_4$ and 2.11 in $FePO_4$ media were recorded the lower solubilization of their respective mineral P source (Table 2).

These results indicate that malic, oxalic and gluconic acids solubilized higher $Ca_3(PO_4)_2$ this is mainly due to acidification of environment. Acidification of media was caused by protonation of organic acids (Iwase, 1992). However, $AlPO_4$ and $FePO_4$ solubilization may be due to complexing ability of organic acids. Malic and oxalic acids may have more complexing ability with Al and Fe respectively. Mineral phosphate solubilization by *P. oxalicum* has been attributed to processes involving acidification and chelation.

Acknowledgements

This work was supported by the Ministry of Agriculture and Forestry through the R & D Promotion Center for Agriculture and Forestry, Korea.

References

Asea P E A, Kucey R M N and Stewart J W B 1988 Inorganic phosphate solubilization by two *Penicillium* species in solution culture and soil. Soil Biol. Biochem. 20, 459–464.

Banik S and Dey B K 1983 Alluvial soil microorganisms capable of utilizing insoluble aluminum phosphate as a source of phosphorus. Zent. Mikrobiol. 138, 437–442.

Basu U, Godbold D and Tayler G J 1994 Aluminum resistance in *Triticum aestivum* L. associated with enhanced exudation of malate. J. Plant Physiol. 144, 747–753.

Beever R E and Burns D J W 1981 Phosphorus uptake, storage and utilizing by fungi. Adv. Bot. Res. 8, 17–23.

Burgstaller W, Strasser H and Schinner F 1992 Solubilization of zinc oxide from filter dust with *Penicillium simplicissimum*; bioreactor, leaching and stoichiometry. Environ. Sci. Tech. 26, 340–346.

Cunningham J E and Kuiack C 1992 Production of citric and oxalic acids and solubilization of calcium phosphate by *Penicillium bilajii*. Appl. Environ. Microbiol. 58, 1451–1458.

Delhaize P R, Ryan P R and Randall P J 1993 Aluminum tolerance in wheat (*Triticum aestivum* L.). II. Aluminum stimulated excretion of malic acid from root apices. Plant Physiol. 103, 695–702.

Gadagi R S and Alagawadi A R 1999 Response of sunflower to the P-solubilizing biofertilizer with different sources and levels of phosphorus. Karnataka J. Agric. Sci. 11, 50–55.

Gadagi R S and Sa T 2002 New isolation method for microorganisms solubilizing iron and aluminum phosphates using dyes. Soil Sci. Plant Nutr. 48, 615–618.

Gaur A C, Madan M and Ostwal K P 1973 Solubilization of phosphatic compounds by native micro flora of rock phosphates. Indian. J. Expt. Biol. 11, 427–429.

Goldstein A H 1995 Recent progress in understanding the molecular genetics and biochemistry of calcium phosphate solubilization by Gram negative bacteria. Biol. Agric. Hortic. 12, 185–193.

Illmer P and Schinner F 1995 Solubilization of inorganic calcium phosphates: solubilization mechanisms. Soil Biol. Biochem. 27, 257–263.

Illmer P, Barbato A and Schinner F 1995 Solubilization of hardly soluble $AlPO_4$ with P-solubilizing microorganisms. Soil Biol. Biochem. 27, 265–270.

Iwase K 1992 Gluconic acid synthesis by the ectomycorrhizal fungus *Tricholoma robustum*. Can. J. Bot. 70, 84–88.

Kucey R M N, Janzen H H and Leggett M E 1989 Microbial mediated increases in plant-available phosphorus. Adv. Agron. 42, 199–228.

Liu S T, Lee L Y, Tai C Y, Hung C H, Chang Y S, Wolfram J H, Rogers R and Goldstein A H 1992 Cloning of an *Erwinia herbicola* gene necessary for gluconic acid production and enhanced mineral phosphate solubilization in *E.* coli HB101:nucleotide sequence and probable involvement in biosynthesis of the coenzyme pyrroloquinoline quinone. J. Bacteriol. 174, 5814–5819.

Ma J F 2000 Role of organic acids in detoxification of aluminum in higher plants. Plant Cell Physiol. 41, 383–390.

McLaughlin M J, Slaston A M and Martin J K 1988 Phosphorus cycling in wheat-pasture rotations: the source of phosphorus taken up wheat. Aus. J. Soil Res. 26, 323–331.

Moksia J, Larroche C and Gros J B 1996 Gluconate production by spores of *Aspergillus niger*. Biotechnol. Lett. 18, 1025–1030.

Molla M A Z and Chawdhury A A 1984 Microbial mineralization or organic phosphate in soil. Plant Soil. 78, 393–399.

Murphy J and Riley J P 1962 A modified single solution method for the determination of phosphate in natural waters. Anal. Chim. Acta 27, 31–36.

Nielson J, Johansen C L and Villadsen J 1994 Culture fluorescence measurements during batch and fed-batch cultivation with *Penicillium chrysogenum*. J. Bacteriol. 38, 51–62.

Pacina R, Bonn G and Burris R H 1984 High performance liquid chromatographic elution behavior of alcohols, aldehydes, ketones, organic acids and carbohydrates on a strong cation exchange stationary phase. J. Chromatogr. 278, 245–258.

Pitt D, Mosley M J and Barnes J C 1983 Glucose oxidase activity and gluconate production during calcium induced conidiation of *Penicillium notatum* in submerged culure. Trans. Br. Mycol. Soc. 81, 21–27.

Pitt J I 1980 The Genus *Penicillium* and Telemorphic Status *Eupenicillium* and *Talaromyces*. Academic Press, New York pp. 273–274.

Reyes I, Bernier L, Simard R R and Antoun H 1999 Effect of nitrogen source on the solubilization of different inorganic phosphates by an isolate of *Penicillium rugulosum* and two UV-induced mutants. FEMS Microbiol. Ecol. 28, 281–290.

Sample E C, Soper R J and Racz G J 1980 Reactions of phosphate fertilizers in soils. *In* The Role of Phosphorus in Agriculture. Ed. F E Khasawneh. pp. 263–310. Am. Soc. Agron, Madison, Wisconsin.

Stumm W and Moragan J 1995 Aquatic Chemistry. Chemical Equilibria and Rates in Natural Waters. 3John Wiley, New York.

Surange S M 1985 Comparative phosphate solubilizing capacity of some soil fungi. Curr. Sci. 54, 1134–1135.

Vassilev N, Baca M T, Vassileva M, Franco I and Azcon R 1995 Rock phosphate solubilzation by *Aspergillus niger* grown on sugar beet waste medium. Appl. Microbiol. Biotechnol. 44, 546–549.

Whitelaw M A, Harden T J and Helyar K R 1999 Phosphate solubilization in solution culture by the soil fungus *Penicillium radicum*. Soil Biol. Biochem. 31, 655–665.

Growth promotion of rice by phosphate solubilizing bioinoculants in a Himalayan location

P. Trivedi, B. Kumar, A. Pandey[1] & L.M.S. Palni

G. B. Pant Institute of Himalayan Environment and Development Kosi- Katarmal, Almora, 263 643, Uttaranchal, India. [1]*Corresponding author**

Received 27 November 2002. Accepted in revised form 2 January 2003

Key words: bioinoculants, growth promotion, phosphate solubilizing microorganisms, rice

Abstract

Three bacterial species, viz. *Bacillus megaterium*, *B. subtilis* and *Pseudomonas corrugata*, originally isolated from temperate locations in the Indian Himalayan region, were examined for their growth promotion ability using both pot and field based assays. A local landrace of rice was used as test crop. The three bacterial species exhibited *in vitro* phosphate solubilizing activity in following order: *P. corrugata* > *B. megaterium* > *B. subtilis*. The bacterial treatments (broth based in pot and charcoal based in field experiments) resulted in improved plant performance. Out of the three treatments, *B. subtilis* gave best performance resulting in 1.66 and 1.55 fold increase in grain yield of rice in pot and field trials, respectively. Inoculations also stimulated the rhizosphere associated bacterial and actinomycetes populations and suppressed the fungal flora. Colonization of roots by mycorrhizal fungi improved in all the treatments. Out of the three bacterial inoculants, *B. subtilis* was the best in affecting these changes. Bacterial treatments also resulted in higher values for phosphorus in shoots and grains in inoculated rice plants. The study indicates that the stimulation of native bacterial flora, including mycorrhizae, in and around roots is one of the important parameters playing indirect role in improving the overall plant growth. The study suggests that charcoal based *B. subtilis* cultures can be developed as an efficient bioinoculant for rice fields in the mountains.

Introduction

Occurrence, importance and use of phosphate solubilizing microorganisms in various ecological niches have been documented (Pandey and Kumar 1989; Chabot et al., 1996; Pandey and Palni, 1998a; Johri et al., 1999; Vazquez et al., 2000). While the mechanism(s) of microbial solubilization of insoluble phosphate has received some attention (Illmer et al., 1995), phosphate solubilization is considered to be an important attribute of plant growth promoting rhizobacteria (PGPRs; Kloepper et al., 1989; Tilak, 1991; Chabot, 1996; Kumar et al., 2001; Peix et al., 2001).

The beneficial effect of PGPR inoculation results from the interaction of three factors, the bacterial strain, the plant cultivar and the ecological niche. In a recent field study on maize conducted at two climatic zones-subtropical and temperate, in Mamlay watershed of Sikkim Himalaya, the bacterial inoculations resulted in statistically significant and improved plant performance at the subtropical zone; the same inoculants did not respond at the temperate site (Pandey et al., 1998). This indicated the value of isolation, identification and screening of native bacteria for selection of PGPR, suitable for use in the mountains.

* FAX No: 05962-241150.
E-mail: pandeyanita1@rediffmail.com

E. Velázquez and C. Rodríguez-Barrueco (eds.),
First International Meeting on Microbial Phosphate Solubilization, 291–299

A number of landraces of rice are cultivated in the rainfed upland farming systems of Uttaranchal Himalaya (Agnihotri et al., 2000). Use of blue green algae, generally recommended in the rice fields at lowlands, is not feasible in the rainfed mountain areas. In the present study, three strains of bacteria with phosphate solubilizing activity, namely- *Bacillus megaterium*, *B. subtilis* and *Pseudomonas corrugata* have been tested as seed inoculants in rice using pot as well as field based experiments. Observations were recorded in respect of rhizosphere micrflora, mycorrhizal infection, phosphorus content, growth and yield parameters.

Materials and methods

Bacterial inoculants

The bacterial inoculants used were soil isolates, all from temperate locations: *Bacillus megaterium* from the rhizosphere soil of pine forest, *B. subtilis* from the rhizosphere of tea, and *Pseudomonas corrugata* from the maize fields (Pandey et al., 1997; Pandey and Palni, 1998b). These were maintained on Tryptone Yeast extract slants at 4 °C.

Phosphate solubilizing activity of bacterial inoculants

The spot inoculation of bacteria was carried out using petridish assay on Pikovskaya agar (Pikovskaya, 1948) at 28 °C. The halo (zone of solubilization) around the bacterial colony and colony diameter were measured following incubation for 7 days. The halo size was calculated by subtracting the colony diameter from the total (halo) diameter. Quantitative estimation of tricalcium phosphate solubilization was done at 28 °C using 100 mL of NBRIP broth in Erlenmeyer flasks (250 mL) (Nautiyal, 1999) inoculated with 1 mL (10^5 cells mL^{-1}) of bacterial suspension; uninoculated medium served as a control. The pH of the broth was adjusted to 7.0 before autoclaving. An aliquot was withdrawn from the medium aseptically from each flask after 7 days, and centrifuged at 10,000 rpm for 20 min. The supernatant was analyzed for P_2O_5 content by chlorostannous reduced molybdophosphoric acid blue colour method (Allen, 1974); the pH of the supernatant was also measured. The data are an average of three replicates.

Pot experiment

Seeds of rice (*Oryza sativa* L.; landrace: *dudil*) were obtained from a local farmer in village Katarmal, District, Almora. Five to eight seeds were sown per earthen pot (11"dia; 12"ht.), containing approximately 8 kg of soil local. Twenty pots were used for each treatment and control; a minimum of 100 plants were considered for each treatment. *B. megaterium* and *B. subtilis* cultures were grown on TY agar while pseudomonas isolation agar was used for *P. corrugata*. Following 48 h incubation at 28 °C, the bacterial growth was harvested in 100 mL of sterile distilled water obtaining an approximate population 10^5–10^6 cfu mL^{-1}. Inoculation was carried out by adding 1 mL of broth culture to the soil around each seed at the time of sowing. Seeds dipped in broth alone were used as control. The pots were kept in the Institute nursery in the open.

Field experiment

The seeds were inoculated with the bacterial inoculants using sterile charcoal as a carrier. Bacterial suspension was prepared as outlined for the pot experiment. The bacterial suspension was mixed with 200 g of seeds, 150 g of charcoal and 10 g of sticker (*gur*-raw sugar, dissolved in sterile water). The mixture was thoroughly stirred to facilitate even coating of rice seeds. Seeds treated with charcoal slurry without bacteria served as control.

Separate plots were prepared, in triplicate, for each treatment (plot size = 2.50 m × 1.25 m). The plots for different treatments were separated from each other by an adequate gap and a raised mud wall (15 cm above ground) in the middle.

The seeds were sown in the second week of May and the crop harvesting was done in the first week of October 2002, both in the pot as well as field experiments. The soil pH, before seed sowing, was found to range between 7.2 and 7.5, and the soil nutrient analyses revealed following values: C = 0.560%, N = 0.094%, P = 0.100% and K = 0.022% (Allen, 1974). The experiments were conducted in the Institute nursery at Katarmal (29°38′10″ N to 79°37′30″ E; 1250 m above mean sea level), District-Almora in the state of Uttaranchal.

Microbial analyses

Three groups of microorganisms, viz. bacteria, actinomycetes, and fungi were enumerated to define the rhizosphere colonization, using the serial dilution technique (Johnson and Curl, 1972; soil samples analyzed in triplicate). Nutrient agar for bacteria, Actinomycetes isolation agar for actinomycetes and potato dextrose agar for fungi were used for these enumerations. Jensen's agar (Jensen, 1954), a nitrogen free medium, was also used for the enumeration of a specific group of bacteria. Following incubation at 28 °C for 1 week, the plates were observed for colonies.

Mycorrhizal colonization

Mycorrhizal colonization was determined on the basis of mycorrhizal roots per plant. The fine roots were separated, rinsed several times in tap water, cut into 1.0 cm pieces and treated with 10% KOH for 12 h at room temperature. These pieces were then bleached in alkaline hydrogen peroxide before staining in trypan blue (0.01%) (Phillips and Hayman, 1970). Microscopic observations were carried out to quantify % infection (colonization). Number of positive root pieces × 100/total number of root pieces observed gave the value for % mycorrhizal infection.

Plant growth, yield, harvest index and phosphorus analyses

Ten plants for each treatment were randomly uprooted from different pots as well as plots. Measurements were recorded for root and shoot length as well as biomass after 150 days of seed sowing. The grain yield was recorded at the time of harvesting. Harvest index was calculated using the following formula (Hall et al., 1993): Harvest index = Economic yield × 100/Biological yield. The pH of the soil after the crop harvest was also estimated. The phosphorus content of different plant parts were analyzed using the oven-dried (80 °C, for 48 h) and powdered (2 mm) samples. Triplicate samples (0.1 g) were digested on a hot plate and analyzed for phosphorous by the molybdophosphoric blue colour method (Allen, 1974).

Data were statistically analyzed as per Snedecor and Cochran (1967).

Results

Phosphate solubilizing activity of bacterial inoculants

All the bacterial isolates exhibited phosphate solubilizing activity and formed clear halo around the bacterial colony on Pikovskaya agar plates. Out of the three bacteria, *Pseudomonas corrugata* exhibited strongest activity, followed by *B. megaterium* and *B. subtilis*. These results were confirmed by quantitative measurements carried out with broth cultures. The bacterial inoculations also resulted in the lowering of pH of the broth indicating acid production with time (Table 1).

Pot experiment

Increase in yield and growth parameters was recorded for treated plants (Table 2). The biomass of different plant components was influenced positively, but differentially by bacterial inoculations. Out of the three bacterial species, *B. subtilis* performed best and resulted in 1.37 fold increase in the total biomass over control.

Table 1. Phosphate solubilization in Pikovskaya and broth cultures, and corresponding lowering of pH following incubation at 28 °C after 7 days

Bacterial inoculants	Halo size (mm) on Pikovskaya agar*	P solubilized in NBRIP broth**(μg/mL)	pH
Bacillus megaterium	2.3	8.0	5.08
Bacillus subtilis	1.7	5.5	5.37
Pseudomonas corrugata	9.7	11.0	4.57

* = Pikovskaya, 1948; **NBRIP (National Botanical Research Institute's phosphate growth medium = Nautiyal 1999.

Table 2. Influence of bacterial inoculation on morphological and yield attributes in rice using pot based assays

Bacterial inoculants	Length (cm)		Biomass production and yield (g dry weight)				Harvest Index	
	Root	Shoot	Root	Straw	Grain	Total		
Control	13.63 ± 2.49	122.53 ± 9.63	4.76 ± 1.07	15.40 ± 4.44	8.46 ± 3.38	28.63 ± 07.44	29.56	
B. megaterium	17.10 ± 3.16	133.20 ± 18.32	5.14 ± 1.30	16.66 ± 4.86	11.30 ± 4.75	33.11 ± 08.61	34.14	
B. subtilis	18.90 ± 4.72	137.40 ± 12.47	6.49 ± 2.42	18.73 ± 5.57	14.07 ± 6.91	39.30 ± 13.26	35.80	
P. corrugata	18.17 ± 5.23	134.21 ± 21.79	5.27 ± 2.61	16.96 ± 7.44	12.22 ± 6.51	34.43 ± 15.66	35.41	
ANOVA table								
Parameters	Length of root				Length of shoot			
Source of variation	df	MS	*F*	*P*-value	df	MS	*F*	*P*-value
Between treatments	3	58.96815	3.2414987	0.034787*	3	406.7321	1.394266	0.262422**
Within treatments	32	18.18889			32	291.7177		
Parameters	Grain yield				Total biomass			
Source of variation	df	MS	*F*	*P*-value	df	MS	*F*	*P*-value
Between treatments	3	59.66398	1.833685	0.160895**	3	230.7732	1.56824	0.216217**
Within treatments	32	32.53774			32	147.1542		

Values are a mean ± SD of ten individual plants.
*Significant at $P < 0.05\%$, ** Not significant.

The increase affected by *P. corrugata* and *B. megaterium* was 1.20 and 1.16 fold, respectively. The proportionate increase in grain yield was maximum due to bacterial inoculation in all the three treatments. The enhancement in grain yield obtained with *B. subtilis*, *P. corrugata* and *B. megaterium* was 1.66, 1.44 and 1.34 fold over control, respectively. The harvest index on per plant basis also recorded, increased irrespective of the treatments. The root length was 13.63 cm in control, 18.90 cm in *B. subtilis*, 18.17 cm in *P. corrugata* and 17.10 cm in *B. megaterium* treatments. The shoot height was also found to increase in the same order.

Field experiment

The data showing the influence of bacterial inoculations on growth and yield of rice are presented in Table 3. The treatments resulted in improvement in biomass, in terms of root, straw and grain weight both on per plant and unit area (m^2) bases. In this experiment also, *B. subtilis* gave the best performance, with an increase of 1.40 and 1.55 fold for total biomass and grain yield, respectively, on per plant basis. For *P. corrugata* treatment the increase was 1.26 and 1.36 fold and for *B. megaterium* it was 1.17 and 1.25 fold, respectively. The harvest index per unit area also recorded an increase in all the treatments as compared to control. There was a positive increase in root length; it was 12.34, 15.60, 16.80 and 15.83 cm in control, *B. megaterium*, *B. subtilis* and *P. corrugata* treatments, respectively. The shoot height was also positively influenced by bacterial treatments in the order: *B. subtilis* > *P. corrugata* > *B. megaterium* > control.

Microbial analyses

Changes in the microbial community in the rice rhizosphere due to bacterial inoculations, under field conditions, are presented in Figure 1. The populations of bacteria (in general and those grown on Jensen's medium) and actinomycetes were found to be stimulated due to inoculations. The maximum stimulation was found in case of *B. subtilis* treatment, where the counts increased by 1.5–2.4 fold for bacteria, 1.5–2.9 fold for actinomycetes and 1.7–3.5 fold for the bacteria on Jensen's medium. In case of *B. megaterium* treatment the counts were found to increase between 1.3–2.1, 1.6–2.4 and 1.7–3.3 fold, and in *P. corrugata* treatment the increase was 1.3–2.1, 1.7–2.2 and 1.1–2.2 fold for bacteria, actinomycetes and for bacteria on Jensen's medium, respectively. The counts were higher in all the three treatments as compared to control, during the entire period of plant growth. The microbial populations were highest during the middle of the growth period, after which a decline was recorded. Contrary to this, the fungal population in the rhizosphere was not stimulated and the counts remained lower than the counts recorded on zero day of sampling. Also during the entire period the plant growth, the fungal counts in all the bacterial treatments remained lower as compared to control. Results were similar for the rhizosphere soil samples collected from the pot experiment (data not presented).

Mycorrhizal colonization

The per cent roots colonized by mycorrhizae increased up to 90 days following seed sowing, after which the per cent colonization remained more or less constant (Figure 2). The maximum root colonization was found in case of *B. subtilis* treatment (88.4%), followed by *B. megaterium* (80.6%), *P. corrugata* (78.4%), and untreated control (76.4%), 90 days after sowing.

Phosphorus content of plant parts

The bacterial inoculation positively influenced the phosphorus (P) content of various plant components; *P. corrugata* treatment was most effective in this respect. The treatments were found to enhance enhance the P content of shoots and grains. The P content of roots was not enhanced (Table 4).

Discussion

The observations recorded on growth and yield related parameters demonstrate the beneficial effects of bacterial inoculation on rice; the best response was obtained with *B. subtilis* treatment. The other effects recorded were in terms of

Table 3. Influence of bacterial inoculation on morphological and yield attributes in rice following a field experiment

Bacterial inoculants	Length (cm)		Biomass production and yield (g dry weight)							
			Per plant basis				Per unit area (m^2) basis			
	Root	Shoot	Root	Straw	Grain yield	Total	Crop residue	Grain yield	Total	Harvest index
Control	12.34 ± 2.66	119.50 ± 11.24	2.85 ± 1.06	12.50 ± 3.40	6.85 ± 3.18	22.21	460.98	205.89	666.87	30.87
B. megaterium	15.60 ± 3.03	127.70 ± 17.54	3.58 ± 1.32	13.74 ± 4.33	8.60 ± 4.23	25.92	520.00	258.10	778.10	33.17
B. subtilis	16.80 ± 4.37	131.17 ± 11.05	4.72 ± 2.17	15.59 ± 5.14	10.76 ± 6.24	31.07	609.43	323.04	932.47	34.64
P. corrugata	15.83 ± 4.51	129.40 ± 21.26	3.53 ± 1.71	15.02 ± 7.47	9.36 ± 7.35	27.91	560.96	281.05	842.01	33.37
ANOVA table										
Parameters	Length of root				Length of shoot					
Source of variation	df	MS	*F*	*P*-value	df	MS	*F*	*P*-value		
Between treatments	3	33.91213	2.304433	0.095595**	3	190.5463	0.79243	0.507124**		
Within treatments	32	14.71604			32	240.4583				
Parameters	Grain Yield				Total biomass					
Source of variation	df	MS	*F*	*P*-value	df	MS	*F*	*P*-value		
Between treatments	3	15002.85	0.604915	0.616622**	3	65861.9	0.653472	0.586658**		
Within treatments	32	24801.57			32	100787.7				

Values are a mean ± SD of ten replicates.
**Not significant.

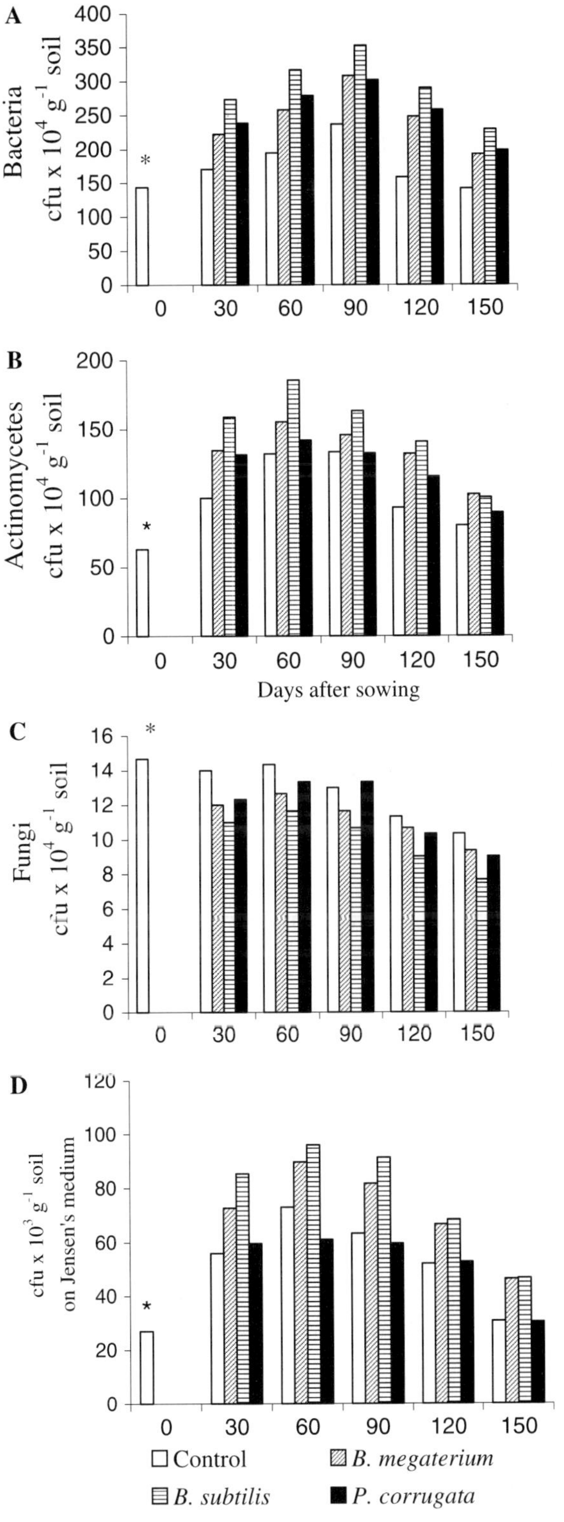

Figure 1. Influence of bacterial inoculation on the microbial communities in the rhizosphere of rice. The LSD values for various microbial communities are: 9.01, 7.78, 10.61, 5.43, 6.21 for bacteria; 8.23, 8.32 7.07 8.41, 7.31 for actinomycetes; 1.24, 2.03, 1.52, 1.70, 1.31 for fungi; and 6.37, 5.74, 4.64, 5.02, 5.42 for the cfu(s) recorded on N-free medium at 30, 60, 90, 120 and 150days, respectively after seed sowing. Cfu = colony forming units. * Bars indicate counts at the time of sowing.

stimulation of rhizosphere associated native bacterial and actinomycetes populations, increase in mycorrhizal colonization of roots and suppression of fungal population in the rhizosphere. Improved phosphorus content of plants was also related to the bacterial inoculations. A number of physiological properties like N-fixation, P-solubilization, production of antifungal and plant growth promoting substances are given importance while selecting effective strains of bioinoculants. Besides these, original habitat of the isolates and their ability to positively influence the native microflora are other parameters of importance. In previous pot as well as field based studies, the beneficial effects of bacterial inoculations have been correlated with the stimulation of native microbial communities in the rhizosphere (Pandey et al., 1998; Pandey et al., 1999). Similar observations have been recorded in this study. Besides the stimulation of general bacterial and actinomycetes flora, root colonization of ectomycorrhizal fungi was also found to be stimulated in all the treatments. The role of ectomycorhizal fungi in improving the phosphorus nutrition of plants is well documented (Lapeyrie

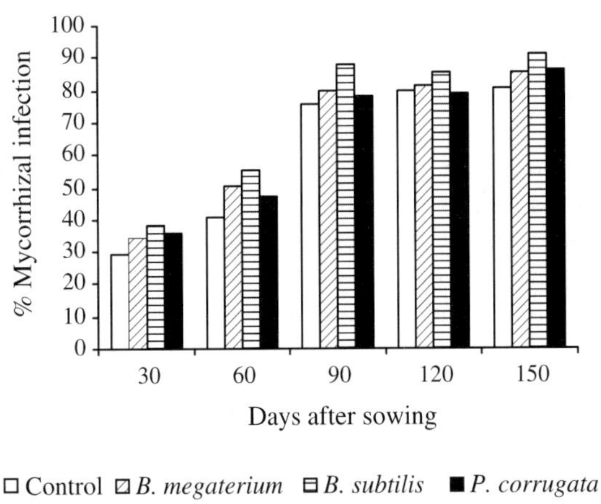

Figure 2. % Mycorrhizal colonization in the roots of rice. The LSD values are 4.88, 3.40, 2.06, 1.81, 3.14 at 30, 60, 90, 120 and 150 days respectively after sowing.

Table 4. Phosphorus content (%) of different parts of rice plant following bacterial inoculation

Bacterial inoculants	Root	Shoot	Grain
Control	0.0221	0.039	0.0307
Bacillus megaterium	0.0207	0.041	0.032
B. subtilis	0.0193	0.050	0.0343
P. corrugata	0.0221	0.059	0.0357

et al., 1991). This group of fungi is well known for a number of other properties associated with plant growth promotion, e.g., improved water status and nutrient uptake, and the protection of root system against phytopathogens (Marschner and Dell, 1994). While the inoculants used in this study possessed phosphate solubilizing property, the overall positive influence obtained may have resulted from a combined effect exerted by the stimulated microbial communities, including mycorrhizae. *B. subtilis*, the weakest phosphate solubilizer among the three bioinoculants, was most effective in stimulating the general microflora, mycorrhizal colonization and the suppression of fungal flora in the rhizosphere. In fact, improvement in the mycorrhizal colonization would seem to be an important attribute of the use of native strains in inoculation trials. The pH of the soil (7.2–7.5) recorded at the time of sowing was found to decline (up to 6.8–6.9) in various treatments after harvest. The decline in the soil pH may be an outcome of the microbial activity in the rhizosphere. Suppression of the general fungal flora in the rhizosphere of treated plants is indicative of antifungal property of the inoculants. (Pandey et al., 1997; Pandey and Palni, 1998b).

Results of the present study represent a step forward of a systematic programme initiated for the isolation of native bacteria, screening for plant growth promoting rhizobacteria, and subsequent selection of suitable inoculants for use in the colder regions of mountains. The programme began with the isolation of a large number of bacteria from the soil samples collected from various temperate/alpine (up to 3600 m above mean sea level) locations. The initial *in vitro* experiments revealed the dominance of species of *Bacillus* and *Pseudomonas* in these soils (Pandey and Palni, 1998a, b). The isolates were screened for various beneficial properties, e.g., ability to solubilize tricalcium phosphate, production of antifungal compounds, intrinsic antibiotic resistance, nitrogen fixing ability, etc., with emphasis on their ability to tolerate lower temperatures (Pandey et al., 1997; Pandey and Palni, 1998a, b; Pandey et al., 2002). Based on the results of above cited studies, the efficient bacterial isolates were tested as inoculants using bioassays and pot assays (Pandey et al., 1999, 2000, 2001). The programme has now progressed to the stage of testing the potential inoculants in field trials using local hill crops. The growth promotion of rice observed in this investigation seems to result from a combined effect of various mechanisms, involved directly or indirectly. Regardless of the mechanism(s) involved, the study suggests the suitability of native bacterial species to be developed as carrier based bioinoculants for use in the colder regions of mountains.

Acknowledgements

The Department of Biotechnology and the Union Ministry of Forests and Environment, Government of India, New Delhi, are acknowledged for financial support.

References

Agnihotri R K, Chandra S, Sharma S and Palni L M S 2000 Genetic variability in photosynthesis and chlorophyll content of various landraces of upland rice. IRRN. 25(2), 13–14.

Allen S E 1974 Chemical Analysis of Ecological Materials. Blackwell Scientific Publications, Oxford.

Chabot R, Antoun H and Cescas Michel P 1996 Growth promotion of maize and lettuce by phosphate solubilizing *Rhizobium leguminosarum* biovar. *phaseoli*. Plant Soil 184, 311–321.

Hall D O, Scurlock J M O, Bolhar-Nordenkamph H R, Leegood R C and Long S P 1993 Photosynthesis and Production in a Changing Environment. A Field and Laboratory Manual. Chapman & Hall, London.

Illmer P, Barbato A and Schinner F 1995 Solubilization of hardly- soluble $AlPO_4$ with P- solubilizing microorganisms. Soil Biol. Biochem. 27(3), 265–270.

Jensen H L 1954 The Azotobacteriaceae. Bacteriol. Rev. 18, 195–214.

Johnson L F and Curl E A 1972 Methods for Research on the Ecology of Soil-Borne Plant Pathogens. Burgess, Minneapolis.

Johri J K, Surange S and Nautiyal C S 1999 Occurrence of salt, pH, and temperature-tolerant, phosphate-solubilizing bacteria in alkaline soils. Curr. Microbiol. 39, 89–93.

Kloepper J W, Lifshitz R and Zablotowicz R M 1989 Free living bacterial inocula for enhancing crop productivity. Trends Biotechnol. 7, 39–44.

Kumar V, Behl R K and Narula N 2001 Establishment of phosphate-solubilizing strains of *Azotobacter chroococcum* in the rhizosphere and their effect on wheat cultivars under green house conditions. Microbiol. Res. 156, 87–93.

Lapeyrie F, Ranger J and Vairelles D 1991 Phosphate solubilizing activity of ectomycorrhizal fungi *in vitro*. Can. J. Bot. 69, 342–346.

Marschner H and Dell B 1994 Nutrient uptake in mycorrhizal symbiosis. Plant Soil 159, 89–102.

Nautiyal C S 1999 An efficient microbiological growth medium for screening phosphate solubilizing microorganisms. FEMS Microbiol. Lett. 170, 265–270.

Pandey A, Durgapal A, Joshi M and Palni L M S 1999 Influence of *Pseudomonas corrugata* inoculation on root colonization and growth promotion of two important hill crops. Microbiol. Res. 154, 259–266.

Pandey A and Kumar S 1989 Potential of azotobacters and azospirilla as biofertilizers for upland agriculture: a review. J. Sci. Indus. Res. 48, 134–144.

Pandey A and Palni L M S 1998a Microbes in Himalayan Soils: biodiversity and potential applications. J. Sci. Indus. Res. 57, 668–673.

Pandey A and Palni L M S 1998b Isolation of *Pseudomonas corrugata* from Sikkim Himalaya. World J. Microbiol. Biotechnol. 14, 411–413.

Pandey A, Palni L M S and Bag N 2000 Biological hardening of tissue culture raised tea plants. Biotechnol. Lett. 22, 1087–1091.

Pandey A, Palni L M S and Coulomb N 1997 Antifungal activity of bacteria isolated from the rhizosphere of established tea bushes. Microbiol. Res. 152(1), 105–112.

Pandey A, Palni L M S and Hebbar K P 2001 Suppression of damping-off in maize seedlings by *Pseudomonas corrugata*. Microbiol. Res. 156, 191–194.

Pandey A, Palni L M S, Mulkalwar P and Nadeem M 2002 Effect of temperature on solubilization of tricalcium phosphate by *Pseudomonas corrugata*. J. Sci. Indus. Res. 61, 457–460.

Pandey A, Sharma E and Palni L M S 1998 Influence of bacterial inoculation on maize in upland farming systems of the Sikkim Himalaya. Soil Biol. Biochem. 30(3), 379–384.

Peix A, Rivas-Boyero A A, Mateos P F, Rodriguez-Barrueco C, Martinez-Molina E and Velazquez E 2001 Growth promotion of chickpea and barley by a phosphates solubilizing strain of *Mesorhizobium mediterraneum* under growth chamber conditions. Soil Biol. Biochem. 33, 103–110.

Phillips J M and Hayman D S 1970 Improved procedure for clearing roots and staining parasitic and vesicular arbuscular mycorrhizal fungi for rapid assessment of infection. Trans. Brit. Mycol. Soc. 55, 158–160.

Pikovskaya R I 1948 Mobilization of phosphorus in soil in connection with the vital activity of some microbial species. Mikrobiologiya 17, 362–370.

Snedecor G W and Cochran W G 1967 Statistical Methods. Oxford & IBH, New Delhi.

Tilak K V B R 1991 Bacterial Fertilizers. Publication and Information Division. Indian Council of Agricultural Research, New Delhi.

Vazquez P, Holguin G, Puente M E, Lopez-Cortes A and Bashan Y 2000 Phosphate solubilizing microorganisms associated with the rhizosphere of mangroves in a semiarid coastal lagoon. Biol. Fertl. Soils 30, 460–468.

Evaluation of the effect of a dual inoculum of phosphate-solubilizing bacteria and *Azotobacter chroococcum*, in crops of creole potato (papa "criolla"), "yema de huevo" variety (*Solanum phureja*)

G. Faccini, S. Garzón, M. Martínez[1] & A. Varela

Facultad de Ciencias, Carrera de Microbiología Industrial, Pontificia Universidad Javeriana, Santafé de Bogotá, Cra. 7a N.° 43-82, Colombia. [1]*Corresponding author**

Received 23 December 2002. Accepted in revised form 2 January 2003

Key words: *Azotobacter chroococcum*, creole potato, papa criolla, phosphate-solubilizing bacteria, *Solanum phureja*

Abstract

Four isolates of PSB (*Pseudomonas cepacia*, *Xanthomona maltophilia*, *Enterobacter cloacae* and *Acidovorans delafieldii*, formerly called *P. delafieldii*) and four strains of *Azotobacter chroococcum*, isolated in a previous work were chosen. They did not show antagonism among themselves, by means of in vitro tests made on GISA medium (PSB-*Azotobacter* modified medium). A dual inoculum was made with the 8 isolates in 4.6 L of sterile GISA broth, which was under continuous air flow. This dual inoculum was taken to a field sample where seeds of "criolla" potato, yema de huevo variety (*Solanum phureja*) were cultivated. After120 days from inoculation, statistical analyses showed that as for stem height, dry weight of the root, number of tubers and soil available phosphorus, there were significant differences among the various treatments. As for all other variables, there were no observable differences among them. With the a posteriori test of Tukey, it was possible to determine that with chemical fertilization – with or without dual inoculum, – the stem height, the fresh weight of plants, fresh weight of leaves and tubers, the results were significantly greater than with the other treatments. The dry weight of roots, and the soil available N, showed better results with the inoculation of 50% of the inoculum plus 50% of chemical fertilizer. The number of tubers showed better results with 100% of fertilizer. A dual inoculum of PSB and *Azotobacter chroococcum* like the one used in this research, will maintain production (ton/ha) of "criolla" potato, Yema de Huevo variety (*Solanum phureja*), at a level matching that of crops with 100% NPK fertilization only, and at the same time, will contribute to the reduction of costs (in nearly 7.4%), a fact that represents favorable implications at both, economical and environmental levels.

Introduction

Cultivation of all varieties of potato requires a balanced and timely fertilization which must be handled in an efficient and cost-effective way. Currently, this kind of fertilization is carried out, using chemical fertilizers.

Fertilizers applied to potato crops basically include elements such as nitrogen and phosphorus. Frequently, these nutrients are distributed within agricultural soils, but are not available for plants without the presence of microorganisms that are able to solubilize and fix such elements.

These microorganisms are part of the soil's native micro-flora, and belong to the *Pseudomonas*, *Enterobacter*, *Bacillus* and *Micrococcus* groups, among others. They have the ability to solubilize phosphates into assimilable compounds, allowing the plant to absorb such nutrient. On the other hand, there are native free-moving bacteria

* E-mail: mmmartin@javeriana.edu.co

E. Velázquez and C. Rodríguez-Barrueco (eds.),
First International Meeting on Microbial Phosphate Solubilization, 301–308

populations, such as the *Azotobacter* class, which can produce growth-promoting substances that enhance the plant's ability to grow.

On the 1980s, the use of chemical fertilizers on potato fields in Colombia amounts, as an average, 20% of the total production investments (Ortega, 1992), increasing the acquisition costs of agricultural products. Therefore, scientists and biotechnologists in particular, while trying to cut on these costs, have used different soil-isolated bacterial groups, in the form of bio-fertilizers, which may eventually reduce chemical fertilizers overuse, as well as all derived serious problems in soil where they are applied.

Therefore, this survey intends to evaluate an inoculum to be used as a bio-fertilizer, which is made up of phosphate-solubilizing bacteria (PSB) and *Azotobacter chroococcum*, and which may be able to solubilize sources of insoluble phosphate and to improve on the use of nitrogen in crops of "criolla" potato, Yema de Huevo variety (*Solanum phureja*). In this way, use of chemical fertilizers could be diminished and crop productivity could be optimized, thus providing enhanced benefits to growers who are directly dependent on these crops for their survival.

Materials and methods

Strains and culture conditions

Twenty isolates of PSB, and 60 of *Azotobacter* were taken from those previously isolated by Cuenca and González (1996); and Martínez and Moreno (1996), respectively. A colony from each isolate was emulsified in 10 ml of nutritious broth, and incubated at 30 °C, from 24 to 48 h, for their recovery. PSB isolates were inoculated in agar, as described by Pikosvskaya (1948), for the determination of solubilization halos formation, after the incubation at 30 °C, for 24–48 h. Afterwards, they were re-isolated in King B agar, and incubated at 30 °C, from 24 to 48 h.

Starting with this culture medium, a colony was taken from each isolate and inoculated three-ways in Pikosvskaya agar, making a groove on the medium. These colonies were incubated at 30 °C, for 72 h, in order to measure halos (in mm), and to choose, through class intervals, the best phosphate-solubilizing isolates.

Recovered *Azotobacter* isolates were seeded in an Ashby medium, as described by Novo et al (1983), and were incubated at 30 °C, from 24 to 48 h. Later, biochemical tests were performed upon them, in order to identify those that belong to the *A. chroococcum* species.

Azotobacter chroococcum isolates were seeded in 10 ml of nutritious broth, and were incubated at 30 °C, for 24 h. After this time had elapsed, serial dilutions 1:10, up to 10^9 in peptonated water (0.1%) were carried out. From these, one ml of each dilution was seeded in depth, in nutritious agar – threeways, – and then, all were incubated at 30 °C, for 48 h. Afterwards, colonies were recounted at the 48th h. In order to verify the isolates purity, a Gram staining was performed.

Antagonism trials

Bacteria from each of the groups undertook antagonism trials in a GISA medium (PSB – modified *Azotobacter*), customized for the growth of both species. By means of the diffusion in agar method (Gauze, 1965). Boxes were incubated at 30 °C, for 24 h. Later on, the presence of an isolate's growth inhibition halo, indicating any antagonism among these microorganisms, was determined.

Preparation of the dual inoculum

A primary inoculum was prepared with a colony from each selected isolate in peptonated water (0.1%), incubated at 30 °C, for 24 h. A concentration ranging between 10^5 and 10^7 ufc/mL of each isolate was seeded in GISA broth, at 30 °C, for 24 h.

Out of this inoculum, two suspensions were prepared: one featuring the mixture of *Azotobacter chroococcum* isolates, and the other one featuring the mixture of PSB isolates. All of them were cultivated in 400 mL of sterile GISA broth under continuous air flow. Incubation was carried out at 30 °C, for 24 h.

Two recipients were taken for the preparation of the final inoculum, and each one contained 4.6 L of GISA broth. Then, 400 mL of the *A. chroococcum* isolates mixture, and 400 mL of the PSB isolates mixture were added to the recipients. These were kept under continuous air flow, and were incubated at 30 °C, for 48 h.

Field test

This was carried out at the "Monte Orión" farm, located at the Tausaquirá town area, municipality of Suesca (Cundinamarca; Colombia). The farm is 3000 m over sea level, and has a yearly average temperature of 8–9 °C (Feged, 1995). Farm soil is clayish, and has a pH of 5.2.

To prepare the terrain, fifteen 1.20 × 25 m plots were used. Seeds of "criolla" potato, Yema de Huevo variety, were sown into them. Each plot was divided into 25 portions, each measuring 1.20 × 1 m, in order to expedite all corresponding statistical analyses.

Experimental design

The design was that of randomized complete blocks, and included the following treatments:

T1: Water (negative control/indicator).

T2: NPK (Abocol 10:30:10) at 100% (750 kg/ha), applied 30 days after the test was started.

T3: NPK (Abocol 10:30:10) at 50% + 20 L/ha of the dual inoculum. The compost was applied after 30 days, and the bacterial suspension was applied on days 0, 30 and 90, after the treatment was started.

T4: 20 L/ha of the dual inoculum. Applied on days 0, 30 and 90 after the treatment was started.

T5: NPK (Abocol 10:30:10) at 50%, applied 30 days after the test was started.

The following variables were evaluated, at 120 days after the inoculum was applied:

Stem height (cm). Fresh weight of the plant (g), dry weight of the root (dried in stove, at 64 °C, for 24 h), fresh weight of leaves (g), number of tubers, nitrogen and phosphorus in leaves (Olsen's method) and available in the soil (Bray II method), weight of tubers (g), persistence of bacterial populations in the soil during the crop cycle (recounts in GISA agar plates) and production (ton/ha).

The total weight of plants and tubers, the stem heights and the bacterial recounts were measured at the agricultural microbiology laboratory of the PUJ University. Nitrogen and phosphorus levels in leaves were determined at Colinagro S.A. and soil available nitrogen and phosphorus at the CIAA-UJTL.

Inoculation

Contents of both 4.6 L recipients, including cultures of both bacterial groups, were mixed in a 20 L-capacity aspersion pump. By means of this pump, inoculation of 60 mL/era (20 L/ha) of in-soil bacteria was carried out at 0, 30 and 90 days.

Results and discussion

Characterization of isolates

From the 20 PSB isolates, 12 were recovered in 10 mL of nutritious broth (60%). From the 12 recovered isolates, 11 showed formation of solubilization halos in Pikosvskaya agar. These 11 isolates were re-isolated in King B agar. Isolates having halos whose diameter ranged between 2.4 and 3.3 mm were selected (9 isolates). These readings correspond to the last two class intervals.

From the 60 *Azotobacter* sp isolates, 36 were recovered (60%). Gram coloration showed that in 20 of said 36 isolates, cyst formation was observed, indicating physiological maturity of the cultures (Moat, 1977; Stanier, 1989). While recounting the 36 isolates in nutritious agar, 14 of such isolates had recounts falling between 10^5 and 10^7 ufc/mL; during biochemical tests, out of these 14, 8 belong to the *A. chroococcum* species.

After 24 h of incubation at 30 °C, in GISA agar, recounts ranging from 10^5 to 10^7 ufc/mL were measured for 6 PSB isolates and 6 *Azotobacter chroococcum* isolates.

During the antagonism trials, the 6 PSB isolates were confronted among themselves, the 6 *Azotobacter chroococcum* isolates among themselves, and PSB and *Azotobacter chroococcum* isolates were confronted to one another. Isolates showing antagonism among themselves (2 PSB and 2 *Azotobacter chroococcum*) were discarded from the survey. Therefore, a total of eight (8) isolates were used to prepare the inocula, and each of these had recounts ranging between 10^5 and 10^7 ufc/mL.

Field experiments

Data gathered from the initial soil analysis are shown in Table 1, as determined by the CIAA-UJTL (Centro de Investigaciones y Asesorias Agroindustriales and Agroindustrial Consulting and Research Center of the Jorge Tadeo Lozano University).

As for the evaluated variables, the following results were obtained:

Stem height
After 90 days, T2 and T3 showed significant differences (P = 1.44E-24; f = 75) when compared with all other experimental treatments, while showing a similar level of efficiency (Table 7, V1). In T3, this may be due to production by the inoculated bacteria of growth-promoting substances that carry out an important role in the stem expansion process, as discovered by Bonner (1961) and Weaver (1980).

These results are corroborated by works by Chabot (1994), who inoculated lettuce and corn plants with *Enterobacter* sp and *Pseudomonas* sp, and observed an increase in plant height, which he believed was due to production of siderophores and auxines by bacteria. Likewise, Dibut et al. (1995) demonstrated the ability to bio-synthesize aminoacids and cytoquinines by an *Azotobacter chroococcum* isolate, which may, if applied, have an impact on bio-stimulation of horticultural crops.

Table 1. Initial soil analysis

Element	Concentration (ppm)
$N\text{–}NH_4$	10.2
$N\text{–}NO_2$	3.4
P	15
K	3.89
Ca	905
Mg	128
Na	77
Fe	94
Mn	1.98
Cu	0.27
Zn	0.5
B	0.28
Al	2.7

On the other hand, the simple addition of a fertilizer containing N and P may promote growth of stems and plants in general (Salisbury and Ross, 1992).

After 120 days, no significant differences were observed among treatments, since the plant reaches its maximum development at the 90-day mark, and does not require or absorb a great amount of nutrients after that.

Fresh weight of plants
The ANOVA shows that there are no differences among treatments having $P = 0.40$ and $f = 1.44$. However, treatments 2 and 3 show heavier fresh weight (Table 7, V2). This suggests that there are elements (N and P) available for the plant's root, and these may be used and incorporated to enhance the plant's weight, as described by Alexander (1980).

Dry weight of the root
The ANOVA showed that there are some differences among treatments ($P = 0, f = 3.30$).

The Tukey trial reveals that treatments 2 and 3 showed really significant minimum differences among themselves, and as compared to all other treatments (Table 7, V3). T3 offered the best results when compared to all other treatments, probably due to the production of growth-promoting substances and vitamins by the inoculated bacteria.

Similar results were reached by Stoyanov and Kudrew (1978), who gained an increase in root dry weight for corn plants, by adding vitamins B1, C6 and B6; and by Mozafar (1994), who showed the importance of the role played by vitamins such as vitamin B12, produced by *Azotobacter* sp, in bean, barley and spinach plants.

Fresh weight of leaves
The ANOVA showed that differences within themselves are bigger than differences among each other, when featuring values $P = 0.15$ and $f = 2.83$.

Treatments 2 and 3 showed a really significant minimum difference, as compared with all other treatments (Table 7, V4).

As for T3, this was probably due to the presence of inoculated bacteria that support absorption, production of vitamins and growth-promoting substances, and solubilization of phosphates. This process has a direct impact on production of ATP and enzymes in charge of stimulating the growth apex (Bonner, 1961; Galston and Mc Cone, 1961).

As for T2, since the radicular system of plants is wide enough (radicular weight), the absorption surface becomes larger, so when 100% of the fertilizer is added, the plant is able to make good use of the micro-nutrients that are to be transported to the leaves, having a positive effect on foliage growth and thus, on foliage weight.

Number of tubers

At the 120-day mark, the ANOVA showed that with values $P = 0.03$ and $f = 2.66$, treatments differ significantly from one another.

The Tukey test showed that there are no significant differences between T2 and T3, nor among T2, T4 and T5. However, a slight increase may be observed in the average tubers for treatments 2 and 3, placing them above all others (Table 7, V5). As per the cultivation needs defined through the initial soil analysis (Table 1), nitrogen and phosphorus levels were optimal. When fertilizing with NPK 10:30:10, amounts of phosphorus and nitrogen are added to the cultivation which exceed the required levels; therefore, plants may use both, these sources and those already existing within the soil. These results are similar to those recorded by Tyler et al. (1983) while experimenting in the sandy soil of Shafter, California, in potato crops where plants treated with 100% of fertilizer showed an increase in the number of tubers, probably due to pre-existing quantities of chemical in the soil, which, in turn, increased the quantities of available N and P.

Nitrogen in leaves and available in soil

Tables 2 and 3 show results on amounts of N in leaves and available in soil, at the 0 and 120-day marks.

When comparing data for nitrogen, as obtained in soil, before and after the survey was performed, it is observed that, in regards to $N–NH_4$, there might have been some accumulation in treatments 2, 3 and 5. This is probably due to the fertilization performed in previous crops and which could be still beneficial to the current crop; or to the nitrogen that, being present in the fertilizer, quickly transforms itself into this ionic form, thus increasing its volume in the soil. As for T4, it was observed that after fertilization there was a decrease in amount of $N–NH_4$, probably because bacteria were able to oxidize a part of the existing compound into nitrate which, in turn, could have been absorbed by the plant, or mineralized by bacteria to form bio-mass. These results are comparable to those published for the First Programmed Series by Colombian-Venezuelan Monomeres ("Primera Serie Programada por Monómeros Colombo-Venezolanos") (1984), which indicate that approximately 25% of the applied N is immobilized in the soil and is used in the coming crops, during the following years. Likewise, it is suggested that the $N–NH_4$, – and even the $N–NO_3$, may experiment chemical reactions and, by means of a physical–chemical association, may form metallic–organic and clay–organic complexes, thus allowing the compounds to be protected from microbial attacks (Burbano, 1989).

Now, as for in-soil nitrate, this decreased in treatments T3 and T4, probably due to good levels of nitrification by bacteria (although the microorganisms in use are not able to nitrify by themselves, they could stimulate populations which effectively are able to, and which are present in the soil).

Table 2. Nitrogen available in soil, before and after treatment application

Treatment		Days	Ammoniacal N (ppm)	Nitrate (ppm)
Before		0	10.2	3.4
	T1	120	12.4	6.0
	T2	120	14.2	13.4
After	T3	120	16.2	0.77
	T4	120	11.5	5.8
	T5	120	24.7	1.1

Table 3. Statistical analysis of foliar nitrogen concentration

Treatments	Average (5 data)	Homogenous groups
T1	3.73	A
T2	4.18	A
T3	4.42	A-B
T4	3.77	A
T5	3.63	A
DMRS	0.58	

DMRS: really significant minimum difference.

At a foliar level, it is evident that, in T3, this could have been the most important event, since this was the treatment that produced the highest concentration of N in leaves. In T4, this was not the case, probably because, at soil level, microorganisms performed an efficient nitrification process and, at the same time, took all that nitrate to form bio-mass, thus competing with the plant and stopping it from taking the nutrient for itself; or, the nitrate that was formed within the soil might have been lost through lixiviation, since this is not absorbed as colloids and is easily disposed of, thus resulting in a decrease at foliar level (Salisbury and Ross, 1992).

In T5, the plant might have taken the nitrate from the soil and incorporated it to its structures, causing the decrease observed in the soil. Also, it might have been lost through lixiviation, since it is an ion that is loosely retained within the soil (Neira, 1992). Tyler et al. (1983), while using different concentrations of nitrified fertilizer in potato fields, concluded that in soils where plants were fertilized 100% with the compound, the nitric form was found in a larger proportion than any other concentration, due to the excessive application of the fertilizer, in amounts that the plant does not really need. This survey corroborates data from T2, in which, as compared with all other treatments, there is a higher quantity of available nitrogen.

Phosphorus in leaves and available in soil

Tables 4 and 5 show results obtained through the phosphorus analysis.

These results show that treatments do not differ among themselves. Phosphorus percentages do not vary among treatments, a fact that can be explained if the element has probably been immobilized at soil level, as described by Data et al. (1982), who performed experiments with PSB that immobilized phosphorus and caused a decrease during the first year of experiments, but that on the following year, offered results that, due to the release of ions, increased significantly the percentage of phosphorus in those treatments with bacteria, derived from a catabolic repression of nitrogenases and phosphatases, which limited the availability of the compounds at soil level. Similar data was gathered by Martínez (1996), while working with sugar cane with PSB.

Table 4. Phosphorus available in soil, before and after inoculation

Treatments		Days	P (ppm)
Before		0	15.0
	T1	120	30.0
	T2	120	32.3
After	T3	120	72.7
	T4	120	20.0
	T5	120	51.6

Table 5. Statistical analysis of phosphorus in leaves

Treatments	Average (6 data)	Homogenous groups
T1	0.22	A
T2	0.23	A
T3	0.27	A
T4	0.27	A
T5	0.21	A
DMRS	0.48	

DMRS: really significant minimum difference

There is another possible fact that could explain the lack of variations in phosphorus percentages among treatments, and that is that all of such treatments were exposed to a constant low temperature (9 °C). According to Nielsen et al. (1961), absorption of phosphorus and other nutrients, such as N, Ca, Mg, and K may depend on periodic increases in temperature, since the higher the temperature, the higher the absorption levels.

Weight of tubers

The ANOVA showed that, in statistical terms, treatments do not differ significantly ($P = 0.50$, $f = 1.12$).

Satisfactory results were reached through treatments 2 and 3 (Table 7, V6), because there was a higher level of availability of N and P when NPK was applied to the soil in concentrations of 100 or 50% compound + 50% dual inoculum. Phosphorus is present in enzymes involved in the ATP synthesis, as needed to carry out vital processes for the plant and to participate in the fixation of N and the conversion of compounds such as CO_2 into glucose. Starting with this monosacharide and inorganic sources of N, the plant will synthesize all the bio-molecules required for its development, such as carbohydrates (starch, among others), lipids, and proteins (Burbano, 1989; Macarulla and Goñi, 1987).

Permanence of bacterial populations in soil during the crop cycle:

Table 6 shows bacterial recounts from soil samples, as performed in GISA agar plates.

In those treatments in which the inoculation was carried out, bacterial populations at the 120-day mark had increased at 10^3–10^5 rates, as compared to day 0. On the other hand, recounts for all other treatments remained constant, indicating that the inoculated bacterial populations that were responsible for the increase were also responsible for changes observed in the plants.

Production of tubers (ton/ha)

Taking into account that the potato field prior to the experiment had been fertilized as a whole with 100% NPK 10:30:10, a comparison among treatment productions was possible, using T2 (NPK 100% 10:30:10) as a control. Therefore, it was observed that T2 and T3 have higher production rates (ton/ha), as compared to all other treatments (Table 7, V7). Nonetheless, when inoculating 100% of NPK, or 50% NPK + 50% of dual inoculum, the same production results may be reached.

Through this survey, we were able to find a culture medium (GISA) that, with the appropriate combination of nutrients, allows the growth of two different bacterial groups (*Azotobacter chroococcum* and PSB, *Pseudomonas*, *Enterobacter* and *Xanthomonas* groups). The dual inoculum prepared in this medium and applied in combination with a chemical fertilizer (50–50) to a crop of "criolla" potato, Yema de Huevo variety (*Solanum phureja*) enhanced the integral development of the plant, due to the bacteria's ability to make the nitrogen and phosphorus compounds assimilable, and to simultaneously produce growth-promoting substances (phyto-hormones).

This suggests that inoculation of bacteria, together with the addition of smaller quantities of those chemical fertilizers that are generally applied, is a real alternative that may increase crop's production and, at the same time, lower costs implied in 100% chemical fertilization processes. Normally, fertilization with chemical products alone demands 750 kg/ha, each kilogram costing COL$360 (that is, COL$270.000/ha); Switching to mixed fertilization (biological–chemical) causes chemical-related expenses to be cut by half, while the other half is supplied by bacteria. Therefore, use of bacteria causes a 7.4% decrease in fertilization total costs (this means that, at a commercially representative value, total costs for

Table 6. Average of triplicates of bacterial recounts in soil, at 0, 90 and 120 days

Time (days)	T1	T2	T3	T4	T5
0	3.3E + 3	3.3E + 3	3.3E + 3	3.3E + 3	3.3E + 3
90	3.4E + 3	2.7E + 3	2.3E + 6	4.6E + 6	2.1E + 3
120	4.0E + 3	4.5E + 3	1.1E + 8	8.2E + 6	3.2E + 3

Table 7. Statistical analysis of variables

Variables treatments	V1 (cm)	V2 (g)	V3 (g)	V4 (g)	V5 (g)	V6 (g)	V7 (ton/ha)
T1. Water	29.7	1019	681	49.26	27.53	616.66	384
T2. NPK 100%	49.97*	2363.3*	20.16	1067.33*	53.8*	1550*	11.53*
T3. 50% NPK + 50% Bacteria	64.76*	2859*	30.1*	1421*	52.13	1602*	11.99*
T4. 100% Bact	29.8	1675.8	9.88	74.4	43.06	758.66	5.84
T5. 50% NPK	29.58	16.26	7.33	72.66	48.53	492.66	3.66

*Treatments that provided the best statistical results.

fertilization of one hectare would amount to COL $175.000).

References

Alexander M 1980 Transformaciones microbianas. *In* Introducción a la Microbiología del suelo. Ed. M Alexander. pp. 355–371. Editorial AGT, México D.F.

Bonner J 1961 Plant Growth Regulation. Fourth International Conference. The Iowa State University Press, Aims, Iowa, USA p. 307

Burbano H 1989 El suelo. Una visión sobre sus componentes biorgánicos. Serie de investigaciones No. 1. Universidad de Nariño, Pasto Colombia p. 131

Cuenca D and González A 1996 Obtención de un biofertilizante a partir del crecimiento de *Azotobacter* en una mezcla en los desechos provenientes de la Industria Licorera de Caldas. Tesis de Grado, p. 72.

Chabot R A, Hani A and Cescas M 1994 Stimulation do la croissance du'mais et de la latitue romain par des microorganismes dissolvant le phosphere inorganique. Can. J. Microbiol. 39, 941–947.

Datta M, Banik S and Grupta R P 1982 Studies of the efficacy of a phytohormona producying by phosphate solubilizing *Bacillus firmus* in aumenting paddy held in acid soils of Nagaland. Plant Soil 69, 365–373.

Dibut B, Acosta M C, Martínez-Viera M and Ljinggren H 1995 Producción de aminoácidos y citoquinas por una cepa cubana de *Azotobacter chroococcum*. Cultivos Tropicales 16, 16–18.

Feged G 1995 Comunicación Personal. Suesca-Cundinamarca, Colombia.

Galston A W, Mc Cone D C 1961 On the mechanics of auxin-induced growth. *In* Plant Growth Regulation. Fourth International Conference. The Iowa State University Press, Aims, Iowa, USA, p. 611.

Gauze G I 1965 Antibióticos elaborados por hongos. Ed. Universitaria. La Habana-Cuba, p. 57.

Macarulla J and Goñi F 1987 Bioquímica Humana. Ed. Reverte. Barcelona, España, p. 198.

Martínez A 1996 Efecto de BFS durante el período de enraizamiento de la caña de azúcar (*Saccharum officinarum*) var. Venezuela 51–71 en el sustrato Grower's Oasis. Tesis de Grado, p. 86.

Martínez A and Moreno A 1996 Relación de cepas de BFS identificadas en el Centro de Investigaciones Microbiológicas (CIMIC) para la Licorera de Cundinamarca.

Moat A 1977 Microbial Physiology. Ed. John Wiley & Sons, New York, USA, p. 558.

Mozafar A 1994 Plan vitamins: agronomic, physiological, and nutritional aspects. CRC Press, U.S.A., p. 65.

Neira P 1992 La papa, el descubrimiento que conquistó al mundo. Fedepapa, p. 47.

Nielsen K E, Halsted R L, Maclean A J, Borget S J and Holmes R M 1961 The influence of soil temperature on the growth and mineral composition of corn, bromegrass. Soil. Sci. Soc. Am. Proc. 25, 369–372.

Novo R, Quintana E and Valdés R 1983 Prácticas de microbiología. Instituto Superior de Ciencias Agropecuarias de la Habana. Facultad de Agronomia, Cuba, p. 47.

Ortega J 1992 La papa, el descubrimiento que conquistó al mundo. Fedepapa, p. 32.

Pikosvskaya R I 1948 Mobilization of phosphorus in soil in connection with the vital activity of some microbial species. Mikrobiologiya 17, 362–370.

Primera de la serie programada por monómeros colombo-venezolanos. 1984 La fertilización en cultivos de clima cálido. Edición Abonos Nutrimon 1.

Salisbury F B and Ross C R 1992 Fisiología vegetal. Segunda edición. Wadsworth Publishing Company Inc.

Stanier R 1989 Microbiología. Segunda edición. Ed. Reverte S A, p. 436.

Stoyanov Y and Kudrew T G 1978 Influence of magnesium and certains vitamins on the restoration of maize plants. New Phytol. 84, 687–694.

Tyler K B, Broadbent F E and Bishop J C 1983 Efficiency of nitrogen uptake by potatos. Am. Potato J. 60, 261 269.

Weaver R J 1980 Reguladores del crecimiento de las plantas en la agricultura. Ed. Trillas. México D.F., p. 91.

Effect of inoculation with a strain of *Pseudomonas fragi* in the growth and phosphorous content of strawberry plants

L. Martín[1], E. Velázquez[1], R. Rivas[1], P.F. Mateos[1], E. Martínez-Molina[1], C. Rodríguez-Barrueco[2] & A. Peix[2,3]

[1]*Departamento de Microbiología y Genética, Facultad de Farmacia, Edificio Departamental, Universidad de Salamanca, Salamanca, 37007, Spain.* [2]*Departamento de Producción Vegetal, Instituto de Recursos Naturales y Agrobiología (IRNA. CSIC), C/ Cordel de Merinas, 40-52, 37008, Salamanca, Spain.* [3]*Corresponding author**

Received 21 October 2002. Accepted in revised form 2 January 2003

Key words: phosphate solubilization, plant growth promotion, *Pseudomonas*, strawberry

Abstract

Within genus *Pseudomonas*, several species are able to solubilize phosphate in plates and some of these species are also able to mobilize phosphorous to plants. In this work we isolated a strain, SAPA2, from the rhizosphere of barley plants growing in a soil from Northern Spain. This strain was able to solubilize phosphates in plates forming great halos of solubilization in 24 h. Moreover, this strain retained its ability to solubilize phosphate after five culture passes. The 16S rRNA sequence of this strain showed a similarity of 99.9% with that of *Pseudomonas fragi*. The inoculation of strawberry plants with this strain was carried out in growth chamber applying 10 ml of a suspension containing 10^8 UFC/ml to each plant. According to the results obtained, the plants inoculated with this strain growing in a soil amended with insoluble phosphate had a phosphorous content significantly higher than uninoculated plants growing in soil with or without insoluble phosphates. Therefore, the strain SAPA2 promotes phosphorous mobilization to strawberry plants. Therefore, the inoculation of plants with suitable phosphate solubilizing bacteria can increase the crop yield and allows a better exploitation of natural soil resources.

Introduction

The phosphorous is an essential element for the plant growth that is added to soil as soluble inorganic phosphate. However, a large portion of inorganic phosphates used as fertilizers is immobilized after application and becomes unavailable to plants (Singh and Kapoor, 1994). Therefore, the solubilization of phosphates is a process that may promotes the plant growth. Many bacterial species have been described as phosphate solubilizers (reviewed by Rodríguez and Fraga, 1999) and some of them may mobilize phosphorous to plants (Chabot et al., 1993; Chabot et al., 1996; Antoun et al., 1998; Chabot et al., 1998; Kim et al., 1998; Singh and Kapoor, 1999; Peix et al., 2001a, b). At present bacilli, rhizobia and pseudomonads are the best studied P-solubilizers groups. Currently the strains belonging to pseudomonads group are dispersed in several genera, families and even subclass of Proteobacteria. For example, *P. cepacia* has been reclassified as *Burkholderia cepacia* (Yabuuchi et al., 1992) that belongs to beta subclass of Proteobacteria, whereas genus *Pseudomonas* belongs to gamma subclass of Proteobacteria. Within current genus *Pseudomonas*, *P. putida* (Kumar and Singh, 2001; Manna et al., 2001; Villegas and Fortin, 2002; Viveganandan and Jaurhi, 2000), *P. aeruginosa*

* FAX No: +34 923 224876.
E-mail: alvarp@usal.es

E. Velázquez and C. Rodríguez-Barrueco (eds.),
First International Meeting on Microbial Phosphate Solubilization, 309–315

(Musarrat et al., 2000), *P. corrugata* (Pandey and Palni, 1998), *P. stuzteri* (Vázquez et al., 2000) *and P. fluorescens* (Deubel et al., 2000; Di-Simine et al., 1998) are the best known as phosphate solubilizers. Nevertheless, many rhizospheric phosphate solubilizing bacterial species remain unknown and their study may be very important to establish their possible role in the P-uptake by plants. Although all plants needed the phosphorous to grow, some of them are specially sensitive to phosphorous fertilization because this element is involved in the colour of their fruits. This is the case of strawberry fruits that are the most important berry used in human nutrition and whose European production is mainly located in Spain.

The cultivation of the strawberry is carried out in two steps that are performed in relatively small areas susceptible to be inoculated with microorganisms. The aim of this study was to analyse the effect of the inoculation of a extremely phosphate solubilizing strain isolated during a study of microbial populations of a soil subjected to monocrop with barley for three years. We identified this strain and analysed the effect of the inoculation on strawberry plants.

Materials and methods

Isolation of strain used in this study

Soil samples for isolation of phosphate solubilizers were taken from a soil (soil 1) in Salamanca (Spain) that was cultivated for several years with a cereal (barley). The inoculation experiments were performed in a sandy soil (soil 2) that has adequate characteristics to be mixed easily with insoluble phosphate and to transplant strawberry plants. Soil samples from plant rhizosphere were taken at a depth of 15–20 cm from three sites in both soils. Soil samples were placed in a cool box for transport, stored at 5 °C, and then used for plant inoculation tests within 2 days of collection. Soil analyses were performed according to the guidelines of the Soil Conservation Service (1972). The soil was classified according to their morphology and analytical data following the U.S. Soil Taxonomy (Soil Survey Staff, 1994). The characteristics of the two soils are shown in Table 1.

Isolation of PSB

To isolate PSB we used the method of Thomas and Shantaram (1986) modified as follows: for each site, the pooled soil was sieved (2 mm) and mixed thoroughly. A 10 g sample from each soil was emulsified in 90 mL of sterile water. Serial decimal dilutions were made from this suspension up to $1{:}10^7$. Five aliquots of 0.1 mL of each dilution were used to inoculate Petri dishes with YED (yeast extract 0.5%; glucose 1% and agar 2%) supplemented with a 0.2% of tricalciumz-phosphate (YED-P). The plates were incubated at 28 °C for 7 days. A strain whose colonies were surrounded by a great solubilization halo was isolated.

Phenotypic tests

The isolated strain was stained according to the Gram procedure and was inoculated in the Hugh–Leifson's medium to test the ability to oxidize or ferment glucose. According to the results obtained we used API20NE (Biomèrieux, France) for phenotypic characterization of this strain.

Amplification and determination of nucleotide sequences of the 16S rRNA gene and analysis of the sequence data

DNA extraction was carried out as previously described (Rivas et al., 2002a). PCR was performed using an AmpliTaq reagent kit (Perkin-Elmer Biosystems, California, USA) following the manufacturer's instructions (1.5 m*M* $MgCl_2$,

Table 1. Characteristics of soil used in this study

Soil	Texture	pH (in water)	Organic matter (%)	Total N (%)	Assimilable P (ppm)	Assimilable K (ppm)
Pedrosillo	Loamy	7.9	1.86	0.08	26	149
Aldearrubia	Sandy	6.6	0.4	0.04	59	153

2 μM of each dNTP and 2 U of Taq polymerase for 25 μL final volume of reaction). The PCR amplification of 16S rDNA was carried out using the following primers: 5′-AGAGTTTGATCTGG CTCAG-3′ (*Escherichia coli* positions 8–27) and 5′-AAGGAGGTGATCCANCCRCA-3′ (*Escherichia coli* positions 1502–1522) at a final concentration of 0.2 μM. PCR conditions were as follows: pre-heating at 95 °C for 9 min; 35 cycles of denaturing at 95 °C for 1 min; annealing at 59 °C for 1 min and extension at 72 °C for 2 min, and a final extension at 72 °C for 7 min. The PCR product (25 microliters) was electrophoresed on 1% agarose gel in TBE buffer (100 mM Tris, 83 mM boric acid, 1 mM EDTA, pH: 8.5) at 6 V cm^{-1}, stained in a solution containing 0.5 μL ethidium bromide mL^{-1}. Standard VI (Boehringer-Roche, USA) was used as a size marker. About 3 μL of 6× loading solution (30% glycerol, 0.25% xylene cyanol and 0.25% bromophenol blue) were added to each sample. The band corresponding to the 16S rDNA was purified directly from the gel by centrifugation in Eppendorff tubes with a special filter (Millipore Co., Illinois, USA) for 10 min at 5000 × *g* at room temperature according to the manufacturer's instructions.

The sequence reaction was performed on an ABI377 sequencer (Applied Biosystems Inc.) using a BigDye terminator v3.0 cycle sequencing kit as supplied by the manufacturer. The following primers were used: 5'-AACGCTGGCGGCR KGCYTAA-3′, 5′-ACTCCTACGGGAGGCAG-CAG-3′, 5′-CTGCTGCCTCCCGTAGGAGT-3′, 5′-CGTGCCAGCAG-CCGCGGTAA-3′, 5′-CA GGATTAGATACCCTGGTAG-3′ and 5′-GAG-GAAGGTGGGGATGACGTC-3′, which correspond to *E. coli* small-subunit rDNA sequence positions 32–52, 336–356, 356–336, 512–532, 782–803 and 1173–1194, respectively. The sequence obtained was compared with those from the GenBank using the FASTA programme (Pearson and Lipman, 1988).

Sequences were aligned using the Clustal W software (Thompson et al., 1997). The distances were calculated according to Kimura's two-parameter method (Kimura, 1980). Phylogenetic trees were inferred using the neighbour-joining method (Saitou and Nei, 1987). Bootstrap analysis was based on 1000 resamplings. The Mega 2 package (Kumar et al., 2001) was used for all analyses. The trees were rooted using *Bradyrhizobium japonicum* as outgroup.

Inoculation of strawberry plants

Experiments for studying the phosphorous mobilization in plants were made on strawberry plants and were conducted in pots containing soil 2. The soil for the following experiments was collected four months after the addition of the fertilizer NPK 20-10-20 (100 kg of nitrogen per Ha) to the soil. Each pot (20 × 60 cm) contained 5 Kg of soil. The pots were placed in a plant growth chamber with mixed incandescent and fluorescent lighting (400 microeinsteins m^{-2} s^{-1}; 400–700 nm), programmed for a 16 h photoperiod, day–night cycle, with a constant temperature varying from 15–27 °C (night–day), and 50–60% relative humidity. The experimental design was performed as follows: treatment 1: uninoculated plants grown in the soil without insoluble phosphate addition. Treatment 2: uninoculated plants grown in the soil with insoluble phosphate (Ca_3PO_4 0.2% w/w). Treatment 3: plants inoculated with strain SAPA2 and treatment 4: plants inoculated with strain SAPA2 grown in soil with addition of insoluble phosphate (Ca_3PO_4 0.2% w/w). The soluble and insoluble phosphates were mixed thoroughly with the soil in a plastic bag before use. Five pots were used for each treatment. Three plants were placed in each pot. The pots were watered with destilled sterile water because the soil was amended with NPK before to be used in this work.

For inoculation, strain SAPA2 was grown in petri dishes with YED-P for 2 days. After this time, sterile water was added to the plates in order to obtain a suspension with approximately 10^8 cells mL^{-1}. We added 1 mL of the suspension of the strain SAPA2 to each plant placed in petri dishes. The inoculation was performed using a micropippete in sterile conditions adding the suspension of the strain on the root.

At harvest (30 days) the dry weight of the aerial part of the plants of strawberry was determined. Plant nitrogen, phosphorous, potassium, calcium and magnesium content was measured according to the A.O.A.C. methods (Johnson, 1990). The data obtained were analyzed by one-way analysis of variance, with the mean values compared using Fisher's Protected LSD (Low Significative Differences) ($P = 0.05$).

Results and discussion

Isolation of PSB

We selected a strain, SAPA2, from the soil 1. The diameter of a clear halo surrounding the colonies of this strain was larger than 18 mm after 4 days of incubation. According to de Freitas et al. (1997) the strains producing a clear halo larger than 15 mm are considered as good P-solubilizers. Moreover, strain SAPA2 retains its ability to solubilize phosphate after five subcultures and therefore we considered it suitable to be used for inoculation of plants.

Phenotypic tests

The strain SAPA2, stained according to the Gram procedure, was a Gram-negative rod. The strain was unable to ferment glucose in Hugh–Leifson medium. As it was a strictly aerobic Gram negative rod, we used the API20NE to characterize this strain. The strain was arginine dehidrolase positive and it was able to grow in arabinose, gluconate, caprate, malate and citrate as sole carbon source, and was identified as *Pseudomonas putida* according to the API20NE system. Nevertheless, other authors have shown that the API20NE gives erroneous identifications when they were used to analyse rhizospheric strains of genus *Pseudomonas* (Behrendt et al., 1999). For that reason, we have confirmed the identification using molecular methods such as 16S rRNA sequencing.

16S rDNA sequence analysis

Strain SAPA2 sequence (accession number AY195842) showed a 99.9% similarity with that of *P. fragi* ATCC 4973^{T}, whereas the similarity with type strain of *Pseudomonas putida* DSM291^{T} was only 96.6%. The phylogenetic analysis of 16S rRNA sequences places the strains from this study in the genus *Pseudomonas* and the closest related species is *P. fragi* (Figure 1). Therefore, the identification using API20NE was good at genus level, but not at species level because *P. fragi*, as well as other non-pathogenic species from this genus, are not included in its database.

Inoculation of strawberry plants

The results of the inoculation assays are shown in Table 2. Four parameters have been evaluated to analyse the differences among the different treatments. According to the results obtained, the plants from treatment 1 have a lower weight of shoots and fruits. Also, the plants from this treatment contain a low phosphorous content. The addition to soil of insoluble phosphate originates a significant increase in the parameters measured, including dry matter and P content, with respect to the control. When the plants were inoculated with strain SAPA2 and insoluble phosphates were not added to soil, the parameters increased with respect to the treatments 1 and 2. The P content in plants from treatment 3 was higher than in plants from treatment 1 but lower than in plants from treatment 2. From these results we can conclude that in plants from treatment 2 the natural phosphate solubilization in the soil lead to an increase of P content in plants which is lower in treatment 3 because in this last treatment insoluble phosphate was not added to soil. It is remarkable that the plants inoculated with strain SAPA2 presented a significantly higher dry weight than those of treatments 1 and 2, indicating that this strain can be considered as a plant growth promoting rhizobacterium with independence of its ability to solubilize phosphate. Nevertheless, the better results were obtained when the plants were inoculated with strain SAPA2 and cultivated in soil added with insoluble phosphate. In this case all the parameters were higher than in control treatments (1 and 2). As can be seen the inoculation with the strain SAPA2 promotes plant growth and the addition of phosphate increases the dry weight of fruits. The plants from treatment 4 showed a P content three times higher than in control plants (treatment 1).

Therefore, from the results obtained we can conclude that the inoculation with P-solubilizing strain SAPA2 in presence of insoluble phosphate enhances the plant growth and increases the P-uptake by the strawberry plants. To our

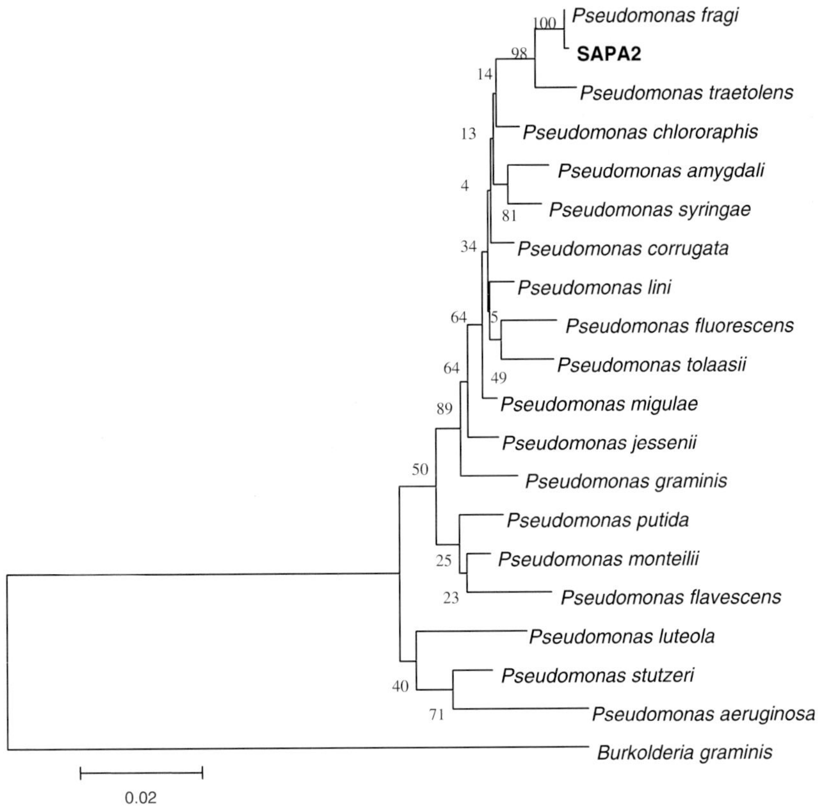

Figure 1. Comparative sequence analysis of 16S rDNA from strain SAPA2 and representative strains from the GenBank. The significance of each branch is indicated by a bootstrap value calculated for 1000 subsets. Bar, 2 nt substitutions per 100 nt. The GenBank accession numbers for the sequences used to generate the phylogenetic tree are the following: strain SAPA2, AY195842, *P.amygdali* ATCC33614^T, D84007, *P. aeruginosa* LMG1242^T, Z76651, *P. chlororaphis* IAM12354^T, D84011, *P. corrugata* ATCC29736^T, D84012, *P. flavescens* B62^T, U01916, *P. fluorescens* IAM12022^T, D84013, *P. fragi* ATCC 4973^T, AF094733, *P. graminis* DSM11363^T, Y11150, *P. jessenii* CIP105274, AF068259, *P. lini* CFBP5739^T, AY035996, *P. luteola* IAM13000^T, D84002, *P. migulae* CIP105470^T, AF074383, *P. monteilii* CIP104883^T, AB0214094, *P. putida* IAM1236^T, D84020, *P. syringae* ATCC19310^T, D840126 *P. stutzeri* CCUG11256^T, U26262, *P. taetrolens* IAM1653^T, D84027 , *P. tolaasii* ATCC33618^T, D84028 and *Burkholderia graminis* C4D1M^T, U96939.

Table 2. Effect of inoculation with *Pseudomonas fragi* SAPA2 on growth and P-uptake by strawberry plants

Treatment	Dry weight (g)	Fruit weight (g/fruit)	Fruits (n°/plant)	Total N (g)	Total P (mg)	Total Ca (mg)	Total K (mg)	Total Ca (mg)
Uninoculated plants without P	4.9^{ab}	2.8^{ab}	3.6^{a}	0.14^{a}	5^{a}	14.7^{a}	49.0^{a}	9.8^{a}
Uninoculated plants with isoluble P	5.5^{ab}	5.6^{cd}	3.0^{a}	0.16^{a}	11^{c}	16.5^{a}	55.1^{a}	11.0^{a}
Plants inoculated with SAPA2	7.5^{c}	3.8^{ab}	4.8^{a}	0.23^{b}	8^{b}	30.0^{b}	75.5^{b}	15.1^{b}
Plants inoculated with SAPA2 added with insoluble P	7.4^{c}	5.3^{cd}	4.7^{a}	0.22^{b}	15^{d}	29.6^{c}	74.2^{b}	14.8^{b}

Values followed by the same letter are not significantly different from each other at $P = 0.05$ according to Fisher's Protected LSD (Low significative differences).

knowledge, there are not works about the beneficial of the inoculation with phosphate solubilizing bacteria in strawberry plants, therefore the results of this work open a new way to increase their production via inoculation with phosphate solubilizing bacteria that promote the plant growth.

Acknowledgements

This work was supported by the Junta de Castilla y León to EV, the DGICYT (Dirección General de Investigación Científica y Técnica) to EM and by CLUE and TLINKS European Research Projects to CRB. We also wish to thank F. Santos and the soil analysis staff of the IRNA for their collaboration.

References

Antoun H, Beauchamp C J, Goussard N, Chabot R and Lalande R 1998 Potential of *Rhizobium* and *Bradyrhizobium* species as growth promoting rhizobacteria on non-legumes: effect on radishes (*Raphanus sativus* L.). Plant Soil 204, 57–67.

Behrendt U, Ulrich A, Schumann P, Erler W, Burghardt J and Seyfarth W 1999 A taxonomic study of bacteria isolated from grasses: a proposed new species *Pseudomonas graminis* sp. nov. Int. J. Syst. Bacteriol. 49, 297–308.

Chabot R, Antoun H and Cescas M P 1993 Grotwh stimulation of corn and romiane lettuce by microorganisms solubilizing inorganic phosphorous. Can. J. Microbiol. 39, 941–947.

Chabot R, Antoun H and Cescas M P 1996 Growth promotion of maize and lettuce by phosphate-solubilizing *Rhizobium leguminosarum* biovar *phaseoli*. Plant Soil 184, 311–321.

Chabot R, Beauchamp C J, Kloepper J W and Antoun H 1998 Effect of phosphorous on root colonization and growth promotion of maize by bioluminiscent mutants of phosphate-solubilizing *Rhizobium leguminosarum* biovar.*phaseoli*. Soil Biol. Biochem. 30, 1615–1618.

de Freitas J R, Banerjee M R and Germida J J 1997 Phosphate-solubilizing rhizobacteria enhance the growth and yield but not phosphorous uptake in canola (*Brassica napus* L.). Biol. Fertil. Soils 24, 358–364.

Deubel A, Gransee A. and Merbach W 2000 Transformation of organic rhizodepositions by rhizosphere bacteria and its influence on the availability of tertiary calcium phosphate. J. Plant Nutr. Soil Sci. 163, 387–392.

Di-Simine C D, Sayer J A and Gadd G M 1998 Solubilization of Zinc phosphate by a strain of *Pseudomonas fluorescens* isolated from a forest soil. Biol. Fertil. Soils 28, 87–94.

Jonhson F J 1990 Detection method of nitrogen (total) in fertilizers. *In* Methods of Analysis of the Association of Official Analytical Chemists. Ed. K Elrich. pp. 17–19. Association of Official Analytical Chemists, USA.

Kim K Y, Jordan D and McDonald G A 1998 Effect of phosphate solubilizing bacteria and vesicular-arbuscular mycorrhizae on tomato growth and soil microbial activity. Biol. Fertil. Soils 26, 79–87.

Kimura M 1980 A simple method for estimating evolutionary rates of base substitutions through comparative studies of nucleotide sequences. J. Mol. Evol. 16, 111–120.

Kumar V and Singh K P 2001 Enriching vermicompost by nitrogen fixing and phosphate solubilizing bacteria. Biores. Technol. 76, 173–175.

Kumar S, Tamura K, Jakobsen I B and Nei M 2001 Molecular Evolutionary Genetics Analysis Software. Arizona State University, Tempe, Arizona, USA.

Manna M C, Ghosh P K, Ghosh B N and Singh K N 2001 Comparative effectiveness of phosphate-enriched compost and single superphosphate on yield, uptake of nutrients and soil quality under soybean-wheat rotation. J. Agr. Sci. 137, 45–54.

Musarrat J, Bano N and Rao R A K 2000 Isolation and characterization of 2,4-dichlorophenoxyacetic acid-catabolizing bacteria and their biodegradation efficiency in soil. World J. Microbiol. Biotechnol. 16, 495–497.

Pandey A and Palni L M S 1998 Isolation of *Pseudomonas corrugata* from Sikkim Himalaya. World J. Microbiol. Biotechnol. 14, 411–413.

Pearson W R and Lipman D J 1988 Improved tools for biological sequence comparison. Proc. Nat. Acad. Sci. USA 85, 2444–2448.

Peix A, Rivas-Boyero A A, Mateos P F, Rodríguez-Barrueco C, Martínez-Molina E and Velázquez E 2001a Growth promotion of chickpea and barley by a phosphate solubilizing strain of *Mesorhizobium mediterraneum* under growth chamber conditions. Soil Biol. Biochem. 33, 103–110.

Peix A, Mateos P F, Rodríguez-Barrueco C, Martínez-Molina E and Velázquez E 2001b Growth promotion of common bean (*Phaseolus vulgaris* L.) by a strain of *Burkholderia cepacia* under growth chamber conditions. Soil Biol. Biochem. 33, 1927–1935.

Rivas R, Velázquez E, Palomo J L, Mateos P, García-Benavides P and Martínez-Molina E 2002a Rapid identification of *Clavibacter michiganensis* subspecies *sepedonicus* using two primers random amplified polymorphic DNA (TP-RAPD) fingerprints. Eur. J. Plant Pathol. 108, 179–184.

Rodríguez H and Fraga R 1999 Phosphate solubilizing bacteria and their role in plant growth promotion. Biotech. Adv. 17, 319–339.

Saitou N and Nei M 1987 A neighbour-joining method: a new method for reconstructing phylogenetics trees. Mol. Biol. Evol. 44, 406–425.

Singh S and Kapoor K K 1994 Solubilization of insoluble phosphates by bacteria isolated from different sources. Environ. Ecol. 12, 51–55.

Singh S and Kapoor K K 1999 Inoculation with phosphate solubilizing microorganisms and a vesicular-arbuscular mycorrhizal fungus improves dry matter yield and nutrient uptake by wheat grown in a sandy soil. Biol. Fertil. Soils 28, 139–144.

Soil Conservation Service 1972 Soil Survey Laboratory Methods and Procedures for Collecting Soil Samples. U.S.D.A, Washington.

Soil Survey Staff 1994 Keys to Soil Taxonomy, 6th edition, Soil Conservation service, USA.

Thomas G V and Shantaram M V 1986 Solubilization of inorganic phosphates for bacteria from coconut plantation soils. J. Plantation Crops 14, 42–48.

Thompson J D, Gibson T J, Plewniak F, Jeanmougin F and Higgins D G 1997 The clustalX windows interface: flexible strategies for multiple sequence alignment aided by quality analysis tools. Nucleic Acid Res. 24, 4876–4882.

Vázquez P, Holguin G, Puente M E, López-Cortez A and Bashan Y 2000 Phosphate solubilizing microorganisms associted with the rhizosphere of mangroves in a semiarid coastal lagoon. Biol. Fertil. Soils 30, 460–468.

Villegas J and Fortín J A 2002 Phosphorous solubilization and pH changes as a result of the interactions between soil bacteria and arbuscular mycorrhizal fungi on a medium containing NO_3^- as nitrogen source. Can. J. Bot. 80, 571–576.

Viveganandan G and Jaurhi K S 2000 Growth and survival of phosphate-solubilizing bacteria in calcium alginate. Microbiol. Res. 155, 205–207.

Yabuuchi E, Kosako Y, Oyaizu H, Yano I, Hotta H, Hashimoto Y, Ezaki T and Arakawa M 1992 Proposal of *Burkholderia* gen. nov. and transfer of seven species of genus *Pseudomonas* homology group II to the new genus, with the type species *Burkholderia cepacia* (Palleroni and Holmes 1981) comb. nov. Microbiol. Inmunol. 39, 897–904.

Effects of phosphate-solubilizing bacteria during the rooting period of sugar cane (*Saccharum officinarum*), Venezuela 51-71 variety, on the grower's oasis substrate

M. Martínez[1] & A. Martínez[1,2]
[1]*Facultad de Ciencias, Pontificia Universidad Javeriana, Carrera de Microbiología Industrial, Cra. 7a N. 43-82, Santafé de Bogotá, Colombia.* [2]*Corresponding author**

Received 20 December 2002. Accepted in revised form 2 January 2003

Key words: *Enterobacter cloacae*, phosphate-solubilizing bacteria, sugar cane, *Xanthomonas maltophila*

Abstract

One hundred samples of rhizosphere taken from 16 crop lands located in the Tolima, Cundinamarca, and Casanare departments were assessed to isolate solubilizing bacteria (PSB) populations. PSB constituted 20% of the total isolated population. Two isolates identified as *Xanthomonas maltophilia* and *Enterobacter cloacae*, were selected for their ability to form solubilization halos within a Pikovskaya medium with $Ca_3(PO_4)_2$. In a greenhouse assay, buds of sugar cane, variety Venezuela 51-71, were planted and fertilized with Fosforita Huila, in the Grower's Oasis rooting medium. Then, an inoculation with the two selected microorganisms was performed on treatments that showed no significant differences in phosphate uptake at 90 days, as compared with the control. However, there were significant differences in plant growth, as the stem length under the phosphate treatment and the mixture of the two bacteria was of 32 cm, while such length was of 17 cm under the positive control. Similarly, stem length with the phosphate treatment and *E. cloacae* was of 9 cm, as compared with 5 cm with the positive control. As for root growth, an increase of 2.57 times was observed, as related to the positive control's dry weight. Root length was increased 0.73 times, being similar to that of the KNO_3 control.

Introduction

Phosphorus is the less available element to sugar cane and to plants in general. This is due to two phenomena occurring when contacting the soil, the first phenomenon is called immobilization, and is carried out by those microorganisms that populate the mineral's deficient regions lacking the nutrients needed to perform their vital processes (Jungk et al., 1993). The second phenomenon is called precipitation or fixation to insoluble complex minerals, and is due to the union of phosphorus with elements such as iron and aluminum in acid soils, and calcium in alkaline soils, denying the plant up to 75% of all soluble phosphorus (Goldstein, 1966; Kucey et al., 1989), and thus, generating a 0.002–0.5% concentration of mineral in the soil (Chabot et al., 1993). This has forced many crop raisers to apply up to four times the required amount of phosphorus to plants. In the case of sugar cane (caña panelera), this amount ranges between 45 and 200 kg of phosphorus per hectarea. This procedure generates an increase in the application of chemical fertilizers and, therefore, an increase in production costs.

Several researchers have demonstrated the existence of bacterial groups in the rhizosphere of various crops, which have the ability to solubilize the insoluble forms of phosphate

* E-mail: mmmartin@javeriana.edu.co

E. Velázquez and C. Rodríguez-Barrueco (eds.),
First International Meeting on Microbial Phosphate Solubilization, 317–323

compounds through the production of organic acids which are better assimilated by the plant (Alexander, 1980). Therefore, production of inoculants could improve the availability of soluble phosphorus, which in turn, would cause a decrease in the use of phosphate fertilizers and thus, a decrease in the crop's production costs, while simultaneously having a positive effect on the environment.

The objective of this survey is to isolate phosphate-solubilizing bacteria (PSB) from various agricultural crops, in order to prepare inoculums that have the ability to solubilize phosphoric rock in the Grower's Oasis rooting medium; to evaluate their effect on root development, plant growth, and phosphorus absorption after 90 days in sugar cane, Venezuela 51-71variety; and, to compare them with chemical fertilization.

Materials and methods

Bacteria isolation

One hundred samples taken from rhizosphere of 16 crops located in the Cundinamarca, Tolima, and Casanare departments were studied.

10 g of sample were diluted in 90 mL of peptonated water at 0.1%, until a 10^4 dilution. Starting with 10^3 and 10^4 dilutions, 0.1 mL were seeded in the surface of an agar described by Pikovskaya (1948), and were incubated for 72 h, at 26 °C. Afterwards, colonies showing transparent halos around them were selected. Isolated colonies were maintained in agar King B and Vinazin (10 g glucose L^{-1}, 0.1 g starch L^{-1}, 0.5 g yeast extract L^{-1}, 0.5 g peptone caseine L^{-1}, 0.1 g NaCl and KH2PO4 L^{-1}).

Phosphorus solubilization test

Each of the selected colonies was seeded on a straight line, in a medium described by Pikovskaya (1948), and was incubated for 72 h, at 26 °C. Colonies showing solubilization halos of 0.5–3.5 mm in diameter were selected. From this, class intervals were created to allow selection of the best solubilizing colonies.

Antagonism trials were performed on the selected bacteria, in order to know which of such bacteria could be a part of a mixture leading to the preparation of a mixed inoculum.

Identification of these bacteria was carried out by means of the methodology described in the Bergey's Manual (1986), and samples were sent to the Universidad de los Andes' CIMIC, for verification purposes.

Scaling

A colony was taken from the selected bacteria and was seeded in 10 mL of modified K.B. (MgSO4, glycerol and peptone at pH = 7.0). Starting with this solution, scaling up to 10, 100 and 400 mL was performed while keeping a 2% inoculum concentration. This generated a bacteria count of 10^6–10^7 ufc mL^{-1}.

Greenhouse trial

Bacteria were inoculated upon sugar cane buds (Venezuela 51-71 variety) in Grower's Oasis rooting medium, during day 0, 20 and 40, combined with commercial fertilizers, together with controls, as follows: TO Water (negative control), T1 phosphorus (positive control), T2 NPK (fertilization control), T3 KNO_3 (rooting control), T4 mixed inoculum (bacteria control), T5 phosphorus + bacteria 1, T6 phosphorus + bacteria 2, T7 phosphorus + mixed inoculum (B1 + B2), thus producing eight treatments with four repetitions each, where every repetition was made up of nine experimental units. These treatments were taken into a greenhouse featuring an average temperature of 25 °C, and were randomly distributed.

Statistical analysis

Five experimental units were taken at random for each repetition, and averages for the following variables were calculated after 90 days:

Rooting, through root length and dry weight; growth, through calculation of the stem length and length of the longest leave; and, phosphorus absorption, through foliar analysis, reading of total amount of phosphorus and of phosphorus available at the rooter.

These data were analyzed by means of a one-way variance analysis, and of a Tukey trial *a posteriori*.

Results and discussion

Bacteria isolation

From the 16 crops under evaluation, six proved to have PSB populations (PSB), in concentrations ranging from 10^3 to 10^6 (that is, 37.5% of the total amount). From the 100 samples that were analyzed, only 20 had this type of bacteria. These results are comparable with those reached by Katnelson et al. (1962), who found out that from 40 to 70% of bacteria isolated from rhizosphere have the ability to solubilize phosphates in laboratory trials.

Solubilization test

From the samples showing PSB counts, 36 strains were selected. Sixteen of those selected lost their ability to solubilize phosphates in a culture medium. Therefore, only 20 strains were left to be evaluated. This reaction is similar to that described by Chabot et al. (1993), in which PSB populations from four soil types found in Quebec (Canada) were reduced after the second passing, from 69 to 31. These authors believed that this decrease is probably due to stress in the laboratory.

A solubilization test was performed upon the selected strains, in order to determine the solubilization halo diameters. From these diameters, ranging between 0.5 and 3.5 mm, class intervals were created to select six (6) strains that featured halos having 2.5 to 3.5 mm diameters.

Table 1 shows recounts of PSB isolated from various crops and their corresponding solubilization halo diameters.

An antagonism trial was performed upon these strains, in order to determine which of those strains could be a part of a mixture for the production of a mixed inoculum. Only two strains proved to be able to be a part of a mixture; these strains were later identified at species level as *Xanthomonas maltophilia* and *Enterobacter cloacae*.

Table 1. Recount of PSB in Pikovskaya agar and of its corresponding solubilization halos

Culture	Recount of PSB	Diameter in mm
Rice	1×10^3	1.5
Rice	2×10^4	0.95
Sugar cane	2×10^5	1.9
Sugar cane	1×10^3	0.8
Sugar cane	2.2×10^3	2.2
Sugar cane	3×10^3	2.5
Sugar cane	1×10^3	0.9
Sugar cane	1×10^4	1.7
Sugar cane	2.2×10^3	2.0
Sugar cane	1×10^4	3.0
Sugar cane	2×10^3	2.0
Sugar cane	1×10^5	0.75
Sugar cane	2×10^3	2.5
Sugar cane	2×10^4	1.2
Onion	1×10^3	2.3
Penny royal	2×10^3	2.0
Sorghum	3×10^6	3.2
Sorghum	2×10^5	3.0
Sorghum	2×10^3	3.3
Carrot	1×10^4	1.3

Rooting

Root length

The variance analysis reflected differences among treatments having $f = 42.41$ and $P << 0.0001$, as shown on the following table and its corresponding statistical analysis.

Applying the Tukey trial, six homogenous groups were conformed. Although such groups did not differ significantly among them, they do differ significantly from the rest (Table 2)

According to the aforementioned results, it may be concluded that treatment 4 is as effective as treatment 3 in the enhancement of the plant's root length. In treatment with KNO_3, fertilization was carried out with a concentration of 200 g of fertilizer for every 4,000 L of water, and adding 2 mL/plant; meanwhile, in treatment with B1 + B2, no fertilization was carried out besides that which contained the rooter; bacteria were only added in the same days in which the fertilizer was added, leading to an advantage: rooting stimulation while using chemical fertilizers.

Likewise, treatments inoculated with bacteria (such as treatments 5, 6 and 7) showed a

Table 2. Root length (in cm) after 90 days

Variable	Mean avr.	Homogenuous groups					
		A	B	C	D	E	F
KNO_3	21.000	*					
B1 + B2	18.750	*					
Phosphorus + B1	16.928		*				
Phosphorus + B1 + B2	16.750		*	*			
Phosphorus + B2	14.500		*	*	*		
Phosphorus	13.750			*	*	*	
H_2O	11.250					*	*
NPK	10.750						*

B1: *X. maltophilia* B2: *E. cloacae.*
Critical value per comparison: 2.5657.
Significance level: 0.005.
Standard error per comparison: 0.7747.

Table 3. Root dry weight (in gm) after 90 days

Variable	Mean avr.	Homogenuous groups			
		A	B	C	D
Phosphorus + B1 + B2	0.4336	*			
Phosphorus + B2	0.4210	*			
B1 + B2	0.4006	*	*		
Phosphorus + B1	0.3553		*		
KNO_3	0.1928			*	
H_2O	0.1745			*	*
Phosphorus	0.1684			*	*
NPK	0.1279				*

B1: *X. maltophilia* B2: *E. cloacae.*
Critical value per comparison: 0.0595.
Significance level: 0.05.
Standard error per comparison: 0.180.

lengthier root, as compared with fertilization controls 1 and 2 having the negative control 0.

Thiman and Schneider (1939) pointed out that there are substances that exercise a primary control on root production. These substances are called auxines. Datta et al. (1982) extracted an auxine named acid 3-indol acetic from a phosphate-solubilizing strains of *Bacillus firmus*. It is possible that microorganisms may have the ability to produce certain substances that act in a similar way; in fact, *Enterobacter* is used in soil recovery processes, due to its ability to transform nutrients and to produce vegetable growth stimulating substances.

While using *Enterobacter* strains in corn seeds inoculation, Chabot et al. (1993) observed root elongation due to production of auxines.

Root dry weight

The statistical analysis reflected significant differences among treatments having $f = 104.20$ and $P << 0.000$.

The Tukey trial (Table 3) conformed 4 homogenous groups that showed no significant differences among them, but significant as compared to the rest.

The best responses were those from treatments inoculated with bacteria, including control 4.

Stoyanov (1978) submitted evidence proving that root dry weight could be increased by adding Vitamins C1, B1 and B6, supporting allegations by Rempe (1973), who stated that such compounds have a positive effect on respiration, protein production and nutrient transfers within the roots. In a survey by Baya et al. (1981), it was concluded that green plant roots cannot produce all those vitamins required to carry out their vital processes. It also pointed out, though, that PSB, isolated from rhizosphere of *Viola tricolor, Diplotaxis muralis, Triticum vulgare* among others, were excellent generators of Vitamin B12, riboflavine and niacine. It is possible that bacteria used in this project may produce this type of substances; likewise, it is also possible that the resulting effect may be directly proportional to production of growth-promoting substances.

Now, the KNO_3 rooter control showed the best root length results, probably due to the fertilizer's presentation: if a fertilizer is dissolved in water, plants are able to absorb faster the available K, which stimulates production of meristems. This causes an increase in cellular divisions, which, in turn, generates root elongation. However, this control had the smallest dry weight, probably due to the fact that this treatment had neither the mineral nor the organic elements (organic matter) required to enable production of radicular hair (Salisbury and Ross, 1978) which might increase dry weight.

Growth

Stem length

The statistical analysis reflected differences among those treatments having f values = 31.74 and $P \ll 0.0001$.

Within the Tukey trial (Table 4), out of the eight treatments, five (5) homogenous groups were conformed in which there were no significant differences among same.

Treatments featuring the best results were those inoculated with bacteria. These results may be correlated with those from Chabot et al. (1993), who inoculated PSB strains, such as *Enterobacter* and *Pseudomonas*, and observed an increase in the height of corn and lettuce plants after 60 days. These authors believed this is due to the production of siderophorus and auxines by such bacteria since, according to Weaver (1980), these substances play an important role in stem expansion.

Leaf length

The variance analysis reflected differences among treatments having f values = 75.46 and $P \ll 0.0001$.

Using the Tukey trail (Table 5), 4 homogenous groups were formed. In these groups, the highest responses were those from treatments inoculated with bacteria, including the control.

According to Weaber (1980), cytocinines induce leaf elongation. And, according to a survey by Barea et al (1976) cytocinines were extracted from a phosphate-solubilizing strain of *Escherichia coli.*

Table 4. Stem length (in cm) after 90 days

Variable	Mean avr.	Homogenuous groups				
		A	B	C	D	E
Phosphorus + B2	9.000	*				
Phosphorus B1 + B2	7.750		*			
Phosphorus + B1	7.500		*			
B1 + B2	7.250		*	*		
KNO_3	6.750		*	*	*	
NPK	6.250			*	*	
H_2O	5.750				*	*
Phosphorus	5.000					*

B1: *X. maltophilia* B2: *E. cloacae.*
Critical value per comparison: 1.0418.
Significance level: 0.05.
Standard error per comparison: 0.3146.

Table 5. Leaf length (in cm) after 90 days

Variable	Mean avr.	Homogenuous groups			
		A	B	C	D
Phosphorus + B1 + B2	32.000	*			
Phosphorus + B2	30.750	*	*		
B1 + B2	28.500	*	*		
Phosphorus + B1	27.500			*	
KNO_3	26.500			*	
H_2O	18.500			*	*
Phosphorus	17.750				*
NPK	17.000				*

B1: *X. maltophilia* B2: *E. cloacae.*
Critical value per comparison: 3.2944.
Significance level: 0.05.
Standard error per comparison: 0.9948.

Phosphorus solubilization

Foliar analysis

The variance analysis showed no differences among treatments having f value = 2.08 and $P = 0.0858$. Results are stated below, in Table 6.

It can be observed that phosphorus absorption levels by plant controls were higher than those by treatments inoculated with bacteria. This is probably due to bacteria performing a certain type of immobilization and, through time, these results may vary because of a slow ionic release action by microorganisms. These results

Table 6. Foliar analysis of P after 90 days (% ppm)

Variable	Mean avr.	Homogenuous groups
H_2O	0.0377	*
Phosphorus	0.0235	*
NPK	0.0185	*
Phosphorus + B2	0.0168	*
Phosphorus + B1	0.0155	*
B1 + B2	0.0152	*
Phosphorus + B1 + B2	0.0120	*
KNO_3	0.0009	*

B1: *X. maltophilia* B2: *E. cloacae.*
Critical value per comparison: 0.0282.
Significance level: 0.05.
Standard error per comparison: 0.0008.

may be correlated with a survey by Datta et al. (1982), who inoculated PSB into rice crops and observed a drastic decrease in phosphorus absorption levels during the first year of experimenting, due to an immobilization process, but on the following year, however, observed a significant increase of the nutrient was observed.

Another possible cause could be that the bud itself stores enough phosphorus contents to cover its needs during the first stage of rooting and, therefore, does not absorb significant quantities.

Calculation of total P in the rooter

As per the variance analysis, differences were observed among treatments having an f value = 40.42 and $P << 0.0001$.

Using the Tukey trial (Table 7), four groups were formed, showing no significant differences among them.

No significant differences were observed for those treatments that were fertilized with phosphorus, except for the treatment that was inoculated with bacteria 1. It is possible that said bacteria has taken a portion of phosphorus, in order to carry out its own vital processes.

As expected, total quantities of phosphorus for treatments that were not fertilized with phosphorus were lower when compared with the rest. Treatment B1 + B2 provided the highest total quantity of phosphorus within this group. This is probably because during nutrient determination, microbial phosphorus was also read.

Table 7. Calculation of total P in the rooter (% ppm)

Variable	Mean avr.	Homogenuous groups			
		A	B	C	D
Phosphorus	32.000	*			
Phosphorus + B1	30.750	*			
Phosphorus + B1 + B2	28.500		*		
NPK	27.500		*	*	
Phosphorus + B1	26.500		*	*	
B1 + B2	18.500			*	*
KNO_3	17.750				*
H_2O	17.000				*

B1: *X. maltophilia* B2: *E. cloacae*.
Critical value per comparison: 201.69.
Significance level: 0.05.
Standard error per comparison: 60.902.

Analysis of P available at the rooter

According to the variance analysis, there are differences among treatments having an f value = 76.70 and $P << 0.0001$.

Using the Tukey trial (Table 8), three homogenous groups were formed which had no significant differences among them. The highest available quantity of phosphorus was that of treatments fertilized with phosphorus, while no significant differences were observed among them. It is worth mentioning that within these treatments, the highest quantity of available phosphorus was that of those inoculated with bacteria.

As for the total quantity of phosphorus, no significant differences were observed among treatments fertilized with phosphorus. However, from all of the aforementioned, that with the highest total quantity of phosphorus was the phosphorus control. This is probably due to the fact that in treatments inoculated with bacteria, a certain type of interaction might be happening with the fertilizer, such as immobilization.

In this case, the solubilizing behavior of bacteria in the rooter is evident, since the rooter shows a higher quantity of available phosphorus when the chemical fertilizer is applied together with the bacteria, than when it is fertilized with 100% of chemical products.

These results match those by Gómez (1984), who mixed phosphoric rock with organic matter, generating a positive effect on solubilization, due to the production of organic acids through the microbial action.

Table 8. Analysis of P available at the rooter (% ppm)

Variable	Mean avr.	Homogenuous groups		
		A	B	C
Phosphorus + B1	126.21	*		
Phosphorus + B1 + B2	125.85	*		
Phosphorus + B2	123.48	*		
Phosphorus	121.45	*		
NPK	110.20	*		
B1 + B2	62.35		*	
H_2O	45.975		*	*
KNO_3	32.715			*

Critical value per comparison: 21.206.
Significance level: 0.05.
Standard error per comparison: 6.4033.

If the results for phosphorus available in rooters are taken into consideration, the plant may be expected to contain, at a foliar level, a similar quantity. However, this did not happen, probably because sugar cane, being a member of the graminean family, does not absorb important quantities of phosphated compounds (Ramírez, 1995). This probably caused an accumulation of the mineral in the roots or in the rooter.

Conclusions

The Pikovskaya medium is a good culture medium for the isolation of PSB. Although these bacteria are present in the soil at low concentrations, they have the tendency to lose their ability to solubilize phosphates in this medium. Therefore, isolation of these microorganisms is thus possible. Such microorganisms, aside from solubilizing phosphates, probably produce substances that facilitate sugar cane growth, as observed in a better plant development and a greater production of radical hair, as compared to the controls. This fact could have an impact on the crops productivity, since the plant's mineral nutrition depends on its roots' absorption ability.

Use of moderate doses of commercial fertilizers combined with inoculums of PSB may be a profitable, enhanced and less-contaminating alternative to the integrated handling of soils and to phosphorus absorption.

References

Alexander M 1980 Transformaciones microbianas. En Introducción a la Microbiología del Suelo AGT editores. Segunda Edición. México. pp. 356–375.

Barea J M, Navarro E and Montoya E 1976 Production of plant growth regulators by rhizosphere phosphate solubilizing bacteria. J. Appl. Bacteriol. 40, 129–134.

Baya A M, Boethling R and Ramos C 1981 Vitamin production in relation to phosphate solubilization by soil bacteria. Soil Biol. Biochem. 13, 527–531.

Bergey H 1986 Gram Negative aerobic rods and cocci. *In* Bergey's Manual of Systematic Bacteriology (2nd edition). Williams & Wilkins, Baltimore, USA. 124 p.

Chabot R, Hani A and Cescas M M 1993 Stimulation do la croissance du'mais et de la latitue romaine par des microorganismes disolvant le phosphore inorganique. Can. J. Microbiol. 39, 941–947.

Datta M, Banik S. and Gupta R P 1982 Studies of the efficacy of a phyotohormona producing by phosphate solubilizing *Bacillus firmus* in augmenting paddy field in acid soils of Nagaland. Plant Soil 69, 365–373.

Goldstein A H 1966 Bacterial solubilization of mineral phosphates: historical perspective and future prospects. Am. J. Altern. Agric. 20, 51–57.

Gómez G 1984 Abonos de uso en Colombia. Ed. ABOCOLICA, Colombia, p. 28.

Jungk A, Seeling B and Gerk J 1993 Mobilization of different phosphate fractions in the rhizosphere. Plant Soil 155/156, 91–94.

Katznelson N H, Peterson E A and Kucey R M 1989 Phosphate solubilizing bacteria and fungi in various cultivated and virgin Alberta soils. Can. J. Soil Sci. 63, 671–678.

Pikovskaya R I 1948 Mobilization of phosphorus in soil connection with the vital activity of some microbial species. Mikrobiologiya 17, 372–370.

Ramírez P 1995 Proceedings: I Congresso Internacional de Microbiología Ambiental. Pontificia Universidad Javeriana, Bogotá, p. 21.

Rempe E K 1973 The role of vitamins syntetized by the root microflora for the plants. Fiziologivarstenii 19, 663–668.

Salisbury F B, Ross C R 1978 Plant physiology (2ª Edicion) Wadsworth publishing company Inc.

Stoyanov U and Kudrew T G 1978 Influencie of magnesium and of certains vitamins on the restoration of maize plants. New Phytol. 84, 687–694.

Thiman K V and Scheneider A 1939 The relative activities of differentes auxins. Am. J. Bot. 25, 627–641.

Weaver R J 1980 Efectos biológicos y mecanismos de acción. Reguladores del crecimiento de las plantas en la agricultura. Editorial Trillas. Segunda Edición. México. 113 p.

Phosphate solubilizing bacteria isolated from the rhizosphere soil and its growth promotion on black pepper (*Piper nigrum* L.) cuttings

K. Ramachandran[1,2], V. Srinivasan[1,3], S. Hamza[1] & M. Anandaraj[1]
[1]*Division of Crop Production, Indian Institute of Spices Research (ICAR), Calicut, 673 012, India.* [2]*P.O. Box 78313, Nairobi, Kenya.* [3]*Corresponding author**

Received 16 December 2002. Accepted in revised form 2 January 2003

Key words: *Azospirillum*, composted coir pith, greenhouse trial, phosphate solubilizing bacteria, *Pseudomonas*, rhizosphere

Abstract

Bacterial isolates from the rhizosphere soil and root cuttings of bush black pepper (*Piper nigrum* L.) (pepper raised by laterals) exhibiting high phosphate solubilizing ability *in-vitro* is described in this paper. Microbial phosphorus solubilization (MPS) trait was analyzed by determining the P solubilization efficiency E (E = Diameter of bacterial growth/Diameter of clearing zone × 100). The highest P solubilization efficiency was demonstrated by the isolate PB-21 followed by the isolate PB-16: which was identified as *Pseudomonas* sp. All isolates under study released inorganic phosphate from tricalcium phosphate (TCP) indicating the potential of these strains to release soluble inorganic phosphates from fixed phosphate sources for plant uptake. The isolate $PIAR_{6\text{-}2}$ was able to solubilize 20.01% of P and also fix atmospheric nitrogen, which was later identified as the nitrogen fixing *Azospirillum* sp. Greenhouse trials using two systems; viz; Soil:Terracare (composted coir pith) and Sand:Soil:FYM with three experimental sets such as rock phosphate (RP as an external P source), PSB isolate in combination with VAM and PSB isolate alone, all against their respective control sets showed very clearly the growth promoting activity of phosphate solubilizing bacteria. Field studies were also carried out using these isolates and some promising results were obtained. Further studies are required to analyze these strains to confirm its plant growth promoting properties. The *in-vitro* analyses and greenhouse studies of these bacteria reflect their potentiality as efficient P solubilizer in black pepper growing soils.

Introduction

The soil environment surrounding plant roots is the zone of intense microbial activity. The existence of soil microbes capable of transforming soil phosphorus and fixing nitrogen from the atmosphere to forms available to the plants has been recorded by many investigators. It has been observed that a high proportion of phosphate solubilizing microorganisms are concentrated in the rhizosphere of plants (Whipps and Lynch, 1986). The recognition in mid-70s that biological nitrogen fixation offers the most promising supplement to chemical nitrogenous fertilizers led to a wide array of studies on associative, free living and symbiotic nitrogen fixation by rhizosphere micro flora. These microbes are commonly found in association with the roots of diverse plants (Brown, 1975).

The main problem in applying P as a plant nutrient in P fixing tropical soils is its conversion

* FAX No: +0091-495-2730294.
E-mail: vsrinivasan@spices.res.in

E. Velázquez and C. Rodríguez-Barrueco (eds.),
First International Meeting on Microbial Phosphate Solubilization, 325–331

to unavailable P in soil up to 85%. Therefore, for making P available to plants, several microorganisms are used as P solubilizers. Isolation of indigenous micro flora capable of phosphorus solubilization and nitrogen fixation is an important procedure when studying their inherent capacity to benefit crops probably because of their superior adaptability to the environment than the introduced strains. The aim of the present study was to record the phosphate solubilizing capacities of some indigenous bacteria isolated from rhizosphere of bush black pepper (*Piper nigrum* L.) raised by rooting lateral shoots and to select promising strains for further experiments to determine the effectiveness of these strains as plant growth promoting rhizobacteria (PGPR).

Materials and methods

Isolation and evaluation of phosphate solubilizing bacteria (PSB)

Tricalcium phosphate (TCP) is regarded as a model compound for measuring the potential or relative rates of microbial solubilization of insoluble inorganic phosphate compounds. In addition, the insoluble calcium phosphate forms a major portion of insoluble phosphate in soil (Devi and Narasimhan, 1978). Solubilization of precipitated TCP in unbuffered solid agar medium plates has been used widely as the initial criterion for the isolation of phosphate solubilizing microorganisms (Pikovskaya, 1948). Microorganisms on precipated calcium phosphate agar produces clear zones around their colonies if they are capable of solubilizing calcium phosphate.

Suitable dilutions (10^{-4}) of serially diluted rhizosphere soil suspension were poured plated on Pikovskaya's Agar (glucose – 10 g, $Ca_3(PO_4)_2$ – 5 g, $(NH_4)_2SO_4$ – 0.5 g, KCl – 0.2 g, $MgSO_4$ – 0.1 g, $MnSO_4$ – traces, $FeSO_4$ – traces, Yeast Extract – 0.5 g, Agar – 15 g, Distilled water – 1 L, pH – 7.0) and the plates were incubated at 30 ± 5 °C for 48–96 h. Phosphate solubilization is indicated by the formation of a solubilization or a clear zone around the bacterial colonies. Single bacterial colonies having clear solubilization zones were isolated separately on to fresh Pikovskaya's agar plates, incubated at 30 ± 5 °C for 10 days and an analysis of the MPS trait was made by measuring the zone of solubilization around the colony growth. The solubilization efficiency (E) of these isolates was calculated based on the relation,

$$\text{Solubilization efficiency(E)} = \frac{\text{Solubilization diameter (S)}}{\text{Growth diameter (G)}} \times 100$$

(Nguyen et al., 1992).

The release of soluble P from TCP was determined by the method described by Jackson (1967). Single colonies of these isolates grown on nutrient agar were inoculated into 0.05 L of nutrient broth and 0.0005 L of these 24 h grown cultures were inoculated into 0.05 L of Pikovskaya's broth and incubated over shaker at 29 ± 3 °C and 100 rpm for 5 days. The drop in pH of the medium was measured by a pH meter. The nitrogen-fixing bacteria isolated from secondary roots of bush pepper, *Azospirillum* sp. $PIAR_{6\text{-}2}$ has also been tested for its P solubilizing efficiency.

To evaluate the P release and growth promoting efficiency of the phosphate solubilizing bacteria PB-21, a nursery experiment on black pepper was conducted at IISR Experimental Farm, Peruvannamuzhi, Calicut, Kerala. The single node black pepper cuttings (cultivar. Karimunda) were planted in 0.20 × 0.15 m poly bags (1.5 kg mixture per bag) with two potting media viz., (1) Potting mixture (1:1:1, Sand:Soil:FYM) and (2) Soil:TC (TC – Terracare Coir pith compost marketed by M/s. Marson Biocare Ltd., Mumbai, India) (80:20 w/w). The treatments imposed were Control (C), Potting media added with rock phosphate (RP), Phosphate solubilizing bacteria (PSB), RP + PSB mixture, VAM and VAM + PSB combined inoculation. The PSB culture PB-21 that is identified as efficient strain from *in-vitro* studies was used for growth promotion studies. The VAM culture isolated and identified (*Glomus fasiculatum*) by the Division of Crop Protection was used to evaluate the growth promotion and P utilization in combination with PB-21.

The VAM culture was multiplied on sorghum and soil with root bits at 50 g per bag containing 500–600 propagules mixed thoroughly with the mixture. The rock phosphate was applied at 0.1% P per bag and mixed thoroughly. PSB was applied at 0.05 L culture concentrate (with 10^6 cfu/mL) per bag. The cuttings were grown for 3 months and the observations on shoot length, total root length, length of finer roots with < 1 mm diameter (GS root, v5.2-GS root is a software used for the root length measurement studies and its version is 5.2), total dry matter and P uptake were recorded. The P availability in the medium was also measured after 3 months (Jackson, 1967). The population of PSB was also monitored.

Results and discussion

On Pikovskaya's agar, strains PB-21, PB-16, PB-19c and PB-13 showed very distinct clearing zone with PB-21 giving a very large, clear and transparent solubilization zone and highest percentage of P solubilization in Pikovskaya's broth (Table 1). Though the isolate $PIAR_{6\text{-}2}$ did not give any detectable solubilization zone in plate assay in Pikovskaya's broth about 20.01% of P was solubilized by the same strain (Figure 1). All isolates released inorganic phosphate from TCP indicating the potential of these strains to release inorganic soluble phosphate from fixed phosphates sources for plant uptake. Highest P solubilization efficiency (E) was demonstrated by PB-21 (Table 1).

An interesting feature of the *Azospirillum* isolate $PIAR_{6\text{-}2}$ is that it is also a phosphate solubilizer bringing about 20% of phosphate solubilization in Pikovskaya's broth containing TCP as phosphate source (Table 1). Katznelson and Bose (1959) found that rhizosphere bacteria has greater metabolic activity and suggested that they might contribute significantly to the phosphate economy of the plant. Laboratory studies reviewed by Kucey et al. (1989) have shown that the microbial solubilization of soil phosphate in liquid medium studies has often been due to the excretion of organic acids as a result of which a decrease in pH was effected.

Table 1. Phosphate solubilization by some bacterial isolates from rhizosphere soil of bush pepper with TCP as P source

Isolate	Fall in pH	Solubilization zone (mm)	% P solubilized	Solubilization efficiency (E)
PB-21	1.56	33	40.43	412.5
PB-16	1.72	12	39.08	133.3
PB-19C	0.30	25	30.43	192.3
PB-13	2.01	15	32.91	125.0
$PIAR_{6\text{-}2}$	1.15	ND[a]	20.01	NC[b]

[a]ND, Not Detectable; [b]NC, Not Calculated.

A few reports have indicated the phosphate solubilizing activity of some nitrogen fixers (Mahesh Kumar et al., 1999: Seshadri et al., 2000). A detailed study of the organism would throw light on the phosphate solubilizing properties that could be incorporated in agriculture as the same strain ($PIAR_{6\text{-}2}$) offers traits for nitrogen fixation also.

Growth promotion

The maximum and significantly high shoot length was observed in VAM treated cuttings on par with VAM + PSB treatment (Figure 2).

The inoculation of PSB alone and PSB + Rock phosphate has also increased the shoot growth (Table 2). The control plants on both the potting media have recorded the lowest shoot length. The dry matter production in 3 months also observed a similar trend where VAM, VAM + PSB inoculations produced highest dry matter of black pepper followed by PSB + RP & RP treatments. Dry matter in inoculation of PSB alone and PSB + RP was on par. Similar studies by Ahmed et al. (1999) also showed that combining *phosphorene* (as a source of phosphate solubilizing bacteria) at 0.1% with phosphate fertilizers had an incrementally beneficial effect on growth and P uptake on olive seedlings. Combined inoculation of organisms has given a better growth of the plant. The total and finer root length measured was significantly highest in VAM and combined inoculation of VAM + PSB, followed by PSB alone (Table 2).

Similar synergistic interaction between VAM and PSB was observed on black pepper (Kandiannan et al., 2000) and tomato (Kim et al., 1998). The inoculation of PSB + RP also recorded significantly high root length over RP application alone. The control plants produced only very less root mass. Production of finer roots (< 1 mm diameter) was maximum in microbial inoculated treatments. All VAM, PSB and combined inoculations recorded significant finer root production over RP and control treatments.

There was not much difference due to the media that was used to raise the cuttings on

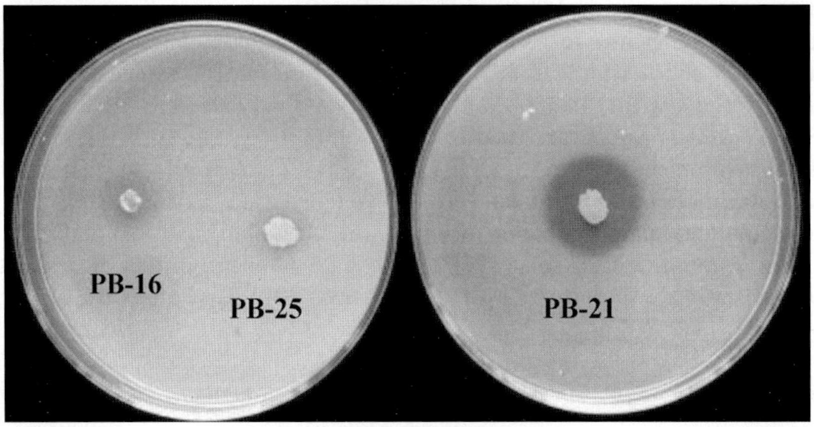

Figure 1. Phosphate solubilizing bacterial isolates from bush black pepper rhizosphere.

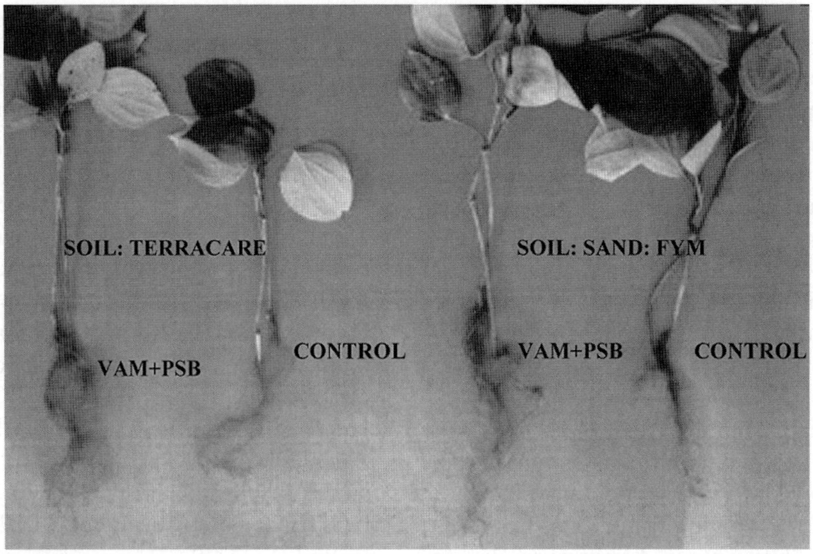

Figure 2. Effect of VAM + PSB isolate PB-21 on the growth of black pepper.

Table 2. Effect of PSB and VAM on P availability, uptake and growth promotion, of black pepper cuttings

Treatment	Shoot length (mm)	Total root length (mm)	Root (< 1 mm dia.) length (mm)	Dry matter (g/bag)	P uptake (mg/plant)	Avail. P (mg/kg)	PSB ($\times 10^4$ cfu/g)
C	11.33	170.5	93.45	4.43	6.22	19.25	8.3
RP	15.56	211.3	170.6	6.58	9.02	40.03	11.5
PSB	14.83	668.3	540.1	7.23	9.39	48.86	17.0
RP & PSB	15.57	575.5	520.2	6.91	8.98	57.16	13.0
VAM	21.13	731.7	561.8	8.17	10.75	30.47	8.2
VAM + PSB	18.83	696.5	567.4	7.98	9.82	43.31	14.0
CD 5%	3.84	39.1	36.2	1.04	1.67	2.97	2.2

shoot length and dry matter production of black pepper cuttings over 3 months. Even though soil:sand:FYM medium recorded highest soil available Bray's P (56.89 ppm) over soil:TC medium (22.81 ppm), the shoot length and dry matter production were on par. But the root length measured was significantly high in soil:TC media as regard to total and finer root lengths (Table 3). As the medium was porous and provided very suitable condition for root penetration, the soil:TC produced maximum finer and total root mass as compared to ordinary potting mixture (Srinivasan and Hamza, 2000). The P uptake by the cuttings grown on different media followed their P availability in respective media. The cuttings from potting mixture (soil:sand:-FYM) have recorded significant uptake of P (0.01104 g/plant) than that from soil:TC media (0.00702 g). This higher uptake might be due to higher P availability from FYM source in the potting mixture grown.

Among treatments, highest P availability was recorded in PSB + RP inoculation followed by PSB inoculation. Treatment with RP alone recorded the next best. P availability from rock phosphate increased when it was co-inoculated with PSB. The inoculation of PSB releases more available P from insoluble source like RP than any water soluble source of P (Chhonkar, 1994). More pronounced effect of PSB in the presence of Mussoorie rock phosphate on P release, uptake and growth promotion was observed in mungbean by Singh and Kapoor (1992). Higher P availability in VAM + PSB inoculation over VAM also reveals the definite role of PSB strain PB-21 in P solubilization. The P uptake by the plant was significantly highest in VAM, PSB and combined inoculations over RP and control. Soil:TC mixture has maintained a population of 14×10^4 cfu/g from the initial population of 0.3×10^4 cfu/g at zero hour of inoculation. FYM based mixture has supported upto 8×10^4 cfu/g. The coir compost base supported good multiplication of PSB culture PB-21 for mass production than FYM base (Figures 3 and 4). PSB population was recorded in all the treatments, whereas

Table 3. Effect of potting media on P availability, uptake and growth promotion of black pepper cuttings

Treatment	Shoot length (mm)	Total root length (mm)	Root (< 1 mm dia.) length (mm)	Dry matter (g/bag)	P uptake (mg/plant)	Avail. P (mg/kg)	PSB ($\times 10^4$ cfu/g)
Soil:Sand:FYM	16.78	371.5	284.0	7.02	11.04	56.89	8.0
Soil:TC	15.63	646.4	533.8	6.75	7.02	22.81	14.0
CD 5%	NS	22.6	20.9	NS	0.97	1.72	1.27

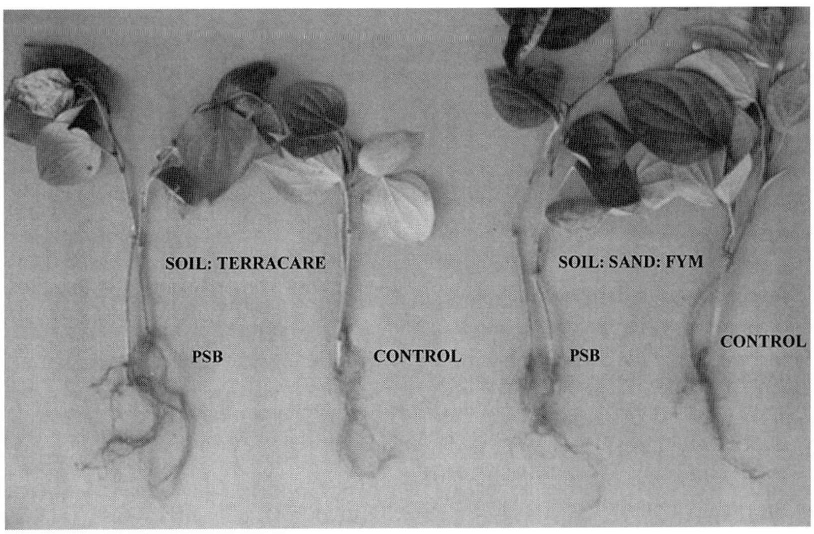

Figure 3. Effect on PSB isolate PB-21 on the growth of black pepper.

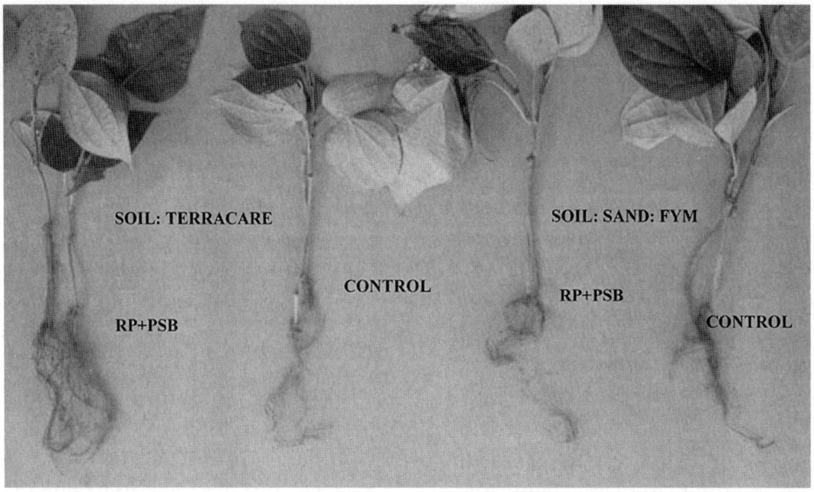

Figure 4. Effect of PSB isolate PB-21 + RP on the growth of black pepper.

it was significantly higher in respective treatments where PSB was added.

Conclusion

In the present study some efficient indigenous phosphate solubilizing bacteria were isolated from bush pepper rhizosphere and their potential worked out. As one of the isolate $P1AR_{6\text{-}2}$, identified as *Azospirillum* sp. has a unique characteristic of phosphate solubilization, the same can be used for nitrogen fertilization and phosphate solubilization. PB-21, the efficient PSB strain has efficiently solubilized and released P from insoluble RP and improved the shoot and root growth of the black pepper cuttings either alone or in combination with VAM. The present findings establish the potential for the use of efficient indigenous phosphate solubilizing bacterial isolate as a reliable alternative in low capital agriculture.

Acknowledgements

The authors gratefully acknowledge the Indian Council of Agricultural Research, New Delhi for providing the financial support in the form of Adhoc scheme and Director, Indian Institute of Spices Research, Calicut, India for providing the necessary facilities to carry out this investigation.

References

Ahmed F F, El Dawwey G M and Papadopoulos A P 1999 Efficiency of phosphorene (a source of phosphate solubilizing bacteria) in enhancing growth and P nutrition of Chemlali olive seedlings. International symposium on growing media and hydroponics, Windsor, Ontario, Canada, 19–26 May 1997 Volume II. Acta Horticult. 481, 701–705.

Brown M E 1975 Rhizosphere microorganisms. Opportunities bandits or benefactors. *In* Soil Microbiology. Ed. N Walker. p. 21. Halsted Press, New York.

Chhonkar P K 1994 Crop response to phosphatic biofertilizers. Fertilizer News 39: 41–43.

Devi M P and Narasimhan R L 1978 Phosphate and lime potentials of some alluvial soils. J. Indian. Soc. Soil. Sci. 26, 33–37.

Jackson M L 1967 Soil Chemical Analysis. Prentice Hall of India Ltd, New Delhi.

Kandiannan K, Sivaraman K, Anandaraj M and Krishnamurthy K S 2000 Growth and nutrient content of black pepper (*Piper nigrum* L.) cuttings as influenced by inoculation with biofertilizers. J. Spices Aromat. Crops 9(2), 145–147.

Katznelson H and Bose B 1959 Metabolic activity and phosphate dissolving capability of bacterial isolates from wheat roots, rhizosphere and non-rhizosphere soil. Can. J. Microbiol. 5, 79–85.

Kim K Y, Jordan D and McDonald G A 1998 Effect of phosphate solubilizing bacteria and vesicular arbuscular mycorrhizae on tomato growth and soil microbial activity. Biol. Fert. Soils 26(2), 79–87.

Kucey R M N, Janzen H and Leggett M E 1989 Microbial mediated increases in plant available P. Adv. Agron 42, 199–228.

Mahesh Kumar K S, Krishnaraj P U and Algawadi A R 1999 Mineral phosphate solubilizing activity of *Acetobacter diazotrophicus*: A bacterium associated with sugarcane. Curr. Sci. 76, 874–875.

Nguyen C, Yan W, Le Tacon F and Lapayrie F 1992 Genetic variability of phosphate solubilizing activity by monocaryotic and dicaryotic mycelia of the ectomycorrhizal fungus

Laccaria bicolor (Maire) P.D. Orton. Plant and Soil 143, 193–199.

Pikovskaya R I 1948 Mobilization of P in soil in connection with vital activity by some microbial species. Microbiologica 17, 362–370.

Seshadri S, Muthukumarasamy R, Lakshminarasimhan C and Ingacimuthu S 2000 Solubilization of inorganic phosphates by *Azospirillum halopraeferens*. Curr. Sci. 79(5), 565–567.

Singh S and Kapoor K K 1992 Solubilization of insoluble phosphates by bacteria and their effect on growth and phosphorus uptake by mungbean. Intl. J. Trop. Agric. 10, 209–213.

Srinivasan V and Hamza S 2000 Use of coir compost as a component of nursery mixture for spices. *In* Spices and Aromatic plants – Challenges and Opportunities in the New Century Centennial Conference on Spices and Aromatic Plants. Eds. K V Ramana et al. pp. 91–96. Sept. 20–23 2000, Calicut, Kerala .

Whipps J M and Lynch J M 1986 The influence of the rhizosphere in crop productivity. Adv. Microbial. Ecol. 9, 187–244.

Immobilization of mercury in soils of Venezuela using phospho-gypsum and sulphate-reducing bacteria

E. Adams[1,5], A. García-Sánchez[1], F. Santos[2], E. Velázquez[3] & M. Adams-Meléndez[4]
[1]*Instituto de Recursos Naturales y Agrobiología (IRNA, CSIC), C/Cordel de Merinas, 40-52, 37008, Salamanca, Spain.* [2]*Facultad de Ciencias Agrarias y Ambientales, Universidad de Salamanca, Salamanca, Spain.* [3]*Departamento de Microbiología y Genética, Universidad de Salamanca, Salamanca, Spain.* [4]*Facultad de Agronomía, Universidad Central de Venezuela-CONICIT, Caracas, Venezuela.* [5]*Corresponding author**

Received 24 December 2002. Accepted in revised form 2 January 2003

Key words: phospho-gypsum, phosphate industry, mercury, sulphate-reducing bacteria, bioremediation

Abstract

The phospho-gypsum is the most important residue from phosphate fertilizer industry and it can be used as emmedant in soil recovery. In the gold mining the mercury, which used by its ability of amalgamation, reaches concentrations that are toxic for plants, animals and man. To recover these soils it is necessary to immobilize the mercury using sorbent substrates, preferably residues of contaminant industries. The main subproduct of phosphate solubilizers industry is the phospho-gypsum. According the results of this work the capability to immovilize mercury is comparable to that of the other inorganic substrates such as kaolin or hematite. Moreover, sulphate-reducing bacteria (SRB) can contribute to the detoxification of soils contaminated with mercury by forming insoluble mercury sulfide. The use of phospo-gypsum and sulfate reducing bacteria together will can allows the integral rehabilitation of the devastated areas. This study shows that the inoculation of a soil with SRB can fail due to the presence of other bacterial species that compete with those inoculated, indicating the need to know the bacterial microbiota of a soil before the inoculation with any bacteria.

Introduction

The industrial process for phosphate obtaining originates a residual product named phospho-gypsum that contains basically calcium sulphate. This product can be used in bioremediation of soils contaminated with metals. Several of these metals, such as mercury, are very toxic for living beings. One of the greater sources of the worldwide contamination by mercury is the use of this element in gold mining activities (Lodenius and Malm, 1998). It is used by its ability of amalgamation, which facilitates the extraction of very fine particles of gold in rocks (once pulverized) as well as in soils and stream sediments.

In Venezuela this artisan mining activity began in the XVIII century at El Callao (State Bolivar) which was the most important centre of the auriferous production, with more than 230 mining sites that operated veins of quartz, mineralized with auriferous pyrite and other sites with native gold (Rodríguez, 1986). Approximately 1 kg of Hg by each kilogram of produced gold is lost (Johnson and Lindberg, 1995) and these mercury unloading go to soils and waters. These facts give an idea of the mercurial contamination level that had been accumulated in the zone throughout the time. Recent studies (Carrasquero, 2001), in three mining zones (of this area), reported concentrations in the soils up to 50 mg g^{-1}. These

* FAX No: +34-923-224876.
E-mail: a109144@usal.es

E. Velázquez and C. Rodríguez-Barrueco (eds.),
First International Meeting on Microbial Phosphate Solubilization, 333–336

Hg contents constitute a potential risk for the environment.

Adsorption of Hg is an important process that affects its mobility in soils. The understanding of these processes and their mechanisms are crucial to the assessment of the potential groundwater contamination and the development of cost-effective soil remediation strategies (Yin et al., 1996).

In soils the mercury exists commonly in form of Hg^{2+} compounds (Figure 1). The reduction to the elementary form, Hg^0, is easily reached by biological and chemical reactions. In anaerobic conditions, the microorganisms can metilate mercury, forming volatile organ-mercury compounds. Nevertheless, at the same time, the anaerobic conditions can precipitate the Hg^{2+} as HgS, that is extremely insoluble ($P_s = 3 \times 10^{-54}$). In aerobic conditions, the state of oxidation (Hg^{2+}) is stable and the mercury originates single complexes with the functional groups phenolic and carboxilic of the organic matter (Mc Bride, 1994).

Inorganic and organic compounds are able to immobilize mercury and therefore may be used as soil conditioners. The phospho-gypsum is usually used as a soil amendment or conditioner for sodic soils (Espinosa, 1996).

Also, the insolubilization of mercury is a way to reduce the mercury levels converting these compounds in insoluble forms (immobilization), such as HgS. This way can be developed by sulphate-reducing microorganisms originating metallic sulphides. The predominant microorganisms related to the reduction of sulphate belong to delta Proteobacteria. These organisms are strictly anaerobic that do not form spores and can produce H_2S from sulphate. This characteristic is used to detect these microorganisms using iron salts in the isolation media because they originates a black colour due to FeS.

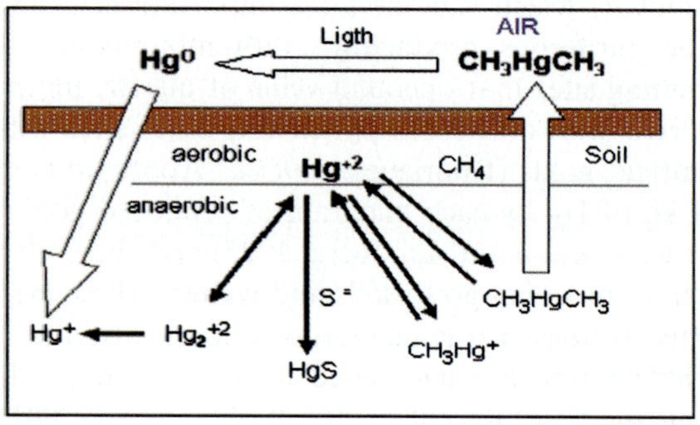

Figure 1. Cycle of mercury in the soil. From Mc Bride (1994).

Therefore, the aims of this work was, to evaluate the capability of several inorganic compounds to immovilize Hg and the isolation of sulphate-reducing bacteria to be used in the future in the integral rehabilitation (physical–chemistry–biological) of the devastated areas.

Materials and methods

Analysis of the capability of different amendment compounds to immobilize Hg

This analysis was performed in several inorganic substrates abundant in soils of Venezuela (Table 1). The phospho-gypsum, a residue of phosphate fertilizer industry, is composed basically by $CaSO_4{\cdot}2H_2O$ with low amounts of P, F, Si, Fe and Al. The kaolin is composed by hydrogenated aluminium silicate. The goethite and hematite are basically composed by ferric oxides. All these materials are abundant in soils from Venezuela. To saturate 20 mg of each substrate 20 ml of a solution of 40 ppm of $Hg(NO_3)_2$ at pH 5 were used. The solution was mixed by constant agitation with each substrate and samples were taken at 30 min, 1, 2, 4 and 6 h. For sampling, each suspension was centrifugated at 4000*g* during 5 min and the supernadants were collected. After that, newly 20 ml of mercurial solution was added to each sample. The content of Hg was measured using an AMA 254 Analizer of Mercury.

Sediment sampling and isolation of sulphate-reducing bacteria

Samples of sediments from 17 mining sites in El Callao area (Venezuelan Guayana) were collected. The samples were taken from the bottom of

Table 1. Amounts of Hg (mg/g) retained by the different inorganic compounds used in this study

Time (h)	Phospho-gypsum	Kaolin	Goethite	Hematite
0.5	7.5	8.5	10.8	7.3
1	8.5	9.5	12.7	7.4
2	10.3	12.1	13.4	7.4
4	10.2	11.7	13.8	7.2
6	9.5	12.5	13.9	7.3

ponds, in plastic containers and were sealed, placed to a temperature of 4 °C and transported to the laboratory.

In each sediment sample an aliquot 10 ml was taken. Then it was inoculate in specific media for isolation of sulphate-reducing bacteria in dilution 1:10 according to the most probable number (MPN) method. The conditions maintained before, during and after preparation of the inocula were reached in an anaerobic chamber with a CO_2 gas atmosphere mixed with N_2. Then the tubes were incubated in anaerobic conditions during 2 weeks at 25 °C.

The tubes in which a black coloration appears after incubation were considered positive due to the SO_4^{2-} reduction caused by the presence of sulphate-reducing bacteria and the iron sulphide formation.

Inoculation in experimental field plots

In El Callao zone an experimental field was designed in order to evaluate the evolution of sulphate-reducing bacteria populations after inoculation together phosphogypsum. This experimental field was installed near an active mining site. The plot with a total area of 4 m^2 was splits in four subplots of 1 m^2 each one. They received different treatments. The first subplot was treated with bacteria (1.2×10^5 cells/g.) + gypsum (300 g), the second a single treatment of gypsum, the third a single treatment of bacteria (1.2×10^5 cells/g.) and the fourth was designated as a control treatment.

Results and discussion

Analysis of the capability of different amendment compounds to immobilize Hg

The results of this analysis are recorded in the Table 1. The Hg retention values obtained in this studies along the time were similar in kaolin and goethite, lightly lower in phospho-gypsum and the lowest values were obtained when hematite was used. Due to the abundance of phosphogypsum, its low cost and the availability of this substrate we decided to test the interaction between this compound and the sulphate reducing populations of bacteria (SRB).

Count of sulphate-reducing bacteria in initial samples and in experimental field plots

The isolation of bacteria in anaerobic conditions was done using liquid media as indicated in the "Materials and methods" section and the count of sulphate-reducing bacteria was made using the MPN method. The number of cells per gram of soil in the initial samples was approximately 1.2×10^5 cells/g. For that reason, the same concentration was used to inoculate the subplots in the present work.

The evolution of sulphate reducing populations of bacteria is recorded in the Figure 2. As can be seen in this figure the evolution of SRB population in the untreated subplot and the plot treated with phospho-gypsum show the same results and the peak observed from 30 days of the treatments can be due to the climatic and environmental conditions. Nevertheless, the number of cells per gram was 10 times higher in untreated soil. In the case of the soil inoculated with bacteria and of the soil treated with phospho-gypsum and bacteria both the evolution of the SRB population and the number of these bacteria rend poor results. However, this result is not due to phospho-gypsum since in the soils treated with bacteria and this compound an increase of SRB is observed at the end of the treatment.

According previous studies, the SRB populations generally range from 10^4 to 10^3 cells per gram, but these values can rise to several millions per gram after near 2 weeks in flooded fields (Alexander, 1980). On the contrary, the results obtained in this work showed that the number of bacteria per gram decreased after field flooding. The cause of the failure in the recovering of SRB bacteria from the soils inoculated with a suspension (directly isolated from soil in liquid medium) of these bacteria is probably due to the presence of other bacteria that compete with the autochthonous SRB present in the soil. Recent results obtained in our laboratory show that the SRB suspensions obtained in selective media incubated in anaerobic conditions contain at least two species of aerobic bacteria that are able to grow quickly when the suspensions are incubated in presence of oxygen. In any case, although more studies are needed, the phosphogypsum has a positive interaction with SRB and

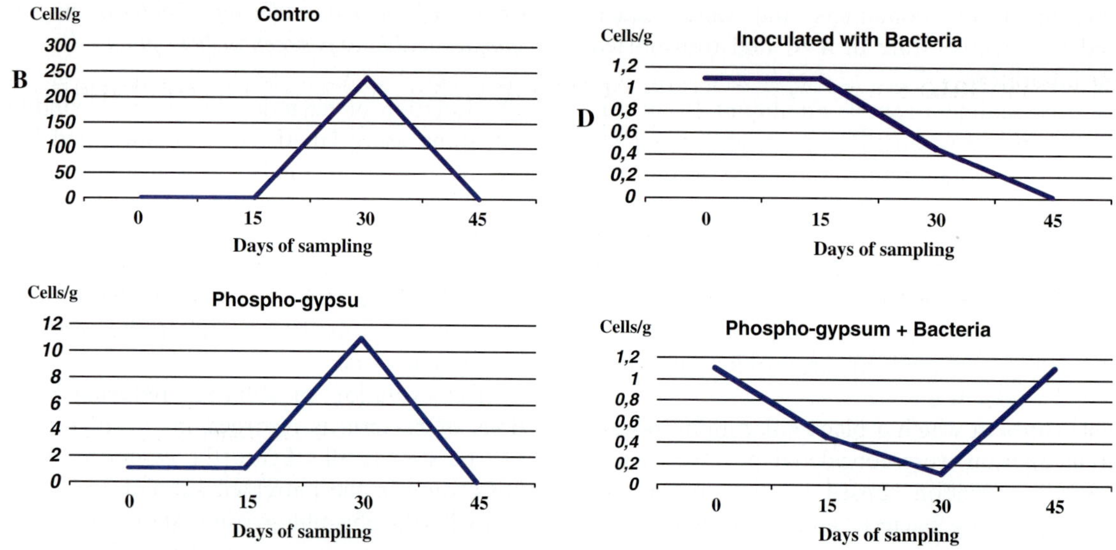

Figure 2. Evolution of the SRB population (cells/g × 10^5) along the time (days) in the different treated plots.

also shows a good capability to immobilize mercury and therefore, this product of phosphate fertilizer industry is suitable for treatments of soils contaminated with this metal.

Acknowledgement

This work was supported by GRANTS to AGS, FS and MAM.

References

Alexander M 1980 Introducción a la Microbiología del Suelo. Traducción de la segunda edición por J.J. Peña – Cabriales. AGT Editor, S.A. México.

Carrasquero A 2001 Study of the biogeochemistry cycle of the Hg. Universidad Central de Venezuela, Postgrado de Ciencias de Suelo, Tesis.

Espinosa J 1996 Efectos de la aplicación de fosfoyeso sobre las propiedades de un suelo Franco-Arenoso del campo experimental de la Facultad de Agronomía, UCV, Tesis de Grado.

Johnson D W and Lindberg S E 1995 The cycling of mercury in forests: For alternative methods quantifying total deposition and soil emissions. Water Air Soil Poll. 80, 1069–1077.

Lodenius M and Malm O 1998 Mercury in the amazon. Rev. Environ. Contam. Toxicol. 157, 25–52.

Mc Bride, M B 1994 Environmental chemistry of soils. Oxford University Press, USA. 406 pp.

Rodríguez S 1986 Mineral resources of Venezuela. Bulletin of Geology Vol. XV No 25, Ministry of Energy and Mines.

Yin Y, Allen H, Li Y, Huang C and Sandres P 1996 Adsorption of mercury (II) by soil: Effects of pH, Chloride, and Organic Matter. J. Environ. Qual. 25, 837–844.

Soil phosphate solubilizing microorganisms and cellulolytic population as biological indicators of iron mined land rehabilitation

S.R. Matias, R. Passos, M.R.M. Scotti[1] & N.H. Sá

Department of Botany, Institute of Biological Sciences, Federal University of Minas Gerais, Avenida Antonio Carlos, 6627, Pampulha, Cep: 31.270-901 Belo Horizonte, Minas Gerais, Brazil. [1]*Corresponding author**

Received 19 December 2002. Accepted in revised form 2 January 2003

Key words: *Bradyrhizobium* sp., arbuscular mycorrhizal fungi, ore iron mine, rehabilitation, soil phosphate solubilizing microorganisms

Abstract

Mining is one of the most important segments of the Brazilian economy. An area of rupestrian fields, which was used as a deposit for iron-ore and where no vegetation was found for 3 years, showed a low soil cellulolytic and phosphate solubilizing microorganisms (PSM) population. The success of rehabilitation of that site was achieved using two native species, *Centrosema coriaceum* (Leguminosae) and *Tibouchina multiflora* (Melastomataceae), inoculated with mycorrhizal fungi (AMF) and/or rhizobia. Efficient strains of *Bradyrhizobium* sp. were isolated from *Centrosema coriaceum* nodules and were used to inoculate seedlings of *Centrosema* grown in deposits of iron-ore soil. Four-month-old seedlings were transplanted to the field where *C. coriaceum* was intercropped with *T. multiflora* using 3 completely randomized blocks with 2 plots/block corresponding the treatments of inoculated and uninoculated plants. After 24 months, the plant growth, mycorrhizal colonization, soil cellulolytic and PSM were evaluated. Inoculated plants of *C. coriaceum* plants showed greater growth and inoculation of both studied species improved the AMF root colonization, PMS and cellulolytic population in root zone in relation to the uninoculated plots or in relation to uncultivated site. The results suggest that the use of efficient strains of *Bradyrhizobium* and AMF associated with native species may be a useful technology for revegetating mined soil. The PSM and cellulolytic population may be used as rehabilitation indicators of iron-ore mined land.

Introduction

The state of Minas Gerais has the largest deposits of iron-ore in Brazil iron ore land. Restoration of mined sites requires a functional microbial community for soil development, biogeochemical cycling and for a successful plant community establishment. Rehabilitation success has been achieved by the use of native species using beneficial microorganisms such as rhizobia (Rao, 2001) and mycorrhizal fungi (AMF) (Rao and Tak, 2001) as described for gypsum mine spoil in India. As a criteria for judging the reclamation success the composition of soil microbiota is an essential biological indicators of soil health (Mummey et al., 2002). The purpose of the present study was to evaluate the benefits of rhizobia and/or AMF for the field establishment and growth of *Centrosema coriaceum* (Leguminosae) and *Tibouchina multiflora* (Melastomataceae) and to evaluate the role of soil phosphate solubilizing

* FAX No: + 31-4992673/225-6627.
E-mail: scottimuzzi@ig.com.br

E. Velázquez and C. Rodríguez-Barrueco (eds.),
First International Meeting on Microbial Phosphate Solubilization, 337–340

(PS) and cellulolytic microorganisms as environmental indicators.

Material and methods

Study site

The study site is located near the town of Itabira – MG, southeast Brazil, in a mine called Brucutu belonging to the Vale Rio Doce Company (19°53′08″ S 43°26′10.9″ W). The natural vegetation is a rupestrian field and the experimental site was used as a deposit for mined iron-ore and no vegetation was found after being abandoned for 3 years. The floristic studies in undisturbed area near to the study site allow us to select 2 species with potential for rehabilitation program: *C. coriaceum* (Leguminosae) and *T. multiflora* (Melastomataceae).

Soil samples

Soil samples (0–10 cm depth) were collected near from each plant, resulting three replicates for each treatment after mixed and sieved (2 mm). Then it was used to chemical analysis.

Inoculum preparation

Rhizobial strain was previously isolated from nodules of *C. coriaceum*, collected from an undisturbed site. Mycorrhizal fungi used were *Gigaspora margarita*, and *Glomus etunicatum*, from our laboratory collection. Seedlings of *C. coriaceum* and *T. multiflora* (Melastomataceae) were planted in pots using two treatments: (I) complete fertilization (Somasegaran and Hoben, 1985) and (II) fertilization without nitrogen plus inoculation with rhizobia strain BHCB-Cc-38 + AMF for *C. coriaceum* and single inoculation of AMF for *T. multiflora*.

Experimental design and plant growth

After 120 days, the plants were transplanted to the field using an experimental design of 3 completely randomized blocks. In each block there were two plots (2 treatments) with 8 plants/species or 16 plants per plot. Plant growth was evaluated in relation to height for *T. multiflora* plants and stoloniferous cover for *C. coriaceum*, during 24 months.

Microorganisms evaluation

After 24 months, root segments were collected, selected randomly, counted and the percent of AMF root colonization was calculated as follows: Colonization (%) = number of AMF positive segments/total number of segments scored × 100. The data were statistically analysed by analysis of variance (ANOVA) and means were compared by the Tuckey test. Soil samples were collected at root zone and analysed in relation to the number of phosphate solubilizing microorganisms (PSM) using Pikovskaia's agar (Pikovskaia, 1948) and also the number of cellulolytic microorganisms (Rougieux, 1964), bacteria (Thorton, 1956) and fungi. The diameter of colony and halo formed by PSM was measured using samples collected at 30 months after transplanting.

Results and discussion

The ability of *C. coriaceum* to cover the disturbed site was significantly increased (5-fold) by the dual inoculation of *Bradyrhizobium* sp. (BHICB-Cc-38) and AMF as shown in Figure 1, especially due to the efficient nitrogen fixing bradyrhizobia strain. Rao (2001) found similar results with *Bradyrhizobium* sp. used to inoculate *Albizia lebbeck* grown in mine spoils, which improved growth. In spite of the high root colonization of inoculated *C. coriaceum* and *T. multiflora* plants (Figure 2), the growth of *T. multiflora* was not increased (Figure 1). The AMF colonization of *C. coriaceum* roots may have had an additive effect on plant growth through the increase of mineral nutrition (Manjunath et al., 1984). These soils showed a general nutrient deficiency with special emphasis for phosphorus (data not shown). Microorganisms that present a nutritional versatility were less affected by soil nutritional deficiency than those that need a specific substrate for their growth. The soil PSM and cellulolytic population were inhibited by soil disturbance while the total bacteria and fungal population was not modified (Figure 3).

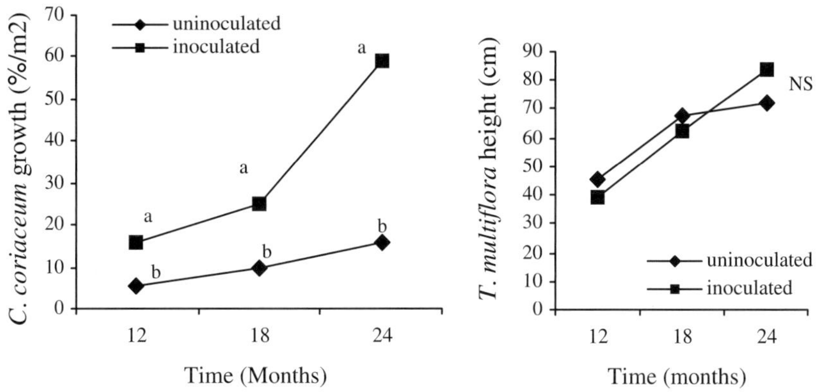

Figure 1. Effect of double inoculation with *Bradyrhizobium* strain (BHCB-Cc38) and AMF on stoloniferous growth of *C. coriaceum* and single inoculation of AMF on height growth (cm) of *T. multiflora* cultivated in experimental mined area after 24 months. Means followed by different letter(s), within each time are statistically significant, $P = 0.05$ (Tuckey's multiple-range test). NS: Not significant.

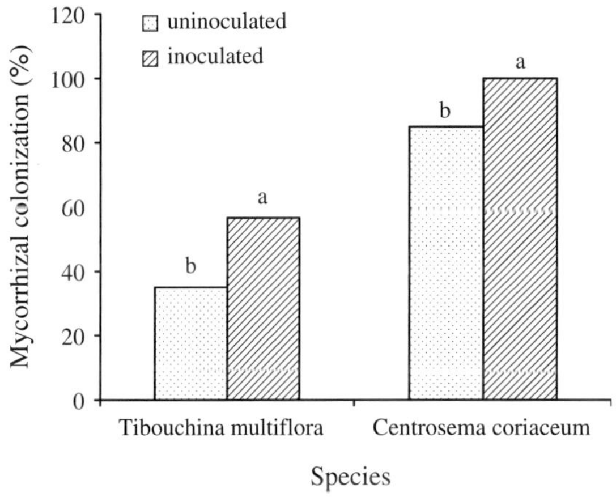

Figure 2. Mycorrhizal colonization (%) of *C. coriaceum* and *T. multiflora* roots inoculated with AMF and 24 months after transplantation in disturbed experimental site. Means followed by different letter(s), within each species, are statistically significant, $P = 0.05$ (Tuckey's multiple-range test). NS: Not significant.

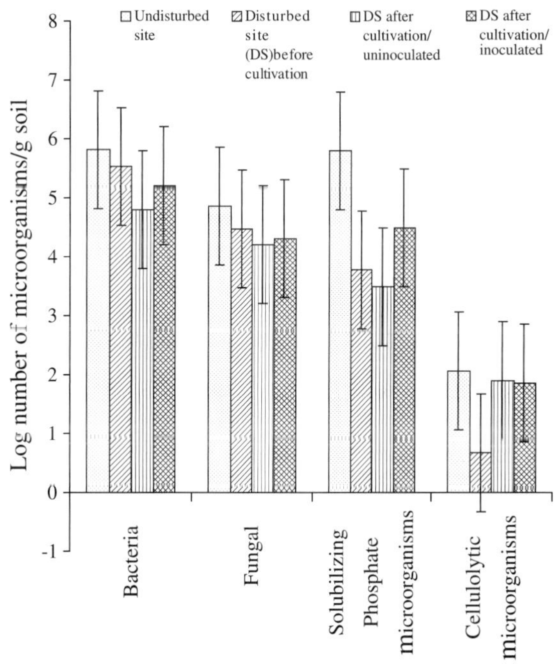

Figure 3. Soil microorganisms population in undisturbed site, before and 24 months after transplantation in experimental site.

The substrate bioavailability is the problem since cultivation stimulated the PSM and cellulolytic microorganisms especially in inoculated plots (Figure 3) approaching the number of microorganisms found in the undisturbed site. These results allow us to use such microorganisms as biological indicators of soil quality able to measure the rehabilitation degree. Microbial life in these mined lands develops only in conjunction with the root systems of plants. Since plants differ in chemical composition, physiology and nutrition parameters they would attract specific groups or species of organisms, which may be beneficial (Katznelson et al., 1948). Therefore, plant exudate elicite a root zone effect which

Table 1. Colony number, halo forming colony and ratio for the size colony and halo of PSM 30 months after transplanting

Treatment	Colony number (log cfu/g)	Number of solubilizing halo forming colony/total cfu	Diameter ratio halo/colony
Preserved area	6.7^{ab}	0.12^{NS}	2.46^{NS}
Disturbed experimental site	5.0^{b}	–	–
Experimental site cultivated with uninoculated *C. coriaceum*	6.0^{ab}	0.09	5.8
Experimental site cultivated with inoculated *C. coriaceum* (rhizobia + AMF)	7.2^{a}	0.12	6.1
Experimental site cultivated with uninoculated *T. multiflora*	6.7^{ab}	0.02	4.9
Experimental site cultivated with inoculated *T. multiflora* (AMF)	7.0^{a}	0.12	8.7

Means followed by the same letter(s), within each time and each soil type, are not statistically significant, $P = 0.05$ (Tuckey's multiple-range test). NS: Not significant.

result in a stimulatory effect of both symbiotic and non-symbiotic microorganisms such as those related to release of nutrients (Rengel, 1997). The PSM population was higher in the root zone of plants inoculated with AMF but the diameter of the halo formed was not significantly improved (Table 1). However, PSM population isolated from uncultivated site did not produce halo. This data suggests a synergistic action of both PSM and AMF population, probably related to phosphate metabolism. Literature confirms that AMF can increase root exudation and through rhizosphere effect stimulate PSM population (Khan and Bhatnagar, 1977).

Conclusions

(1) The cultivation model demonstrates that the production and soil cover had increased. The selected native species were successfully reintroduced. (2) The double inoculation with bradyrhizobia (BHICB-Cc-38) and AMF had played a critical role in plant growth, being considered a useful technique for mine reclamation. (3) The soil PSM and cellulolytic microorganisms can be considered as soil quality indicator of disturbed mine 2 sites under rehabilitation process (Figure 3).

Reference

Katznelson H, Lochhead A G and Timonin M I 1948 Soil microorganisms and the rhizosphere. Bot. Rev. 14, 543–586.

Khan J A and Bhatnagar R M 1977 Studies on solubilization of insoluble phosphates by microorganisms. Fertil. Technol. 14, 329–333.

Manjunath A, Bragyaraj D J and Gopala-Gowada H S 1984 Dual inoculation with V.A. mycorrhiza and *Rhizobium* is beneficial to *Leucaena*. Plant Soil 78, 445–448.

Mummey D L, Stahl P D and Buyer J S 2002 Microbial biomarkers as an indicator of ecosystem recovery following surface mine reclamation. Appl. Soil Ecol. 21, 251–259.

Pikovskaia R L 1948 Mobilization of phosphorus in soil in connection with vital activities by some microbial species. Mikrobiologia 17, 362–370.

Rao A V 2001 Effect of rhizobial inoculation on *Albizia lebbeck* and its rhizosphere activity in mine. Arid Land Res. Manag. 15, 157–162.

Rao A V and Tak R 2001 Influence of mycorrhizal fungi on the growth of different tree species and their nutrient uptake in gypsum mine spoil in India. Appl. Soil Ecol. 17, 279–284.

Rengel Z 1997 Root exudation and microflora populations in rhizosphere of crop genotypes differing in tolerance to micronutrient deficiency. Plant Soil 196, 255–260.

Rougieux G 1964 Técnicas de microbiologia agricola. Acribia. Zaragoza. Espanha. p. 181.

Somasegaran P and Hoben H J 1985 Methods in legume-Rhizobium Technology. Ed. Niftal USAID, Hawaii. p. 367.

Thornton H G 1956 The ecology of microorganisms in soil. Proc. R. Soc. (London) B.145, 364–374.

Effect of certain phosphate-solubilizing bacteria on root-knot nematode disease of mungbean

M. R. Khan[1], S. M. Khan, F. A. Mohiddin & T. H. Askary

Department of Plant Protection, Aligarh Muslim University, Aligarh, 202 002, India. [1]*Corresponding author**

Received 22 November 2002. Accepted in revised form 2 January 2003

Key words: *Bacillus* spp., *Pseudomonas* spp., *Meloidogyne incognita*, root nodules, crop yield

Abstract

Bacillus subtilis, B. polymyxa, Pseudomonas fluorescens and *P. stutzeri* were cultured on nutrient broth and were applied on the seeds of mungbean at 5 mL/kg of seeds along with the rhizobium. The seeds were sown in microplots (3 × 1.5 m) where nematode suspension (1500 second stage juveniles of *Meloidogyne incognita*/kg soil) had already been added or not added. Three microplots which were randomly distributed in the field were maintained for each treatment. In the plots without nematodes, plants showed better growth with significantly improved yield, especially with *P. stutzeri*. Root-nodulation was significantly increased due to the treatments of *P. stutzeri* or *B. polymyxa* compared to uninoculated control. Inoculation with the nematode caused severe galling on roots and decreased the yield of mungbean by 23.8%. Application with *P. fluorescens* or *B. subtilis* suppressed the gall formation, reproduction and soil population of *M. incognita*. The nematicidal effect of the PSMs was considerably less than the fenamiphos treatment (at 2 g/kg seed). Rhizosphere population of all PSMs tested was significantly increased over time irrespective of nematode presence or absence. Application of *P. fluorescens* suppressed the nematode pathogenesis and increased the yield of mungbean by 30.9% that was greater than the nematicide treatment (16.7%).

Introduction

Interest is growing to use biofertilizers based on plant growth-promoting organisms particularly phosphate-solubilizing microorganisms in place of inorganic fertilizers. PSMs such as *Azotobacter chroococcum, Beijerinkia indica, Bacillus subtilis, B. polymyxa, Pseudomonas fluorescens, P. striata, P. stutzeri* etc. solubilise phosphorus in the soil and supply it to plant roots in much greater amount than otherwise (Gaur, 1990; Rao, 1990). PSMs can also be used in nematode disease management. Nematode suppression may occur through two ways. Greater availability of phosphorus may make the host plant strong enough to tolerate pathogen attack. Phosphorus itself plays vital role in building self defence of plants against nematodes (Kirkpatrick et al., 1964). PGPOs produce toxins such as bulbiformin by *B. subtilis* (Brannen, 1995) and phenazin (Toohey, 1965), pyoleutorin (Howell and Stipanovic, 1980), pyrrolnintrin (Burkhead and Geoghegan, 1994) by *P. fluorescens* which may inhibit the infection by nematode. Evidences exist which have shown that soil application of *B. subtilis, P. stutzeri* (Khan and Tarannum, 1999), *P. fluorescens* and *B. polymyxa* (Khan and Akram, 2000) suppressed the disease severity of root-knot of tomato caused by *Meloidogyne* species.

Mungbean, *Vigna radiata*, L. Wilczek is an important pulse crop in India, and an excellent source of high quality protein. Root-knot is one of the prevalent diseases of mungbean which causes around 28% or more yield loss to the crop

* E-mail: mrkhan777in@yahoo.co.in

E. Velázquez and C. Rodríguez-Barrueco (eds.),
First International Meeting on Microbial Phosphate Solubilization, 341–346

(Bridge, 1981). The present study was aimed to examine and ascertain the feasibility of applying seed treatment with *B. subtilis, B. polymyxa, P. fluorescens* and *P. stutzeri* to manage root-knot disease of mungbean caused by *M. incognita* under field condition. Effects of the microorganisms were also examined on root nodulation, biomass production and yield of mungbean and on the reproduction and soil population of the nematode, and rhizosphere population of PSMs.

Materials and methods

Thirty microplots, each 3 × 1.5 m dimension were prepared in a field of 24 × 12 m. Twelve treatments were made to examine the effect of 4 strains of PSMs on the root-knot disease and other variables of mungbean. A treatment with the nematicide, fenamiphos at the rate of 2 g of active ingredient/kg seed was maintained to compare efficacy of the PSMs. The 12 treatments are listed in Table 3. Each treatment was replicated three times and randomly distributed in the field. The experiment was conducted over two consecutive years.

Culture of *M. incognita* (Kofoid & White) Chitwood was prepared from egg masses, incubated on tissue paper on a wire gauze (coarse sieve) placed in a Baermann funnel inside a B.O.D. incubator at 25 °C for 6–8 days. The nematode suspension was diluted to 1500 second stage juveniles (J_2) in 1L water, which was added to soil in an area 15 cm deep and wide, at 36 locations in a microplot (12 locations/row, 3 rows/microplot). The inoculation was made 1 day prior to sowing seed of mungbean cv. LOC.

Pure cultures of *Bacillus polymyxa* (Prazmowski) Mace, *B. subtilis* Cohn *emend.* Prazmowski, *P. fluorescens* (Threvesan) Migula and *P. stutzeri* (Lehmann and Neumann) Sijderius were obtained from the Institute of Microbial Technology, Chandigarh, India. Mass-cultures of these microorganisms were prepared in nutrient broth in conical flasks. The 2–3-day-old cultures (CFU $\times 10^8$/mL) were applied on seeds at 2 mL/kg along with the rhizobium (commercial formulation). After application the seeds were sown at the 36 sites in each microplot where nematode suspension had been added or not added. Two weeks after sowing, the plots were irrigated, and thinned to one seedling at each site to maintain 12 seedlings in a row (36 seedlings/microplot). Four months after sowing, 10 plants from each microplot were randomly selected for harvest and galls, egg masses/root system, fecundity (eggs/egg mass), dry matter production and yield (weight of grains/plant) were determined. Nodulation in terms of number of functional, non-functional and total nodules/root system were counted on 2-month-old plants. Soil population of *M. incognita* juveniles was estimated from 10–15 cm deep rhizosphere soil collected from 10 plants at sowing, mid-age (2 months) and harvest (4 months)

Table 1. Effects of seed treatment with certain phosphate-solubilizing bacteria on the galling, reproduction and soil population of *Meloidogyne incognita*

Treatment	Number/Root system		Fecundity	Soil population/kg soil			LSD ($P = 0.05$)	F-value (df = 2)
	Galls	Egg mass	Egg/egg mass	Pre-plant	Mid-age (2 months)	Harvest (4 months)		
Control	73	66	253	1500	2474	2919	121.6	59.4**
B. subtilis	68	58*	236	1500	2238*	2704	139.5	47.2**
B. polymyxa	71	67	241	1500	2401	2852	141.4	46.5**
P. fluorescens	58*	46*	219*	1500	2169*	2415*	172.0	38.3**
P. stutzeri	69	66	246	1500	2394	2877	155.8	49.0**
Fenamiphos	48*	42.0*	174	1500	1872*	2080*	91.5	61.8**
LSD ($P = 0.05$)	6.38	4.7	29.7		215.3	295.6		
F-value (df = 5)	12.5**	34.2**	21.5**		28.4**	10.7**		

Each value is mean of two replicates.
*Significantly different from the control at $P = 0.05$.
**Significant at $P = 0.05$.

Table 2. Population of phosphate-solubilizing bacteria in the rhizosphere of mungbean inoculated with *Meloidogyne incognita* or not inoculated

Phosphate-solubilizing bacteria	Nematode inoculated	Rhizosphere population × 10^6 CFU/g soil				
		Pre-plant	Mid-age (2 months)	Harvest (4 months)	LSD (P = 0.05)	F-value (df = 2)
Bacillus subtilis	500	6.59	44.7*	9.22*	0.83	419**
	00	6.61	43.5	9.17*	0.81	407**
B. polymyxa	1500	5.45	46.2*	8.50*	0.66	552**
	00	5.43	46.8*	8.15*	0.62	541**
Pseudomonas	1500	9.10	65.3*	19.71*	0.91	507**
fluorescens	00	9.12	63.9	19.07*	0.93	482**
P. stutzeri	1500	8.87	50.8*	18.15*	0.70	523**
	00	8.90	52.3	18.31*	0.69	512**

Each value of mean of two replicates.
*Significantly different from the control at P = 0.05.
**Significant at P = 0.05.

in each microplot inoculated with the nematode using Cobb's decanting and sieving method (Southey, 1986). Rhizosphere population of the PSMs from nematode-inoculated and uninoculated microplots was also determined simultaneously using dilution plate method.

Thirty observations taken for each variable of a treatment (10 observations/microplot) were averaged and considered as one replicate. Hence, there were two replicates as the experiment was replicated over time (two consecutive years). The standard error calculated on two replicates for various variables ranged to ±1.2 2.5. The data on biomass, yield and nodules were subjected to a two-factor analysis of variance (ANOVA), and galls, egg masses, fecundity and soil population were analysed by a single factor analysis of variance. Least significance difference (LSD) was calculated for each variable at the probability level of P = 0.05 (Dospekhov, 1984).

Results and discussion

Inoculation with *M. incognita* caused extensive galling on roots of mungbean. Treatments with the *P. fluorescens* suppressed gall formation by 14.8% (Table 1). Decline in gall formation due to the fenamiphos treatment was 29.4% compared to inoculated control. Nematicide treatment resulted to maximum decline in the egg mass production (27.6%), followed by *P. fluorescens* (20.7%) and *Bacillus subtilis* (12.1%). Fecundity of the nematode (number of eggs/egg mass) was decreased by 31.2 and 13.4% due to nematicide and *P. fluorescens* treatments, respectively (Table 1). Soil populations of root-knot nematodes at mid-age (2-month-old plants) were decreased due to application of the PSMs, being significant for fenamiphos (24.3%), *P. fluorescens* (12.3%) and *B. subtilis* (9.5%) compared to the control. At harvest, significantly low population of the nematode was recorded for the nematicide (28.7%) or *P. fluorescens* (17.3%). Out of the four tested strains of PSMs, *P. fluorescens* and *B. subtilis*, in particular the former produced definite nematicidal effects on root-knot nematode suppressing all the variables of parasitism examined. This effect may have resulted due to toxins such as phenazin (Toohey et al., 1965), pyrrolnintrin etc. (Burkhead and Geoghegan, 1994) produced by *P. fluorescens*. Rhizosphere populations of the PSMs were drastically increased over time irrespect of inoculation with root-knot nematode (Table 2). At mid-age, greatest increase in the population was recorded for *B. polymyxa* (748–761%), followed by *P. fluorescens* (600–615%), *B. subtilis* (558–578%) and *P. stutzeri* (473–487%) compared to respective preplant populations. The harvest populations were increased by 109–117% (*P. fluorescens*), 99–109% (*P. stutzeri*), 50–56% (*B. polymyxa*) and 38–40% (*B. subtilis*).

Table 3. Effects of seed treatments with certain phosphate-solubilizing bacteria on the dry biomass, yield and nodulation of mungbean inoculated with *Meloidogyne incognita* or not inoculated

Treatments	Nematode inoculation (Juveniles/kg)	Dry plant weight (g)	Weight of grains per plant (g)	Number of nodules/root system		
				Functional	Non-functional	Total
Control (Uninoculated)	00	34.7	69.2	16	5	21
Bacillus subtilis	00	35.5	68.3	16	5	21
B. polymyxa	00	38.4*	74.2	18*	5	23*
Pseudomonas fluorescens	00	36.9	72.9	17	5	22
P. stutzeri	00	37.8*	77.3*	19*	5	24*
Fenamiphos	00	34.9	70.5	15	6*	21
Control (inoculated)	1500	29.4*	52.7*	11*	7*	18*
B. subtilis	1500	30.5	53.5	12	6*	18
B. polymyxa	1500	34.0*	59.6*	12	6*	18
P. fluorescens	1500	35.1*	69.0*	13*	6*	19
P. stutzeri	1500	33.7*	60.9*	12	6*	18
Fenamiphos	1500	33.1*	61.5*	12	7*	19
LSD (P = 0.05)	2.90	6.19	1.24	0.39	0.69	1.83
F-value	Treatments (df = 4)	5.3**	NS	6.5**	NS	7.4**
Nematode (df = 1)	9.2**	12.5**	8.2**	6.1**	5.8**	
Treatments × Nematode (df = 4)	6.9**	7.2**	NS	NS	NS	

Each value of mean of two replicates.
*Significantly different from the control at P = 0.05.
**Significant at P = 0.05.
NS – Not significant at P = 0.05.

A non-significant difference in the rhizosphere populations of PSMs in the presence and absence of *M. incognita* indicates that all the four microorganisms established in soil because of their own adaptibility and reproductive potential, and the nematode presence did not influence their normal multiplication rate (Table 2).

Seed treatment with *B. polymyxa* or *P. stutzeri* significantly promoted the plant dry weight of uninoculated plants (Table 3). Yield of mungbean was, however, significantly increased due to the treatment with *P. stutzeri* only compared to uninoculated control. Yield enhancement induced by other PSMs were not significant. Nematode infection alone suppressed the dry plant weight (15.3%) and yield (23.8%) compared to uninoculated control. Application of PSMs except *B. subtilis* significantly increased the dry plant weight and yield of mungbean compared to the inoculated control (Table 3). Greatest increase in the dry plant weight and yield occurred due to the treatments with *P. fluorescens* (14.6 and 30.9%), followed by *B. polymyxa* (15.6 and 13.1%), *P. stutzeri* (14.6 and 15.6%) and fenamiphos (12.6 and 16.7%), respectively. Plant growth and yield enhancements were apparently due to solubilization of phosphates in soil and making available greater amount of utilizable form to plants (Gaur, 1990; Rao 1990). Yield promotion of nematode-inoculated plants due to *P. fluorescens* might had occurred cumulatively due to phosphate solubilization and toxin production. Whereas *B. polymyxa* or *P. stutzeri* increased the yield through their plant growth promoting effects only. Similar effects of *Pseudomonas stutzeri* and *P. fluorescens* on the root-knot disease of tomato have been recorded (Khan and Akram, 2000; Khan and Tarannum, 1999).

Seed treatments with *P. stutzeri* and *B. polymyxa* promoted the nodulation induced by *Rhizobium* sp., leading to 18.8 and 14.3% and 12.5 and 9.5% increase in the number of functional and total nodules/root system of uninoculated plants, respectively (Table 3). Nematode infection caused 31.2 and 14.3% decrease in functional and total nodules/root system, respectively compared to the uninoculated control, whereas non-functional nodules were increased by 40%. A similar adverse effect of root-knot nematodes on nodule formation has been recorded on chickpea (Khan et al., 1996), cowpea (Khan and Khan, 1996), soybean (Kabi, 1983) and pigeonpea (Taha, 1993). Root-knot nematodes also invade nodules and form galls on them (Melakeberhan and Webster, 1993). Such nodules degenerate and are shed much earlier than the normal ones (Ali et al., 1981). In the present study, 5–10% nodules/root system were invaded by the nematode as evidenced by the presence of a small gall on nodules. Probably for this reason, lower number of nodules were present on nematode-infected roots. Application of the microorganisms, however, enhanced the number of functional nodules/root system by 9–18% on nematode-inoculated plants, the increase was greatest with *P. fluorescens* in comparison to nematode-inoculated control. This increase was apparently due to significant decrease (14.3%) in the non-functional nodules resulted by the application of all PSMs invariably. For this reason total nodule count of nematode-inoculated plants treated with PSMs had remained equal to the inoculated control.

The study has demonstrated a potential scope of use of phosphate-solubilizing microorganisms in root-knot disease management. Seed treatment with *P. fluorescens* suppressed the disease severity, nematode reproduction and promoted yield of mungbean (30.9%) greater than the nematicide (16.7%) compared to the control.

References

Ali M A, Trabulsi J Y and Abd-Elsamea M E 1981 Antagonistic interaction between *Meloidogyne incognita* and *Rhizobium leguminosarum* on cowpea. Plant Dis. 65, 432–435.

Brannen R 1995 Production of antibiotics by *Bacillus* subtilis and their effect on fungal colonists of various crops. Trans. Br. Mycol. Soc. 65, 203.

Bridge J 1981 Nematode. *In* Nematode Pest Control in Tropical Grain Legumes. Eds A Ward, S L Mercer and V Howe. pp. 111–125. Center for Overseas Pests Research, London.

Burkhead K and Geoghegan M J 1994 Antibiotics. *In* Soil-borne Plant Pathogens. Ed. K Burkhead. p. 368 Macmillan, New York.

Dospekhov B A 1984 Field Experimentation. Mir Publishers, Moscow, Russia p 352.

Gaur A C 1990 Phosphate Solubilising Microorganisms as Biofertilizers. Omega Scientific Publishers, New Delhi p 176.

Howell C R and Stipanovic R D 1980 Suppression of *Pythium ultimum*-induced damping-off of cotton seedlings by *Pseudomonas fluorescens* and its antibiotic pyoluteorin. Phytopathology 70, 712–715.

Kabi M S 1983 Rhizobial interaction in the rhizosphere of soybean for nodule sites. Zentralbla. Mikrobiol. 138, 269–275.

Khan M R and Akram M 2000 Effect of certain antagonistic fungi and rhizobacteria on wilt disease complex caused by *Meloidogyne incognita* and *Fusarium oxysporum* f.sp. *lycopersici* on tomato. Nematol. Mediterr. 28, 139–144.

Khan M R and Khan M W 1996 Development of root-knot nematodes on cowpea as influenced by sulphur dioxide. Nematol. Mediterr. 24, 33–35.

Khan M R, Khan M W and Singh G K 1996 Effects of nickel and root-knot nematode on the growth and protein content of chickpea. Nematol. Mediterr. 24, 87–90.

Khan M R and Tarannum Z 1999 Effect of field application of various microorganisms on the root-knot disease of tomato. Nematol. Mediterr. 27, 33–38.

Kirkpatrick J D, Mai W F, Parker K G and Fisher E G 1964 Effect of phosphorus and potassium nutrition of sour cherry on the soil population of fire plant parasitic nematodes. Phytopathology 54, 706–712.

Melakeberhan H and Webster J M 1993 The phenology of plant–nematode interaction and yield loss. *In* Nematode Interactions. Ed. M W Khan. p. 26–41. Chapman and Hall, London, U.K.

Rao A B 1990 Role of microorganisms in plant nutrition under acid conditions. *In* Biofertilizers. Ed. L L Vyas. p. 67–84. Scientific Publ, Jaudpur.

Southey J F 1986 Plant Nematology. Her Majesty's Stationary office, London, U.K, p 440.

Taha Y A 1993 Nematode intractions with root-nodule bacteria. *In* Nematode Interactions. Ed. M W Khan. p. 170–202. Chapman and Hall, London, U.K.

Toohey J I, Netson C D and Krotkov G 1965 Isolation and identification of two phenazines from a strain of *Pseudomonas aureofacions*. Can. J. Bot. 43, 1055–1062.

Two strains isolated from tumours of *Prunus persica* are able to solubilize phosphate *in vitro*

J.L. Palomo[1], P. García-Benavides[1], P.F. Mateos[2], E. Martínez-Molina[2,4] & E. Velázquez[3]

[1]*Centro Regional de Diagnóstico de la Junta de Castilla y León, Salamanca, Spain.* [2]*Departamento de Microbiología y Genética, Centro Hispano Luso de Investigaciones Agrarias, Universidad de Salamanca, Salamanca, Spain.* [3]*Departamento de Microbiología y Genética, Edificio Departamental, Universidad de Salamanca, Salamanca, Spain.* [4]*Corresponding author**

Received 30 November 2002. Accepted in revised form 2 January 2003

Key words: phosphate solubilizing bacteria, plant tumour, *Agrobacterium*, *Rhizobium*

Abstract

Bacteria belonging to genus *Agrobacterium* are able to induce tumours in plants. *A. radiobacter*, the only non-pathogenic species of this genus, has been reported as phosphate solubilizer. In this work, we have tested the ability to solubilize phosphate of the type strains of *Agrobacterium* species and that of several isolates from tumours of different plants using bicalcium phosphate as P source. These strains have been already identified using LMW RNA profiles as *A. tumefaciens*, *A. radiobacter* and *A. rhizogenes*. The results obtained showed that the type strains of *A. rhizogenes* and *A. tumefaciens* solubilize phosphate. From the strains isolated from tumours two strains, 163C and 97/546, able to originate tumours in tomato plants were able to solubilize bicalcium phosphate in plate. These strains showed a 100% homology with that of the type strain of *A. rhizogenes*.

Introduction

The genus *Rhizobium* has been extensively studied from the point of view of solubilization and mobilization of phosphate (Antoun et al., 1998; Chabot et al., 1998; Halder et al., 1990). However the genus *Agrobacterium* belonging to the same family (Rhizobiaceae) has been few studied (Belimov et al., 1995, 1999).

A. tumefaciens, *A. rubi*, *A. vitis* and *A. larrymoorei* originate tumours in plants. *A. radiobacter* is a non-pathogenic species phylogenetically closed with *A. tumefaciens*. *A. rhizogenes* produces hairy roots and is more closely related with genus *Rhizobium* than with genus *Agrobacterium* (de Lajudie et al., 1998, Velázquez et al., 1998). This fact has lead to a premature reclassification of this genus into genus *Rhizobium* (Young et al., 2001), because more studies are necessary to clarify the taxonomy of family Rhizobiaceae. Although the type strain of *A. rhizogenes* originates hairy roots in plants, other strains from the same species can induce tumours in plants (Sawada et al., 1993).

At present all strains studied as phosphate solubilizers belong to the non-pathogenic species *A. radiobacter*. Nevertheless, since the ability to solubilize phosphate cannot be related with the symptom in plants, the aim of this work was to analyze the ability to solubilize phosphate of several strains isolated from plant tumours identified in a previous work (Velázquez et al., 2001a).

Material and methods

Strains used in this study

We included in this study several strains isolated from tumours that in some cases did not produce

* FAX No: +34-923-224876.
E-mail: emm@usal.es

E. Velázquez and C. Rodríguez-Barrueco (eds.),
First International Meeting on Microbial Phosphate Solubilization, 347–349

symptoms in plant. Also, we have included the type strains of *Rhizobium* species closely related to *A. rhizogenes* (Table 1).

16S rRNA gene sequencing and analysis

DNA extraction was carried out as previously described (Rivas et al., 2001). The amplification of 16S rDNA and its sequencing was performed according to the method already described (Rivas et al., 2002). The sequence obtained was compared with those from the GenBank using the FASTA program (Pearson and Lipman, 1988).

Evaluation of phosphate solubilization of strains used in this study

The ability to solubilize bicalcium phosphate was tested in petri dishes containing YED (yeast extract 0.5%; glucose 1% and agar 2%) supplemented with a 0.2% of bicalcium phosphate (YED-P). A suspension of each strain was inoculated in this medium and the plates were incubated for 7 days until the solubilization zone surrounding the colonies was observed.

Table 1. Phosphate solubilization by the strains of this study

Strains	Host	Phosphate solubilization
A. radiobacter AT1	*Phaseolus vulgaris*	–
A. radiobacter AT2	*Phaseolus vulgaris*	–
A. radiobacter AT14	*Beta vulgaris*	–
A. tumefaciens AT8	*Beta vulgaris*	–
A. tumefaciens AT9	*Beta vulgaris*	–
A. tumefaciens AT10	*Beta vulgaris*	–
A. tumefaciens AT11	*Beta vulgaris*	–
A. rhizogenes 163C	*Prunus persica*	+
A. rhizogenes 97/546	*Prunus persica*	+
A. tumefaciens ATCC 23308^T	*Malus* sp.	+
A. radiobacter ATCC 19358^T		–
A. rhizogenes ATCC 11325^T		+
A. rubi ATCC 13335^T	*Rubus* sp.	–
A. vitis CECT 4799^T	*Vitis vinifera*	–
R. leguminosarum bv. viceae ATCC 10004^T	*Pisum sativum*	–
R. leguminosarum bv. trifolii ATCC 14480	*Trifolium repens*	w
R. etli CFN42^T	*Phaseolus vulgaris*	–
R. gallicum R602^T	*Phaseolus vulgaris*	–
R. mongolense USDA 1844^T	*Medicago ruthenica*	–
R. tropici IIB CIAT899^T	*Phaseolus vulgaris*	–

Results

16S rRNA gene sequence analysis

All strains from this study, as mentioned above, have been identified in a previous study (Velázquez et al., 2001a). The strains 163C and 97/546 display the same LMW RNA profile than type strain of former species *A. rhizogenes* and therefore must be included in the same species (Velázquez et al., 2001b). Nevertheless, the strains 163C and 97/546 are tumour inducing whereas the type strain of the species induces hairy roots. Although these symptoms are codified in plasmids and the identification at species level is not dependent on the extrachromosomal content, to confirm the identification, the 16S rDNA complete sequence of strain 163C was obtained (Accession number AY206687). The results obtained showed that it has 100% homology with that of type strain of former species *A. rhizogenes* and that therefore the identification of strains 163C and 97/546 using LMW RNA was correct. As mentioned above, the species *A. rhizogenes* is clearly related with the species of genus *Rhizobium*, but a complete study of polyphasic taxonomy is needed to reclassify this species into genus *Rhizobium*.

Evaluation of ability to solubilize phosphate

The results of the solubilization of phosphate in the strains of this study are recorded in Table 1. In disagreement with the literature data (Belimov et al., 1995), the strains belong to *A. radiobacter*, including the type strain, were not able to solubilize phosphate. The type strains of other tumourigenic strains such as *A. rubi* and *A. vitis* were not able to solubilize phosphate.

On the contrary, the type strain of *A. tumefaciens* was able to solubilize phosphate. However, the other strains from this study belonging to this species do not show this ability.

The type strain of *A. rhizogenes* was able to solubilize phosphate. The strains 163C and 97/546 that were previously identified as *A. rhizogenes* also showed phosphate solubilization in YED-P plates.

As the sequence of strain 163C was identical to those of type strain of *A. rhizogenes* and this species is phylogenetically related with species of *Rhizobium leguminosarum* group, we have analyzed the ability to solubilize bicalcium phosphate of type strains of these species. According to the results, a weak ability was found only in *Rhizobium leguminosarum* bv trifolii ATCC14480. These results are in agreement with those previously reported using tricalcium phosphate as P source (Peix et al., 2001), however when other compounds were used, such as phytate, we have observed a high ability to solubilize these substrats in several *Rhizobium* species (Yanni et al., 2001).

In conclusion, the strains belong to *A. rhizogenes* used in this study, including the type strain, were able to solubilize phosphate. From these strains, two are tumour inducing and the type strain originates hairy roots; therefore, the ability to solubilize phosphate is not related to the pathogenicity of these bacteria.

Acknowledgements

This work was supported by the Junta de Castilla y León (Spain).

References

Antoun H, Beauchamp C J, Goussard N, Chabot R and Lalonde R 1998 Potential of *Rhizobium* and *Bradyrhizobium* species as growth promoting rhizobacteria on non-legumes: effect on radishes (*Raphanus sativus* L.). Plant Soil 204, 57–67.

Belimov A A, Kunakova A M, Vasilyeva N D, Gruzdeva E V, Vorobriev N I, Kojemiakov A P, Khamova O F, Postavskaya S M and Sokowa S A 1995 Relationship between survival rates of associative nitrogen-fixers on roots and yield response of plants to inoculation. FEMS Microbiol. Ecol. 17, 187–196.

Belimov A A, Serebrennikova N V and Stepanok V V 1999 Interaction of associative bacteria and an endomycorrhizal fungus with barley upon dual infection. Microbiology, UK 68, 104–108.

Chabot R, Beauchamp C J, Kloepper J W and Antoun H 1998 Effect of phosphorous on root colonization and growth promotion of maize by bioluminiscent mutants of phosphate-solubilizing *Rhizobium leguminosarum* biovar *phaseoli*. Soil Biol. Biochem. 30, 1615–1618.

De Lajudie P, Laurent-Fullele E, Willems A, Torck U, Coopman R, Collins C D, Kersters K, Dreyfus B and Gillis M 1998 *Allorhizobium undicola* gen. nov., sp. nov., nitrogen-fixing bacteria that efficiently nodulate *Neptunia natans* in Senegal. Int. J. Syst. Bacteriol. 48, 1277–1290.

Halder A K, Mishra A K and Chakrabartty P K 1990 Solubilization of phosphatic compounds by *Rhizobium*. Indian J. Microbiol. 30, 311–314.

Peix A, Rivas-Boyero AA, Mateos PF, Rodríguez-Barrueco C, Martínez-Molina E and Velázquez E 2001 Growth promotion of chickpea and barley by a phosphate solubilizing strain of *Mesorhizobium mediterraneum* under growth chamber conditions. Soil Biol. Biochem. 33, 103–110.

Pearson W R and Lipman D J 1988 Improved tools for biological sequence comparison. Proc. Natl. Acad. Sci. USA 85, 2444–2448.

Rivas R, Velázquez E, Valverde A, Mateos P F and Martínez-Molina E 2001 A two primers random amplified polymorphic DNA procedure to obtain polymerase chain reaction fingerprints of bacterial species. Electrophoresis 22, 1086–1089.

Rivas R, Velázquez E, Willems A, Vizcaíno N, Subba-Rao N S, Mateos P F, Gillis M, Dazzo F B and Martínez-Molina E 2002 A new species of *Devosia* that forms a nitrogen-fixing root-nodule symbiosis with the aquatic legume *Neptunia natans* (L.f.) Druce. Appl. Environ. Microbiol. 68, 5217–5222.

Sawada H, Ieki H, Oyaizu H and Matsumoto S 1993 Proposal for rejection of *Agrobacterium tumefaciens* and revised descriptions for the genus *Agrobacterium* and for *Agrobacterium radiobacter* and *Agrobacterium rhizogenes*. Int. J. Syst. Evol. Microbiol. 43, 694–702.

Velázquez E, Cruz-Sánchez J, Mateos P and Martínez-Molina E 1998 Analysis of stable low-molecular weight RNA profiles of members of the family Rhizobiaceae. Appl. Environ. Microbiol. 64, 1555–1559.

Velázquez E, Palomo J L, Lastra B, Mateos P F, García Benavides P and Martínez-Molina E 2001a Rapid identification of *Agrobacterium* species by staircase electrophoresis of low molecular weight RNA profiles. Eur. J. Plant Pathol. 107, 931–938.

Velázquez E, Trujillo M E, Peix A, Palomo J L, García-Benavides P, Mateos P F, Ventosa A and Martínez-Molina E 2001b Stable low molecular weight RNA analyzed by staircase electrophoresis, a molecular signature for both prokaryotic and eukaryotic microorganisms. Syst. Appl. Microbiol. 24, 490–499.

Yanni Y, Rizk R, Abd-El Fattah F, Squartini A, Corich V, de Bruijn F, Rademaker J, Maya-Flores J, Ostrom P, Vega-Hernández M, Hollingsworth R, Martínez-Molina E, Mateos P, Velázquez E, Wopereis J, Triplett E, Umali-García U, Rolfe B, Ladha J K, Hill J and Dazzo F B 2001 The beneficial plant growth-promoting association of *Rhizobium leguminosarum* bv. *trifolii* with rice roots. Aust. J. Plant Physiol. 28, 845–870.

Young J M, Kuykendall L D, Martínez-Romero E, Kerr A and Sawada H 2001 A revision of *Rhizobium* Frank 1889, with an emended description of the genus, and the inclusion of all species of *Agrobacterium* Conn 1942 and *Allorhizobium undicola* de Lajudie et al. 1998 as new combinations: *Rhizobium radiobacter, R. rhizogenes, R. rubi, R. undicola* and *R.vitis*. Int. J. Syst. Evol. Microbiol. 51, 89–103.

Inorganic phosphate solubilization by two insect pathogenic *Bacillus* sp

S. Seshadri[1], S. Ignacimuthu[2], M. Vadivelu[3] & C. Lakshminarasimhan[4]
[1]*Shri AMM Murugappa Chettiar Research Center (MCRC), Taramani, Chennai, 600 113, India.* [2]*Entomology Research Institute, Loyola College, Chennai, 600 034, India.* [3]*Main Biocontrol Research Laboratory, Tamil Nadu Co-operative Sugar Federation, Chengalpattu, India.* [4]*P. R. Group of Institutions, Thanjavur, 613 007, India.* [1]*Corresponding author**

Received 31 December 2002. Accepted in revised form 2 January 2003

Key words: *Bacillus*, insect pathogens, phosphate solubilization

Abstract

Two strains (*Bacillus thuringiensis* (*kurstaki*) and *Bacillus sphaericus*) having ability to solubilize inorganic phosphates were screened from the local isolates tested to control lepidopteran pests. In the phosphate solubilization and pH reduction properties investigated, both were found to solubilize phosphate in both Sperber's medium and Pikovskaya's medium. In plate assays *B. thuringiensis* (*kurstaki*) showed very poor solubilization in Sperber's medium and *B. sphaericus* did not show any sign of solubilization in both the media used. However, in broth cultures, maximum phosphate solubilization was seen in Pikovskaya's medium than Sperber's medium. P-solubilization was found to be associated with pH reduction.

Introduction

Although phosphorus is one of the most important macronutrients for biological growth and development of all living organisms, many processes of the P cycle in nature are still obscure (Illmer and Schinner, 1992). Many soil fungi and bacteria are capable of solubilizing insoluble inorganic and organic phosphates (Katznelson et al., 1962; Goldstein and Liu, 1987; Vora and Shelat, 1996). *Bacillus thuringiensis* and *Bacillus sphaericus* have received considerable attention as microbial control agents. Around 60% of the isolates from five continents showed to be toxic for lepidopetrans and dipterans. They infect insects through mouth and digestive tract. They produce exo and endotoxins that play a role in the invasion of bacteria through the digestive tract (Tanada and Kaya, 1993). *Bacillus* sp. is aerobic or facultative aerobe and usually produces catalase (Claus and Berkeley, 1986). Within the digestive tract, they produce enzymes (e.g., lecithin ase, proteinase and chitinase) that act on the midgut cell and enable the bacteria to enter the hemocoel (Smirnoff and Valero, 1977). *Bacillus* sp. also produces a thermolabile enzyme phospholipase C or lecithinase that is toxic to insects. The phospholipase C is believed to act by penetrating the cell membranes and attacking the cell substrate (Heimpel, 1967). *In vitro*, the enzyme inhibits cell growth and causes cellular deformation and internal deterioration (Ikezawa et al., 1989). The present investigation has been undertaken to evaluate the performance of insect pathogenic *Bacillus sphaericus* and *Bacillus thuringiensis* (*kurstaki*) (Sally Gloriana et al., 2001) on phosphate sources.

Materials and methods

Plate assays

Individual strains of *Bacillus sphaericus* (MTCC 511) and *B. thuringiensis* (*kurstaki*) (MTCC 868)

* E-mail: tsseshadri@rediffmail.com

E. Velázquez and C. Rodríguez-Barrueco (eds.),
First International Meeting on Microbial Phosphate Solubilization, 351–355

were grown on the Sperber's medium (1957) and modified Pikovskaya's medium (1948). Media prepared, autoclaved and poured in sterile petri-plates were used for plate assays. The strains were transferred to the plates from actively growing culture using a stab. The plates were incubated at room temperature (30 ± 2 °C) for 1 week to study the clearing zone brought out by the bacteria. The results are recorded as solubilization efficiency (*E*) (Nguyen et al., 1992).

P-solubilization studies

A solution of 100 ml of Pikovskaya's broth was distributed in a conical flask (250 ml) and sterilized at 15 lb for 20 min. The flasks were inoculated with 100 μl of overnight grown culture of *B. sphaericus* and *B. thuringiensis* (*Kurstaki*). Uninoculated medium in flasks was treated as control. All treatments and control were incubated at room temperature (30 ± 2 °C) for 20 days. Estimation of phosphate was carried out by paramolybdate blue method of Olsen and Sommers (1982). For analysis the cultures were harvested once in 2 days, centrifuged at 10,000 rpm for 15 min and then subjected for P estimation. pH of the media was noted simultaneously. All the experiments were conducted in triplicate.

Results and discussion

B. sphaericus did not produce any clearing zone in both the media used. However, *B. thuringiensis* (*kurstaki*) showed a very poor clearing zone (Solubilization efficiency = 188.89 ± 19.25) after 96 h in the Sperber's medium and no clearing zone was found in Pikovskaya's medium.

In contrast, both the strains performed moderately in broth culture studies. Analysis of P in the medium showed that both the strains released Phosphate in Sperber's medium as well as Pikovskaya's medium. Results are shown in Figures 1, 2, 3 and 4. Both the strains differed significantly in solubilization. *B. thuringiensis* (*Kurstaki*) was efficient in solubilization from second day itself with better solubilization efficiency till the end of experiment. In *B. sphaericus* the action initially was very slow and proceeded gradually. Soluble phosphate level was between 1072.63–1382.77 and 315–1398.79 μg/mL in *B. thuringiensis* and *B. sphaericus* inoculated Pikovskaya's medium, respectively. In Sperber's medium the phosphate level was 688.63–817.09 and 304.52–760.45 μg/mL, respectively. The P concentration in solution showed few ups and down over incubation. This is in conformity with the results reported earlier (Morris and Allen, 1994; Seshadri et al., 2000) where the microorganisms show no clearing zone in plates but produced a good oxalate and phosphate metabolism in broth. Both the media were found to be preferred by both the bacteria for P-solubilization, indicating the part played by substrate in triggering the action (Illmer and Schinner, 1992; Seshadri, 1995).

In broth, pH drop was noticed in both the media used. The pH change was gradual where it slowly went down upon incubation. *B. thuringiensis* (*kurstaki*) reduced the media pH more than *B. sphaericus*. Earlier reports also indicate the reduction in pH during P-solubilization by microbes (Agnihothri, 1971; Illmer and Schinner, 1992; Cunningham and Kuiack, 1992; Seshadri, 1995). Although the pH drop was noticed from 6.5 to 4.5, no positive correlation could be established between the P-solubilization and pH.

P-increase in solution by both microbes in broth was attributed to (1) acidification or (2) proton extrusion and/or (3) cation dissociation and exchange reaction processes described by various workers (Roos and Luckener, 1984; Thomas et al. 1985; Leyval and Berthelin, 1989; Salih et al., 1989; Seshadri et al., 2000). *Bacillus sphaericus* does not use sugars as a carbon source for growth (Russel et al., 1989). In the present study, in the presence of glucose as carbon source, the bacteria could be able to increase the P concentration in solution with pH reduction. Eventhough the possibility of induction of multicomponent gene system (regulon) that includes bacterial alkaline phosphatase, to hydrolyze exogenous phosphates to soluble phosphates under starvation conditions as reported earlier (Russel et al., 1989) is there, it requires further studies for confirmation.

Reports are also available on the P-solubilization potential and pH reduction by *Bacillus* sp. in broth (Robinson and Syers, 1990; Yadav and Singh, 1991; Gaind and Gaur, 1999). Enzymes viz. phospholipase, phytase, nuclease and lecithinase, produced by different bacteria and fungi were also reported to act on the phosphate sources (Nahas, 1996). Studies undertaken earlier showed that the maximum solubilization of

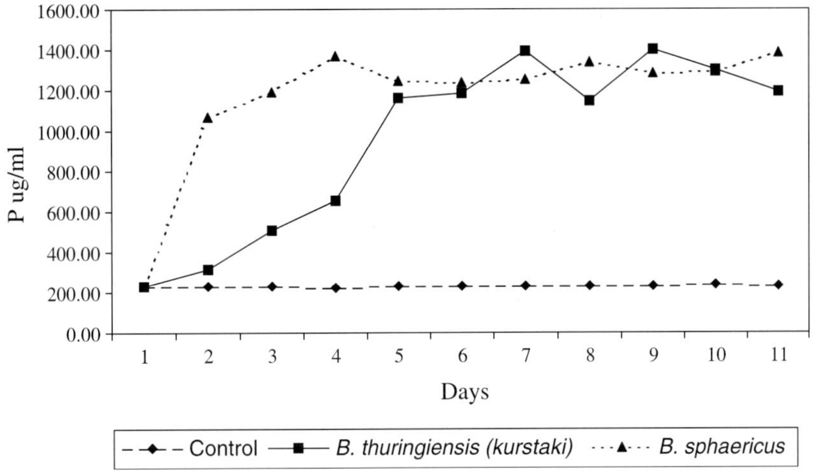

Figure 1. Tricalcium phosphate solubilization by *B. sphaericus* and *B. thuringiensis* (kurstaki) in Pikovskaya's medium.

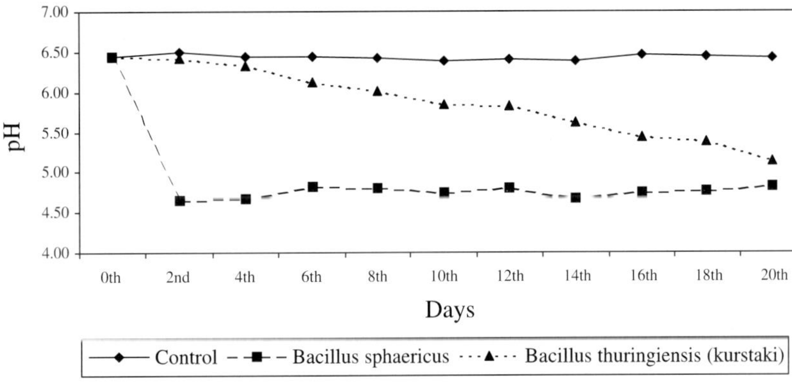

Figure 2. pH changes during Tricalcium phosphate solubilization by *B. sphaericus* and *B. thuringiensis (kurstaki)* in Pikovskaya's medium.

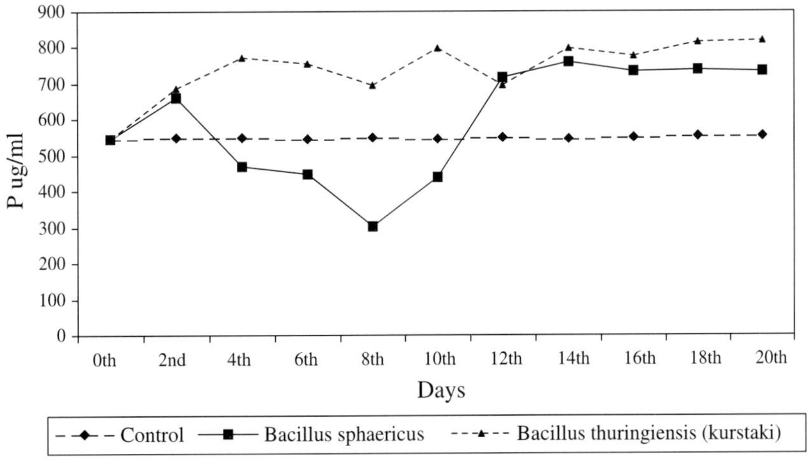

Figure 3. Phosphate solubilization by *B. sphaericus* and *B. thuringiensis (kurstaki)* in Sperber's medium.

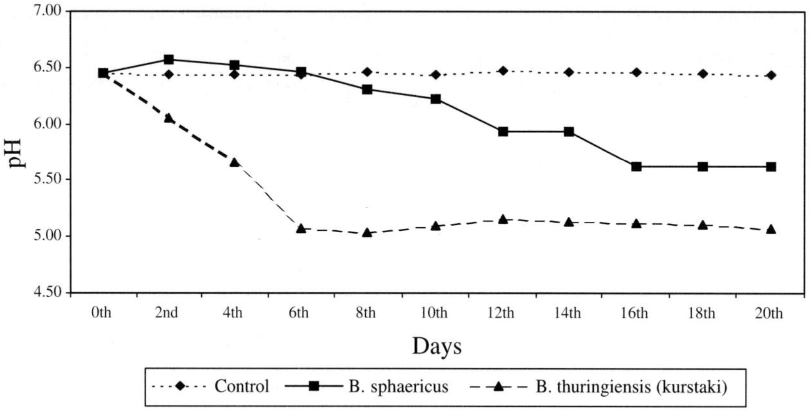

Figure 4. pH changes during Phosphate solubilization by *B. sphaericus* and *B. thuringiensis* (*kurstaki*) in Sperber's medium.

phosphate occurred within 2 weeks period under controlled conditions (Goswami and Sen, 1962; Wani et al., 1979; Gaur, 1990). In the present study there was a gradual solubilization process that lasted up to 20 days, indicating the slow action of the microbes used.

From the above results it is established that both *B. thuringiensis* (*kurstaki*) and *B. sphaericus* have P-solubilizing capacity in broth. *Bacillus* sp. is reported colonize the phyllosphere of plants (Sudarsan et al., 1994). A further study would enrich our knowledge on the role played by these bacteria in phyllosphere and add additional knowledge in understanding the insect-plant-microbe interrelationship.

References

Agnihothri V P 1971 Solubilization of insoluble phosphate by some soil fungi isolated from nursery seedbeds. Can. J. Microbiol. 16, 877–880.

Claus D, Berkeley R C 1986 *In* Bergey's Manual of systematic Bacteriology. Ed. P H A Sneath. Vol. 2. pp. 1105–1141. Williams and Wilkins, Baltimore, USA.

Cunningham J E and Kuiack C 1992 Production of citric and oxalic acids and solubilization of calcium phosphate by Penicillium bilaii. Appl. Environ. Microbiol. 58, 1451–1458.

Gaind S and Gaur A C 1999 Microbial phosphate solubilization as influenced by sodium chloride. Indian J. Expt. Biol. 47, 209–210.

Gaur A C 1990 Phosphate solubilizing microorganisms as Biofertilizers. Omega Scientific Publishers, 240 pp.

Goldstein A H and Liu S T 1987 Molecular cloning and regulation of a mineral phosphate solubilizing gene from *Erwinia herbicola*. Bio/Technology 5, 72–74.

Goswami K P and Sen A 1962 Solubilization of calcium phosphate by three strains of phosphobacterin. Indian J. Agri. Sci. 32, 96–101.

Heimpel A M 1967 A critical review of *Bacillus thuringiensis* var. *thuringiensis* Berliner and other crystalliferous bacteria. Ann. Rev. Entomol. 12, 287–322.

Ikezawa H A, Hashimoto R, Taguchi T, Nakabayashi T and Himeno M 1989 Effect of Bacillus thuringiensis β endotoxin on larval maturation adult logevity, fecundity, and egg viability in several species of lepidoptera. Toxicon 27, 637–645.

Illmer P and Schinner F 1992 Solubilization of inorganic phosphates by microorganisms isolated from forest soils. Soil Biol. Biochem. 24, 389–395.

Katznelson H, Peterson E A and Rouatt J W 1962 Phosphate dissolving microorganisms on seeds and in the root zone of plants. Can. J. Bot. 40, 1181–1185.

Leyval C and Berthelin J 1989 Interactions between *Laccaria laccata*, *Agrobacterium radiobacter* and beech roots: influence on P, K, Mg and Fe mobilization from minerals and plant growth. Plant Soil 117, 103–110.

Morris S J and Allen M F 1994 Oxalate metabilizing microorganisms in sagebrush steppe soils. Biol. Fertil. Soils 18, 255–259.

Nahas E 1996 Factors determining rock phosphate solubilization by microorganisms isolated from soil. World-J. Microbiol. Biotechnol. 6, 567–572.

Nguyen C, Ya. W, Le Tacon F and Lapeyrie F 1992 Genetic variability of phosphate solubilizing activity by monocaryotic and dicaryotic mycelia of the ectomycorrhizal fungus Laccaria bicolor (Maire)P.D.Orton. Plant Soil 143, 193–197.

Olsen S R and Sommers L E 1982 Phosphorus. *In* Methods of soil analysis. Part 2: Chemical and microbiological properties. Ed. A L Page. pp. 403–430. American Society of Agronomy, Soil Science Society of America, Madison, WI, US.

Pikovskaya R E 1948 Mobilization of phosphorus in soil in connection with vital activity of some microbial species. Mikrobiologiya 17, 362–370.

Robinson J S and Syers J K 1990 A critical evaluation of the factors influencing the dissolution of Gafs phosphate rock. J. Soi. Sci. 41, 597–605.

Roos W and Luckener M 1984 Relationship between proton extrusion and fluxes of ammonium ions and organic acids in Penicillium cyclopium. J. Gen. Microbiol. 130, 1007–1010.

Russel B L, Jelley S A and Yousten A A 1989 Carbonhydrate metabolism in the mosquito pathogen *Bacillus sphaericus* 2362. Appl. Environ. Microbiol. 55, 294–297.

Salih H M, Yahya A I, Abdul-Rahem A M and Munam B H 1989 Availability of phosphorus in a calcareous soil treated with rock phosphate or superphosphate as affected by phosphate dissolving fungi. Plant Soil 120, 181–185.

Sally-Gloriana A, Raja N, Seshadri S, Janarthanan S and Ignacimuthu S 2001 Pathogenicity of entomopathogens, *Bacillus thuringiensis* subsp. *Kurstaki* and *Beauveria bassiana*, to the larvae of *Spodoptera litura* (F.) and *Pericallia ricini* (F.). Biol. Agric. Hortic. 18, 235–242.

Seshadri S 1995 Phosphate solubilizing fungi from Thanjavur Delta, South India. Ph.D. Thesis. Bharathidasan University, Tiruchirappalli, India.

Seshadri S, Muthukumarasamy R, Lakshminarasimhan C and Ignacimuthu S 2000 Curr.Sci. 79, 565–567.

Smirnoff W A and Valero J 1977 Susceptibility of population of Indian meal moth and almond moth *Bacillus thuringiensis* isolates. J. Invertebr. Pathol. 30, 265–266.

Sperber J I 1957 Solution of mineral phosphates by soil bacteia. Nature (London) 180, 994–995.

Sudarsan N, Suma N R, Vennison J S and Sekar V 1994 Survival of a strain of *Bacillus megaterium* carrying a lepidopteran-specific gene of *Bacillus thuringiensis* in the phyllospheres of various economically important plants. Plant Soil 167, 321–324.

Tanada Y and Kaya H K 1993 In Insect Pathology. Academic Press, New York, 666 pp.

Thomas G V, Shantaram M V and Saraswathy N 1985 Occurrence and activity of phosphate solubilizing fungi from coconut plantation soils. Plant Soil 87, 307–311.

Vora M S and Shelat H H 1996 Isolation and screening of microorganisms for dissolving tricalcium phosphate. Madras Agric. J. 83, 353–354.

Wani P V, More B B and Patil P L 1979 Physiological studies on the activity of phosphorus solubilizing microorganisms. Indian J. Microbiol. 19, 23–25.

Yadav K and Singh T 1991 Phosphorus solubilization by Microbial Isolate from a Calcifluvent. Indian Soc. Soil Sci. 39, 89–93.

Effect of certain fungal and bacterial phosphate solubilizing microorganisms on the fusarial wilt of tomato

M.R. Khan[1,2], S.M. Khan[1] & F.A. Mohiddin[1]
[1]*Department of Plant Protection, Aligarh Muslim University, Aligarh, 202 002, India.* [2]*Corresponding author**

Received 22 November 2002. Accepted in revised form 2 January 2003

Key words: *Aspergillus* spp., *Pseudomonas* spp., *Fusarium oxysporum* f. sp. *lycopersici*, tomato yield, rhizosphere population

Abstract

A study was conducted under field condition in microplots to explore the feasibility of using certain fungal and bacterial phosphate solubilizing microoganisms (PSMs) through soil application and root-dip treatment to manage wilt of tomato caused by *Fusarium oxysporum* f. sp. *lycopersici*. *Aspergillus awamori* and *A. niger* were cultured on potato dextrose broth, whereas *Pseudomonas fluorescens* and *P. striata* were cultured on nutrient broth. Inoculum of the wilt fungus was prepared on sorghum seeds and was added to soil at 2 g/kg soil. For soil application of PSMOs pure culture on the broth was diluted in ample water and added to soil at 2 ml/kg soil. For root-dip treatment, roots of tomato seedlings, *Lycopersicon esculentum* cv. Pusa Ruby were dipped in the pure broth culture of the microorganisms for 30 min. The seedlings were planted in the microplots (3 × 2 m) in the presence or absence of wilt fungus. The pathogenic fungus caused characteristic symptoms of wilt and significantly decreased the shoot dry weight and the yield by 36.6% and 34.7%, respectively. Soil application of *A. awamori* or *P. fluorescens* improved the yield of tomato (without wilt fungus) by 22% and 9%. Root-dip treatment with the former also improved the yield significantly. Application of PSMs, in general, suppressed the wilt fungus to a varied extent and decreased the wilt severity being greatest with *A. awamori* (29–37%). Fungicide treatment (carbendazim) decreased the severity of wilt by 8–19%. Greatest increase in the yield of infected plants was recorded due to soil or seed treatment with *A. awamori* (15–20%), followed by *P. fluorescens* (11–16%) and carbendazim (12–19%) compared to inoculated control. Soil population of wilt fungus (CFUs/g soil) in untreated plots was increased during first 2 months. Whereas in bioagent or fungicide treated plots, the population was gradually and significantly decreased during the 4 months of plant growth period. Greatest decrease in the population was recorded with *A. awamori*, followed by *A. niger*, *P. fluorescens* and carbendazim. On overall basis, soil application was found 5–10% more effective than root-dip treatment.

Introduction

Tomato is an important vegetable crop and is extensively grown throughout the world. The crop is frequently attacked by the wilt caused by *Fusarium oxysporum* f. sp. *lycopersici* in both temperate and tropical climates (Woltz and Jones, 1981). The fungus causes tremendous yield loss to tomato in India depending on the cultivar and the prevailing environmental conditions (Khan and Akram, 1997). Several measures are taken by growers to control the disease. Use of chemical fungicides is avoided for their various adverse effects. Biological control offers a potential alternative to chemical fungicides (Parker et al., 1985). Mycoparasites and other antagonists have been extensively explored (Papavizas, 1985), but none of them provide control of plant diseases

* E-mail: mrkhan777in@yahoo.co.in

E. Velázquez and C. Rodríguez-Barrueco (eds.),
First International Meeting on Microbial Phosphate Solubilization, 357–361

satisfactory in all respects. In fact, majority of the researches on biocontrol have engaged mycoparasites or a limited number of other antagonists. Least efforts have been made to test potential of rhizobacteria, especially phosphate solubilizing microorganisms in disease management. These organisms have a specific significance in the nutritional economy of soil. Proliferation of PSMs in the root zone results to an increased availability of utilizable form of phosphorus to plants, subsequently the plant growth and yield increase considerably (Sakthivel et al., 1986). In addition, metabolites of certain PSMs may possess fungicidal property. *Pseudomonas fluorescens* produces phenazin (Toohey et al., 1965), which may suppress soil-borne fungal pathogens such as *F. oxysporum* (Leeman et al., 1995). Certain species of *Aspergillus* such as *A. awamori*, *A. niger* etc. are active solubilizers of phosphate. They synthesize organic acids like formic, acetic, propionic, lactic, glycolic, fumaric, succinic acids etc. (Arora and Gaur, 1979) and other metabolites or toxins, which may suppress plant pathogens (Dalla, 1986; Nair and Burke, 1988). A field trial was conducted to examine the effect of soil application or root-dip treatment with *A. awamori*, *A. niger*, *P. fuorescens* and *P. striata* on the fusarial wilt of tomato caused by *F. oxysporum* f. sp. *lycopersici*. Effects were also determined on the biomass production and yield of tomato.

Materials and methods

Pure culture of the wilt fungus, *F. oxysporum* Schlecht. f. sp. *lycopersici* (Sacc.) Snyder and Hansan was obtained from IARI, New Delhi, India. The fungus was mass cultured on sorghum seeds. For field inoculation, fungus colonized seeds were blended in distilled water (1:50) in an electric blender. Hundred ml fungus suspension (1.7×10^3 CFU/ml) containing 2 g colonized seeds/plant was added to soil (15 cm wide and deep) as spot application, where seedlings were to be planted a day later. Strains of phosphate solubilizers viz., *A. awamori* and *A. niger* van Tighem were procured from the Division of Microbiology, IARI, New Delhi and were cultured in potato dextrose broth. The mycelial mat was removed from the broth, weighed and blended in a known volume of distilled water (1:1 w/v) to get a homogenous mixture. *Pseudomonas fluorescens* (Threvesan) Migula and *P. striata* were obtained from the Institute of Microbial Technology, Chandigarh, India. The bacteria were cultured on nutrient broth. For soil application, 1 ml fungus suspension or bacteria broth was diluted with 99 ml distilled water, and was added to soil (15 cm deep and wide) at 30 locations in a microplot (10 locations/row, 3 rows/microplot) at the time of seedling planting. For root-dip treatment, seedling roots were dipped for 15 min in 100 ml suspension containing 50 ml pure fungal or bacterial suspension. The treated seedlings were planted without any rinse. To compare effectiveness of the PSMs, treatments with an efficacious fungicide, carbendazim were maintained for soil (300 g of active ingredient/ha) and root application (0.1% solution for 15 min).

A field of 44×22 m dimension was properly prepared and was divided into two equal plots (22×11 m). In each plot 36 microplots each of (3×2 m dimension) were prepared. Twelve treatments each for soil and root-dip application were maintained in the two plots separately. The treatments are indicated in Table 2. Each treatment was represented by three microplots, and arranged in a complete randomized block design. Seedlings of tomato cv. Pusa Ruby were planted at 30 sites in each microplot, where wilt fungus had already been inoculated or not inoculated. Three rows, each with 10 plants were maintained in a microplot. Seedlings were watered immediately after planting. During the growth period, mature fruits were collected periodically. Five plants from each row were harvested randomly at 4 months age to determine dry shoot weight and yield of tomato. Wilt severity (%) was recorded on two and a half month old plants. Rhizosphere population of wilt fungus was estimated monthly during November to February using dilution plate method.

The experiment was conducted during two consecutive years in different fields under identical agro-climatic conditions. Mean of 45 plants harvested randomly from three microplots (15 plants/microplot) was considered as one replicate (2–5% variance). Hence, there were two replicates, as the study was replicated over time (SE ± 1.5–2.9). The data on shoot weight and yield were subjected to a two-factor analysis of variance (ANOVA). The data on wilting and rhizosphere population were analysed by a single factor ANOVA. Least significance difference (LSD) was calculated for each parameter at $P = 0.05$ to isolate significant effects of different treatments (Dospekhov, 1984).

Results and discussion

F. oxysporum f. sp. *lycorpersici* caused characteristic wilt symptoms that appeared on around one and a half month old plants. The wilting gradually became intensified. Lower leaves developed the wilting first, which extend to upper leaves. Plants, however, partially recovered the wilting after irrigation but at later stage the wilting became permanent. Such leaves, twigs or the whole plant turned brown and later dried and died. Similar symptoms due to infection by *F. oxysporum* f. sp. *lycorpersici* have been recorded under local conditions (Khan and Akram, 2000). Mild wilting developed on the plants applied with PSMs, especially *A. awamori*; its soil application or root-dip treatment decreased the wilting by 37% and 29.5%, respectively compared to the control (Table 1). Certain organic acids produced by *A. awamori* were apparently responsible for suppression of wilt severity (Arora and Gaur, 1979). Corresponding values for carbendazim were 29.5% and 8.7%, respectively.

Soil application of *A. awamori* or *P. fluorescens* increased the dry weight of shoot by 19.0% and 9.5%, respectively, compared to un-inoculated control; the root-dip treatments resulted to 15.8% and 8.5% increase. Fruit formation was increased by 15.6% and 12.5% due to soil or root-dip application of *P. fluorescens*, respectively. Total weight of fruits/plant was increased by 21.7% and 14.9% due to applications of *A. awamori* compared to the control. Soil application of *P. fluorescens* also resulted to significant increase in the yield (Table 1). Phosphorus is a major plant nutrient and essentially required for plant growth and vigour. Proliferation and activity of *A. awamori* and *P. fluorescens* in the root zone of tomato had made greater availability of utilizable form of phosphorus to roots which reflected as improved growth and yield.

Inoculation with the wilt fungus, *F. oxysporum* f. sp. *lycopersici* caused significant decline in shoot dry weight (36.6%), number of fruits (18.6%), mean fruit weight (19.8%) and weight of fruits/plant (34.6%) of tomato compared to the control (Table 2). Application of *A. awamori* checked the suppressive effect of the pathogen and induced greatest increase in the biomass and yield variables of the infected plants compared to the inoculated control (Table 2). Soil or root-dip treatment with *A. awamori* significantly increased the shoot dry weight (25.3% and 17.6%), number (14.4% and 10.6%) and weight of fruits/

Table 1. Effects of soil application or root-dip treatment with certain phosphate solubilizing microorganisms on the wilt severity and rhizosphere population of *Fusariumoxysporum* f. sp. *lycopersici* on tomato cv. Pusa Ruby

Treatments	Wilt Severity %*	Rhizosphere population × 10^6 CFUs/g soil					
		November	December	January	February	LSD ($P = 0.05$)	F-value (df = 3)
Control	18.3/18.3*	2.1/2.1	2.7/2.6	3.1/3.2	3.3/3.3	0.59/0.59	$10.7^b/11.9^b$
Pseudomonas fluorescens	$16.6^a/16.9$	2.1/2.1	$1.9^a/2.1^a$	$1.7^a/2.0^a$	$1.7^a/1.8^a$	0.39/0.42	NS/NS
P. striata	17.6/17.8	2.1/2.1	$2.0^a/2.6$	$2.0^a/3.0$	$1.9^a/3.2$	0.23/0.42	NS/NS
Aspergillus awamori	$11.5^a/12.9^a$	2.1/2.0	$1.6^a/2.0^a$	$1.4^a/1.8^a$	$1.3^a/1.7^a$	0.34/0.41	14.8^b/NS
A. niger	$15.7^a/16.9^a$	2.1/2.0	$1.8^a/2.0^a$	$1.7^a/1.9^a$	$1.6^a/1.8^a$	0.38/0.41	NS/NS
Carbendazim	$15.0^a/16.7$	2.0/2.1	$0.6^a/2.1^a$	$1.6^a/1.9^a$	$1.8^a/2.0^a$	0.32/0.42	51.6^b/NS
LSD ($P = 0.05$)	1.9/2.1		0.27/0.32	0.57/0.32	0.29/0.28		
F-value (df = 5)	NS/NS		44.0^b/NS	$9.1^b/19.8^b$	$42.1^b/34.8^b$		

Each value is mean of two replicates
[a]Significantly different from the control at $P = 0.05$.
[b]Significant at $P = 0.05$.
NS: Not significant at $P = 0.05$.
*Soil application/root-dip treatment.

Table 2. Effect of certain fungal and bacterial phosphae solubilizing microorganisms on dry shoot weight and yield of tomato plants inoculated with *Fusarium oxysporium* f. sp. *lycopersici* or not inoculated

Phosphate solubilizing microorganisms	Wilt fungus (g/kg)	Dry shoot weight (g)		No. of fruits per plant		Mean fruit weight (g)		Weight of fruits per plant	
		Soil	Root-dip	Soil	Root-dip	Soil	Root-dip	Soil	Root-dip
Control (uninoculated)	00	67.3	67.5	74.3	74.3	28.3	28.3	2102.7	2099.9
Pseudomonas fluorescens	00	73.7[a]	73.0[a]	78.5	77.0	29.1	28.7	2284.4[a]	2209.9
P. striata	00	68.3	68.9	75.9	75.2	28.6	28.4	2170.7	2135.7
Aspergillus awamori	00	80.1[a]	78.6[a]	85.9[a]	83.5	29.8	28.9	2559.8[a]	2413.2[a]
A. niger	00	70.7	69.4	77.4	75.1	29.0	29.0	2244.6	2177.9
Carbendazim	00	67.0	66.7	73.1	70.8	28.2	28.3	2061.4	2003.6
Control (inoculated)	2.0	42.7[a]	42.7[a]	60.5[a]	60.5[a]	22.7[a]	22.7[a]	1373.4[a]	1373.4[a]
P. fluorescens	2.0	46.9[a]	45.1	66.9[a]	63.5	23.9	24.0	1589.9[a]	1524[a]
P. striata	2.0	45.1	44.0	63.0	62.2	23.0	22.9	1449	1424.4
A. awamori	2.0	53.5	50.2[a]	69.2[a]	6.9[a]	23.9	23.7	1653.9[a]	1585.5[a]
A. niger	2.0	45.3	43.6	65.5	63.1	23.5	23.0	1539.3[a]	1451.3
Carbendazim	2.0	48.0[a]	44.5	70.4[a]	67.0	23.2	23.0	1633.3[a]	1541[a]
LSD $P = 0.05$		5.6	5.2	6.3	5.9	3.4	3.6	173.8	149.5
F-value Fungus	(df = 1)	39.5[b]	17.8[b]	12.6[b]	6.7[b]	24.8[b]	23.5[b]	89.5[b]	73.8[b]
Treatments	(df = 5)	5.5[b]	12.6[b]	6.9[b]	3.5[b]	NS	NS	8.3[b]	5.2[b]
Combined	(df = 5)	8.6[b]	6.7[b]	NS	NS	NS	NS	4.3[b]	NS

Each value is mean of two replicates; Values in parenthesis are angular transformed values.
[a]Significantly different from the control at $P = 0.05$.
[b]Significant at $P = 0.05$.
NS: Not significant at $P = 0.05$.

plant (20.4% and 15.4%) of inoculated tomato plants, respectively. Soil application of *P. fluorescens* significantly increased the number and weight of fruits/plant by 10.6% and 16.4%, respectively. Increase in the yield by root-dip treatment with *P. fluorescens* was also significant. Soil or root-dip treatments of carbendazim increased the fruit formation (16.4% and 10.7%) and yield (18.9% and 12.2%) of infected plants compared to the inoculated control. Application of PSMs promoted mean fruit weight, but it was not significant for any treatment. Soil application had resulted to spread of cells/spores of PSMs in a relatively larger area providing an opportunity to solubilize greater amount of phosphorus compared to root-dip treatment. For this reason, soil application of PSMs was some 5–10% more effective in suppressing the wilt fungus and enhancing tomato yield.

Rhizosphere population of *F. oxysporium* f. sp. *lycopersici* was gradually increased during December (24–29%), January (48–52%) and February (57%) compared to November population. Application of PSMs caused significant decrease in the population. Greatest decrease in the CFU count of the wilt fungus/g soil was recorded due to soil or root-dip treatments with *A. awamori* during December (24.7% and 23.1%), January (54.8% and 43.8%) and February (60.6% and 48.5%) compared to respective month controls. Next were *A. niger* and *P. fluorescens* which decreased the rhizosphere population by 33–55% and 23–46% and 30–49% and 19%-due to soil and root-dip treatments, respectively. Carbendazim treatments decreased the fungus population by 46–78% and 19–41% in comparison to the inoculated control.

The study has demonstrated that besides being plant growth promoter, PSMs possess potential of being good antagonists of wilt fungus, *F. oxysporum* f. sp. *lycopersici*. *A. awamori* was found to be the most effective PSMs tested and suppressed the wilt severity by 37.1% and 29.5% and increased the yield by 20.4% and 15.4% of infected tomato plants due to soil and root-dip treatments compared to 18% and

8.7% and 18.9% and 12.2% by carbendazim treatments, respectively.

References

Arora D and Gaur A C 1979 Microbial solubilization of different inorganic phosphates. Indian J. Exp. Biol. 17, 1258–1261.

Dalla G C 1986 Esperienze di lotta biologicacontrola fusariosi vascolare del garofano. [Trials on the biological control of vascular wilt of carnations]. Ann. dell'Instituto Sper. Flori. 17(1), 3012.

Dospekhov B A 1984 Field Experimentation. Moscow Russia Mir Publishers. 352 p.

Khan M R, Akram M 1997 Effect of certain bacterial and fungal bioagents on a disease complex caused by *Meloidogyne incognita* and *Fusarium oxysporum* f. sp. *lycopersici* on tomato. Abstracts, Intern. Conf. Integ. Pl. Dis. Manag.Sust. Agr., New Delhi, India. 252 p.

Khan M R and Akram M 2000 Effect of certain antagonistic fungi and rhizobacteria on wilt disease complex caused by *Meloidogyne incognita* and *Fusarium oxysporum* f. sp. *lycopersici* on tomato. Nematol. Mediterr. 28, 139–144.

Leeman M, van Pelt J A, Hendrickz M K, Scheffe R J, Bakker P A H M and Schippers B 1995 Biocontrol of *Fusarium* wilt of radish in commercial green house trials by seed treatment with *Pseudomonas flourescens* WCS 374. Phytopathology 85, 1301–1305.

Nair M G and Burke B A 1988 A new fatty acid methyl ester and other biologically active compounds from *Aspergillus niger*. Phytopathology 27(10), 3169–3173.

Papavizas G C 1985 *Trichoderma* and *Gliocladium*: Biology, ecology and potential for control. Ann. Rev. Phytopathol. 23, 23–54.

Parker C A, Rovira A D, Moore K J and Wong P T W 1985 Ecology and Management of Soil-borne Plant Pathogens. APS Press, St. Paul, Minnesota, USA.

Sakthivel N, Sivamani E, Unnamalai N and Gnanamanickam S S 1986 Plant growth promoting rhizobacteria in enhancing plant growth and suppressing plant pathogens. Cur. Sci. 55(1), 22–25.

Toohey J I, Netson C D and Krotkov G 1965 Isolation and identification of two phenazines from a strain of *Pseudomonas aureofacions*. Can. J. Bot. 43, 1055–1062.

Woltz S S and Jones J P 1981 Requirements of *Fusarium oxysporum*: Basis for a disease control system. *In* Fusarium Disease, Biology and Taxanomy. Eds. P E Nelson, T A Toussoun and R J Cook. pp. 340–349. Pennsylvania State University Press, USA.